The Physics of Music and Color

Leon Gunther

The Physics of Music and Color

Sound and Light

Second Edition

Springer

Leon Gunther
Department of Physics and Astronomy
Tufts University
Medford, MA
USA

ISBN 978-3-030-19221-1 ISBN 978-3-030-19219-8 (eBook)
https://doi.org/10.1007/978-3-030-19219-8

This Springer imprint is published by the registered company Springer Nature Switzerland AG.
The registered company address is: Gewerbestrasse 11, 6330 Cham, Switzerland

Dedicated to my mother, Esther (Weiss) Gunther Wand,
who nurtured me with a deep appreciation
of music and the beauty of nature,
and to my wife, Joelle (Cotter) Gunther,
who sustains me with her love and wisdom

Preface to the Second Edition

Over the course of the years since the first edition was published in 2011, I have come across many fascinating phenomena that involve or are related to sound and light. As a result, I have felt the desire to publish a second edition.

Highlights in this second edition are the addition of sections on the following subjects:

- On the **beats of the sound of a bell**
- **AM and FM radio**
- On the **spectrogram**, which can display the flow of the frequency of a wave in time. It can capture the essential characteristics of a wave, often of sound waves—with applications in diagnosing medical problems with the voice and in identifying bird calls visually—but also of electronic signals such as those produced by **EKGs** and more recently even of **gravitational waves** emitted by binary neutron stars that collapsed billions of years ago.
- On the **diopter**, which is a parameter central to expressing the power of a lens.
- On the **luminance** of a beam of light, which is related to brightness. I feel that this subject is quite complex but is not treated in a clear and accessible way by a multitude of sources on the web. This book presents an essential explanation of luminance without excessive complexity.

I have added two new **Appendices**:

- **On Numbers**: dealing most importantly with *order of magnitude*, *significant figures*, and *relative changes*, all of which are important in making calculations in this book. Included are tricks ("shortcuts" might be a better word) for simplifying the calculation of the relationship between the relative change of one parameter and the relative change of another parameter upon which the first depends. As examples of using a trick, we will be able to easily determine the effect of a change in the tension of a violin string on its pitch or the effect of a change in temperature on the pitch of a flute.
- **On Photons**: dealing with the **mysterious, incomprehensible behavior of photons** in a clear way. This behavior indicates that quantum theory requires a major change in the conduct of the discipline of Physics and changes the assumptions as to what Physics has to say about "reality." It is the opinion of this author that Physics has nothing to say about reality and that the term "reality" falls only within the realm of Philosophy.

I have added numerous problems at the end of the chapters. And finally, I have clarified much of the material in the first edition.

The second edition calls for me to express my deep gratitude to my friend and colleague Ivo Klik for reading through the new sections in this edition, providing me with numerous suggestions for improved wording and correcting typos.

Preface to the First Edition

This textbook has its roots in a course that was first given by Gary Goldstein and me at Tufts University in 1971. We are both theoretical physicists, with Gary focusing on the study of elementary particles and my focusing on condensed matter physics, which is the study of the fundamental behavior of various types of matter—superconductors, magnets, fluids, among many others. However, in addition, we both have a great love and appreciation for the arts. This love is fortunately also manifested in our involvement therein: Gary has been seriously devoted to oil painting. I have played the violin since I was seven and played in many community orchestras. I am also the founder and director of a chorus. Finally, I am fortunate to have a brother, Perry Gunther, who is a sculptor and my inspiration and mentor in the fine arts.

It is common to have a course on either the Physics of Music or the Physics of Color. Numerous textbooks exist, many of which are outstanding. Why did we choose to develop a course on both music and color? There are a number of reasons:

1. The basic underlying physical principles of the two subjects overlap greatly because both music and color are manifestations of wave phenomena. In particular, commonalities exist with respect to the production, transmission, and detection of sound and light. Our decision to include both music and color was partly due to the fact that some wave phenomena are relatively easy to demonstrate for sound but not for light; they are experienced in every day life. Examples include diffraction and the Doppler effect. Thus, the study of sound helps us understand light. On the other hand, there are some wave phenomena—common to both sound and light—that are more easily observed for light. An example is refraction, wherein a a beam of light is traveling through air and is incident upon a surface of glass. Refraction causes the beam to bend upon passing into the glass. Refraction is the basis for the operation of eyeglasses. And finally, there are wave phenomena that are easily observable for both sound and light. Interference is an example.

   Two stereo loudspeakers emitting a sound at the same single frequency produce dead (silent) regions within a room as a result of the interference between the two sound waves produced by the two loudspeakers; the colors observed on the CDs of the photo on the cover page are a result of the interference of light reflected from the grooves within the CDs.

2. The production of music and color involves physical systems whose behavior depends upon a common set of physical principles. They include vibrating mechanical systems (such as the strings of the violin or the drum, vibrating columns of air in wind instruments and the organ), electromagnetic waves such as light, the rods and cones of the eye, and the atom. All manifest the existence of **modes** and the phenomena of excitation, resonance, energy storage and transfer, and attenuation.
   CDs 'produce' sound through a series of processes that involve many distinct physical phenomena. First, the CD modulates a laser beam that excites an electronic device into producing an electrical signal. The laser light itself is a manifestation of electric and magnetic fields. The electrical signal is used to cause the cone of a loudspeaker to vibrate and produce the motion in air that is none other than the sound wave that we hear.
3. The course that led to the writing of this book offered us the opportunity to study a major fraction of the basic principles of physics, with an added important feature: Traditionally, introductory physics courses are organized so that basic principles are introduced first and are then applied wherever possible. This course, on the other hand, is based upon a motivational approach: Because of the ease of observing most phenomena that is afforded by including both light and sound, we are able to introduce the vast majority of topics using class demonstrations.
   We challenge ourselves by calling for a physical basis for what we observe. We then turn to basic principles as a means of understanding the phenomena. A study of both subjects involves pretty nearly the entire gamut of the fundamental laws of classical as well as modern physics. (The main excluded areas are nuclear and particle physics and relativity.)

Ultimately, our approach helps us appreciate a central cornerstone of physics—to uncover a minimal set of concepts and laws that is adequate to describe and account for all physical observations. Simplification is the motto. We learn to appreciate how it is that because the laws of physics weave an intricate, vast web among physical phenomena, physics (and science generally) has attained its stature of reflecting what some people refer to as "truth" and, much more significantly, of having an extraordinarily high level of dependability.

The prerequisites for the associated course were elementary algebra and a familiarity with the trigonometric functions. The only material in the textbook that requires a higher level of mathematics is the appendix on the Transformation of Color Matching Functions (Appendix M) from one set of primaries to another—the analysis requires a good understanding of matrices. I have never included this appendix in my course; it is available for those who might be interested in it. The level of the textbook is such as to produce questions as to whether a student without inclinations to major in the sciences can handle the material. It has been my experience in teaching the associated course at Tufts University for over 35 years, that very few such students have failed to do well in the course. In the Fall, 2009 semester, in particular, the fifteen students who took the course were all majoring in the Arts, Humanities or Social Sciences or as yet had not declared a major. The average score on the Final Exam was a respectable 73%, with a range from 61% to 94%.

Note on Problems and Questions

Whether you are reading this book in connection with a course you are taking or reading it on your own, I strongly urge you to take the questions and problems in the book very seriously. To test your understanding, to measure your level of understanding, you have to do problems. In all my more than 50 years of studying physics, I have never truly appreciated a new subject without doing problems.

There are many fine books already available that cover either the physics of sound and music or the physics of light and color. Some of these books go into great depth about a number of the subjects, way beyond the depth of this book. For example, you won't find details on the complex behavior of musical instruments in this book. The book by Arthur Benade, listed in Appendix E of references is a great resource on this subject, even though it is quite dated. And, you won't find in-depth coverage of the incredibly rich range of light and color phenomena that is treated in the wonderful book by Williamson and Cummins. Their section on oil paint is outstanding. Instead, you should look on this book as a resource for gaining an in-depth understanding of the relevant concepts and learning to make simple calculations that will help you test hypotheses for understanding phenomena that are not covered in this book. You will be able to read other books and articles on the web, empowered with an understanding that will help you appreciate the content.

One of the problems raging today (2011) is the proliferation of information. Ah yes, you can look up on the Web any topic in this book. Unfortunately, a huge fraction of the information is incorrect or unreliable.[1] How can you judge what you read? The only solution is for you to accumulate knowledge and understanding of the basics and to criticize what you read.[2]

[1]Recently, the Sharp Corporation announced that it was going to make available a color monitor and TV that has **four primary colors** among the color pixels, in contrast to the three primaries currently used. As a result, it claimed that the number of colors available would approach one-trillion. [See their website: http://www.sharpusa.com/AboutSharp/NewsAndEvents/PressReleases/2010/January/2010_01_06_Booth_Overview.aspx.] Yet you will learn in Chap. 15 that human vision can differentiate only about ten million colors. Therefore, even if the Sharp monitor were able to produce one trillion colors, the viewers wouldn't be able to benefit from this great technology. We can still ask what can possibly be the gain in adding a yellow primary? Is their chosen color yellow for the fourth primary the best one to choose in order to improve our color vision? See Chap. 15 for information on this question. Websites abound dealing with the significance of Sharp's new technology; this book will help you analyze and judge what you read.

[2]What applies to information on science applies to all subjects. If you are given a multitude of conflicting **expert** opinions on a subject, you will tend to choose one expert who is closest to your point of view or you will want to throw all the sources out the window with the conclusion that reliable information not only can't be found but has no meaning. The fascinating book by Neil Postman—*Amusing Ourselves to Death: Public Discourse in the Age of Show Business* [Penguin Books, N.Y, 1986]—discusses some related problems connected with this proliferation of information.

Acknowledgments First and foremost, I am indebted to Gary Goldstein, who was a co-developer of the original course on The Physics of Music and Color. Gary's contributions in teaching a number of the subjects in a clear way were invaluable. Most noteworthy were his ideas for teaching color theory. I am grateful to my daughter, Rachel Gunther, for producing the first word processed draft of the book. I am deeply indebted to Stephan Richter, one of my graduate students, who was a driving force and indefatigable in producing a LaTex copy of the book, worked over numerous figures, and is responsible for the layout of the book. I had a number of teaching assistants over the years who made very valuable contributions in teaching the course, most notably Stephan Richter and Rebecca Batorsky. Both Stephan and Rebecca are gifted teachers and frequently shared productive advice for me. My long time friend and violin teacher, Wolfgang Schocken, was a well-known teacher of the violin. He was also extremely knowledgeable about the numerical issues involved in intonation, which he shared with me. In spite of my familiarity with resonances and overtones of a vibrating string, it was he who taught me to listen carefully to the resonant vibration of unbowed strings in order to vastly improve my intonation. My son, Avi Gunther, who got his Bachelor's degree from the Berklee College of Music in Boston with a major in Music Production and Engineering, was often extremely valuable in advising me on many aspects of music and on sound production.

I benefitted greatly from two readers of this book: the first reader was my personal opthalmologist, Dr. Paul Vinger, who pointed out numerous typos and provided me with questions that he suggested be addressed in the book. My second reader was a student of mine, Bryce Meyer, who did an incredibly dedicated job reading carefully through the book—finding typos and making countless suggestions for improving the clarity of the various passages in the text. Bryce also helped me with some figures.

Many individuals have helped me in one way or another toward the writing of this book. I list the following, with apologies to those who should be here but have been inadvertently omitted: Paavo Alku, Anandajoti Bhikku, Bruce Boghosian, Andrew Bregman, Andrew Clarke, David Copenhagen, Tom Cornsweet, Russ Dewey, Marcia Evans, Oliver Knill, Paul Lehrman, Ken Lang, Jay Neitz, Donna Nicol, Ken Olum, Charles Poynton, Jeffrey Rabin, Brian Roberts, Judith Ross, Eberhard Sengpiel, George Smith, and Raymond Soneira. This book would not have been published were it not for the strong support and help of my editors, Christopher Coughlin and HoYing Fan. I want to bring special attention to Kaća Bradonjić, who refined over one hundred figures with great finesse, especially those in Chap. 5 that are based upon my hand drawings. I am grateful for the secretarial and administrative help and support of Gayle Grant and Shannon Landis in the Department of Physics at Tufts University.

This book has been a work in progress for more than 35 years. It has had many drafts. I need to share with you my deep appreciation for my loving wife, Joelle, for supporting me in this effort. Whenever I needed encouragement to sustain my spirits and energy, Joelle was there for me.

Medford, MA, USA Leon Gunther

Questions Discussed in This Book

1. Why does a flutist have to retune his/her flute a while after having begun playing?
2. Can a soprano really break a glass goblet?
3. How does the brain determine the direction of a source of sound?
4. What is noise?
5. Why does the trumpet sound different from the violin?
6. How is it that sound can bend around corners and pass through spaces of an open door?
7. Does light bend around corners?
8. What simple mathematical relationships form the bases of the musical scales of most of the world's cultures? Are these relationships unique?
9. How can audio signals be transmitted using radio waves even though audio signals have frequencies thousands of times smaller than radio waves?
10. How can the ear provide us with a sense of pitch?
11. Can a fish hear a fisherman talking from offshore?
12. How can a person hear a clock ticking at a frequency of one tick per second while it is said that the lowest frequency that can be heard is about 20 cycles per second?
13. How can we estimate the speed of an airplane flying overhead from the sound it emits?
14. How does the vibrato of a violin help improve our perception of consonance among groups of notes?
15. Why is the sky blue and the setting sun red?
16. How is the mathematics of the interference of a light beam that passes through two slits related to the location of a gunshot in an inner city?
17. How can we obtain a visual representation of a bird song by using a spectrogram?
18. The pitch of a sound is essentially determined by the frequency of the sound wave. On the other hand, the frequency of a sound wave is associated with a sinusoidal wave that has an infinite duration in time. How long does a note have to last for us to determine its pitch?
19. Why does it become more difficult to perceive a sense of pitch as we play ever lower-pitched notes on a piano?
20. How is it that when we view sunlight that has passed through the atmosphere, the color is red, while if we are deep in the ocean and view sunlight that has entered from above the ocean's surface, the color is blue?

21. How can we hear sounds that are not in the air? How is this phenomenon related to the blue color of the ocean?
22. How does the rainbow get its colors?
23. How is the eye like a camera?
24. Are there three primary colors?
25. What is the physiological basis for the existence of a set of three primaries that can be mixed to produce the vast majority of color sensations?
26. Approximately how many colors can a person with normal vision distinguish?
27. What is color blindness?
28. What are the colors white, black, gray, and brown?
29. How is it that all light is a mixture of the colors of the rainbow, yet the color brown is not simply a mixture of these colors?
30. How is it that the ear can perceive two distinct musical tones, yet the eye always perceives a mixture of many colors as a single color?
31. How can we perceive color from purely black and white images?
32. What is a "mirage"? Is it a real image?
33. How do color prints, color slides, and color TV work?
34. What is the range of visible colors (the "gamut" of visible colors) that a color monitor can display?
35. Why does the addition by the Sharp Corporation of a fourth primary **not** increase the gamut of visible colors significantly?
36. Smart phones are ever-increasing the number of their pixels. At what point does that number reach the limit of resolution of the eye, so that there is no gain in a further increase?

Medford, MA, USA Leon Gunther

Contents

1 Introductory Remarks

Why should someone be attracted to a book on the Physics of Music and Color? For people who are well versed in both the sciences and the arts, the question would very likely not arise. But for those who are well versed in but one of these areas, the relationship between the two is probably unclear, if not a total mystery.

To respond to this question, I would like to begin by sharing a personal experience with you. In the fall of 1983 I had the privilege of meeting Wolfgang Schocken, who was to become my violin teacher. I have a memory of my standing there in Wolfgang's elegant living room, in front of a music stand, with my violin in hand, a bit anxious on beholding Wolfgang's aristocratic, tall, and erect stature that was nevertheless accompanied by a soft kind smile. I learned that Wolfgang had been born in Berlin in 1908 and had studied under Karl Flesch and Ottokar Sevčik, two of the most world renowned violin pedagogues. He had left Germany for Palestine in 1933 when Hitler came to power because his father was Jewish. Worse still was that his former very close non-Jewish friends began to shun him. I learned that Wolfgang had known two of the most world renowned violinists, Itzhak Perlman and Pinchas Zuckerman, in Israel, when they had begun their studies as children.

Wolfgang began my lesson by asking me what I would like to focus on regarding technique. I assumed that I was to choose between my bowing and my **intonation**[1] I immediately told him that I felt that my intonation was fine so that I wanted to begin by working on my bowing. Wolfgang quietly responded by suggesting that we first examine and check on my intonation. He asked me to play a certain note associated with the placement of my index finger on a string. Note that the position of the finger determines the **frequency** of the sound and hence the **pitch**.[2] Then he asked me to move that finger a bit to examine any changes in what I heard. He urged me to listen very carefully. Knowing that I was a physicist, he asked me whether I could hear any **resonant** sounds produced by other strings that have no fingers on them—what we refer to as strings that are open, that is, **open strings**. When there is **resonance**, another open string responds to the vibrations of the string that is bowed by vibrating and producing its own sound. Indeed I sometimes did hear a resonance, but only when my finger was very close to a specific position. The sound was beautiful when "I hit the spot" on the string. Before meeting Wolfgang I had not paid attention to these resonances! And so began many years of listening to resonances so as to improve and maintain my intonation. I still do so forty-five years later.

[1] As we will discuss in detail in Chap. 12, "intonation" refers to what the layperson would refer to as "playing in tune."

[2] This subject is discussed in detail in Chaps. 2 and 12.

L. Gunther, *The Physics of Music and Color*,
https://doi.org/10.1007/978-3-030-19219-8_1

What is the physical basis for a resonance and when does it occur between two strings? These questions will be discussed in this book. It is well for the reader to know that resonances play a role not only in vibrating strings but also in a multitude of domains in physics—of sound, light and of electricity, of elementary particles (not discussed in this book), molecules, and even the blue color of the ocean! When I hear a resonance between two strings, I delight not only in the sound I hear and my success in playing in tune, but also in imagining the "dancing of the strings" that produces this sound. Resonances involve numerical relationships.[3]

1.1 Two Contrary Attitudes About Science

Let us consider two contrary attitudes to the role the study of physics can make with regard to our sense of the world about us. One is by the great poet Walt Whitman, the other by the renowned physicist Richard Feynman (Fig. 1.1).

Here is **Walt Whitman's** attitude towards Astronomy. His poem "When I Heard the Learn'd Astronomer" is sardonic:

> When I heard the learn'd astronomer,
> When the proof, the figures, were ranged in columns before me,
> When I was shown the charts and diagrams, to add, divide, and measure them,
> When I sitting heard the astronomer where he lectured with much applause in the lecture-room,
> How soon unaccountable I became tired and sick,
> Till rising and gliding out I wander'd off by myself,
> In the mystical moist night-air, and from time to time,
> Look'd up in perfect silence at the stars.

I wonder whether Whitman would have reacted the same way to the documentary film on the work of **Louis Leakey**, who discovered the remains of **Australopithecus bosei**, a prehistoric form of man that was dated to have existed about 1-3/4 million years ago. Leakey has been described as having worked persistently but unrewardingly for 28 years at the site, before the discovery was made.

There is a scene wherein Leakey is standing on a hilltop overlooking the **Olduvai Gorge** in Kenya. The terrain is devoid of greenery, in fact, lifeless in appearance. Still, Leakey passionately paints word images of the life of the prehistoric people who lived and died in that valley as if they were alive that very day the filming took place. Upon what information were these images based? Merely upon dry pieces of bone and artifacts, most of which would barely be noticed by the average passerby.

The same can be said of the work of astronomers, astrophysicists, and cosmologists. They have provided us with images of our solar system, our galaxy, and our Universe, revealed the detailed workings of the stars and charted their life history, and deduced a possible history of the Universe starting with the Big Bang theory—but only after painstaking patient mathematical analysis of astronomical data, an activity that is fueled by irresistible curiosity, and by egos too!

Still, one need not know any physics to be a successful professional musician or artist, although currently, many artists are making use of physics in their work. The musician must understand the relationships among the various elements that make for a great musical composition, such as musical notes. The musician understands that in some, oftentimes mysterious way, our perception of the specific relationships among these elements exists at various levels, from the subconscious to the conscious levels, so as to produce a sense of esthetic beauty and a variety of emotional responses. There is an obvious underlying degree of order among these elements. The same can be said for the visual artist with respect to a great work of art.

What turns some people off from science? Is it boredom with the subject matter or boredom that is due to an inability to appreciate the content of science? Is there a fear that science will remove the element of mystery, upon which much of our pleasure of music and art is based? Consider the viewpoint of the great physicist **Richard**

[3]Wolfgang's interest in the importance of numerology in music is probably connected with his having published two books on the numerical relationships involved in the Mayan calendar.

Fig. 1.1 Left: Walt Whitman (source: https://upload.wikimedia.org/wikipedia/commons/5/54/Whitman_at_about_fifty.jpg); Right: Richard Feynman (source: https://en.wikipedia.org/wiki/Richard_Feynman#/media/File:Richard_Feynman_Nobel.jpg)

Feynman, as quoted from his book *What Do You Care What People Think?*:

> I have a friend who's an artist, and he sometimes takes a view which I don't agree with. He'll hold up a flower and say, "Look how beautiful it is", and I'll agree. But then he'll say, "I, as an artist can see how beautiful it is. But you, as a scientist, take it all apart and it becomes dull." I think he's nutty.
> First of all, the beauty that he sees is available to other people—and to me, too, I believe. Although I might not be refined aesthetically as he is, I can appreciate the beauty of a flower. But at the same time, I see much more in the flower than he sees. I can imagine the cells inside, which also have a beauty. There's beauty not just at the dimension of one centimeter; there's also beauty at a smaller dimension.
> There are the complicated actions of the cells, and other processes. The fact that the colors in the flowers have evolved in order to attract insects to pollinate it is interesting; that means that insects can see the colors. That adds a question: Does this aesthetic sense exist in lower forms of life? There are all kinds of interesting questions that come from a knowledge of science, which only adds to the excitement and mystery and awe of a flower. It only adds. I don't understand how it subtracts.

The fact is that in many ways, the work of the physicist is similar to that of the impressionistic painter. While people marvel at the visual relationships in art, physicists marvel, in addition, at conceptual relationships in theories that describe natural phenomena as revealed by experimental and theoretical analysis.

1.2 A Scene vs. a Painting

Consider the Langlois Bridge at Arles, France, as shown in the photograph in Fig. 1.2. As seen in the photograph, the bridge would normally not attract much attention to people. Yet Googling this bridge results in quite a number of hits. Many people take the trouble to go out of their way to visit this bridge. Why is this so? Because the painter van Gogh produced a number of paintings of this bridge. A print of one of these paintings is shown in Fig. 1.2. The photograph indicates that the bridge and its surroundings have probably deteriorated quite a bit since van Gogh produced his paintings. Thus, we cannot expect a photographer to be dishonest in presenting a bridge without the color it once had. However, there is an important reason for the interest and attraction in van Gogh's painting.

I suggest the following as a modest response to this question: The human mind cannot absorb and integrate all the information that is transmitted to it by the senses. Nature is too complicated. Van Gogh chose certain elements of the visual field and emphasized them with well-chosen strokes of the brush. Viewing the painting helps you to become more sensitive to and more aware of these elements so that once you have been "impressed" by the painting, bridges and streams will forever appear very different to you,

(a) (b)

Fig. 1.2 Langlois Bridge in Arles, France: (**a**) Photograph thereof. (**b**) Van Gogh painting (photocredit: **stock.xchng** http://www.sxc.hu/browse.phtml?f=download&id=501544)

(a) (b)

Fig. 1.3 A sunflower by Barry Guthertz: (**a**) digital color photograph, (**b**) digitally artistically modified BW photograph

certainly more alive and vibrant. Thus, I expect that my having appreciated impressionistic paintings for many years has reduced the difference between the visual reality and the painting.

Here is an experiment that I recommend for the reader that confirms this idea for me: Stare at the photograph in Fig. 1.2 for about 15 s. Then close your eyes and work to picture the photograph in your mind. Do the same for the painting. When I do so, I find that I can much more easily visualize the painting than I can visualize the photograph, indicating that the reduced focused information in the painting might well be the reason for this experience. And, the particular reduced information selected by the artist makes an intense "impression" upon us that the photograph cannot provide.

NOTE: My comments are not at all intended to demean the art of photography! Note that the photograph of Feynman at the beginning of this chapter is an example of how a good photographer can capture a moment in time. One look at this photograph leaves you with a permanent memory of a significant nature of Feynman's appearance and personality.

Here is an example as to how a photograph can be digitally modified by a photographer so as to accomplish what a painter such as van Gogh accomplished for the Langlois Bridge. In Fig. 1.3 we see the printed images of two photographs[4] of a sunflower. The first photograph (a) is the original digital color photograph of the sunflower.

[4]Courtesy of the photographer, Barry Guthertz.

Fig. 1.4 Strings in a piano

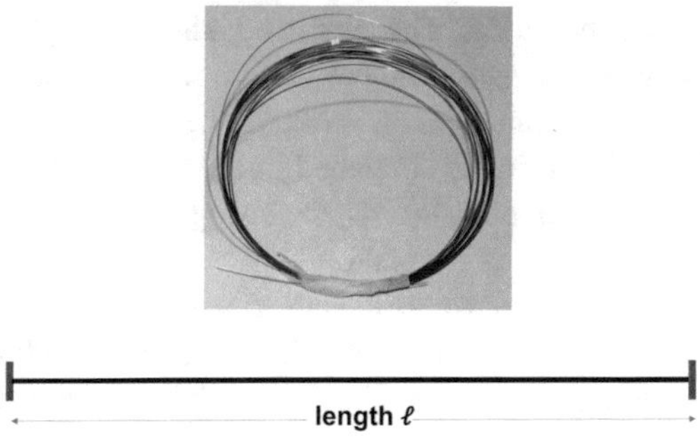

Fig. 1.5 Piano wire and a piano string

The second image (b) is produced by digitally modifying the image file using a computer. The resulting image is incredibly captivating.

How was the original image modified? According to the photographer, the image was converted to a BW (black and white) image, followed by the addition of **contrast** and **grain**; "the result is that the viewer looks more closely at the full picture of the sunflower rather than focusing on the yellow of the sunflower head."

1.2.1 The Joy of a Physicist in Looking at a Mathematical Equation

How can we understand how a physicist could have joy in looking at a mathematical equation, a mere set of symbols? Let us see how the physicist approaches an object or system and **focuses attention** on only some of the object's characteristics as does a painter or a photographer. In Fig. 1.4[5] we see some strings in a piano.

Suppose that we need to replace a string in a piano. A length ℓ of piano wire is cut from a spool, as shown in Fig. 1.5. Its two ends are attached to the piano.

In order to analyze the behavior of the string as it is used in a piano, the physicist focuses on **just three of its attributes**: its **mass** m, its **length** ℓ, and the **tension** T in the string produced by attaching its two ends and turning a **tuning pin** in order to produce the tension that is associated with the desired **musical pitch**. Neglected are other characteristics of the wire, such as its diameter, its stiffness, and its shiny surface. The laws of mechanics lead to a mathematical equation that describes the behavior of the string. The reader is not at all expected to understand this equation; I am displaying the equation merely to allow the reader to see the set of symbols in the equation. The equation is an example of a **wave equation** and describes the behavior of waves along the string.[6] It involves the three parameters m, ℓ, and T.

$$\frac{m}{\ell}\frac{\partial^2 y}{\partial t^2} = T\frac{\partial^2 y}{\partial x^2} \tag{1.1}$$

Essentially the equation describes the **displacement** of each point along the string at various times. The symbols are: y for the displacement of the string up and down, that is, transverse (perpendicular) to the length of the string, x for the position along the string, and t tor the time.

Suppose that you take a photograph of a vibrating string at some instant in time. From the photograph, you can produce a **graph** of the displacement of the string vs. the position x

[5]https://upload.wikimedia.org/wikipedia/commons$/3/35$/DuplexScaling.jpg.

[6]Wave equations describe the behavior of **waves** of various sorts. Examples are sound waves, light waves, X-ray waves, nerve signal waves along neurons, and waves moving along the surface of a pond or the ocean. The reader will learn details about the propagation of waves in Chap. 2.

along the string. The above equation describes how such graphs will evolve in time. The entire content of information about the behavior of the vibrating string is contained in the mathematical equation. In other words, the string's behavior is **mapped** onto the equation.

The above wave equation can be used to derive the existence and behavior of the **modes of vibration** of a string, which will be discussed in Chap. 2. Each mode has a frequency; these frequencies are referred to in music as **harmonics**. The patterns of vibration of the five modes with the lowest frequency are shown in Fig. 2.18. The addition of a mathematical theorem derived by **Joseph Fourier** allows us to describe mathematically the **general behavior** of a vibrating string, understand quite a bit about the sound that a piano can emit due to the vibration of its strings, and how the strings can **interact** with each other. Finally, we will be able to understand how a musician can make use of **resonance** to improve intonation.

It is the relation between a beautiful, simple looking mathematical equation, the wave equation, and the observed rich behavior of a string that gives joy and delight to the physicist.[7]

Before closing this section I would like to present the reader with the following questions: Aren't the words you are reading in this book just minute amounts of dried ink? The text is quite meaningless when isolated. Yet you have learned to map text onto words that you hear in your minds and onto ideas, thoughts, and feelings. Children are taught to read and make such mappings. Are mathematical symbols necessarily any more meaningless than ordinary text when isolated? The answer depends upon the reader. Keep in mind that analogously, physicists have learned how to derive the mathematical equations of physics as mappings of observations; they are thus quite meaningful.

[7]As another example: Their exist a few mathematical equations that contain all the information about many of the physical properties of a specific material, such as a sample of copper. Examples of properties are its density, its hardness, how easily you can stretch it or bend it, and how much its temperature will increase if we add a specific amount of heat to the copper—all of these properties are mapped onto a few mathematical equations.

1.3 Two Views on the Creation of a Musical Composition vs. the Discovery of a Law of Physics

It is interesting to consider how **Albert Einstein** viewed the relationship between science and art or music[8]:

"All great achievements of science must start from intuitive knowledge. I believe in intuition and inspiration.... At times I feel certain I am right while not knowing the reason." Thus, his famous statement that, for creative work in science, "Imagination is more important than knowledge" But how, then, did art differ from science for Einstein? Surprisingly, it wasn't the content of an idea, or its subject, that determined whether something was art or science, but how the idea was expressed. "If what is seen and experienced is portrayed in the language of logic, then it is science. If it is *communicated* (my emphasis) through forms whose constructions are not accessible to the conscious mind but are recognized intuitively, then it is art."

We can briefly express Einstein's description of Science as follows: The physicist analyzes data from experimental observations. Through analysis, inspiration, and imagination the physicist arrives at a **theory** that summarizes observations in terms of mathematical relationships—mathematical equations—among observables such as force, momentum, time, and displacement. These relationships display a beauty to the physicist because of their relative simplicity in appearance and in their ability to summarize and account for experimental observations. The construction of the laws of Physics is a highly conscious process (Fig. 1.6).

[8]Based upon the journal article, Physics Today, March 2010 issue, with quotes from Alice Calaprice's *The Expanded Quotable Einstein*.[Princeton University Press, Princeton, N. J., 2000].

Fig. 1.6 Maurice Ravel (composer) and Albert Einstein (physicist) (photo credits: Ravel—http://en.wikipedia.org/wiki/File:Maurice_Ravel_1912.jpg; Einstein—http://commons.wikimedia.org/wiki/File:Albert_Einstein_violin.jpg)

Musicologists and composers would well disagree with Einstein with respect to the absence of logical organization in a great piece of music! Consider, for example, an exchange between the composer, **Maurice Ravel** and the French violinist Andre Asselin who asked Ravel about the role of inspiration in Ravel's Sonata for Violin and Piano. Ravel replied as follows: "Inspiration—what do you mean? No—I don't know what you mean. The most difficult thing for a composer, you see, is choice—yes, choice".[9] For me, "choice" represents logical analysis in musical composition—analysis that is necessary for composing a great original piece of music. The creation of a musical composition involves quite a bit of perspiration in addition to inspiration.[10]

Given the musical score, with notes displayed on a sheet of paper, musicians perform the piece of music, creating musical sound through analysis and a choice of mechanical techniques (such as fingering, bowing, and vibrato on a violin), so as to shape the sound in time, dynamics, and in tonal quality. And finally we have the listener who, upon hearing the piece performed, has a musical experience that will be for the most part, subconscious and might not be analytical. However, a person well educated in musical theory can add greatly to the musical experience by being aware of the unique musical–theoretical relationships in the composition. This last aspect is analogous to the physicist's experience of the mathematical beauty of a theory.

The disagreement between Einstein and Ravel is to some extent expressed in terms of the relative roles of **inspiration** and **perspiration**. The celebrated author Umberto Eco said that genius is 10% inspiration and 90% perspiration.[11] The Hollywood screenwriter, Jeremy Iacone, most noted for his screenplay "Bone Collector," starring Denzel Washington concurs: "Most anyone can have ideas and inspiration. It takes craftsmanship to produce a work of art [in literature or music]."

It is fair to ask what the difference is between inspiration and perspiration. I believe that it is fair to say that inspiration probably involves a lower degree of conscious awareness than perspiration. Moreover, the perspiration stages involve a much

[9]Taken from *A Ravel Reader: Correspondence, Articles, Interviews*, by Arbie Orenstein (Columbia University Press, New York, 1990).

[10]Readers are strongly encouraged to listen to Leonard Bernstein's analysis of the Claude Debussy's musical composition "Afternoon of a Fawn" through the following link: https://www.youtube.com/watch?v=vOlzpfE8bUk. Even a reader who doesn't know musical jargon can appreciate Bernstein's revelation of the profound thinking that went into composing this piece.

[11]*How to Write a Thesis*, MIT Press, March 2015.

Fig. 1.7 Schenkerian analysis of a segment of Beethoven's Piano Sonata No. 30 (source: https://upload.wikimedia.org/wikipedia/commons/7/7b/BeethovenOp109.png)

higher degree of conscious analysis. Whether in the arts or the sciences, the perspiration phase usually consumes by far the greater time. In the future, a much clearer characterization of the components of the creativity process will be available to us as a result of the research of cognitive scientists and neuroscientists.

There is an interesting relationship between the above question of the relative contribution of inspiration and perspiration in producing a musical composition on the one hand and the relative importance of intrinsic talent vs. hours of practice in learning how to play a musical instrument. There is increasing opinion that hours of study and practice are much more important than talent. The book *Outliers*, by **Malcolm Gladwell** claimed that 10,000 hours of practice would enable a person to master a skill—such as playing a musical instrument. Recently, **Anders Ericsson** refined this claim by requiring **the practice to be focused and deliberate**. The basis for this claim is under debate.

We finally come to Einstein's focus on the difference between the arts and science with respect to *communication*. It is probable that Einstein meant that it is possible to appreciate a piece of art without analysis while such is not possible in the case of science. If so, we should recognize that non-scientists might be able to appreciate somewhat the beauty of a scientific theory without understanding the mathematical essence of the theory. Furthermore, modern music has a genre that some musicologists, such as Milton Babbitt, suggest cannot and, moreover, need not be appreciated by the non-educated in musical theory.[12]

Do the laws of Physics have their counterpart in Western music? **Heinrich Schenker**, who lived from 1868 to 1935, produced a highly regarded method of musical analysis that is in use to this day. On his tombstone are inscribed in German the following, in English translation: "Here lies he who examined and revealed the laws concerning the soul of music like none other before him." Below we exhibit a Schenkerian analysis of a segment of Beethoven's Piano Sonata No. 30, Op. 109.

Schenkerian analysis can help a composer compose and help a musician perform a score in a deeply musical way. While you might not be able to read music and therefore might not be able to get any deep understanding of the analysis, you can see that Schenkerian analysis has reduced the multitude of notes in the passage quite considerably and connected notes with solid as well as dotted curves. You can appreciate that a person knowledgeable in Schenkerian analysis will understand how a musical score can give a listener great musical pleasure (Fig. 1.7).

In order to appreciate the difference between science and art, consider the following: Imagine yourself standing next to a stream of water

[12]See "Who Cares If You Listen," http://www.palestrant.com/babbitt.html.

in the woods. Consider how we observe a stream flowing with our eyes. We can observe waves moving along the surface of the water. The painter provides us with focused static content. The physicist seeks to determine the **relationships** that connect all physical phenomena; it is the revelation of these relationships that excites a physicist.

Look at the photo of the Langlois Bridge as a reference. The physicist would seek to understand questions like:

- How does light produce the image of the trees and the bridge on the water?
- What is the nature of the water waves on the stream? How can we characterize their shape and how they evolve, move, and disappear? Given that waves are often produced by breezes and wind, what is the relationship between the wind characteristics, such as the wind velocity, and the waves and surface textures produced? In fact, there are mathematical equations that describe in detail the behavior of waves along the surface of a river, a lake, or an ocean and how they are produced by wind.
- What determines the apparent color of any object and whether the surface of the object is shiny or dull?
- Regarding the Langlois Bridge: What tension must there be in the cables and stresses in the wood in order to keep the sections at rest. This information can lead to information about how the cable is responding to the tension and how the wood is responding to these stresses. We can compare this study to the interest we have as to how various psychological stresses affect one's emotional state. The scientific study of the wood gives the wood a life of its own.

These are not questions that would necessarily bore an artist. If one learns to synthesize one's knowledge, analysis, through a familiarity with Physics, can only add to one's appreciation of nature.

Often people are turned off by the heavy mathematical analysis that dominates Physics and is its essential language. Yet music and mathematics have been inseparable throughout history. Most significantly, it was recognized long ago that pleasurable music is connected with **numbers**, specifically ratios of small integers. This fact is exemplified by the ancient Chinese Legend of the **Huang Chung** (meaning "yellow bell"), the earliest known account of which is due to **Leu Buhwei** (226 BC). This Legend is believed to be over 3,000 years old.

The Legend of the Huang Chung: The Ancient Recognition of the Connection Between Numbers and Music

According to the Legend of the Huang Chung, Emperor Huang Ti one day ordered Ling Lun to make pitch pipes. Ling Lun needed a mathematical recipe for their construction both in order to end up with pleasing sounds and to be able to have an instrument that could be played along with other instruments. So Ling Lun went from the West of the Ta Hia country to the north of Yuan Yu mountain. [See Fig. 1.8.] Here Ling Lun took bamboos from the valley Hia Hi. He made sure that the sections were thick and even and he cut out the nice sections. Their length was 81 lines, that is, about 9 inches.

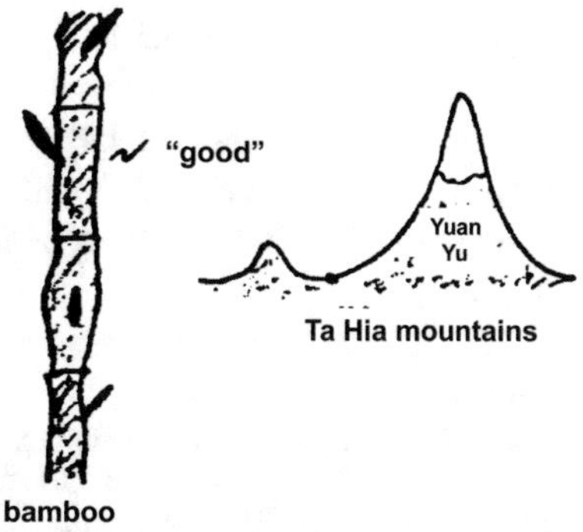

Fig. 1.8 Bamboo from the Ta Hia country

He blew them and made their tone the starting note, the huang chung, of the scale. (The huang chung had the same pitch as Ling Lun's voice when he spoke without passion.)

He blew them and said: "That is just right." Then he made 12 pipes. With what notes? Well, he heard Phoenix birds singing at the foot of the Yuan Yu mountain. From the male birds he heard six notes and from the female birds he heard six notes. These are the lengths of the pipes. [Our current notation for the notes is added for reference purposes.][13]

Male Pipes:	F	G	A	B	C#	D#
	81	72	64	57	51	45

Female Pipes:	C	D	E	F#	G#	A#
	54	48	43	38	34	30

- F is the "huang chung" (yellow bell)
- G is the "great frame"
- A is "old purified"
- C is the "forest bell"
- D is the "southern tube"

What is the basis for these numbers? Here is the recipe for the Chinese scale as recorded in China:

"From the three parts of the 'huang chung generator' reject one part, making the 'inferior generator' (hence equal to 2/3 of the huang chung generator). Next, take three parts of the new (i.e. inferior) generator and **add** one part, making the 'superior generator' (hence equal to 4/3 of the inferior generator)....," and so forth.

The lengths of the pipes are based on repeated applications of the factor 2/3 and 4/3 on the basic length of the huang chung generator (Fig. 1.9). Thus:

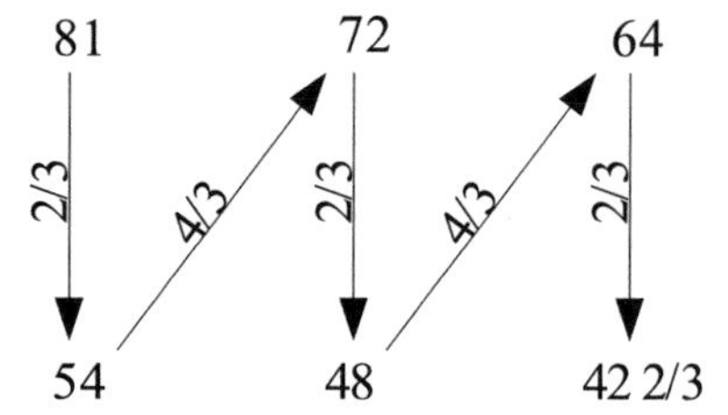

Fig. 1.9 Generating the Chinese scale from the huang chung generator

The coincidence between what was considered esthetically pleasing musically and the role of ratios of small integers and hence mathematics, or as the sixth century AD Roman philosopher **Boethius** put it, the coincidence between "sensus and ratio" (senses and reason) had a significant, meaningful effect on people. The pre-Socratics began a tradition of lack of trust in the senses as not providing truth about reality. The understanding was that **TRUTH** is obtained from **THOUGHT**. Thus, for example, one should not trust the senses to produce an acceptable rendition of the musical interval called the "fifth"; one should use an exact ratio of 3:2 of string lengths or pitch pipe lengths.[14] It should not be surprising that people would be very curious as to why the two—mathematics and music—should be connected. The answer must necessarily lie in mathematics and physics, but the physics necessary to understand the relationship was not known until the past few hundred years. With some background physics, we will try to provide answers to this question in this book. In particular, in the context of the Legend of the Huang Chung, we will learn how pipe lengths are related to frequencies and, in turn, how frequencies are related to pitch. In Chap. 12, we will demonstrate that within the framework of the level of complexity of the classical music of these past few hundred years, the desire for an omnipresence of ratios of small integers, which is connected with consonant musical intervals,

[13]The reader who is versed in music can determine that the full set of twelve notes, comprising the full **chromatic scale**, that is F, F#, G, G#, A, A#, B, C, C#, D, D#, and E, allows a person to play three different **pentatonic scales**, with the root notes, respectively, being C, C#, and D. The scales are, respectively, C, D, F, G, A; C#, D#, F#, G#, A#; D, E, G, A, B. We will see that this is so in Chap. 12. The pentatonic scale is the standard musical scale used in traditional Chinese music.

[14]How interesting it is that in recent times, a large fraction of society abhors the possible squelching of the senses by excessive thought.

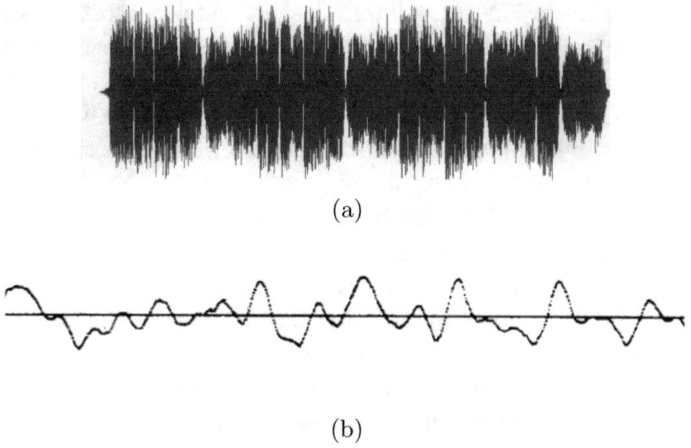

Fig. 1.10 Waveform of *Adon Olam*, by Salomon de Rossi. (**a**) Segment (2 min. 38 s) of waveform. (**b**) One-tenth second segment from the above

cannot possibly be satisfied for purely mathematical reasons.

I needed my teacher Wolfgang Schocken to remind me to use the physical principles that I had known for many years! They involve resonance among the strings. What is the physical basis for a resonance and when does it occur between two strings? They include numerical relationships that are relevant in the Huang Chung. These questions will be discussed in the book.[15] When I hear a resonance between two strings, I delight not only in the sound I hear and my success in playing in tune, but also in imagining the "dancing of the strings" that produces this sound.

In our study of the Physics of Music and Color, we will study the nature of sound and light. Analysis will be our focus. Many people find too detailed an analysis destructive to our ability to appreciate music and art. Interestingly, analysis within the framework of music and art proper seems to be acceptable. Fortunately, analysis leads to a richer synthesis. I hope that the reader will discover that analysis within the framework of Physics enriches our experience and need not be destructive.

In order to analyze sound and light, we must learn how to characterize sound and light. The sound of music is by far the easier of the two because it is characterized by a series of events in time. The sound that strikes our ears can be represented simply by a graph. We see in Fig. 1.10a below a graph of the wave of a short piece of music, 2:38 minutes in duration, composed by the Italian Renaissance composer **Salomone de Rossi** for five voices. It is difficult to see the details of the graph because of the extreme compression. To appreciate the content, Fig. 1.10b provides us with a magnification of an excerpt lasting about one-tenth of a second.[16]

Such a graph might seem to trivialize human experience. Alternatively, one should be amazed at how such a simple graph can fully represent something so powerful! A musical experience involves complex neuronal processing in the brain that maps the neuronal signals sent to the brain by the ears.[17] Clearly, a person would have to be extremely highly trained to tell the difference between a poor performance of a piece of music and an excellent one just by examining the difference in their respective waveforms. The human mind is wonderful.

Art is by far more complex and varied. Typically, it is two- or three-dimensional (2-D and 3-D) as well is static in time. Modern art includes dynamic visual works too. In this book, our study

[15]It is well for the reader to know that resonances play a role not only in vibrating strings but also in a multitude of domains in physic—of sound and light and of electricity, of elementary particles, molecules, and the blue color of the ocean!

[16]The graph represents the output of a single loudspeaker; for stereophonic sound we would simply need two such graphs.

[17]*This is Your Brain on Music*, by Daniel Levitin [Penguin Group, New York, 2006].

of the place of Physics as it relates to art will be extremely limited. We will study the nature of light and its relation to our perception of color. We will not go much beyond 2-D images, with a focus on simple patches of uniform color and interactions between neighboring patches. A 2-D image on a plane can be characterized by specifying the color at each point on the plane. The color can be specified in terms of what is referred to as the **spectral intensity**. We will learn that the spectral intensity gives more information than is necessary. A simpler though **incomplete characterization** of color makes use of a **three-primary representation**. One must specify the intensity of each of three primaries at each point on the image.

1.4 Outline of the Book

Both sound and light propagate as a wave. Therefore, the book will begin with Chap. 2, which will teach us what a wave is and how it propagates. The subsequent Chap. 3 will explain the nature of sound, how a sound is produced, and how it propagates. Chapter 4, covers the various forms of **energy**, the principle of conservation of energy, **power**, **intensity**, the **decibel** measure of intensity, **attenuation**, and **reverberation time** for acoustics in a room. Chapter 5 will teach us the nature of light, being simply a special combination of an **electric field** and **magnetic field** known as an **electromagnetic wave**, that it can be **polarized**, how it is produced, and how it propagates. Chapter 5 will also explain how various audio devices, such as a microphone and an audio loudspeaker, function. Chapter 6 will introduce us to quantum theory, which is the basis for how **atoms** and **molecules** emit and absorb electromagnetic waves as a discrete process involving **photons**. In fact, an electromagnetic wave consists of a stream of photons.

Once we understand the nature of sound waves and light waves and how they are produced, we will learn what happens when there is more than one source, such as when we have two loudspeakers or light that has passed through two slits. This subject is covered in Chap. 7. Chap. 8 deals with complex waves, most importantly **beats, AM/FM radio** and **polarized light**. Chapter 9, provides us with an analysis and description of what happens to a wave when it encounters obstacles in their path. For example, sunlight is scattered by the molecules of a gas such as our atmosphere leading to a bright blue sky. Or, a wave might strike an opaque object or wall that entirely blocks its path so that there will be a combination of absorption and reflection or strike a wall with holes of various sizes and shapes that leads to a reflected wave and a wave on the opposite side of the wall with **diffraction**, or strike a transparent obstacle such as glass leading to **transmission**, **absorption**, **reflection**, and **refraction**. A special important example of a transparent object is a **lens**, which has well-known use. A wave will be affected by the motion of its source and what an observer receives depends upon its motion. This phenomenon is referred to as the **Doppler effect**. The chapter closes with a study of the fascinating behavior of **polarized light** that can be manipulated using **polarizers**.

The next section of the book deals with how a person's sensors, our ears and our eyes, respectively, respond to incident sound and light. In Chap. 10, we describe how our ears amplify a sound, provides a degree of frequency discrimination, and send neuronal signals to the brain. Chapter 11, discusses the response of the ear to sound—how loudness relates to frequency and intensity, **pitch discrimination**, and "ghost sounds" produced by the inner ear known as **combination tones**. Chapter 12, reviews and analyzes musical scales used in current Western musical compositions—their frequencies and corresponding pitches. The significance of precise numerology in music will become evident.

We next move on to the field of light. Chapter 13, discusses how the eye functions much like a camera. Chapter 14 discusses how we characterize a light beam by its **spectral intensity**. **Color** is characterized technically by two parameters: **hue** and **saturation**. For example, a red hue can be ultra-saturated, like the deep red of blood, while **pink** is a low saturated red. **Color filters** function by selectively filtering out light according to the dependence of their transmittance on frequency. When patches of light produced on a surface from more than one source of light are overlapped, the observed color is a result of **additive mixing**. **Subtractive mixing** of colors is produced by passing light through a set of overlapping filters or by the selective absorption of light by **pigments**. The book closes with an extensive treatment of **color vision** in Chap. 15. We will learn how our eye–brain maps spectral intensities onto numerically characterized colors, referred to as **chromaticities**. While there are an infinite number of possible spectral intensities and there are an infinite number of distinguishable of chromaticities, a given chromaticity can be produced by an infinite set of distinct spectral intensities referred to as **metamers**. Each color or chromaticity can be assigned a set of three numbers, the **color coördinates**, labeled r, g, and b. The perceived color is determined by two independent fractions among them, r/(r+g+b) and g/(r+g+b). A central subject of the chapter explains how we can determine the color coordinates of a light source from the spectral intensity of a source of light, using a table of chromaticity coördinates. We will also learn how the color coördinates map onto the color of a color monitor—the familiar numbers R=0. . . 255; G=0. . . 255; B=0. . . 255.

1.5 What About the Sentient Perception of Music and Color?

Consider one of the simplest, pleasing combinations of musical notes, the **major triad**—for example, the C-major triad, consisting of the notes C, E, and G displayed in Fig. 1.11. We will learn in Chap. 11 that the most pleasing ratio of respective frequencies for a major triad is probably 4:5:6. Why does hearing this triad give us pleasure? Can science explain why we have this pleasure?

This book takes us on a journey to learn the scientific basis of sound, its nature, its

Fig. 1.11 C-major triad

propagation, its detection by the ear, and the transmission of neuronal signals to the brain. Neuroscientists[18] have taken us further: They have been able to map a triad of notes onto a specific excited location in the brain, the 3D analog of a color monitor's **pixel**, referred to as a **voxel**. There are three voxels in the brain that are excited by the corresponding three notes of the C-major **triad**, C, E, and G. In addition, these three excitations lead to the excitation of a single voxel in the brain that reflects the sensation of the triad as whole. Thus, the brain has evolved so that it recognizes the special place of a triad in our musical experience.

It is reasonable to expect that neuroscientists will ultimately be able to map out the excitations in the brain that occur in connection with a person's profound experience in listening to a piece of music. However, I wouldn't expect the details of the mapping in the brain to be contained in the sound entering the ear. The mapping is much more complex than the sound itself. Furthermore, even if neuroscientists accomplish this feat, in my opinion, we wouldn't then be able to answer the question as to nature of the **feeling** in purely

[18]See https://en.wikipedia.org/wiki/Cognitive_neuroscience_of_music; https://www.sciencedirect.com/topics/neuroscience/music-cognition; Fujisawa et al, "The Percep- tion of Triads" (2011)—https://link.springer.com- contentpdf-10.1007%2Fs11682011-91165.pdf. webloc; Janata et al, "The Cortical Topography of Tonal Structures" (2002) –science.sciencemag.org/content/sci//298/5601/2167.full.pdf?casa_tokenV̄w8LW-d5RLsAAAAA:XQfei-fBtEp7szKns5muZwFsohSJh/TnCkfbTPcfCu9Tajy_s-mO9Lr6ky_QXnJMq0L6p7fJMTp5Z88.

physical terms. Any question is expressed in a language and must be answered in a language. I don't believe we have a language to deal with the question of feeling. The same is true with respect to the meaning of **consciousness**. In my opinion, the questions of the precise meaning of the terms *feeling* and *consciousness* are not in the domain of **Physics**, they are in the domain of **Philosophy**.

Note I strongly recommend that you the reader review the opening section of this book entitled "Questions Discussed in this Book" so as to get a sense of the subjects that will be covered. Will this text enable you to better account for the esthetic pleasures of music and art? I believe so. I will be satisfied if your study of the Physics of Music and Color reveals new vistas of sound and light, so that your world experience of music and color will be greatly enriched. My goal is for you to become more observant. I look forward to your becoming aware of phenomena that would pass your notice without the benefit of this book. When you notice something, you should try to explain its basis using the relationships you will learn. Test your explanation by making an actual calculation using the basic laws of physics: See whether the estimated values of the physical parameters are reasonable. If they are related by a physical law you have studied in this book, determine how well the law is obeyed by your estimates. This book will give you countless examples of this process.

1.6 Questions and Problems for Chap. 1

1. Discuss briefly why the composer Maurice Ravel was annoyed by the implication of a violinist that Ravel's musical talent was founded essentially on inspiration. What did Ravel feel was not recognized by the violinist?
2. Calculate the lengths of the pipes labeled B and F# using the recipe for the Chinese scale in the Legend of the Huang Chung.

2 The Vibrating String

The subject of this text is music and color. Music is produced by musical instruments, some occurring naturally—such as the songs of birds—and others produced by man-made instruments such as stringed instruments, wind instruments, and the percussive instruments of drum sets. Color is produced by sources of light such as natural sunlight and by man-made sources such as the floodlights for a stage.

Essentially, music and color are *subjective* manifestations of the corresponding *objective* physical phenomena—sound and light, respectively. Both sound and light are manifestations of **wave phenomena**. If we can understand the nature of **waves**, we will increase our awareness of the richness of our human experiences with sound and light and hence of music and color. We can observe the wave nature of some types of waves with our own eyes—such as waves along a vibrating string or waves on the surface of the ocean. On the other hand, the wave nature of many important waves is invisible; examples are sound waves and light waves. Moreover, given that the string is the fundamental component of all stringed musical instruments, it is reasonable for us to begin our study with the **visible waves along a string under tension**.

2.1 Waves Along a Stretched String

Suppose that we have a long string, we stretch it so that it develops **tension** and we attach it to two points, A and B, as shown in Fig. 2.1. Tension is not a trivial subject and will be discussed in detail later in this chapter. Our string will be free to vibrate between two points, A and B. The string is the basic element of all stringed instruments, such as the violin, the piano, or the guitar.

The string is characterized quantitatively by three parameters: the **tension** T of the string, the **mass** m of the string, and the length ℓ of the string. This book will involve mathematical calculations that require the specification of the values of these parameters. The reader should therefore read Appendix A to review the notions of **significant figures**, **order of magnitude**, and **relative value**.

On a stringed instrument it is easy to appreciate how tension can be maintained through the attachment of the string at its ends. We will begin our study by assuming that there is tension and will neglect the ends, imagining the string to be infinite in length.

L. Gunther, *The Physics of Music and Color*,
https://doi.org/10.1007/978-3-030-19219-8_2

Fig. 2.1 A string under tension attached to two points A and B

In Fig. 2.2 below the string is depicted initially as the uppermost solid line . The **tension** in the string keeps the string straight. Next, we disturb the string by pulling the string upward a bit at a particular point along the string. The shape of the disturbance is a small triangle. What will happen next? The disturbance will move along the string as shown in the figure at one milli-second (1-ms) intervals: We set the time t equal to (1-ms), (2-ms),(3-ms), (4-ms), and (5-ms). Each of the vertical dotted lines marks a position along the string at a sequence of one-meter (1-m) intervals. We note that after each one-second interval, the disturbance progresses a distance 1-m to the right. Thus, the disturbance moves at a speed of one meter per milli-second. This value is equivalent to 1000-m/s. Note that this speed is quite large; in common units it is one kilometer per second (1-km/s), which is equivalent to 0.6 miles per second. Nevertheless, this value is close to the speed of a disturbance moving along a typical violin string.

A localized disturbance of this sort is called a **pulse** and is a simple example of **wave propagation**. The speed of the pulse is called the **wave velocity**. Later on in this chapter we will investigate what determines the wave velocity for a stretched string.

We can easily show that the string itself does not move at a speed of 1-km/s, or 1000-m/s. Nor does the string itself move to the right. In order to see this, suppose we focus our attention on a single point along the string, say the point marked with a dot, shown in Fig. 2.3. We note that while the pulse is moving to the right this point along the string has moved *downward*! We say that the wave is **transverse**, here meaning *perpendicular*. Suppose next that the height of the pulse is one millimeter (1-mm) [not drawn to scale above]. Then the average speed of this point is 1-mm/s, a value much less than the wave velocity.

How can we account for the motion of the pulse? Think of the old familiar "telephone game," wherein we have a string of people. The first person whispers a message to the second person. The second person whispers the perceived message on to the third person, and so on. The last person announces the message received and the first person reveals the original message. One hopes that the message will not be garbled!

In the case of the string, the initial material of the pulse along the string pulls upwards on the neighboring string material. The neighboring material pulls upwards on its neighboring material, and so on, leading the propagation of the pulse.

How does this description relate to other types of waves? The most important wave in the context of music is, of course, a sound wave—the focus of Chap. 3. Sound waves can propagate through a variety of media—such as air or water or a solid. Let us try to produce such a wave: Imagine what would happen if you were to move your hand forward suddenly. You would compress the air immediately in front of your hand. That compressed region of air would compress the air immediately in front of it. This process will continue as in the case of a pulse propagating along a stretched string. You will have produced a **sound pulse**. The wave is said to be **longitudinal**, meaning that the motion of the air is along the same direction as the direction of propagation of the disturbance. Unfortunately, you cannot move your hands fast enough to hear this pulse.

If you were to be able to move your hand forward and backward at a rate that exceeds 20 times a second, you would in fact produce an audible sound. Your hand would be acting essentially like a loudspeaker, as shown below in Fig. 2.4. At the left we see the gray cone of the loudspeaker moving forward and backward. There are two positions shown—one as a pair of solid brown curves, the other as a pair of dotted brown curves. The sequence of three dotted pseudo-vertical curves represents the sound wave traveling through the air.

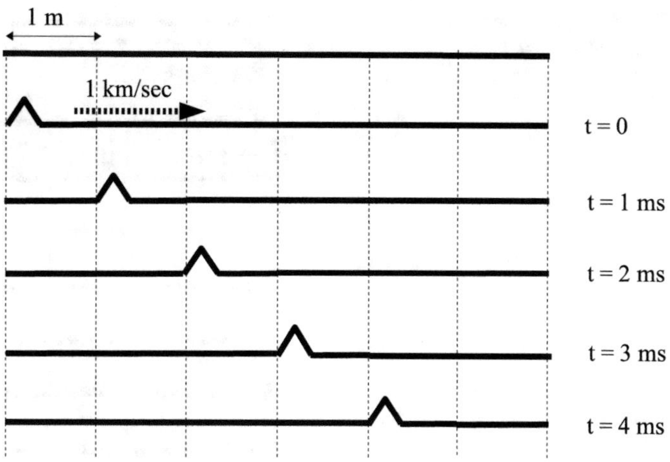

Fig. 2.2 A pulse traveling down the length of the stretched string

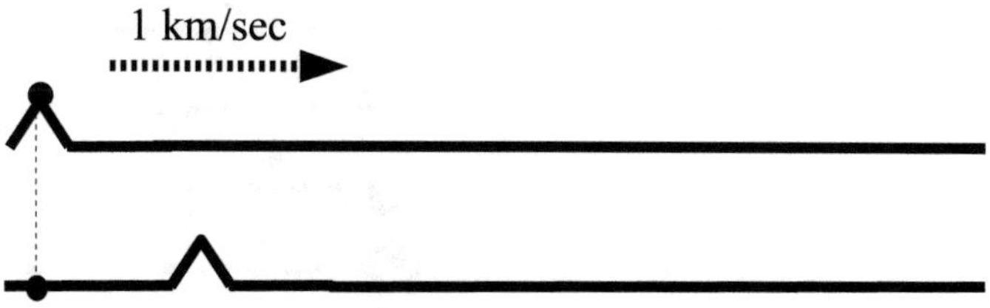

Fig. 2.3 The motion of a point—marked by a dot along the string

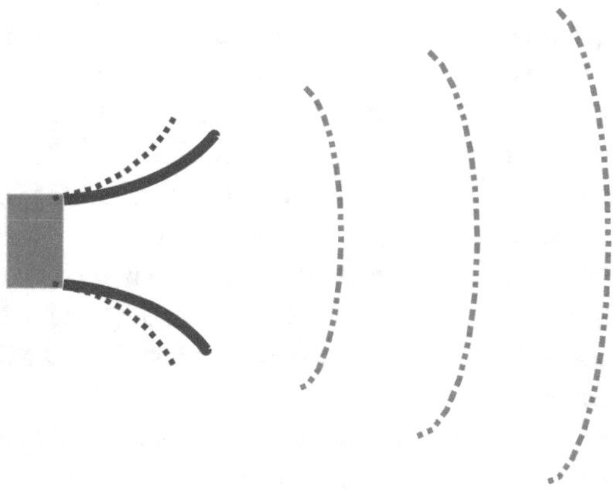

Fig. 2.4 Schematic of a loudspeaker

2.2 A Finite String Can Generate Music!

Consider now a guitar string strung on a guitar. The string considered in the previous section was assumed to be infinite; this string is finite with ends that are held fixed. See the uppermost line segment in Fig. 2.5, where we represent a string of length l =80-cm. We will assume that the wave velocity is v=400-m/s. Imagine what would happen to a pulse that is sent down the string, starting at one end, as in Fig. 2.5. The width of the pulse is exaggerated—the width is understood to be much less than a centimeter, so that it can be ignored in the calculations below.

Let us determine how long it will take for the pulse to reach the opposite end. We will use the relation

$$\text{speed} = \frac{\text{distance}}{\text{time}} \quad \text{OR} \quad \text{time} = \frac{\text{distance}}{\text{speed}} \tag{2.1}$$

We will carry out the calculation using symbols—t for time, l for distance, and v for speed. We must be careful when we are given

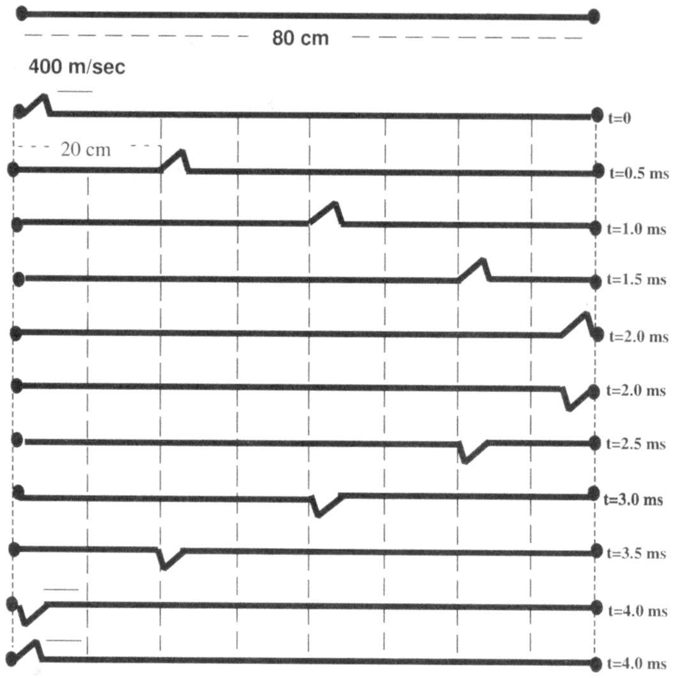

Fig. 2.5 A pulse traveling back and forth along a string with fixed ends

quantities that use different units for a given quantity. This issue is exemplified by the current situation, where we have a distance of 80-cm and a speed of 400-m/s. Thus both the centimeter and the meter are used for the dimension of length. In order to use Eq. (2.1) we must use the same unit of length for both quantities. We will choose to use the meter for both, recognizing that we could also use the centimeter for both without any error.

Since one meter = 100 centimeters, the distance is 0.80-m. We then obtain

$$t = \frac{l}{v} = \frac{0.80\text{-m}}{400\text{-m/s}} = 0.0020\text{-s} = 2.0\text{-ms} \quad (2.2)$$

We note in the figure that the pulse reaches the opposite end in 2.0-ms. The pulse is then reflected back to the left along the string.

Look closely at the shape of the reflected pulse. Notice that the shape of the pulse is "reversed" in two ways: First, the original pulse approached pointing upwards; the reflected pulse is pointing downwards. Second, notice that the original pulse is steeper on the right side compared to the left side; on the other hand, the reflected pulse is steeper on the left side.

What will happen next? The pulse will reach the left end and be reflected back to the right. The same reversals as above will take place once again. The pulse is reversed from pointing downward to pointing upward; the steeper edge is reversed from being steeper on the left side to being steeper edge on the right side. The end result is a pulse that is exactly the same as the original pulse! The time for the round trip will be 2× 2.0-ms = 4.0-ms.

Such a round trip is generally referred to as a **cycle**. Ultimately, the pulse will move back and forth, with one round trip every 4.0-ms. This time interval is called the **period**, with the symbol T. Thus,

$$T = \frac{2l}{v} = \frac{2(0.80)}{400} = 4 \times 10^{-3}\text{-s} = 4\text{-ms} \quad (2.3)$$

The number of cycles per unit time is called the **frequency**, with the symbol f. In the current case, we have

$$f = \text{one cycle per 4 ms} = \frac{1\text{cycle}}{4 \times 10^{-3} - s}$$
$$= 250\text{cycles per sec} \equiv 250\text{-cps} \quad (2.4)$$

An alternative term for the cycle per second as a unit of frequency is the **Hertz,**[1] which is abbreviated as Hz. Thus, **one cycle per second = 1-cps = 1 Hertz = 1-Hz.**

Note that the frequency and the period are inverses of each other:

$$f = \frac{1}{T} \tag{2.5}$$

In the above case, 250-Hz = 1/(4-ms).
One should note that there are many ways that the string could be excited. The most important example for a guitar is the **pluck**, which is shown in Fig. 2.6. The pluck is produced by pulling the string aside at one point and then releasing it from rest. The figure shows the subsequent motion of the string.

We note that the time for a full cycle, the period T, is again 4-ms. The corresponding frequency is 250-Hz.

2.3 Pitch, Loudness, and Timbre

If you pluck a string, a sound is produced. You can identify several attributes of that sound. There is a definite **pitch**. Pitch designates the musical note to which the string is tuned. For example, the so-called G string of the violin (which is tuned to the G below middle C on the piano) produces the pitch G. If you loosen the string, by turning the tuning peg, you will immediately notice that the pitch will change—it will become lower. If you tighten the string, the pitch will become higher.

A second attribute of the sound is its **loudness**. By giving the string a bigger pull when you pluck it, you can produce a louder sound. Furthermore, the loudness decreases after the initial pluck, until the sound is inaudible.

The third attribute is what we identify with the quality of the sound produced by the particular instrument—the **timbre**. Timbre is one of the factors that enables you to distinguish the G played on the violin from an equally loud G played on a piano or a trumpet or any other instrument. You can vary the timbre of the plucked string itself by changing the point at which you pluck as follows: First pluck the string near its center and listen carefully to the quality of the sound. Then pluck the string very near one end, trying to produce the same loudness. The pitch will be the same but there will be a slightly different timbre to the sound. When plucked near the end the resulting sound has a slight high-pitched ring or "twang," which is not present in the sound produced by plucking near its center. Similarly, if a narrow pulse is cycling back and forth along the string, a sound will be produced having the same pitch but different timbre. A bowed string produces a wave that moves back and forth the length of the string with a different characteristic shape; yet again, we will hear a sound with the same pitch.[2]

We have not been very precise, at this point in defining pitch, loudness, and timbre. To be more precise you must first understand what physical phenomena give rise to the "perceptual" qualities we have discussed.

2.4 The Relation Between Frequency and Pitch

Recall that in discussing pitch, we said that if the string being plucked were loosened, the pitch would become lower. Imagine loosening the string of Fig. 2.6 and then plucking it, so that at the moment of release it has exactly the same shape as that in the first frame of the figure. However, it will take more time to complete one cycle. The period will increase, with a consequent decrease in the number of oscillations per second; that is, the frequency will decrease. This is in agreement with Eq. (2.5).

Let us suppose that the string is loosened just enough to increase the period to 5-ms. Then the

[1]Named after Heinrich Hertz (1857–1894). Hertz was a great physicist who first demonstrated the existence of electromagnetic waves, which will be discussed later in this book.

[2]The sound of the violin is strongly affected by the other physical components of the instruments, along with their respective vibrations.

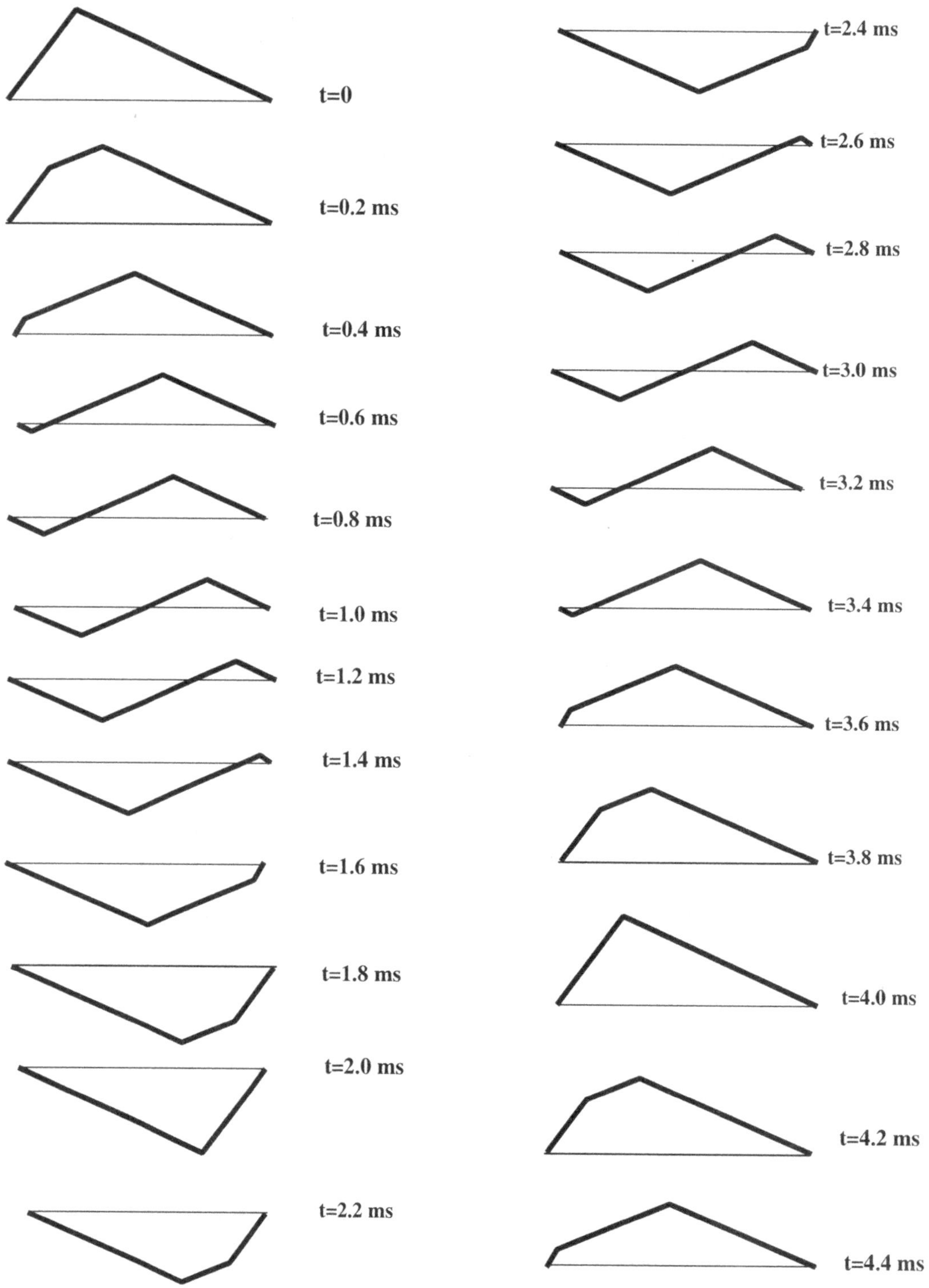

Fig. 2.6 The progressive wave along a plucked string

new frequency is $f = 1/0.005$-s per cycle = 200-cps = 200-Hz.

How much loosening does this change require? To answer this question, we need a quantitative measure of the "tautness" or tension of the string and how that tension is related to frequency. We will return to this question in Sect. 2.8. What we want you to consider at the moment is the qualitative result of this little experiment.

Loosening the string decreases the frequency of the oscillations, as it lowers the pitch. Correspondingly, tightening the string increases the frequency and raises the pitch. So there is a relation between the physical quantity, frequency, and the psychological attribute, pitch. This relationship is the basis for the tuning of musical instruments. In Chap. 11 we will note that the loudness of a note also affects the sense of pitch.

The strings of a piano are tuned to a definite set of frequencies. First one sets the A above the middle key on a piano—referred to as "middle-C"—at a frequency of 440-Hz. The "middle C" on a piano is set at a frequency of approximately 262-Hz. The lowest C is set correspondingly to a frequency of approximately 33-Hz, and so on.

We see that one way to produce different frequencies is to vary the tension. Are there other ways? If we combine Eqs. (2.3) and (2.5) we obtain the relation

$$f = \frac{v}{2\ell} \tag{2.6}$$

We will see later in the chapter that an increase in the tension produces an increase in the wave velocity. As a consequence, according to Eq. (2.6) the frequency will increase and so will the pitch. We will also see later that changing the nature of the string itself will change the wave velocity. Finally, we see that decreasing the length of the string will increase the frequency. All three factors are used to produce the huge range of frequencies of a piano—from 27.5 Hz to $\sim$ 4186 Hz.[3]

Various stringed instruments are tuned accordingly. For example, in order that the A string on the violin be in tune with the corresponding A string on the piano, their frequencies should be equal.

Why the particular frequency of 440-Hz is chosen for the "A" is a matter of history. In fact, this frequency has been rising steadily over the past two-hundred or more years, so much so that in Bach's time it is believed to have been about 415-Hz. Why the notes of the Western scale have the frequencies to which we have just alluded will be the subject of Chap. 12. The development of scales is a fascinating story of the interdependence of scientific understanding and esthetics.

2.5 The Wave Motion of a Stretched Rope

It is difficult to study the motion of the strings of musical instruments without special equipment because the wave velocities and the frequencies are very large. It is possible to check the relation (2.6) by performing a simple, but illustrative experiment. Get a long piece (2 or 3 meters) of heavy rope or clothesline or a long tightly wound spring (as used to close screen doors). Secure one end to a fixed point—say a doorknob on a closed door. Pull the free end so that the rope is stretched loosely to its full extent, as shown in Fig. 2.7a. Estimate the length, ℓ, of the stretched rope.

You are going to set up wave motion of the rope by shaking the held end up and down while using your wrist as a pivot, as shown in Fig. 2.7b. By shaking very slowly at first and gradually increasing the rate of shaking, you will soon reach a rate that sets up a wave of the form shown in Fig. 2.7c. The whole rope will be oscillating up and down at that rate. Notice that once you set up that motion it is easy to maintain the motion. It is as if the system has "locked in" to that mode of oscillation.

[3]The lower frequency is precisely four octaves (a factor of $2^4 = 16$ below 440-Hz), while the latter frequency corresponds to tuning according to equal temperament. (See Chap. 12).

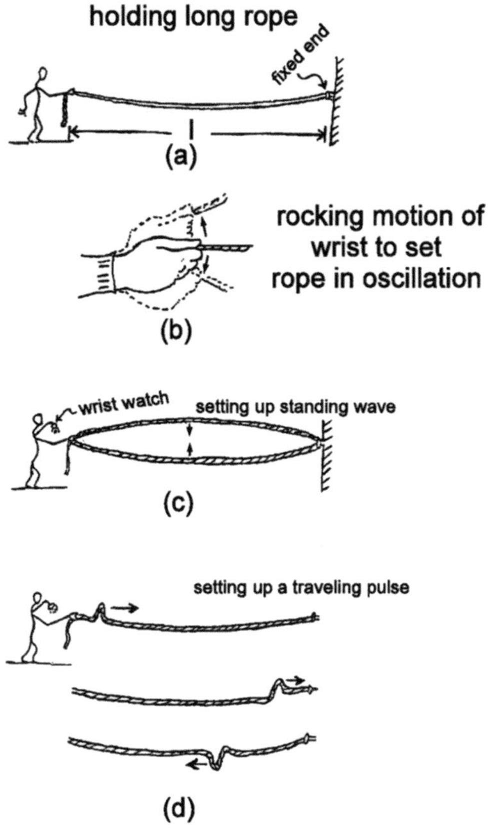

Fig. 2.7 Exciting a long rope [drawing by Gary Goldstein]

While maintaining the motion of Fig. 2.7c, use the seconds hand on your wrist watch to determine the period. (You might have a friend assist you.) This can be done easily by counting, say 10 cycles and observing how many seconds have elapsed. Remember that a cycle is completed when the rope has returned to some initial configuration, so whenever it reaches the lowest point in its motion it has completed a cycle. If, for example, the rope completes 10 cycles in 8 s, the period would be $8/10$ s. The frequency would be $10/8 = 5/4$ Hz.

Next, let the rope return to rest. It is important not to vary the tension, so do not change your position. Now you are going to set up a disturbance in the rope of the form shown in Fig. 2.7d. This is accomplished by very quickly jerking your hand up and down while quickly returning to the starting position. It is best to keep your hand as rigid as possible. Observe what happens. The short disturbance or **pulse** moves rapidly to the end of the rope, is reflected, and returns to your hand upside down. If your hand remains rigid, the pulse will reflect at your hand, turn right side up, and move to the far end again. The pulse might make many round trips before it disappears. You have set up a **traveling wave**. (A pulse travels across the string.)

Note that any particular segment of rope material moves up and down, while the wave pattern, the pulse in this case, moves down the length of the rope. These two directions are perpendicular to each other. The waves are **transverse**.

Now time the pulse by measuring the time required for the pulse to complete several round trips. For example, if the pulse makes 5 round trips in 4 s, then the time for a single circuit would be $4/5$ s or 0.8 s. If you are careful, you will find that the time required for one round trip is the same as the period of oscillatory wave motion that you determined before, for the standing wave.

Measuring the length of the stretched rope will then enable you to determine the velocity of propagation for the traveling wave. In our example the round trip time and the period were 0.8 s. If the rope were 2 meters long, a round trip would be 4 meters and the velocity of propagation would be $4-m/0.8-s = 5-m/s$. Determine the velocity for your rope, using Eq. (2.3).

2.6 Modes of Vibration and Harmonics

One might ask whether the string can be excited so as to produce a vibration that does not have a frequency of 250-Hz. The answer is yes. In the course of demonstrating this fact, we will describe what are referred to as the modes of vibration of the string.

Now that you have become familiar with working the rope you can learn how to excite its modes of vibration. Start by exciting the same standing wave that you did before (Fig. 2.7c). Count the cycles rhythmically while the rope is oscillating. That is, say the numbers out loud

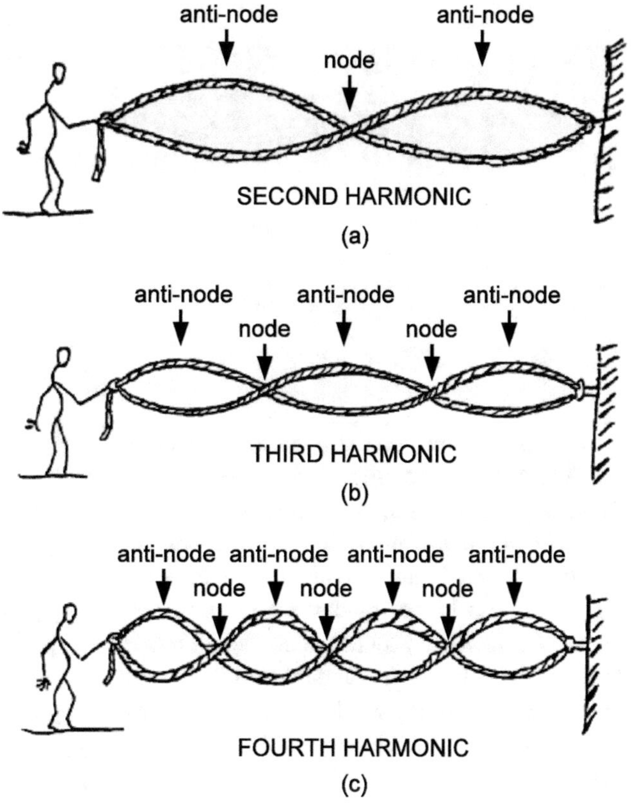

Fig. 2.8 Higher harmonics of the vibrating string [drawing by Gary Goldstein]

every time the rope reaches bottom—"one-two-three-four-one-two…". Now start shaking your hand at twice the original tempo. You will have doubled the rate of oscillations and hence the frequency. The rope will "lock into" a different mode of oscillation. It will now appear as in Fig. 2.8a. The period of this oscillation is one-half the period of the preceding oscillation, wherein a pulse is traveling back and forth along the rope, as in Fig. 2.7.

For the sake of identification we call the mode of Fig. 2.7c the **fundamental mode** or the **first harmonic** of the string. The mode of oscillation you are now producing is called the **second harmonic** (Fig. 2.8a). While the rope is oscillating in this mode notice that near the midpoint the rope is hardly moving at all. This point at which no motion occurs is called a **node**. For the second harmonic there is one node between the end points, whereas the fundamental mode (Fig. 2.7c) had no nodes between the end points.

Observe also that there are two points along the rope which achieve the greatest displacement from equilibrium (either above or below), one at about $1/4$ the distance from your hand, the other at $3/4$ the distance. These points along the rope at which the maximum displacement occurs are called **anti-nodes**. The second harmonic has two anti-nodes, whereas the fundamental mode has one anti-node at the midpoint of the rope (see Fig. 2.7 again).

Now, by shaking your hand at triple the rate for the fundamental mode you can excite the mode shown in Fig. 2.8b, the third harmonic. This is somewhat harder to excite than the preceding mode, but once you get near the right rate of shaking, the rope will respond very strongly and will "lock in" to that mode. The third harmonic has three times the frequency of the fundamental. You will observe that there are two nodes in this mode—one at $1/3$ the distance to the fixed end, the other at $2/3$ that distance. There are 3 anti-nodes.

Table 2.1 Harmonics

Mode	Frequency	No. of nodes	No of anti-nodes
Fundamental = 1st harmonic	f_1	0	1
2nd harmonic	$f_2 = 2f_1$	1	2
3rd harmonic	$f_3 = 3f_1$	2	3
4th harmonic	$f_4 = 4f_1$	3	4
5th harmonic	$f_5 = 5f_1$	4	5
6th harmonic	$f_6 = 6f_1$	5	6
nth harmonic	$f_n = nf_1$	$n - 1$	n

You should now see the pattern. By exciting the fourth harmonic (Fig. 2.8c), which has a frequency four times the fundamental frequency, you will produce a mode having 3 nodes and 4 anti-nodes (It is appreciably harder to excite this mode; the higher modes are progressively more difficult.) The fifth harmonic would have a frequency five times the fundamental frequency, and the wave pattern would have 4 nodes and 5 anti-nodes. We summarize this information in Table 2.1.

The frequencies of the harmonics are written as multiples of the fundamental frequency f_1. We have included a general mode, the nth harmonic, where n symbolizes any integer (1, 2, 3, 4, ...). Letting $n = 7$, for example, tells you that the 7th harmonic has frequency $7f_1$, $(7 - 1) = 6$ nodes, and 7 anti-nodes.

From all of the preceding you now see that the rope, or a stretched string, has many different modes of vibration. These modes of vibration have frequencies which are integral multiples of the fundamental frequency—the modes are harmonic. Then the periods for each of the modes will be different from one another. Recall, however, that the time required for a traveling pulse to make a round trip (Fig. 2.7d) was equal to the period of oscillation of the rope in the fundamental mode (Fig. 2.7c). Therefore the relation between wave velocity, length, and frequency (Eq. (2.6)) should be rewritten to show explicitly that the fundamental frequency is involved.

We have

$$f_1 = \frac{v}{2\ell} \tag{2.7}$$

For the other harmonics, the frequencies are multiples of the fundamental frequency, so

$$\begin{aligned} f_2 &= 2f_1 = 2 \times \frac{v}{2\ell} \\ f_3 &= 3f_1 = 3 \times \frac{v}{2\ell} \\ f_4 &= 4f_1 = 4 \times \frac{v}{2\ell} \end{aligned} \tag{2.8}$$

and so on. This series is summarized by writing the frequency of nth harmonic, f_n, as

$$f_n = n\ f_1 = n \times \frac{v}{2\ell} \tag{2.9}$$

The fact that the rope or stretched string can be set into oscillation in many different modes will be of continual importance. It forms the basis for much of the subsequent discussion.

Notice that the wave patterns for these modes don't move, either to the right or to the left. We refer to such a wave as a **standing wave**. In contrast, the wave described initially in this chapter that moves along an endless string is called a **traveling wave**. We will discuss such waves more fully in the next section.

We close this section by introducing other widely used terms—the **overtone** and the **partial**. By definition, the first overtone is the second harmonic; the second overtone is the third harmonic; and so on. The term **partial** is used to refer to one of the modes of a musical instrument **whether or not** the frequencies form a harmonic series. An example is the sound of a gong, whose mode frequencies do not form a harmonic series.

2.7 The Sine Wave

The shape of the pattern along a string that is vibrating in one of its modes is very specific—being the curve produced by plotting the trigonometric **sine function**. In addition, if we plot the displacement of any point along the string vs. time, we will obtain a graph of the sine function. In fact, of all periodic curves in nature, the sine curve is very unique in its physical ramifications, as we will see many times in the course of our study of sound and light and therefore of music and color. Thus, we now turn to an examination of the sine curve.

You probably have paid attention to how various radio stations are identified. For example, a popular radio station for classical music in the Boston area is WCRB 102.5FM. The number 102.5 stands for a frequency of 102,500,000-Hz. [The letters "FM" stand for "frequency modulation," which is a special means of transmitting information using waves; it will be discussed later in the text.] Or, as another example, you might have heard that most current symphonic orchestras are tuned to a frequency of 440 cycles per second. What these numbers fully represent is the subject of this section.

Let us begin by returning to the long stretched string. Suppose that you were to take hold of the string and move it up and down repeatedly at a constant rate in time. If the pattern of your motion is repeated again and again, we say that the pattern is periodic. As an example let us display the pattern of motion for the most important such motion; it is called a **sine wave** pattern. See Fig. 2.9.

The graph displays the displacement of the hand as it varies in time. Note that there is a pattern that extends over a one-second interval. It is repeated three times over the entire three second interval. This interval is called the **period** or the motion. The maximum displacement is 1-cm and is called the **amplitude** of the motion. The pattern is **sinusoidal** and represents the **sine function** of trigonometry. Let us review the nature of the sine function.

You might recall that the sine of an angle is the ratio of the "side opposite" to the hypotenuse of a right triangle. Thus,

$$\sin\theta = \frac{b}{c} \tag{2.10}$$

We can produce a graph of the sine function by a simple method involving the constant circular motion of the seconds hand of a clock. The seconds hand sweeps around, making a full circle every 60 s. Let us measure the vertical position of the tip of the hand as it sweeps around. We do this by first drawing a base line across the face passing through the center and the 3 and 9 o'clock marks, as shown in Fig. 2.10. Suppose the hand extends 5-cm from the center. Then the *vertical position* of the tip, relative to the base line, will vary from the lowest point (at the 6 o'clock mark) of −5-cm to the highest point (at the 12 o'clock mark) of +5-cm. When the hand points at 11 o'clock, for example, as shown in Fig. 2.10, the vertical position will be +4.3-cm.

Now we will plot the vertical position as the hand sweeps around, starting at the 9 o'clock position when the vertical position is 5.0-cm. Five seconds later the hand will be at 10 o'clock

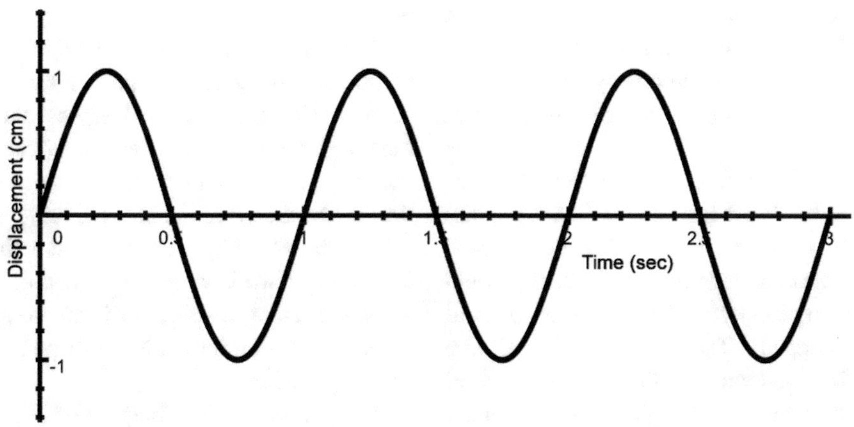

Fig. 2.9 Sine wave of displacement vs. time

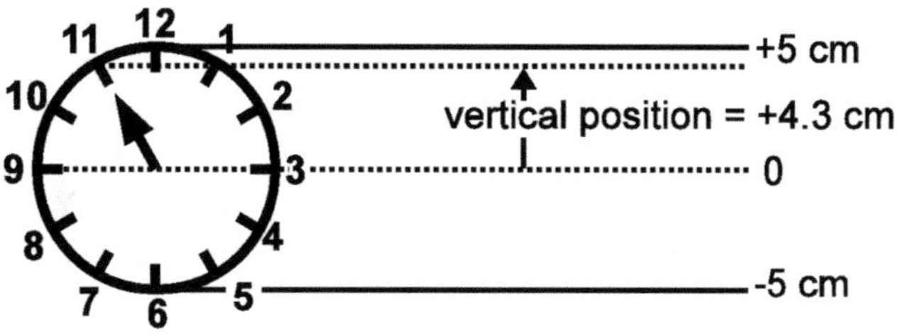

Fig. 2.10 Sweeping hand of a clock defining a sine wave

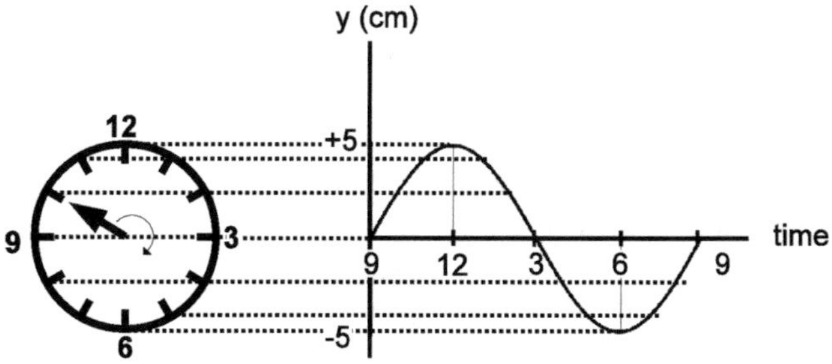

Fig. 2.11 Clock defining one cycle of a sine wave

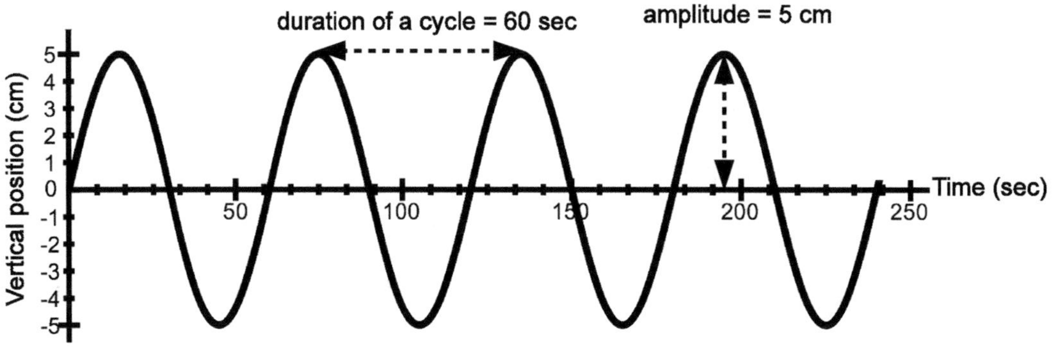

Fig. 2.12 Vertical position of clock hand vs. time elapsed

and the vertical position will be +2.5-cm. Then in 5 s more, the hand will be at 11 o'clock and the vertical position will be +4.3-cm, and so on. The procedure of plotting the vertical position as a function of time elapsed is illustrated in Fig. 2.11, for the first 60 s; we obtain one cycle of the sine wave. Continuing this plotting gives the curve in Fig. 2.12.

The form of the curve repeats exactly every 60 s, when the hand has returned to its initial position. If we continued plotting the vertical position of the hand indefinitely, the curve would continue on repeating itself indefinitely (or until the clock stopped). The full curve is the sine function. It is **periodic** in that it repeats itself indefinitely. Note that the angle changes steadily, going through a full circle of 360 degrees in 60 s. Thus, the rate of change of the angle is 360°/min, or 6°/s. Then, the value of the sine function after 5-s will be $\sin 30^\circ = 0.5$. The result is a vertical position equal to 0.5 (5.0) = 2.5-cm.

Although we have obtained this curve by a particular procedure, its significance is far more general. Being a graphical representation of a function, it represents a mathematical prescription: Given some numerical value of the variable, the sine of that variable has a definite numerical value. The variable may represent a time (as in the example we have used), or it may represent a position along a string, or it may represent an angle. What is important is the shape of the curve and its periodicity.

The particular sine function of Fig. 2.11 can be characterized by two numbers. The first of these is the maximum height of the curve, called the **amplitude**, which is 5-cm for this case. The second is the length of one cycle, which is a time interval of 60 s for this example. When the variable is time, the length of one cycle, its duration, is the **period** of the motion. We will soon consider examples of sine functions representing some vertical position as a function of distance rather than time. In that circumstance the length of one cycle will be a distance, and is called the **wavelength**.

2.8 The Simple Harmonic Oscillator

While the sine curve is central to the behavior of the modes of a vibrating string, it shows itself in a simpler way the behavior of a **simple harmonic oscillator** (**SHO** for short). The SHO is a system that is fundamental for understanding all vibrating systems and therefore deserves significant attention. It consists of a spring having negligible mass that is attached at one end to some fixed support and a rigid object that has the essential mass of the system—referred to as the **mass** of the SHO—at the other end. See Fig. 2.13 below.

When isolated, an SHO will come to rest at its **equilibrium state**, in which the spring is neither stretched nor compressed. To displace the mass from its equilibrium position, a force must be applied. That force F is proportional to the displacement y from the equilibrium position of the mass, as shown in Fig. 2.13b, in which a **downward** displacement corresponds to positive y, while an **upward** displacement corresponds to a **negative** y. Because the graph of y vs F is a straight line, we say that the relation between y and F is **linear**. We have

$$\textit{displacement} \propto \textit{Force} \tag{2.11}$$

Mathematically we write

$$y = \frac{1}{k}F = \frac{F}{k} \tag{2.12}$$

where the constant k is known as the **spring constant** or the force constant. The relation is known as **(Robert) Hooke's Law**.

If the force is measured in lbs and the displacement in inches, the spring constant is expressed in lbs-per-in, which will be written as "lbs/in." In this text, the force will often be expressed in Newtons (abbreviated as N) [one Newton is about 4.5-lbs] and the displacement will be expressed in meters. Then the spring constant will be expressed in Newtons per meter, or N/m.

Sample Problem 2.1 Suppose it takes a force of 5-N to stretch a given spring 2-m. Find the spring constant.

Solution The spring constant is given by

$$k = \frac{F}{y} = \frac{5}{2} = 2.5\text{-N/m} \tag{2.13}$$

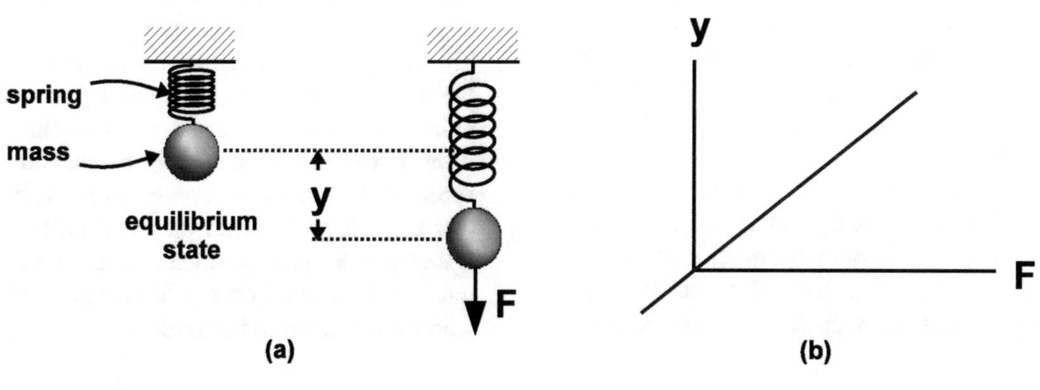

Fig. 2.13 The simple harmonic oscillator

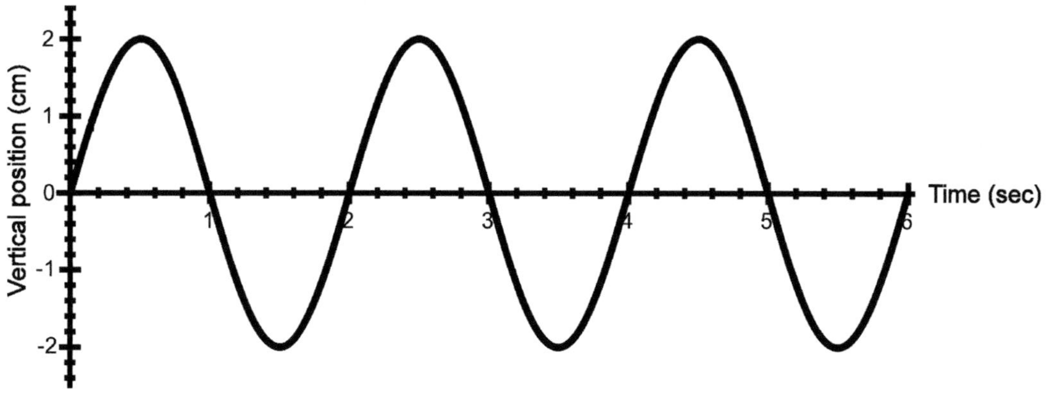

Fig. 2.14 Displacement of an SHO vs. time

Note that if that same spring is stretched by a force equal to 10-N, the displacement will be

$$y = \frac{F}{k} = \frac{10}{2.5} = 4\text{-m} \qquad (2.14)$$

Doubling the force leads to a doubling of the displacement.

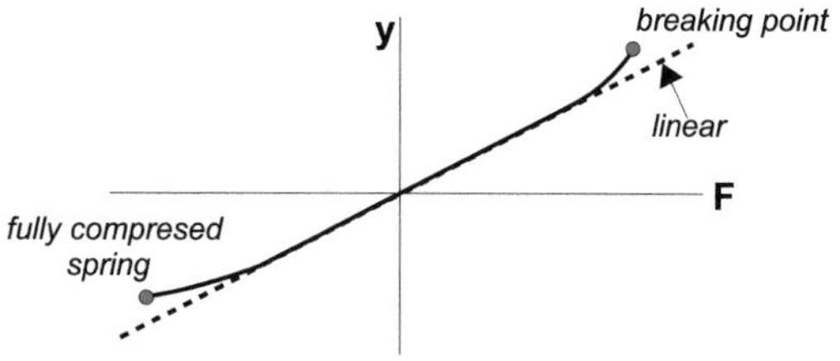

Fig. 2.15 Displacement vs. force for a real spring

If the mass is pulled from its equilibrium position as in Fig. 2.13 and released, it will oscillate in time at a certain frequency.

The *linear* relation between y and F is **unique** in leading to two characteristics:

1. a **sinusoidal** displacement in time, as shown in Fig. 2.14.
2. a frequency that is **independent of the amplitude of oscillation**.

Let us imagine suspending a mass to a spring and letting it oscillate. We see in Fig. 2.14 that the displacement of the mass exhibits a sine wave pattern. Its amplitude is 2-cm. The period is 2 s. The corresponding frequency is f=1/T=1/2=0.5-Hz.

In fact, the two characteristics of oscillatory motion automatically imply a linear relation. Real springs do not obey **Hooke**'s Law precisely, as shown in Fig. 2.15. However, they do so approximately for small enough displacements.

2.8.1 The Physical Basis for the Oscillation of a Simple Harmonic Oscillator

We can understand why an SHO oscillates by studying the various parts of one cycle of oscillation:

1. Start the oscillation by pulling the mass upward a bit and releasing it from rest.
2. The mass will then move downward because it is pulled by the stretched spring.
3. The mass will not come to rest at the equilibrium position because it is moving at that point and cannot come to rest instantaneously. Instead, it moves down further, stretching the spring, which brings it to rest. It can be shown that the rest position has the same distance from the equilibrium position as did its maximum position above the equilibrium position, where it was released from rest.

4. The stretched spring will pull the mass upward so that the mass will move upwards, past the equilibrium position and on to the maximum height it had when it was first released from rest.

   This completes one cycle, which is repeated again and again.

2.8.2 The Vibration Frequency of a Simple Harmonic Oscillator

The spring constant and the mass of an SHO determine its vibration frequency. We will see later that the fundamental frequency of a vibrating string is proportional to the square root of the ratio of a restoring force to a mass. This qualitative relationship holds for an SHO too. It can be shown that the frequency of vibration of the SHO is given by[4]

$$f = \frac{1}{2\pi}\sqrt{\frac{k}{m}} \qquad \text{Frequency of SHO} \tag{2.15}$$

The spring constant reflects the restoring force.

Using this formula, we are not free to express the units of k and mass independently. The choice of units must be consistent. Thus, if we express the spring constant in N/m, the mass must be expressed in kilograms (abbreviated as "kg"). Then the frequency we obtain is expressed in Hz.

Sample Problem 2.2 Suppose that an SHO has a spring constant equal to 25-N/m and a mass of 500-g. Find the frequency and period of vibration of the SHO.

Solution We must first express the mass in kg: 500-g = 0.500-kg. Then

$$f = \frac{1}{2\pi}\sqrt{\frac{k}{m}} = \frac{1}{2\pi}\sqrt{\frac{25}{0.500}} = 1.1\text{-Hz} \tag{2.16}$$

Correspondingly, the **period** of vibration is given by

$$T = \frac{1}{f} = 2\pi\sqrt{\frac{m}{k}} \qquad \text{Period of SHO} \tag{2.17}$$

so that T = 1/1.1 = 0.9-s.

We note that generally, the frequency increases if the spring constant increases and/or the mass decreases. However, if the spring constant is doubled or the mass is halved, the frequency is not doubled. Instead, it is increased by a factor of $\sqrt{2}$: Taking off from the previous numerical example, suppose that the spring constant is 50-N/m and the mass is 0.500-kg. A simple calculation leads to a frequency of 1.6-Hz, which equals $\sqrt{2}$ multiplied by the frequency of the previous example.

We can obtain an estimate of the amplitude of the velocity—the maximum speed—of an SHO over the course of one oscillation. It is on the order of the average speed. The latter is simply the total distance traveled by the mass in one cycle divided by the period. Thus, with A = amplitude

$$\text{average speed} = \frac{4A}{T} = 4Af \tag{2.18}$$

The actual velocity amplitude is $2\pi Af$, which is a bit greater, as it should be.

2.9 Traveling Sine Waves

The modes of vibration of a string of fixed length are such that there is a pattern that remains stationary except for oscillations in the overall amplitude. The pattern doesn't move to the right or the left. To the contrary, a wave that progresses

[4] This expression for the frequency as well as the fact that the motion is sinusoidal can be derived rigorously mathematically by making use of a combination of Hooke's Law $F = kx$ and Newton's Second Law of Motion $F = ma$, where a is the acceleration of the mass. Eliminating the force leads to a direct relation between the displacement and the acceleration: $x = ma/k$. This subject is discussed in Appendix F. In Appendix G, you can see how the sinusoidal motion evolves by using a technique of Numerical Integration.

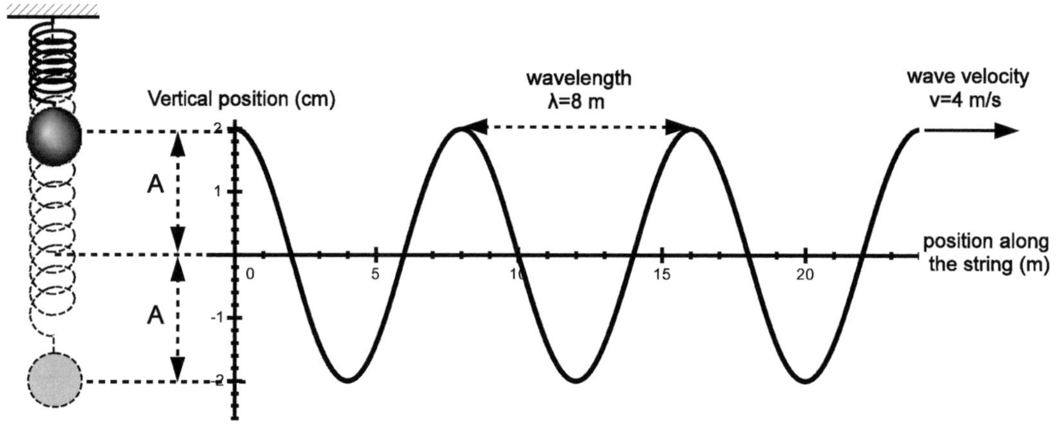

Fig. 2.16 Traveling wave produced by an SHO

in one or the other direction is referred to as a **traveling wave**. Here is a simple way to produce a traveling sine wave.

Suppose that we attach the mass of an SHO to the end of a long stretched string. The mass is set into oscillation with an amplitude of 2-cm, as above. In Fig. 2.16 we see the mass at an instant when it has its maximum upwards displacement of 2-cm. To its right, we see the sinusoidal pattern of the wave along the string. Note that this curve represents the actual material of the string at this instant in time. Furthermore, **whereas the displacement of the oscillator is sinusoidal in *time*, the pattern of the string is sinusoidal in *space*.**

We have assumed that the wave velocity is 4-m/s. As a consequence, during one cycle of oscillation lasting 2 s, the wave moves along the string a distance $x = vt$=4(2)=8-m. Each cycle that has its left end in contact with the mass is replaced by another such cycle. The three cycles along the string were produced by three cycles of oscillation of the mass. The period in space is called the **wavelength** and is here equal to 8-m.

We see that the sinusoidal wave in space is characterized by the following four parameters:

1. the **amplitude** A—here equal to 2-cm,
2. the **wave velocity** v—here equal to 4-m/s,
3. the **wavelength** λ—here equal to 8-m,
4. the **period** T—here equal to 2 s.

Clearly, we have a simple relation among the velocity, the wavelength, and the period:

$$v = \frac{\lambda}{T} \tag{2.19}$$

The frequency and period are inverses of each other: $f = 1/T$. Therefore, we have the relation

$$\lambda f = v \tag{2.20}$$

Notice that this equation can be rewritten as $\lambda = v/f$. As a consequence, for a given velocity, the wavelength decreases as the frequency increases. Alternatively, if the frequency is constant and the velocity decreases, the wavelength must decrease.

This latter result can be understood in terms of the following traffic situation. Suppose a line of cars is traveling along a one lane road over a long time so that the traffic flow is stationary. The number of cars passing a given point must then be constant so that there is neither a pile up of cars someplace nor any buildup of empty space. Then, if the cars speed up, they must be further apart. Similarly, if the cars slow down, the space between the cars must decrease. A consequence of this decrease in space and the need for safety is that the cars usually slow down even more. Figure 2.17 illustrates how the spacing between neighboring cars decreases when the speed decreases. The rate at which cars pass point P is $30/15 = 2$ cars per second, which is equal to the

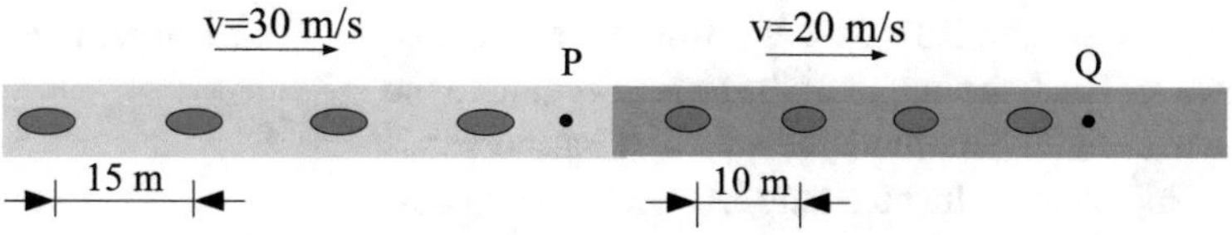

Fig. 2.17 Spacing vs. speed of cars in traffic

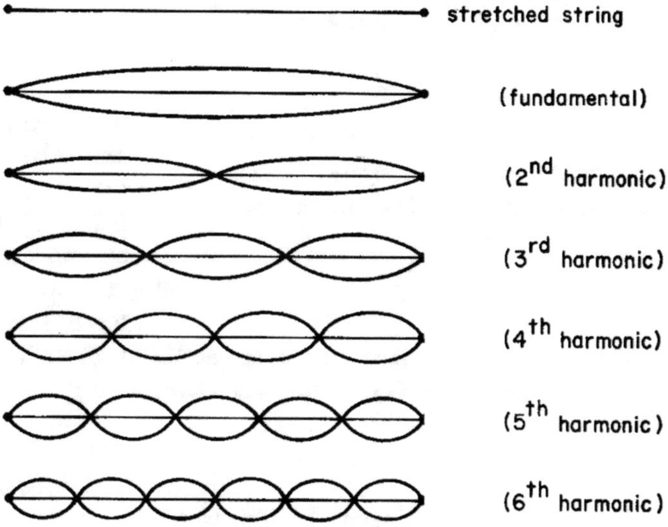

Fig. 2.18 Modes of vibration of the stretched string

rate at which cars pass point Q, that is, $20/10 = 2$ cars per second.

Applications

1. In the case of the standing wave, we recall that $\lambda = 2\ell$. Then, as we have already shown.

$$v = 2\ell f_1 \quad \text{or} \quad f_1 = \frac{v}{2\ell}$$

2. The audible range of frequencies is from 20-Hz to 20,000-Hz. In the case of a traveling sound wave in air, with v = 340-m/s, the corresponding range of wavelengths is

$$\max\lambda = 340/20 = 17\text{-m} \ \textbf{to}$$

$$\min\lambda = 340/20,000 = 0.017\text{m} = 1.7\text{-cm}.$$

3. As we will see in Chap. 5, light is a visible electromagnetic wave having a range of frequencies from 4.0×10^{14}-Hz to 7.0×10^{14}-Hz. In the case of a light wave traveling in vacuum, the wave velocity is given the symbol c and is equal to 3.0×10^8-m/s. The corresponding range of wavelengths is

$$\max\lambda = \frac{3.0\times10^8}{4.0\times10^{14}} = 7.5\times10^{-7}\text{m} = 750\text{-nm}$$

to

$$\min\lambda = \frac{3.0\times10^8}{7.0\times10^{14}} = 430\text{-nm}.$$

2.10 Modes of Vibration—Spatial Structure

The modes of vibration of a stretched string are related to traveling sine waves. Later on in this chapter we will see that when two sine waves of the same wavelength head towards each other, they superpose to produce a standing wave with the same wavelength. The standing waves of the modes of vibration are portions of sine waves. An examination of their shapes reveals the relationship between these shapes and the corresponding wavelengths. Below in Fig. 2.18 we show the first six harmonics of the vibrating string. The patterns display the extreme shapes at two times, one-half cycle apart.

The second harmonic is a full cycle of a sine wave. Thus, the length of the string is equal to the wavelength. The fundamental (first harmonic) is a half cycle, so that the length is equal to one-half of a wavelength.

The wavelength for the fundamental is

$$\lambda_1 = 2\ell \qquad (2.21)$$

For the second harmonic, the shape of the string encompasses a full cycle of a sine wave, so the second harmonic has a wavelength

$$\lambda_2 = \frac{\lambda_1}{2} = \ell \qquad (2.22)$$

In the third harmonic mode, the string has the form of one and one-half cycles of a sine wave. Thus

$$\lambda_3 = \frac{2}{3}\ell \text{ or } \frac{3}{2}\lambda_3 = \ell \qquad (2.23)$$

Lastly, the fourth harmonic has the wavelength

$$\lambda_4 = \frac{1}{2}\ell \qquad (2.24)$$

This sequence of wavelengths can be rewritten in a way that allows us to generalize these examples:

$$\begin{aligned} \lambda_1 &= \frac{2\ell}{1} = 2\ell \\ \lambda_2 &= \frac{2\ell}{2} = \ell \\ \lambda_3 &= \frac{2\ell}{3} = \frac{2}{3}\ell \\ \lambda_4 &= \frac{2\ell}{4} = \frac{1}{2}\ell \end{aligned} \qquad (2.25)$$

Written in this way it is obvious that the fifth harmonic will have wavelength $\lambda = 2\ell/5$, and so on. For the nth harmonic, then, we will have the relation

$$\lambda_n = \frac{2}{n}\ell \qquad (2.26)$$

Now recall that according to Eq. (2.9) the frequency for the nth harmonic is given by $nv/2\ell$. The formula for the frequency is then

$$f_n = \frac{v}{\lambda_n} \qquad (2.27)$$

We can write this equation also as

$$\lambda_n f_n = v \qquad (2.28)$$

The equation becomes identical to Eq. (2.10), which applies to the two sine waves that when added together produce the standing wave.

We can appreciate this process if we realize that when we excite a standing wave by shaking one end of the string, we send a sine wave down the string. This wave is reflected off the other fixed end. Upon reflection we have a second sine wave with the identical wavelength traveling back towards the hand. This sine wave adds together with the sine wave that we are sending with our hand to produce the standing wave.

2.11 The Wave Velocity of a Vibrating String

It is well known to players of string instruments that the pitch and hence fundamental frequency of a string increase with increasing tension. This fact is connected with the increase of the wave velocity with increasing tension. Similarly, one notices that for given string material, the pitch of a string decreases with increasing string thickness. This fact is connected with the decrease of the wave velocity with increasing mass of string, for given length of string. This section is concerned with the parameters that determine the wave velocity and the precise relationship among them.

It turns out that the wave velocity depends upon two parameters that characterize the string. First we have the **Tension**, with the symbol T.

Note It is important to distinguish the symbol for the tension T from the symbol for the period *T*.

The tension acts to restore the string to its equilibrium shape and favors a greater wave velocity. The second parameter is the **mass per unit length**, with the Greek letter μ as a symbol. This parameter is also called the **linear mass density**. The mass of an SHO reflects its resistance to having its velocity change—that is, being

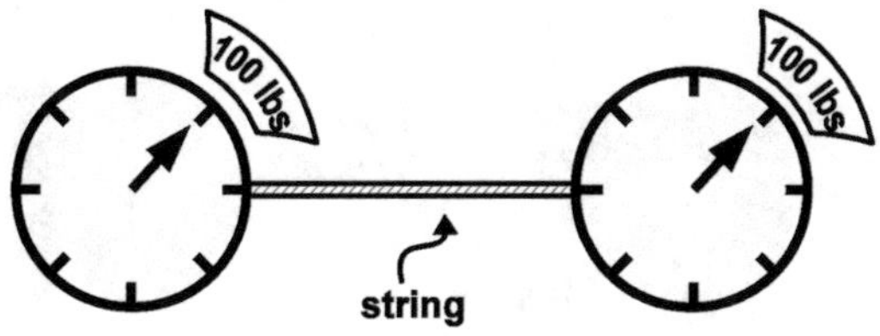

Fig. 2.19 String pulled from two directions

accelerated. Similarly, the linear mass density of a string reflects the string's resistance to having any point along the string undergo a change in velocity. This set of changes is what constitutes a wave. Just as an increase in the mass of an SHO decreases its vibration frequency, an increase in the linear mass density decreases the wave velocity.

We will now discuss these two parameters in greater detail.

Let us turn our attention to tension. This parameter is measured in units of force such as the "pound" (*lb*) or the **Newton** (named after Isaac Newton), abbreviated as N. (The two are related as follows: $1 - lb$ = 4.5-N.) A common device for measuring tension is a "spring scale".[5] On the average, a string of a stringed instrument is under a tension of about 50-lbs (thus about 200-N). We will later show that the total tension of the strings of a piano is on the order of 70,000-lbs!

The physical parameter **force** has direction as well as magnitude. Thus, for example, the gravitational force of the earth on a person is downward, towards the center of the earth. On the other hand, *tension has no directionality*. This fact is illustrated in Fig. 2.19, wherein a string is being pulled on by two spring scales, one to the right and one to the left.

Both scales read a force of 100-lbs. This implies that the right scale pulls on the string with a force of 100-lbs to the right, while the left scale pulls on the string with a force of 100-lbs to the left. These two forces cancel each other out, so that the net force acting on the string is zero and the string can remain stationary in this situation.

[5] If you attach a spring to one end of a string that is under tension, the tension is proportional to the consequent displacement of the spring.

The resulting tension, T, in the string might not be obvious: It might seem that the two applied forces add to produce a tension of 200-lbs. In fact the tension is 100-lbs. We write

$$T = 100 - lbs \tag{2.29}$$

Note

The Nature of Tension

How can we understand the above enigma, that the tension is not 200-lbs? We will be able to answer this question by examining what the tension represents. So, let us imagine two people facing each other with their right arms outstretched and clasping each other's hand. Call them, Richard and Lisa. Richard's shoulder pulls his arm with a force of 10-lbs and so does Lisa's shoulder pull her arm with a force of 10-lbs. Richard's hand pulls Lisa's hand towards him with a force of 10-lbs and correspondingly, Lisa's hand pulls Richard's hand towards her with a force of 10-lbs. We say that there is a tension of 10-lbs at the point where the two hands are clasped.

Now let us turn to the string in Fig. 2.19. We consider two segments of this rope, labeled L and R, respectively, shown at the top of Fig. 2.20a, whose boundary is marked by the letter "P." At the bottom of Fig. 2.20a we focus on segment L, noting the two forces acting on this segment that balance each other—100 lbs by the scale to its left and 100 lbs by the R segment to its right. [In fact, the L segment pulls on the R segment towards L with a force of 100-lbs and correspondingly, the R segment pulls on the L segment towards R with a force of 100-lbs.] Since the point P is arbitrary, the tension is uniform all along the string and is said to be 100-lbs.

This is not so for a string with weight that is hanging from a support, as shown in Fig. 2.20b. At any point along the string, the tension equals the weight of string below that point. Thus, the tension vanishes at the bottom end and equals the total weight of the string at the top end, where the string is supported.

Question: If the string weighs 8oz, what is the tension at the midpoint of the string?

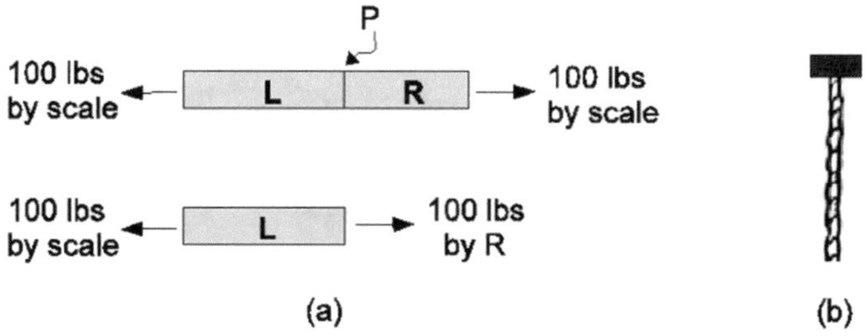

Fig. 2.20 (**a**): the rope viewed as two segments; (**b**): a hanging rope

We now turn to the **linear mass density**. Suppose we have a spool of string and cut off a meter length of string and find that it has a mass of 5-g. Then the linear mass density of the string in the spool is 5-g/m =0.005-kg/m. A two-meter length of such a string would have a mass of 10-g. As a result the linear mass density will be 10-g/2m=0.005-kg/m. The result is that there is no change in the linear mass density. Generally, for a length of string ℓ with a mass m, we have the relation

$$\mu = \frac{m}{\ell} \tag{2.30}$$

The linear mass density of the string of the spool is independent of the length of the string.

We are now ready to reveal the relation between the wave velocity and the two parameters, the tension T and the mass density μ. It is given by

$$v = \sqrt{\frac{\mathrm{T}}{\mu}} \tag{2.31}$$

which is the wave velocity along a string.

Note that the wave velocity involves a square root of a force-like parameter (here T) divided by a quantity which reflects mass or inertia (here μ):

$$\text{wave velocity} = \sqrt{\frac{\text{force-like parameter}}{\text{mass-like parameter}}} \tag{2.32}$$

The force-like parameter is usually referred to as the **restoring force**.

This result is universal for all types of waves.

In this case, tension is the restoring force.

Now we can see fully how the fundamental frequency depends upon the three parameters of a vibrating string—the length, the tension, and the linear mass density. Using Eq. (2.6) and (2.31) we obtain

$$f = \frac{v}{2\ell} = \frac{\sqrt{\frac{\mathrm{T}}{\mu}}}{2\ell} \tag{2.33}$$

Thus, we have found that **the fundamental frequency is proportional to the square root of the tension, inversely proportional to the square root of the linear mass density, and inversely proportional to the length.**

Sample Problem 2.3 Suppose that a violin string has a length of 33-cm and a linear mass density of 6-g/m, and has a fundamental frequency of 440-Hz. Find the wave velocity, the mass of the string, and the tension in the string.

Solution

$$v = 2f\ell = 2(440\text{Hz})(0.33\text{m}) = 290\text{m/s}$$

$$m = \mu\ell = 6 \times 10^{-3} \times 0.33$$
$$= 1.98 \times 10^{-3}\text{-kg} = 1.98\text{-g}$$

To obtain the tension is a bit more complicated because it appears within a square root:

$$v = \sqrt{\frac{\mathrm{T}}{\mu}}$$

so that

$$v^2 = T/\mu$$

and

$$\mathrm{T} = \mu v^2 = (0.006\text{kg/m})(290)^2 = 506N$$

Application of the Above Relations to the Piano

We will now review in a bit of detail the methods whereby the huge range of frequencies (and hence pitches) of piano strings, from 27.5-Hz to 4,186-Hz, can be obtained:

To increase the pitch, one can

- increase the tension,
- decrease the linear mass density, or
- decrease the length.

We can obtain an idea of the **total tension** on the block of metal that supports the strings as follows:

From the relation

$$v = \sqrt{\frac{T}{\mu}}$$

we obtain $T = \mu v^2$.

In order to estimate the average wave velocity we will use the relation $v = 2Rf$. Since the average frequency is about 500-Hz and the average length of the strings is about 0.5 meter, we obtain $v \sim 2(0.5)(500) = 500$-m/s.

Now the linear mass density, expressed in kg/m, is the mass in kg of a one meter length of string. That mass is the product of the mass density (8-g/cc = 8,000-kg/m^3 for the steel of most piano strings) and the volume of 1m of string. From observation, the strings have an average radius of about 0.5-mm. The above volume V is thus one meter times the area of a circle of radius 0.5-mm. Since 1-mm $=10^{-3}$ m, we obtain

$$V = (1\text{m})\pi R^2 = (1\text{-m})(0.5 \times 10^{-3}\text{-m})^2$$
$$= 8 \times 10^{-7}\text{-m}^3.$$

Hence, $\mu \sim 8{,}000\ (8\times 10^{-7}) = 6.4 \times 10^{-3}$-kg/m^3 = 6.4-g/m^3.

The average tension per single string is then

$$T = \mu v^2 \sim (6 \times 10^{-3})(500)^2$$
$$= 1500\text{-N} \sim 300\text{-lbs}.$$

Most piano keys have a few strings (i.e., there is more than one string per note. As a result, while there are 88 keys the total number of strings is about 230. Our estimate for the total tension is then $230\times\ 300 = 69{,}000$-lbs.

We have seen that the wave velocity of a vibrating string involves a characteristic force parameter (the tension) and a mass parameter (the linear mass density). We will next see how the behavior of an SHO can help us understand the expression for the wave velocity of a vibrating string. The SHO has a force parameter (the spring constant) and a mass that together determine its vibration frequency.

2.12 The Connection Between an SHO and a Vibrating String

What has the SHO in common with a vibrating string? To simplify our analysis, we will examine the vibration of a string that is pulled aside at its midpoint and then released and allowed to vibrate.

We have seen that an SHO is characterized simply by a mass that is displaced and experiences a restoring force proportional to its displacement. The entire length of string is the corresponding mass. There is variable displacement all along the length of the string, so the system is more complicated. As an approximation we will let the displacement of the midpoint correspond to the displacement of the SHO. [See Fig. 2.21 below.]

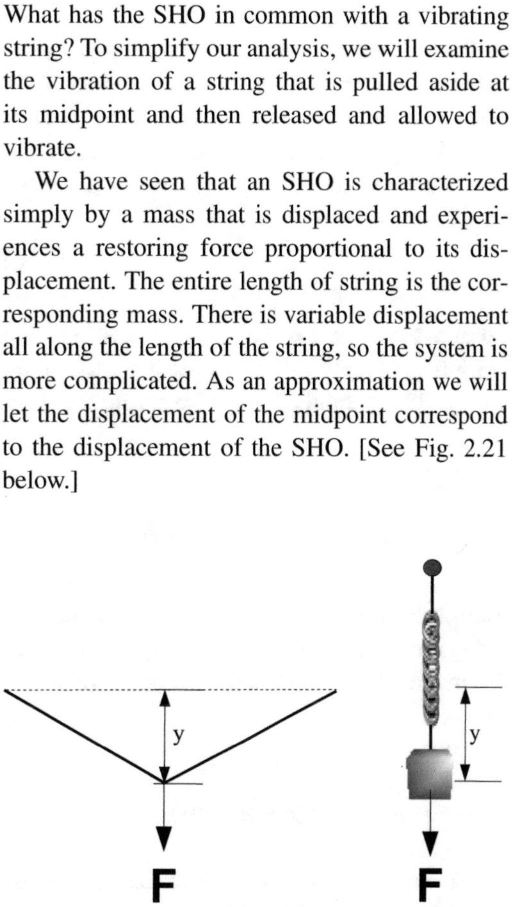

Fig. 2.21 The plucked string vs. the SHO

It can be shown that for small displacements, the restoring force on the string is proportional to the displacement. "Small" means displacement much less than the length of the string. A simpler description of this restriction is that the slope of the string during vibration must be very small. The result is

$$F = \frac{4y}{\ell}\mathrm{T} = \frac{4\mathrm{T}}{\ell}y \quad (2.34)$$

That is, the restoring force is given by the tension T reduced by a factor $(4y/\ell)$, which is typically much less than unity. [For example, a guitar string of length 650-mm might vibrate with an amplitude of but a few mm.] Alternatively, we can express this relation as

$$F = \frac{4\mathrm{T}}{\ell}y \quad (2.35)$$

We see that the restoring force is proportional to the displacement y, as in the case of an SHO. This is the essential reason that a string vibrates sinusoidally. The effective spring constant is defined by the relation $F = ky$. Thus it is given by

$$k = \frac{4\mathrm{T}}{\ell} \quad (2.36)$$

Sample Problem 2.4 Suppose that a string has a tension equal to 200-N and a length equal to 33-cm (=0.33-m). Find the effective spring constant.

Solution

$$k = \frac{4\mathrm{T}}{\ell} = \frac{4(200)}{0.33} = 2400\text{-N/m}$$

Sample Problem 2.5 Suppose that the previous string is displaced at its midpoint by a distance of 1-mm (=1/1000-m). Find the restoring force.

Solution

$$F = kx = (2400)(0.001) = 2.4\text{-N}$$

We can now combine the exact expression (2.17) for the period of an SHO with our expression (2.36) for the spring constant of the plucked string so as to obtain an expression for the **period of a plucked string**. [Beware of the two symbols: T for the period, T and for the tension.]

$$T_{\mathrm{SHO}} = 2\pi\sqrt{\frac{m}{k}} \quad \text{along with} \quad k \approx \frac{4\mathrm{T}}{\ell} \quad (2.37)$$

to obtain

$$T_{\mathrm{SHO}} \sim 2\pi\sqrt{\frac{m}{4\mathrm{T}/\ell}} = \pi\sqrt{\frac{m\ell}{\mathrm{T}}} \quad (2.38)$$

Finally we are ready to obtain an approximate expression for the wave velocity along a string:

$$v_{\mathrm{approx}} = \frac{2\ell}{T_{\mathrm{SHO}}} = \frac{2\ell}{\pi\sqrt{\frac{m\ell}{\mathrm{T}}}} = \frac{2}{\pi}\sqrt{\frac{\mathrm{T}\ell}{m}} \quad (2.39)$$

Substituting μ for the ratio m/ℓ, we obtain

$$v = \frac{2}{\pi}\sqrt{\frac{\mathrm{T}}{\mu}} \quad \text{wave velocity along a string} \quad (2.40)$$

Why the difference between the two equations for the wave velocity, (2.31) and (2.40), amounting to a ratio of $2/\pi \sim 0.6$? The spring constant is the ratio of force to displacement. The displacement of an SHO is well defined with a specific value. In contrast, for a string, the displacement varies from zero at the ends of the string to its maximum value at the string's center. As a consequence, we have over-estimated the average displacement and underestimated the effective spring constant. This leads to an overestimate of the fundamental period.[6]

2.13 Stiffness of a String

So far, we have assumed that the vibrating string is completely flexible. No force is necessary to

[6]The exact expression for the fundamental period of a plucked string is

$$T = 2\sqrt{\frac{m\ell}{\mathrm{T}}}. \quad (2.41)$$

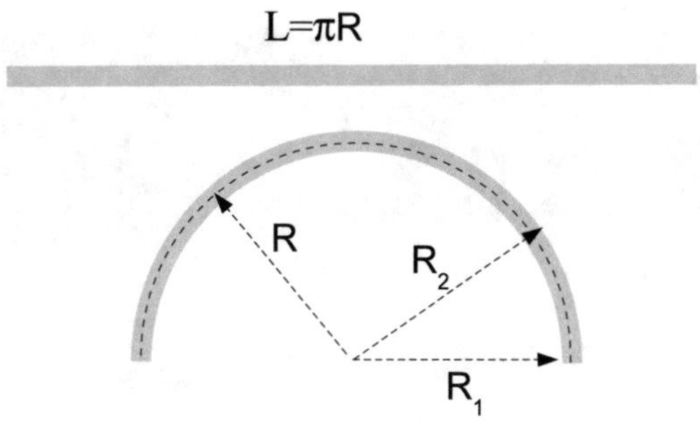

Fig. 2.22 A thick string bent into a semi-circle

bend the string. The term **stiffness** is used to characterize the force necessary to bend a string.

The physical parameter that determines stiffness is the same as that which determines the force necessary to stretch a string. It is called **Young's modulus**. Why is this so? Because when a string is bent, one side of the string is stretched while the other side is compressed. This can be seen in Fig. 2.22, where a string of length $L = \pi R$ is bent into a semi-circle. The thickness of the string is $R_2 - R_1$. The outer perimeter has a length πR_2, while the inner perimeter has a length πR_1. While the outer perimeter is stretched by an amount $\pi(R_2 - R)$, the inner perimeter is compressed by an amount $\pi(R - R_1)$ Thus, the difference in the perimeters is $\pi(R_2 - R_1)$, or π times the thickness.

Note: The shape of the wave for the vibrating stiff string in the nth partial is sinusoidal even in the presence of stiffness. It is a portion of a sine wave having a wavelength

$$\lambda = 2\ell/n \tag{2.42}$$

While we will not discuss Young's modulus because the subject is beyond the scope of this text, we can see qualitatively what effect stiffness might have on the wave velocity along a string and more importantly on the frequency spectrum of the modes.

First, we expect that stiffness contributes to bringing the string back from a curved shape towards a straight shape. It is a restoring force. Therefore, we expect that the wave velocity will increase. Next, as the wavelength decreases, the degree of bending increases. Therefore, we expect the wave velocity to increase with decreasing wavelength—or alternatively, to increase with increasing frequency.[7] Finally, since the wavelength is inversely proportional to n, the effect of stiffness increases with increasing n. In fact, it can be shown that the frequency of the nth partial is given by

$$f_n = n\frac{\sqrt{\frac{T}{\mu}}}{2\ell}\sqrt{1 + \mathcal{B}n^2} \tag{2.45}$$

$-\mathcal{B}$ is a number that is inversely proportional to the square of the length of the string.[8] Thus, the longer the string, the smaller the effect of stiffness. Relatively, the longer a string, the easier it is to bend it. Therefore, for a given mode, longer strings have less of an effect due to stiffness. Also, the constant is proportional to the fourth power of the radius of the string. As a result, thicker strings are stiffer, as we would expect. Typically, the constant $\mathcal{B}$ is much less than unity, so that the effect is small for $n = 1$. On the other hand, for large n the effect will be much more significant. Alternatively, we can write

$$f_n = nf_0\sqrt{1 + \mathcal{B}n^2} \tag{2.46}$$

---

[7]Mathematically, the wave velocity can be expressed as

$$v = \sqrt{\frac{\mathrm{T}}{\mu} + \frac{\overline{\mathcal{B}}}{\rho\lambda^2}} \tag{2.43}$$

where $\overline{\mathcal{B}}$ is a constant. If the tension is absent, as is the case for a suspended rod of metal—e.g., one prong of a tuning fork—the speed of transverse vibrations is given by

$$v = \sqrt{\frac{\overline{\mathcal{B}}}{\rho\lambda^2}} \tag{2.44}$$

We see from this equation that the ratio $\overline{\mathcal{B}}/\lambda^2$ is the restoring force. Sometimes this force is referred to as the **bending force**. This is the equation for the speed of **transverse waves** along a solid rod. The rod also exhibits **longitudinal sound waves**, which will be discussed in the next chapter.

[8]Explicitly, $\mathcal{B} = (\pi Y/T)(\pi a^2/2\ell)^2$, where a is the radius of the string and Y is **Young's modulus**. Young's modulus determines how much of an elongation $\Delta\ell$ results from tension. Thus, if a string is under a tension T, the relative change in its length is given by $\Delta\ell/\ell = T/(\pi a^2 Y)$.

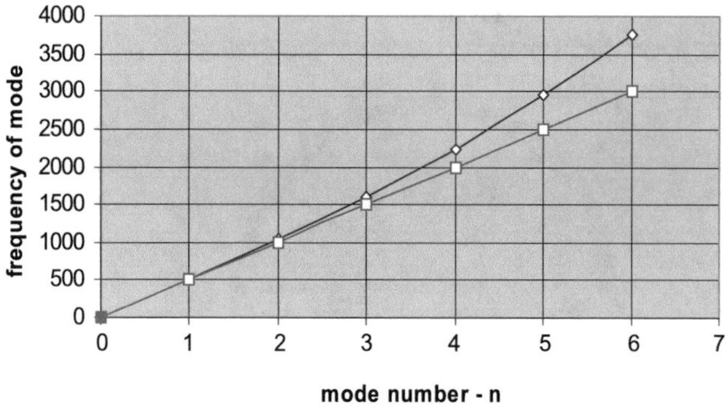

Fig. 2.23 Frequency of modes for string with (dark curve) and without stiffness (magenta curve)

where the parameter $f_0 = \sqrt{T/\mu}/(2\ell)$ is the fundamental frequency in the absence of stiffness. Note that in the presence of stiffness, f_0 is not the fundamental frequency. Instead, the fundamental frequency is $f_1 = f_0\sqrt{1+\mathcal{B}}$. The equation for f_n shows us plainly that the frequency spectrum is no longer a harmonic series.

To gain a sense of the order of magnitude of the constant $\mathcal{B}$, we find that for a steel wire of radius 1-mm and a length of one-meter, under a tension of 100-Newtons, $\mathcal{B} = 0.008$. For the fundamental mode, the correction to the frequency is less than one-percent. However, as the mode number increases, the correction increases too, and not proportionately. We can see the effect dramatically in the graph of Fig. 2.23. The dark curve represents the spectrum with stiffness, while the straight line in magenta represents the spectrum without stiffness included.

Note that the effect on the first two modes is not great. For the third mode, the corresponding frequencies are 1500-Hz and 1600-Hz, respectively—a significant difference. For the fifth mode, the difference is dramatic: 2500-Hz vs. 3000-Hz.

2.14 Resonance

Consider the process of exciting a mode of vibration of a string. It requires that you move the end of the string up and down at a frequency equal to the frequency of that mode. If your hand moves at a frequency that is different from a mode frequency, the degree of excitation of the mode will not be great. However, the closer the frequency match is, the greater the ultimate amplitude of the excited mode.

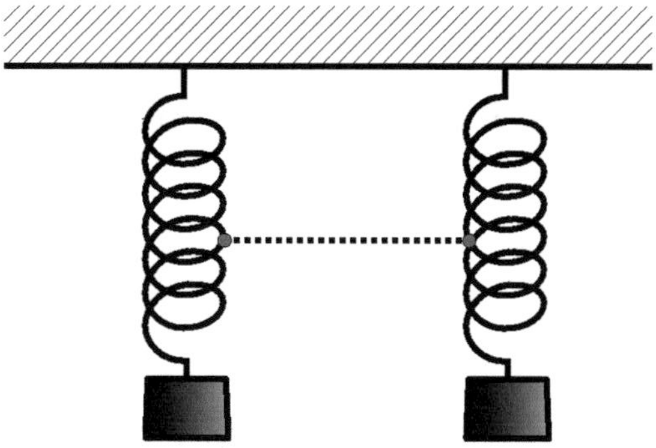

Fig. 2.24 Resonance between two coupled SHOs

We say that there is a **resonance** between the two systems—your hand and the string—when there is a frequency match, or practically speaking, close to a frequency match, so that there is a high degree of excitation.

For a simple example of resonance, consider two SHOs having an identical frequency f. We connect them with a fine string as shown in Fig. 2.24. Clearly, either SHO can excite vibrations in the other through the coupling between them.

Digression on the Modes of Two Coupled SHOs

Suppose that both SHOs are released from rest with the same initial displacement. Clearly, they will oscillate up and down at their common mode frequency f, because the coupling between them will be **inactive**. We say that the SHOs oscillate

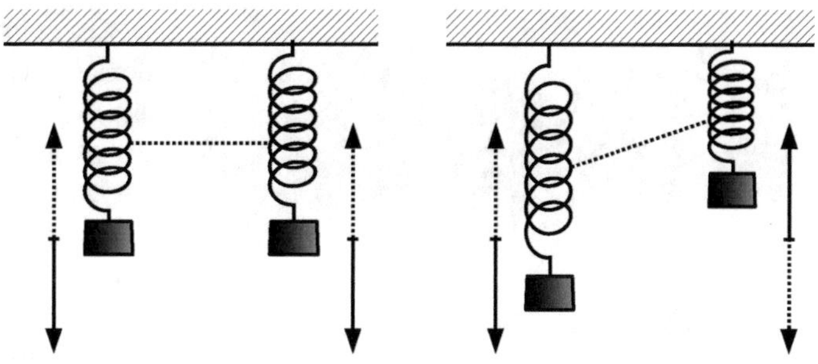

Fig. 2.25 Two modes of two coupled SHOs

in phase. Now, suppose that the two SHOs are released from rest with the same initial amplitude but NOW with **displacements in opposite directions**. The two SHOs will oscillate up and down, always in opposite directions. We say that they oscillate **out of phase**. In this case, there will be strong coupling between the two SHOs.

What we have described above are the two modes of a pair of coupled SHOs. They are depicted in Fig. 2.25. The IN-PHASE mode has a frequency $f_{\text{in}} = f$, while the OUT-OF-PHASE mode has a frequency f_{out} that is slightly larger. The *difference* Δf between these two frequencies increases with increasing coupling between the strings and vanishes in its absence.

It can be shown that any particular motion of the two SHOs can be expressed as a sum of the two modes. A very interesting example is the following:

Suppose that both SHOs are released from rest, with the **left** SHO displaced downward by an amount A from its equilibrium position, while the **right** SHO is kept in its equilibrium position. We see that the initial condition is a sum of the initial conditions described above for the two modes. As a consequence, the subsequent motion is a sum of the in-phase and the out-of-phase modes, with equal amplitudes $A/2$ of each in the sum. The resulting subsequent motion is quite interesting. [We neglect attenuation, for simplicity.]

The left SHO will begin oscillating with an amplitude A. In time, that amplitude will decrease, while the right SHO will begin to oscillate. The amplitude of the left SHO will eventually momentarily vanish, while at the same time, the amplitude of the right SHO will equal A: The left SHO will have passed its energy [initially potential energy] entirely onto the right SHO!

Subsequently, the roles will be reversed, with the right SHO passing its energy back to the left SHO. Ultimately, the two SHOs will exchange energy sinusoidally at an **exchange frequency** f_{ex} that is exactly equal to the frequency Δf! The time dependence of the displacements of the two oscillators is shown in Fig. 2.26 below. The black curve represents the oscillation of the left SHO while the blue curve represents the oscillation of the right SHO. Two cycles of exchange are shown. Note how the left SHO comes to rest at the time 0.5 units, where the right SHO has its maximum oscillation.[9]

2.15 General Vibrations of a String—Fourier's Theorem

Suppose that you don't move a string up and down at exactly any of the mode frequencies. The string will vibrate; however, the pattern of vibration will not resemble any one of the modes unless there is a near frequency match. Furthermore, a plucked string doesn't vibrate with the pattern of any of the modes, even though the

[9]Note that a musical instrument can have a few vibrating components, such as does the violin—the two components being a vibrating string and a vibrating wooden plate. It can be desirable to have mode frequencies of the components match: the original source of vibration—as from a bowed string—might not transfer the vibration into the air efficiently. The second component—here the wooden plate—might be able to do so efficiently. With efficient transfer of vibration to the air, the second component will not fully return the vibration back to the original source and the above annoying phenomenon will be reduced.

Fig. 2.26 Exchange of oscillation between two coupled SHOs

pattern vibrates periodically at the fundamental frequency! **How is the general vibration of a string related to the various modes of vibration?** The answer lies in a mathematical theorem due to **Jean Baptiste Joseph Fourier**, a French mathematician who lived from 1768 to 1830. Here is **Fourier's Theorem** in words:

> Any pattern can be expressed mathematically as a sum of sine waves, the amplitudes of each sine wave in the sum being unique.

This theorem is analogous to one of the most important theorems in number theory, that any number can be expressed as a product of prime numbers. [For example, $60 = 2 \times 2 \times 3 \times 5$. The representation of the number by such a product is unique. In particular, the number of times that any prime number appears in the product is unique.] And so it is with the amplitudes of the various sine waves in the sum which represents the pattern.

Given a particular pattern, there are mathematical as well as electronic means for obtaining the unique mixture of sine waves associated with the pattern, a process known as **Fourier analysis**. Each individual sine wave is referred to as a **Fourier component**. To specify a Fourier component we need to know three factors: its **frequency**, its **amplitude**, and its **relative phase**.

The set of frequencies in the mixture of sine waves is called the **Fourier spectrum** or simply the **frequency spectrum**. The amplitudes of the sine waves in the sum are called **Fourier amplitudes**. The "sum" is obtained by a straightforward graphical sum of the curves representing the waves. For example, in Fig. 2.27 we exhibit the sum of two Fourier components A and B, with the resultant SUM.

The **relative phase** refers to the relative positions of the waves. To appreciate the significance and importance of the relative phase, in Fig. 2.28 we exhibit the sum of the same two Fourier components as Fig. 2.27, except that component B has been shifted by a quarter of a wavelength so as to produce component C.

The reverse process of adding the given mixture of sine waves corresponding to a specific pattern so as to produce that pattern is called **Fourier synthesis**.

Generally, the frequency spectrum will include all frequencies, from zero to infinity. This is not the case for a finite vibrating string. Here, Fourier's Theorem leads to the result that any pattern of vibration is a sum of the modes of vibration of the string, with a unique set of amplitudes for each mode in the sum. Hence, in this case the Fourier spectrum is a harmonic series.

There is a **corollary** to Fourier's theorem that is central to understanding the basis for obtaining a sense of pitch from musical instruments:

The Fourier spectrum of any periodic wave—and hence any sound wave that has a well-defined single pitch—must be a harmonic series with a fundamental frequency equal to the frequency of the periodic wave.

As a consequence, the frequency spectrum for the modes of vibration of a musical instrument that produces a well-defined pitch must be a harmonic series. In contrast, the frequency spectrum of a gong, a tuning fork, or a drum is not a harmonic series. Unless one excites one mode alone of these instruments, the sound produced will be perceived to have more than one sense of pitch. This fact is exhibited in the spectra of some musical instruments. In Fig. 2.29 we see the wave and spectrum of a short segment of sound from a **violin**. We can see the peaks at the harmonics with a fundamental frequency of about 280-Hz. Note the variation in the envelope of the wave, corresponding to varying loudness. More interesting are the numerous spikes surround the main peak. These partially reflect the **vibrato**, which is briefly discussed below.

In contrast, we see in Fig. 2.30 the wave and spectrum of a segment of sound from a **flute**.

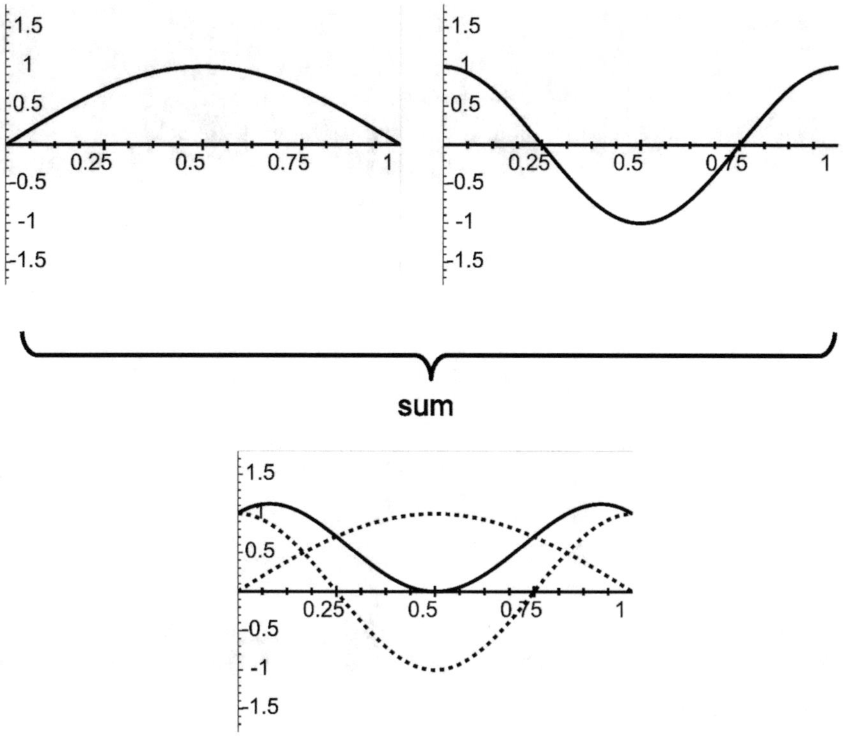

Fig. 2.27 Summing two sine waves

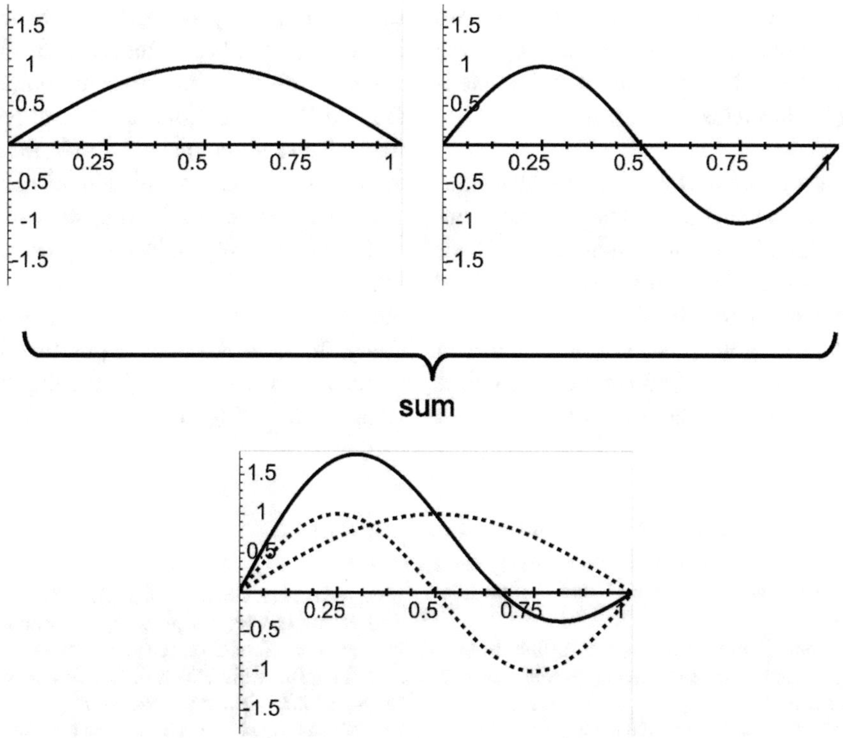

Fig. 2.28 Summing two sine waves with different phase relation from above

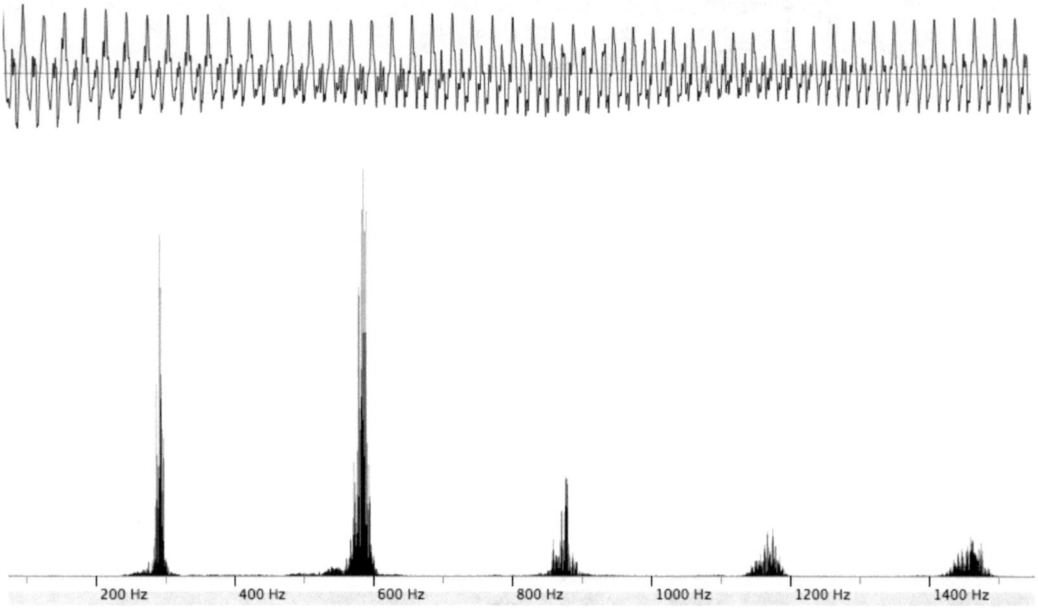

Fig. 2.29 Violin wave and frequency spectrum

The absence of much contribution from harmonics above the second is evident.[10] **NOTE:** We must remember that the spectrum for any given instrument varies, depending upon how a note is played by the musician.

Note The sound of a violin exhibits characteristics that are analogous to the two most important modes of communicating audio signals with radio—AM (**amplitude modulation**) and FM (**frequency modulation**).[11]

Consider the central fundamental frequency of the above violin wave—280-Hz. Let us suppose that the envelope of the wave oscillates at a frequency of 2-Hz. This is the **amplitude modulation frequency**. An AM radio wave of WEEI in Boston, MA, has a frequency of the **carrier wave** of 550-kHz. This frequency corresponds to the 280-Hz of the violin. If the station wants to transmit an audio signal of 280-Hz, the amplitude modulation frequency would be 280-Hz.

Now let us turn to the frequency modulation, which is strongly produced by **vibrato**[12]: The fundamental frequency of the violin is determined by the position of a finger on the violin string that restricts the length that is free to vibrate. If the violinist rocks the finger back and forth on the string at a frequency of 5-Hz, the

[10]The waves and spectra were produced using mp3s of instrumental sounds downloaded into the program **AmadeusPro**.

[11]See the extremely informative applet on this website (2-15-2011): http://engweb.info/courses/wdt/lecture07/wdt07-am-fm.html#FM_Applet
You can vary the modulation frequency as well as the amplitude of modulation and observe the changing waveform as well as the resulting frequency spectrum.

[12]We realize that vibrato is what gives the violin its sweet tone. However, vibrato is probably a very important factor in a number of other specific ways: For example, see Sect. 11.7 for a discussion of the fusion of harmonics and Sect. 12.7, wherein we discuss the important role that vibrato certainly plays in allowing us not to be affected by the impossibility of performing combinations of musical pure tones that are consistently consonant.

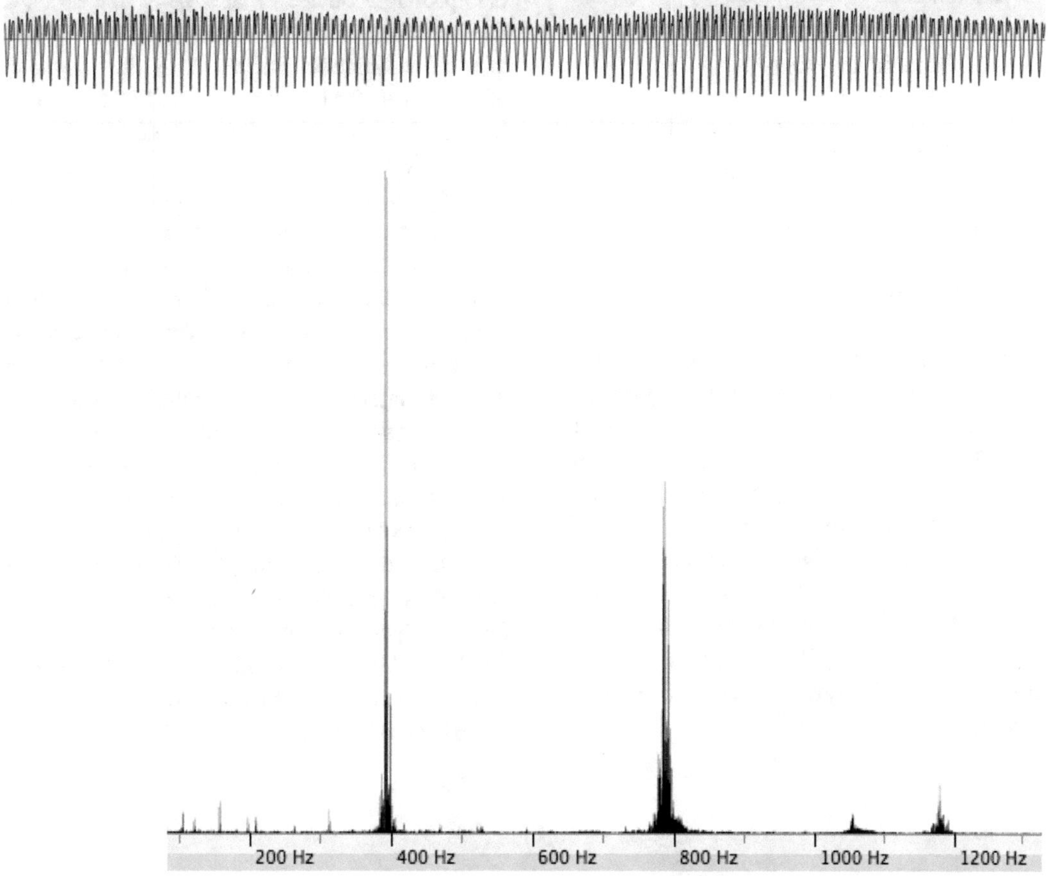

Fig. 2.30 Flute wave and frequency spectrum

length that is free to vibrate will oscillate at this frequency and the sound will be frequency modulated at this frequency.[13] In the case of FM radio, an **FM radio wave** from WCRB-FM in Boston would have a **carrier frequency** of 89.7 -MHz; the audio signal of 280-Hz would be the **FM modulation frequency**.

[13]It can be shown that the resulting frequency spectrum consists of a central peak at 280-Hz along with side peaks at frequencies, $280 \pm 5 = 275---\text{and}-285$, -$280 \pm 10 = 270 - \text{and} - 290$, and $280 \pm 15 = 265 - \text{and} - 295$. The weight of these side frequencies falls off as we move to greater distances from the fundamental and depends upon the amplitude of the rocking motion.

NOTE: The general term for the frequency of a mode of vibration of a system is the **partial**. The first partial is always equivalent to the fundamental frequency. For a vibrating string without stiffness, the set of partials forms a harmonic series.

In order to illustrate the results of a Fourier analysis of a vibrating system, consider the vibration of a string that is plucked at its midpoint, as shown previously in Fig. 2.15. It can be shown using mathematical analysis that the pattern of vibration is a sum of all the odd modes of vibration of the string. Suppose that at some instant the amplitude is $\pi^2/8$ at its midpoint. The shape of the string is triangular. The amplitudes A_1,

A_2, A_3, ... of the sine waves that reproduce this pattern are given by:

$$\text{fundamental: } A_1 = 1$$

$$\text{3rd harmonic: } A_3 = -\frac{1}{9}$$

$$\text{5th harmonic: } A_5 = \frac{1}{25}$$

$$\text{general odd harmonic: } A_n = \frac{(-1)^{(n-1)/2}}{n^2},$$

where n=1, 3, 5, ...

We illustrate this result in Fig. 2.31, where we show how adding sine waves produces a triangular wave pattern. The black curve is the desired triangular wave. The blue curve is the first harmonic—$n = 1$ or the function $\sin(\pi t)$. The red curve is the sum of the first and third harmonics—$n = 1$ and $n = 3$—that is, the function $\sin(\pi t) - (1/9)\sin(3\pi t)$. We see that these two terms alone are within about 10% of reproducing the triangular wave. The green curve is the sum of the 1st, 3rd, and 5th harmonic.

2.15.1 Frequency of a Wave with Missing Fundamental

It is probably obvious that a wave that includes the fundamental in its spectrum has the frequency of the fundamental. For example, the frequency of a mixture 100-Hz, 200-Hz, and 500-Hz is 100-Hz. However consider the mixture 200-Hz and 500-Hz. What is the frequency in this case?

Suppose we focus our attention on the displacement of a specific point along a vibrating string. According to Fourier's Theorem, its wave pattern in time is a sum of sine waves, all of which are members of the harmonic series of the string's mode spectrum. It can be shown that the pattern is always **periodic, with a frequency equal to the largest common denominator (LCD) of all the frequencies in the Fourier spectrum**. For this last example, the LCD is 100Hz, which is the frequency of wave, even though the fundamental $100Hz$ is missing. We can appreciate this result as follows: During one cycle of oscillation over the period of $1/100 = 0.01$-s there will be exactly two cycles of the 200-Hz component and five cycles of the 500-Hz component.

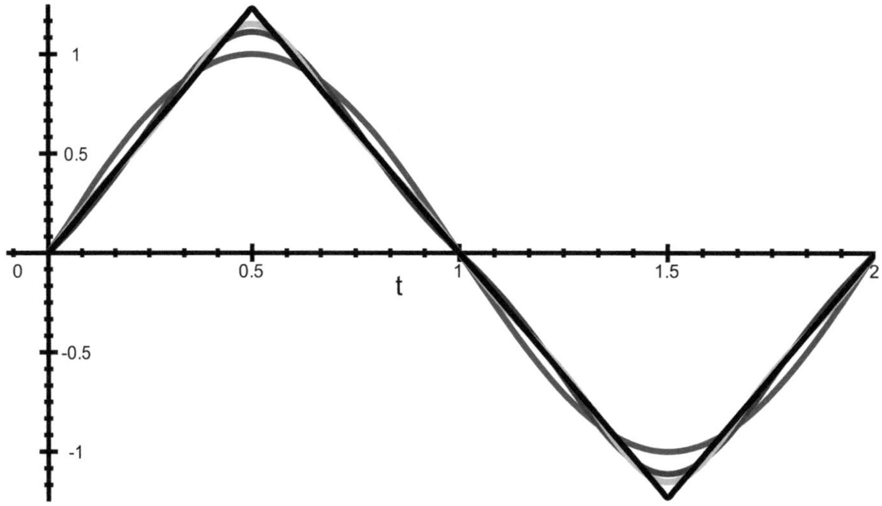

Fig. 2.31 Sum of Fourier components to produce a triangular wave

2.16 Periodic Waves and Timbre

We can now appreciate a major factor that distinguishes the timbre of one musical instrument from another: Two instruments that are producing the same *steady* musical note are producing periodic patterns having the same frequency. It is this frequency that determines the *pitch* of the note. However, the two sets of relative amplitudes of the Fourier components are different. This difference is one of the important factors that distinguishes the timbres of musical instruments.

There are two other factors that contribute to our ability to distinguish one instrument from another when the notes are not steady but have a beginning and end; they are the **attack** and the **decay** parts of the note, which are depicted in Fig. 2.32. The variation of the amplitude—defining with the growth and final decay—is called the **envelope**. It is given by the pair of dashed curves in the figure.

Experiments have shown that in the absence of differing envelopes, it is often difficult to distinguish the sounds of different instruments.

2.17 An Application of Fourier's Theorem to Resonance Between Strings

When a string is disturbed, generally a mixture of modes is excited. The Fourier amplitudes depend upon the manner in which the string is excited. This fact has important ramifications with regard to resonance between strings: Consider two strings, one tuned to 440-Hz and the second to 660-Hz. (These two frequencies correspond to the **fundamental** frequencies of the respective strings.) The frequency spectra are

440-Hz string: 440, 880, 1320, 1760, . . .

660-Hz string: 660, 1320, 1980, 2640, . . .

We see that the third harmonic of the 440-Hz string and the second harmonic of the 660-Hz string have the same frequency. Thus, a general excitation of the 440-Hz string can strongly excite the second harmonic of the 660-Hz string or, a general excitation of the 660-Hz string can excite the third harmonic of the 440-Hz string.

Can you find a **second** matching pair of frequencies for the above strings? Discuss resonance between a 440-Hz string and a second string tuned to its octave at 880-Hz.

Resonance among the strings of a stringed instrument enriches tone quality. It therefore provides us with a partial explanation as to why good intonation of a string player—that is, playing notes "in tune"—improves tone quality. Conversely, by becoming more aware of the resonant response among strings, a string player can improve intonation.

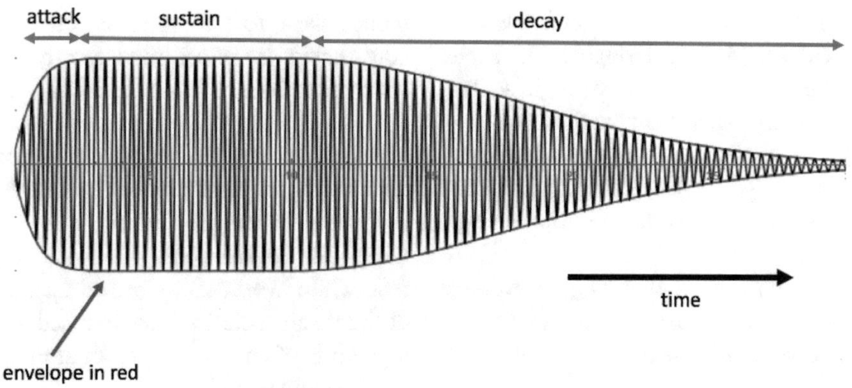

Fig. 2.32 The attack, decay and envelope (in red) of a wave

Fig. 2.33 Piano keyboard

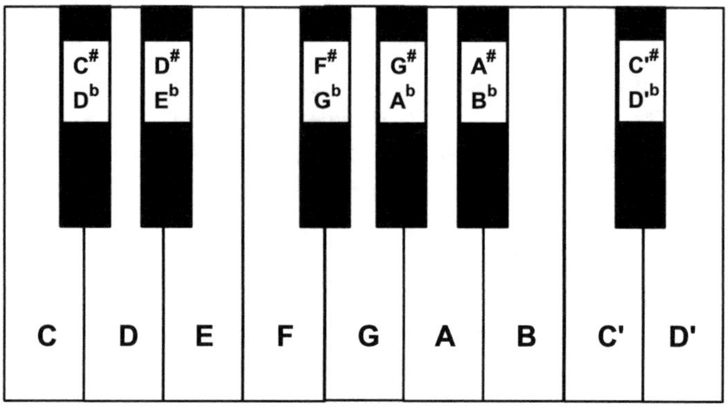

Home Exercise with a Piano

If you have a piano available, you can observe the resonances discussed above as follows. Let us refer to the **piano keyboard** depicted in Fig. 2.33.

You will note that there is a pattern of the **piano keys** that repeats itself. Each cycle of keys is called an octave, with the white keys labeled from *A* through *G*. Focus on the key labeled *C*. This *C* is called *"middle-C."* The *A* above middle-C is the key that is first tuned by a piano tuner, presently usually at a frequency of 440-Hz. We will call this key $A - 440$. The *A* above it, being one octave above, is tuned at double this frequency. The *E* above $A - 440$ is tuned at a frequency that is close to 660-Hz. [See Chap. 12, for more details on the choice of frequencies.]

Now, hold the $A - 880$ key down so as to free the string from a damper which prevents it from vibrating. Next, give the $A - 440$ key a sharp, "staccato" blow, so that the $A - 440$ tone will sound long enough as to excite the $A - 880$ string, but short enough so that you can eventually hear the sound of the $A - 880$ string. To confirm that you are hearing a sound produced by the $A - 880$ string, release the $A - 880$ key so as to dampen that string's vibration.

Repeat the above by exchanging the roles of the two strings. Next, repeat all of the above with a second pair of strings—say, the $A - 440$ and $E - 660$ strings.

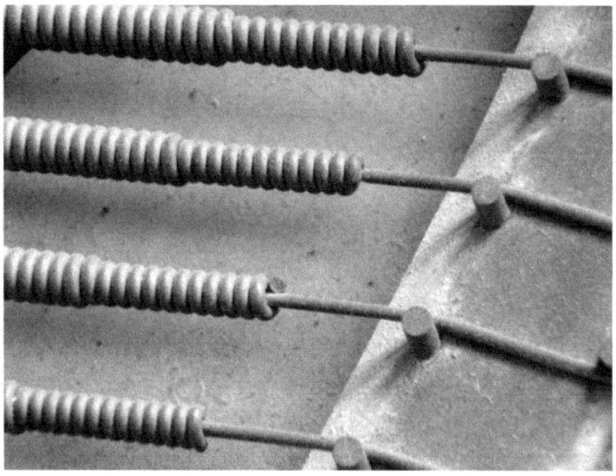

Fig. 2.34 Wound Piano strings

Note Let us now recall that stiffness causes the frequency spectrum of a string not to be a harmonic series. The effect is strongest for the thick strings of a piano at the low end, where increased thickness is necessary to produce the low frequencies. As a result, stiffness reduces the degree of resonance among these strings. In order to reduce the effect of stiffness, these strings are constructed out of a central core of steel that is surrounded by a coil, as seen in Fig. 2.34 above.

In an effort to reduce the mismatch of common harmonics, pianos are **stretch tuned**—a feature discussed further in a problem of Chap. 12. Furthermore, it is interesting to note that while we tend to regard resonance as a desirable characteristic, the reduced resonance in a piano is often regarded as an attractive feature of the sound of a piano.

2.18 A Standing Wave as a Sum of Traveling Waves

A standing wave is not a traveling sine wave since it is moving neither to the right nor to the left. According to Fourier's Theorem, a standing wave must be a sum of sine waves. In fact, it is a sum of two sine waves having the same wavelength and amplitude but traveling in opposite directions. This fact is depicted in Fig. 2.35.[14] Diagrams (a) through (e) one-eighth of a cycle apart. Each diagram depicts the position of the two component sine waves and their sum. The figure reflects a special property of sine waves: **The sum of two sine waves having the same wavelength is a sine wave having the same common wavelength. The amplitude of the resultant sine wave depends upon the amplitudes of the components and their relative phase.** In our case, the components have exactly the same amplitude. In (a), the components are in phase and the resultant wave has an amplitude that is double that of the components. In (c), the components are out of phase so that the components cancel each other. Note that while the displacement vanishes everywhere in (c), the string does have an instantaneous velocity. This situation can be compared to the SHO whose mass is passing through the equilibrium position, where the displacement is zero and the velocity is a maximum.

We can now understand how we are able to set up a standard wave along a string of finite extent: With our hand we propagate a sine wave down the string. The reflected wave is a sine wave traveling in the opposite direction, which when added to the original wave, forms a standing wave! [Of course, the observed standing wave is only a portion of an infinite standing wave.]

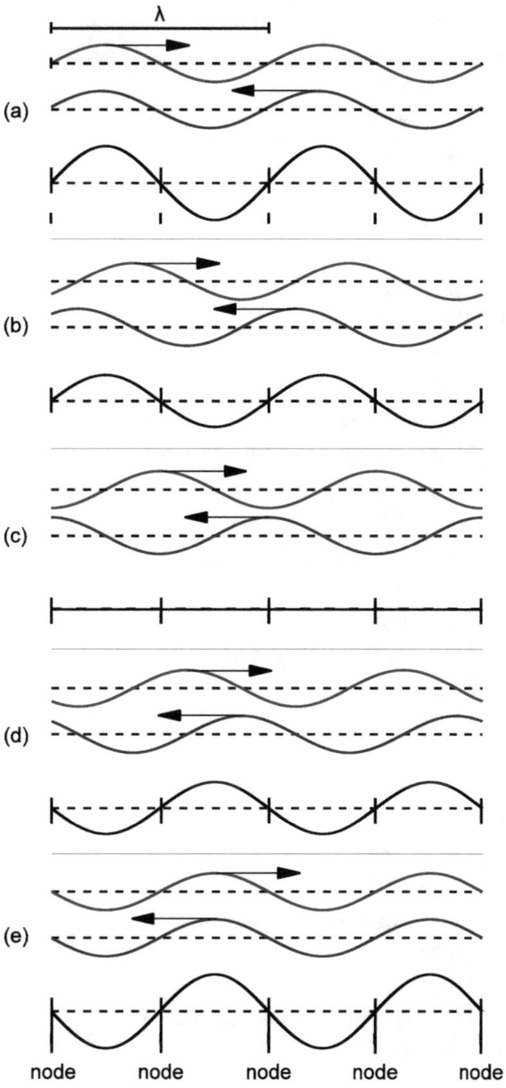

Fig. 2.35 Standing wave from two traveling waves

2.19 Terms

- Amplification
- Amplitude analyzer
- Anti-node
- Attenuation
- Bending force
- Centi–10^{-2}
- Chladni plate
- Cycle
- Damping
- Direction of propagation
- Dispersion
- Dispersive
- Displacement
- Dissipation
- Equilibrium state
- Excitation

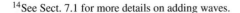

[14] See Sect. 7.1 for more details on adding waves.

- Force constant (or "spring constant") **k**
- Fourier analysis
- Fourier component
- Fourier spectrum
- Fourier synthesis
- Fourier theorem
- Frequency **f**
- Fundamental mode
- Fusion of harmonics
- Giga–10^9
- Gong sound
- Harmonic
- Harmonic series
- Hertz (Hz) [a unit of frequency]
- Integral multiples
- Kilo–10^3
- Largest common denominator
- Linear mass density μ
- Longitudinal wave
- Medium for wave propagation
- Mega–10^6
- Micro–10^{-6}
- Milli–10^{-3}
- Nano–10^{-9}
- Newton (N) [a unit of force]
- Nodal line
- Node
- Octave of notes oscillation
- Overtone
- Period **T**
- Periodic wave (in time or space)
- Phase relation
- Pitch
- Pluck
- Pulse
- Resonance
- Restoring force
- Simple harmonic oscillator (SHO)
- Sinusoidal
- Sonometer
- Spectrum
- Standing wave
- Stiffness
- Stretch tuning
- Stroboscope
- Tension T
- Timbre or tone quality
- Transverse wave
- Traveling wave
- Tuning fork
- Wave propagation
- Wave velocity v

2.20 Important Equations

$$x = vt \tag{2.47}$$

$$f = \frac{1}{T} \tag{2.48}$$

$$f_1 = \frac{v}{2\ell},\ f_2 = 2\frac{v}{2\ell} = \frac{v}{\ell},\ f_3 = 3\frac{v}{2\ell},\ \ldots \tag{2.49}$$

$$\lambda f = v \tag{2.50}$$

linear mass density:

$$\mu = \frac{m}{\ell} \tag{2.51}$$

$$v = \sqrt{\frac{T}{\mu}} \tag{2.52}$$

Hooke's Law:

$$F = ky \tag{2.53}$$

frequency of a Simple Harmonic Oscillator:

$$f = \frac{1}{2\pi}\sqrt{\frac{k}{m}} \tag{2.54}$$

general form of the wave velocity:

$$v = \sqrt{\frac{\text{restoring force}}{\text{mass density}}} \tag{2.55}$$

2.21 Problems for Chap. 2

1. Suppose you see a flash of lightning and then hear the sound of its thunder 5 s later. Assuming a speed of light equals 3×10^8-m/s and a speed of sound equal to 340-m/s:
   (a) Estimate the distance between you and the lightning flash.

   (b) How long did it take for the light of the lightning flash to travel from the lightning to you?
2. Consider a violin string of length 31.6-cm. Waves travel on this string with a velocity of 277-m/s.
   (a) What is the largest period a wave can have if it is to be accommodated by the string as a standing wave?
   (b) Waves of other periods can also exist as standing waves on the string. What are some of these periods?
3. Suppose that there are two strings, with one string excited at the fundamental and the other excited at the second harmonic. Both are vibrating at 800-Hz with the same inter-nodal distance of 0.45-m.
   (a) Draw a diagram depicting the two strings at the same scale.
   (b) Show that the wave velocity v is the same for the two strings and calculate its value.
4. Let us assume that a musical instrument has strings of length 76-cm. The player can press the string against the "fingerboard" so as to reduce the length of string that is free to vibrate. Suppose that when the string is fingered a distance of 8-cm from one end, leaving a length of 68-cm free to vibrate, the string vibrates with a frequency of 300-Hz. Where would the finger have to be placed to obtain a frequency of 311-Hz?
5. How many anti-nodes does a string vibrating in its fifth *overtone* have?
6. What is the fundamental frequency of a string that has five anti-nodes when vibrating at 450-Hz?
7. Express the wavelength of a standing wave in terms of the distance between nodes. Now do so in terms of the distance between anti-nodes.
8. **Ernst Chladni**, who lived from 1756 to 1827, studied the vibrations of a metal plate by sprinkling sand on its surface and exciting one of its modes of vibration. A mode is distinguished by having **nodal lines**, along which the displacement of the metal plate vanishes. The frequency spectrum is *not* a harmonic series. Generally, vibrations of the plate are composed of superpositions of many modes. Excitation of a single mode is facilitated by bowing the plate with a violin bow at some position along the edge and holding the plate at another position along the edge. If the plate is held horizontal, the sand particles dance around as the surface vibrates, being tossed into the air wherever there are vibrations and ultimately settling close to nodal lines. [Where you bow cannot be a node. Why so?] Such plates—in the context of its modes—are called **Chladni plates** (Fig. 2.36).

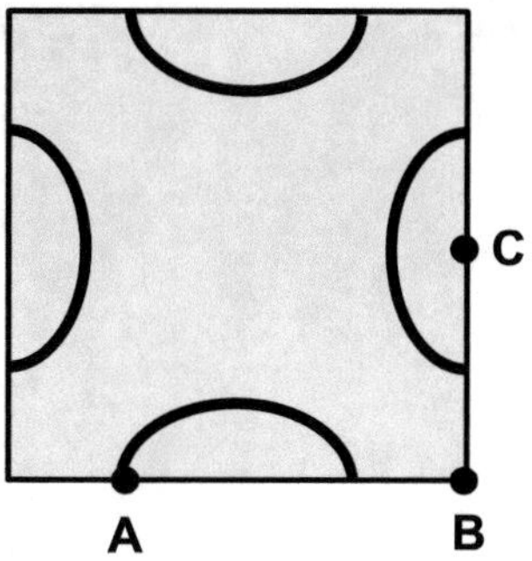

Fig. 2.36 Chladni plate

   A pattern of sand on a square Chladni plate is shown in Fig. 2.36. This pattern could result from
   (a) Bowing the plate at A and holding it at B.
   (b) Bowing the plate at C and holding it at B.
   (c) Bowing the plate at A and holding it at C.
   (d) Bowing the plate at C and holding it at A.
   (e) Bowing the plate at B and holding it at C.
9. (a) Find the wavelength of a sound wave in water (with a wave velocity of 1400-m/s) that has a frequency of 10-kHz.
   (b) Find the frequency of a light wave in vacuum that has a wavelength of 0.5μ-m ($= 5 \times 10^{-7}$-m).

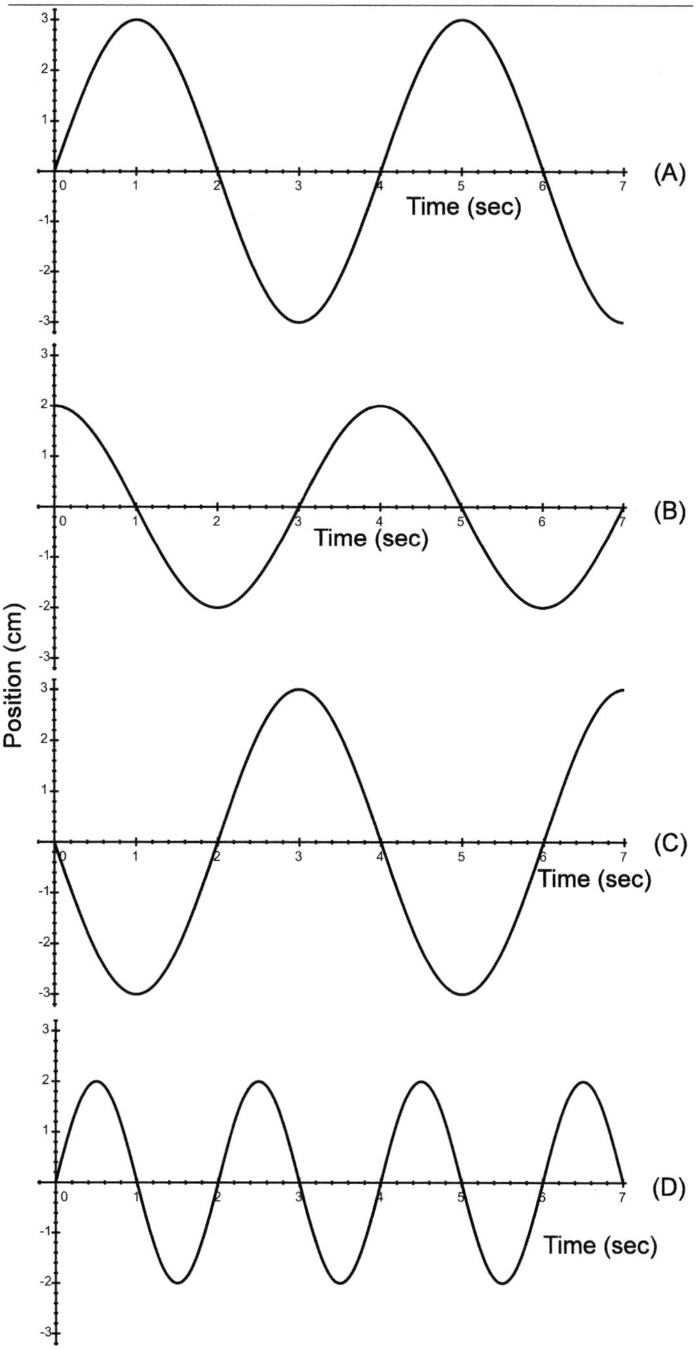

Fig. 2.37 Four different sinusoidal waves

10. A pendulum swings back and forth at 20-Hz. Find its period and frequency.
11. (a) What characteristic of the relation between the displacement of an SHO and an applied force is central to its behavior and distinguishes it from other oscillators?
    (b) Specify **at least two** characteristics of the oscillation of a SHO that make it **unique**.
12. (a) Suppose an SHO has a spring constant of 32-N/m and mass of 500-g. Find its vibration frequency.
    (b) How must the mass be changed so as to double the frequency; to halve the frequency?
    (c) How must the **spring constant** be changed so as to double the frequency; to **halve** the frequency?
13. A simple harmonic oscillator (SHO) has a period of 0.002273 second. What is the frequency of the oscillator?
14. A piano wire of length 2-m has a mass of 8-gm and is kept under a tension of 160-N.
    (a) Find the wave velocity along the wire.
    (b) To double the wave velocity:
        i. The tension can be changed to ______.
        ii. The mass can be changed to ______.
15. A telephone wire electrician needs to determine the tension in a 16-m segment of wire that is suspended between two poles. The wire is known to have a linear mass density of 0.2-kg/m. He plucks the wire at one end of the segment and finds that the pulse returns in 8 s.
    (a) Find the wave velocity.
    (b) Find the tension.
16. (a) A tightrope walker tends to avoid walking at a pace equal to a multiple of the fundamental frequency of the tightrope. **Explain why.** Describe what would happen if he were to do so. Now suppose that the tightrope is 25-m long, has a mass per unit length of 0.2-kg/m, and has a tension of 2,000-N.
    (b) Calculate the speed of wave propagation along the rope.
    (c) Calculate the rope's fundamental frequency and its two **lowest** overtone frequencies.
17. (a) What is stiffness in a string?
    (b) Does it exist in the absence of tension?
    (c) How does stiffness affect the frequency spectrum of a vibrating string?
    (d) Because of their longer strings, grand pianos need less stretch tuning than upright pianos. Give **two** reasons why this is so.
18. (a) Suppose that a guitar string is plucked at its midpoint. Which Fourier components **CANNOT** be excited?
    (b) Repeat the previous question when the string is plucked at a point 1/3 from one end.
19. Explain how a vibrating 440-Hz string can cause a 550-Hz string to vibrate.
20. Find the frequency and period of a periodic wave whose **only** Fourier components are equal to the following:
    (a) 500-Hz, 1,000-Hz;
    (b) 500-Hz, 800-Hz, and 1,000-Hz.
21. What is special about the Fourier frequency spectrum of a periodic wave?
22. Using Eqs. (2.26), (2.27), and (2.43), derive Eq. (2.45).
23. Figure 2.37 depicts the displacement versus time of four wave patterns, labeled A to D:
    (a) What are the respective amplitudes of patterns A and B?
    (b) Which two waves differ **only** in phase?
    (c) What is the frequency of pattern A?

The Nature of Sound

3

While the vibrating strings of guitars and violins are plainly visible, the sound that they produce in air is invisible. We often associate sound with air because we are used to hearing sounds that reach our ears from the air. We also learn that in the absence of air, sound cannot propagate—movies with sound propagating in outer space, notwithstanding. The fact that air is so transparent is not the issue here: Sound travels through liquids such as water and solids such as steel, as well as other gases such as air; nevertheless, we cannot see sound propagating through liquids or solids either. So, what is sound? That is the first subject of this chapter. Once we understand the nature of sound, we will go on to study the modes of vibration of air that is contained in pipes, that is, air columns. These are the basic components for all wind instruments, such as the recorder, flute, and trumpet.

3.1 The Air of Our Atmosphere

Many people think that we cannot see air, not realizing or forgetting that the sky is blue because the air of the atmosphere scatters sunlight. [See Chap. 9, for more details on the origin of the blueness of our sky.] Yet we know that air reveals its presence in the force of blowing wind. To gain a better understanding of the physics and power of invisible and tenuous air, we will begin our study by characterizing air under normal conditions at sea level—in particular, at "standard temperature and pressure" (which is abbreviated as "STP"), that is a temperature of 0°C and a pressure of "one atmosphere."

- Air has weight, represented by a **mass density** ρ of 1.3-kg/m^3. Thus, a room having dimensions of 5-m$\times$5-m$\times$2.5-m, corresponding to a volume of 62.5-m^3, has a mass of $62.5 \times 1.3=$ 81-kg, corresponding to a weight of 81-kg$\times$ 2.2-lbs/kg = 180-lbs.
- Air consists of a mixture of various molecules, mostly nitrogen ($\sim$ 79%), oxygen ($\sim$ 20%), water vapor (of varied percentage in relation to the relative humidity), and carbon dioxide. These non-spherical molecules have a mean diameter of about 6-**Ångstroms**, abbreviated as 6-$\AA$. [1-$\AA$ $=$ 10^{-8}-cm or, equivalently, 1-cm = 100-million- $\AA$.] In one cubic centimeter (cc) there are about 2.7×10^{19} molecules. We say that the number of molecules per unit volume (also referred to as the **number density**) is about 2.7×10^{19}per cc (also written as $2.7 \times 10^{19}/cc$). In contrast, water has about 10^{23} molecules/cc. [For the sake of comparison, the number of stars in the entire observable universe is estimated at 10^{22}.]
- The average distance between a molecule of air and a nearest neighbor molecule is about 34-$\AA$. Its molecular diameter ($\sim$3.5 $\AA$) is much smaller by a factor of about ten. As a result, it can be shown that on the average

L. Gunther, *The Physics of Music and Color*,
https://doi.org/10.1007/978-3-030-19219-8_3

a molecule has to travel a relatively great distance, ~ 670-Å, about a 100 times its own diameter, before it collides with another molecule. This distance is referred to as the **mean free path**. Thus, most of air is empty![1]

It is difficult to imagine the minuscule dimensions of molecules. Therefore, in order to get a sense of the proportions involved, suppose that a molecule has a diameter on the order of a football player—say a sphere one yard in diameter. Now imagine a checkerboard laid out without boundaries, with each side being 10-yards; there would be an infinite number of squares. Next, place one football player randomly within each square. On average, the nearest neighbor to a football player will then be 10-yards, just 10 times the diameter of one player. Finally, imagine that one football player starts running in a straight line and in a random direction. How far would the player have to run on the average till he runs into another player? The answer is 100-yards—which is the square of their nearest neighbor distance apart divided by the diameter of a football player.[2] For a gas in a three-dimensional volume, with all molecules moving about randomly, the mean free path is about (distance between nearest neighbors)^3/(diameter)^2.[3]

- Gas molecules are in constant random motion, with an average speed of about 300-m/s (~ 1100-km/hr ~ 1000-ft/s). This speed happens to be close to the speed of sound in air.

[1]Another way to appreciate this observation is to note that if we were to take a volume of air and compress it so that all the molecules are just touching each other, the volume would be reduced by a factor of $10^3 = 1000$.

[2]The following website [12-26-2010) http://comp.uark.edu/~jgeabana/mol_dyn/] has an animation that shows a collection of particles moving in a square chamber. You can choose the number and size of the particles. You can also run the animation slow enough to be able to follow a single molecule in order to see how far it travels before colliding with another molecule.

[3]The exact expression for the mean free path is $(\sqrt{2}\,\pi \times \text{number density} \times \text{diameter}^2)^{-1}$. Also, note that if we were to distribute football players in a three-dimensional array of boxes, each side being 10-yards, the mean free path turns out to be about 1000-yards or a bit over one-half of a mile!

3.1.1 Pressure

- If most of the space in air is devoid of molecules, how then can the air sustain a sound wave that requires the propagation of density variations through the air? What is the level of interaction between molecules of air. One measure is the rate of collisions between molecules: In one cubic centimeter of air there are on the order of 10^{28} collisions between molecules per sec.
- We barely notice the presence of air but easily note that we need its oxygen to survive. Yet air is not inert even when the wind is not blowing. There is **air pressure**: A value of **one atmosphere pressure**, abbreviated **atm**, corresponds to a pressure of 10^5-N/m^2 ~15-lbs/sq-in; that is, every square inch of flat surface experiences a force of 15-lbs, **whatever its orientation**, be it horizontal, vertical, or otherwise. Mathematically, we write

$$p = \frac{F}{A} \tag{3.1}$$

Alternatively, we write

$$\text{force} = \text{pressure} \times \text{area} \quad \text{or} \quad F = pA \tag{3.2}$$

Thus, for example, consider the palm of my hand, which has an area of ~30-sq-in. See Fig. 3.1. When held horizontal, the top of my palm experiences a force of 30×15 = 450-lbs downwards, while the bottom experiences a force of 450-lbs upwards. These two forces cancel each other, so that I don't have to put any effort into preventing my hand from being moved by the air.

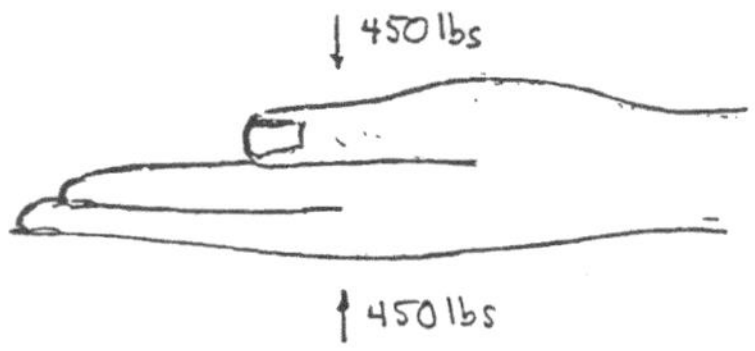

Fig. 3.1 Force by air on hand

Summary of Units of Pressure

lbs/in^2
one **atmosphere (atm)**
N/m^2 $\equiv$ one **Pascal** (Pa)

Conversions:

$$\boxed{\text{one atm} = 15\text{-lbs/in}^2 = 10^{-5}\text{Pa}} \quad (3.3)$$

- My abdomen has dimensions on the order of 15″ by 12″, corresponding to an area of 180 sq-in. The force of the air on my abdomen is therefore about $180\times 15 = 2700$-lbs! Why don't I feel this tremendous force? Why doesn't my hand get crushed? The reason is that there is pressure within the tissues of my hand, and within my nerve cells in particular, that prevents any crushing from taking place. Furthermore, my nerves function in such a way that this normal background pressure of one atmosphere does not produce any sensation. Imagine if it were otherwise!! We will see that there is a background noise impinging upon our ears, but we are simply insensitive to it.
- What is the source of the force associated with air pressure? It is the rapid rate at which molecules of air collide with a surface, as depicted in Fig. 3.2. In fact, there are about 10^{23} collisions per second on each square centimeter of area. To help you sense the nature of this force, imagine rain drops from a torrential rainstorm striking your hand at a rapid rate. You would feel the force they exert, but that force would normally be much less than 15-lbs/sq-in: While the drops have much greater mass than does a molecule, the collision rate of the rain drops would be much less than that for the air molecules.
- At a given temperature, the pressure is proportional to the density. In Fig. 3.3, we depict a chamber of air fitted with a piston. Outside the chamber is air at one atmosphere pressure. In Fig. 3.3a, the air in the chamber has a pressure of one atmosphere.
  The force on the piston from the outside air to the left of the piston then balances the force of the air inside the chamber. In Fig. 3.3b, the air in the chamber has a pressure of 1.3 atmosphere. The density of air in this chamber is 30% greater than the density of air in the chamber of Fig. 3.3a. (Note the difference in density of shading.) In this second case, the force on the piston from the inside is greater than the force from the outside and the piston would tend to move to the left, corresponding to a reduction in the air density and hence the pressure within the chamber. The **difference** in pressure acts as a **restoring force**.[4]

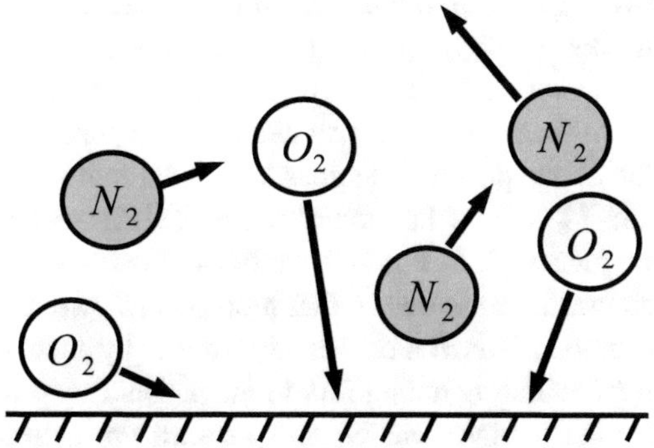

Fig. 3.2 Molecules colliding with wall

3.1.2 Generating a Sound Pulse

Consider again the chamber as in Fig. 3.3a. What would happen if the piston were moved **suddenly** to the right and held fixed in position? [See Fig. 3.4.]

The air in front of the piston will be compressed locally, thus increasing the pressure lo-

---

[4]Here is a beautiful animation that displays the collisions of molecules on a piston. [12-26-2010 http://wilsonspirit.com/] You can vary the number of molecules in the chamber. You can move the piston so as to change the volume so as to change the collision rate with the piston as well as the pressure. Finally, you can vary the temperature so that the speed of the molecules varies. A shortcoming of the animation is that it omits intermolecular collisions.

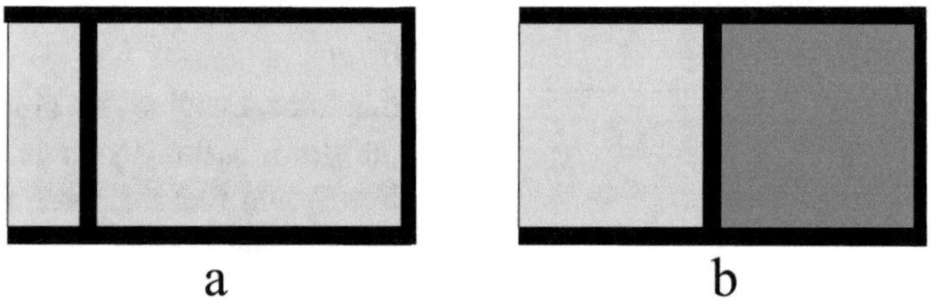

Fig. 3.3 (**a**) A chamber of air fitted with a piston. (**b**) The piston compresses the air in the chamber

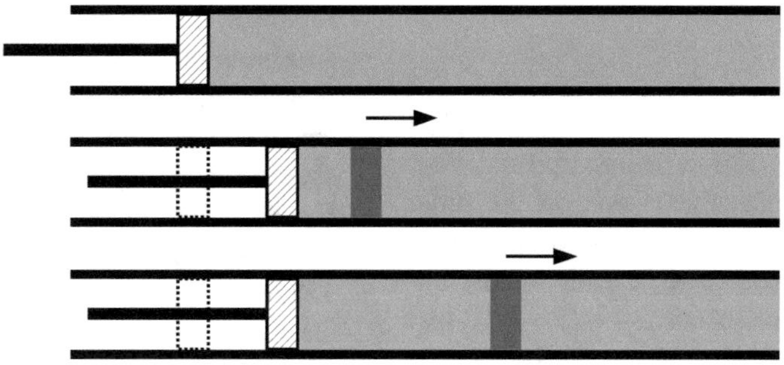

Fig. 3.4 A sudden move of the piston creates a pulse

cally. The non-uniformity in pressure will lead to a tendency for the compressed region to expand outwardly to the right. The net result will be that a compression pulse will move to the right as depicted in Fig. 3.4 to the right. This pulse will strike the opposite end of the chamber and be reflected back, followed by repeated cycles back and forth. In time, because of attenuation [see Sect. 4.9], the pulse will spread out in width and diminish in amplitude, eventually disappearing. The final state will be a chamber with air at a uniform density and pressure, corresponding to the current reduced volume of the chamber.

3.1.3 Digression on Pushing a Block of Wood

When you start pushing a block of material at one of its ends, you might assume that all of the material begins to move at once. The fact is that it takes time for the entire piece of material to respond to your push. Your initial push produces a compression pulse that travels through the block at the speed of sound. Eventually, this pulse is spread out, so that the block moves as a whole. If the block is made of a hard wood such as oak and the pulse is traveling in a direction parallel to the fibers, the speed of sound and hence the pulse is $\sim$ 4,000-m/s. If the block is 10 cm long, the pulse takes 0.10/4,100 = 0.00003-s = 30-microseconds (30-μs) to travel the length of the block.

3.2 The Nature of Sound Waves in Air

We can now understand what the essence of a sound wave is: **A sound wave represents the propagation in a material of a non-uniformity in the density**. Non-uniform pressure provides the restoring force towards the equilibrium state of uniform density. As an example, suppose that the above piston undergoes sinusoidal motion—that is, simple harmonic motion (SHM)—at a frequency f. In Fig. 3.5, we depict the resulting sinusoidal sound wave that propagates down the chamber. This motion can be produced by attaching a vibrating tuning fork to the piston. Regions having a high density and pressure are called **condensations**, while regions with low density and pressure are called **rarefactions**.

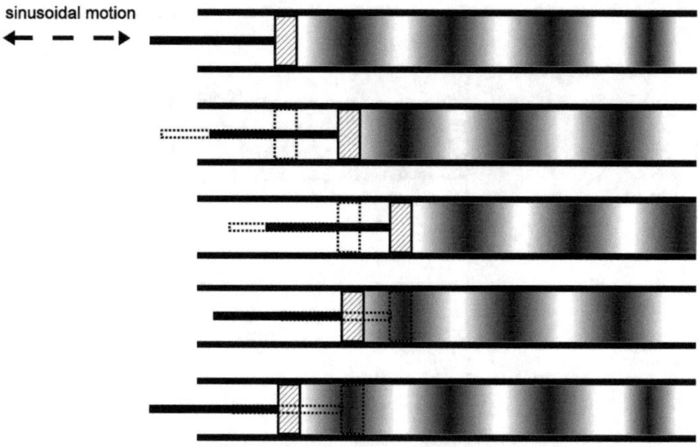

Fig. 3.5 A sinusoidal wave in air produced by a piston oscillating sinusoidally

Note that the action of the piston is similar to that of the cone of a loudspeaker, which vibrates in some pattern time. If the cone vibrates sinusoidally, it produces a sinusoidal wave in space. Generally, the pattern in time produces a sound wave with a corresponding pattern in space.

Since the motion of air in a sound wave is in the same direction as the direction of propagation of the wave. The wave is said to be **longitudinal**, in contrast to the transverse waves along a stretched string.

Indicated too in the figure are the wavelength λ and the waveforms at various fractions of the period $T = 1/f$. After a time interval of one period, the wave progresses a distance of one wavelength.

Equation (2.23) and (2.24) of Chap. 2 hold

$$v = \frac{\lambda}{T} = \lambda f \tag{3.4}$$

The normal person can hear pure sinusoidal sounds having a frequency ranging from about 20-Hz to 20,000-Hz.[5]

Let us calculate the corresponding range of wavelengths of audible sound traveling in air. We will use the velocity of sound at a temperature of 15°C—340-m/s. For the minimum frequency of 20-Hz we obtain

$$\lambda = \frac{v}{f} = \frac{340}{20} = 17\text{-m} \tag{3.5}$$

The wavelength corresponding to a frequency of 20,000-Hz is just 20,000/20 = 1000 times smaller, that is 17-mm = 1.7-cm.

3.3 Characterizing a Sound Wave

A sound wave can be characterized by the **change in pressure** from the ambient equilibrium pressure, which is usually about one atmosphere. The amplitude of the variation of the change in pressure is called the **sound pressure**, symbolized by p_s. (See Fig. 3.6 below.) The softest sound that can be heard has a sound pressure of only 2×10^{-10} atmosphere, symbolized by p_0—a pressure that reflects the ear's incredible sensitivity. A sound can produce pain if the sound pressure is one-ten-thousandth (10^{-4}) of an atmosphere. Thus, the ratio of the largest sound pressure tolerable to the smallest discernible is $10^{-4}/10^{-10} = 10^6$, or a million to one![6] Alternatively, a sound wave can be characterized by the change in mass density from the equilibrium density. We then refer to the **sound density** of the wave.

One can also characterize a sound wave by the **displacement** of the air, although this is not

[5]As a person gets older and/or subjects himself/herself to loud sounds such as rock music, the upper limit goes down. In the author's testing of students from 1973 to 2000, the limit for most students has dropped from about 22,000-Hz to about 18,000-Hz.

[6]It is rare to find a man-made device that can handle such a huge range of inputs.

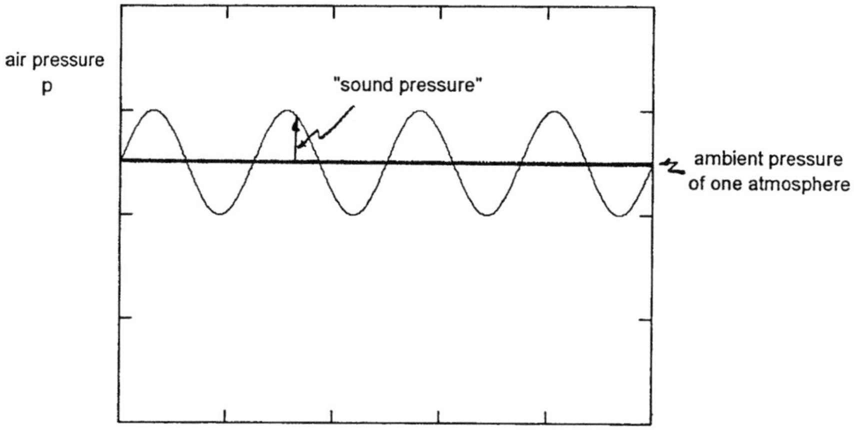

Fig. 3.6 Sound pressure

done in practice. The corresponding range of displacements is incredibly small—ranging from 0.1Å (about 1/30th the size of an atom (!)) to 1/100th of a millimeter (about twice the length of a bacterium).

Note Sound waves in liquids and solids are essentially the same as sound waves in a gas such as air. The major difference lies in the density of material: Solids and liquids are typically about 1000 times more dense than air at STP. Nevertheless, we will see that the velocity of sound is about the same order of magnitude for all materials.

3.4 Visualizing a Sound Wave

Suppose that we have a sound wave and we want to be able to visualize its corresponding wave pattern. A neat way to do so is summarized in Fig. 3.7.

The figure begins on the left with a source of sound—the **signal source**—that is produced by a generator of an electric signal that corresponds to the sound wave. One possible signal source is a microphone into which we might sing. Another can be a **wave generator**, which is an electronic device that produces periodic electric signals having a frequency that we can control. The periodic pattern of the signal is usually a sine wave. [Other patterns include a **sawtooth wave** and a **rectangular wave**. The names have

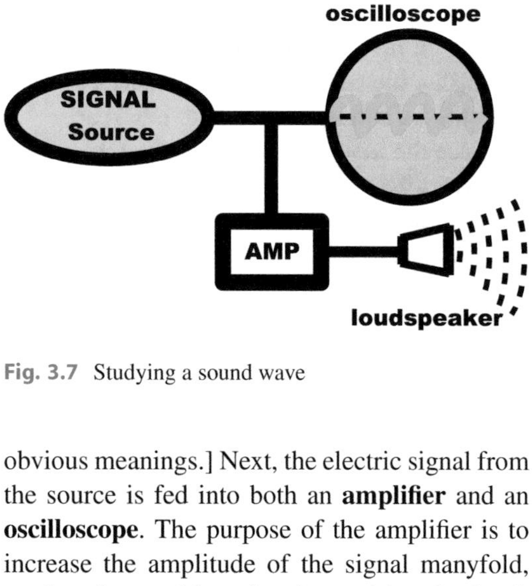

Fig. 3.7 Studying a sound wave

obvious meanings.] Next, the electric signal from the source is fed into both an **amplifier** and an **oscilloscope**. The purpose of the amplifier is to increase the amplitude of the signal manyfold, so that the resulting signal can drive the loudspeaker. The oscilloscope gives us a visual image on the screen of the pattern of the periodic electric signal. This visual image ideally represents the wave pattern of the sound wave that we hear produced by the loudspeaker.

3.5 The Velocity of Sound

What determines the velocity of sound? Recall that generally, the wave velocity of all types of waves depends upon the square root of the ratio of an effective restoring force and an effective mass. In the case of a sound wave, that effective mass is the mass density ρ, which is expressible in kg/m^3. [For example, the mass density of water is 1-g/cc = 1000-kg/m^3 while the mass density of air at STP is 1.3-kg/m^3.] The restoring force for a sound wave is known as the **bulk modulus**, which has the symbol B: This parameter tells us the relative **decrease** in volume of a sample in response to an **increase** in the pressure. In some situations, the temperature remains fixed during the process; in others, the sample is insulated and the temperature will usually rise. For small relative changes in volume, the relative change in volume $\Delta V/V$ is generally proportional to the change in pressure Δp. [This is the analog of **Hooke's Law**: displacement of spring $\propto$ force.] We write

$$\frac{\Delta V}{V} = -\frac{\Delta p}{B} \tag{3.6}$$

The **minus** sign is inserted because an **increase** in pressure leads to a **decrease** in volume. The ratio $\Delta V/V$ is referred to as the **relative change in the volume**. By multiplying by 100 we obtain the percent change.

As the volume increases at fixed amount of matter, the mass density decreases. Correspondingly the relative change in mass density ρ, given by $\Delta\rho/\rho$, is

$$\frac{\Delta\rho}{\rho} = +\frac{\Delta p}{B} \tag{3.7}$$

We see that the $(-)$ sign of Eq. (3.7) is simply replaced by a $(+)$ sign.

Note In the case of air, the bulk modulus is about equal to the pressure itself, so that $\Delta V/V \sim -\Delta p/p$. Thus, a one percent increase in pressure leads to about a one percent decrease in volume. On the other hand, the bulk modulus of liquids and solids is much greater than that for gases: It requires a much greater increase in pressure to produce a given reduction in the volume of a liquid or solid; air is much more compressible. For example, the value of the bulk modulus for water is $20,000$ atmospheres, while for steel it is $1,400,000$ atmospheres! Thus, in order to produce a 1% decrease in the volume of a sample of water requires an increase in pressure of $20,000 \times 0.01$ = 200-atm, while for a sample of steel a pressure of $1,400,000\times 0.01$ = 14,000-atm is required!

Down at the deepest depths of ocean, the pressure is about 1000-atm, so that the density of the water is increased by a factor of $\Delta p/B = 1000/20,000 = 1/20$, corresponding to 5%.

One can show that generally the velocity of sound is given by

$$v = \sqrt{\frac{B}{\rho}} \tag{3.8}$$

Using this equation it is essential to express both B and ρ in terms of a consistent set of units, such as N/m^2 and kg/m^3, respectively.

Recall from Chap. 2 that the wave velocity for a stretched string is the square root of the ratio of a force parameter (that restores the string to its straight line shape) and a mass parameter. Here, the bulk modulus B is associated with the increase in pressure when the air is compressed; it reflects the effective restoring force. The mass density ρ is the mass parameter.

Let us then calculate the speed of sound in water. We recall that the mass density of water is 1-g/cm^3 = 1000-kg/m^3 and that 1-atm = 10^5-N/m^2 , so that B = 20,000$\times$10^5-N/m^2. Thus,

$$v = \sqrt{\frac{20,000 \times 10^5}{1,000}} = 1{,}400\text{-m/s} \tag{3.9}$$

You can carry out a similar calculation for air (B = 1.4-atm, ρ = 1.3-kg/m^3) and for steel (B = 1,400,000-atm and ρ = 7,900-kg/m^3).

3.5.1 Temperature Dependence of Speed of Sound in Air

The speed of sound in a gas increases with increasing temperature. For an **ideal gas**, the speed

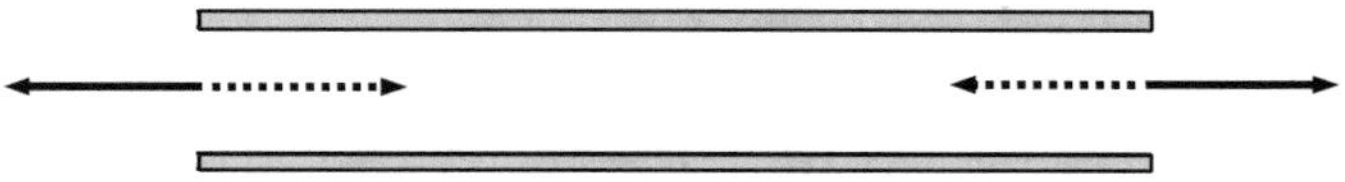

Fig. 3.8 Motion of air in a pipe for the fundamental

of sound in an ideal gas is on the order of average of the speed of a molecule.[7]

$$v \propto \sqrt{\frac{\text{absolute temperature}}{\text{mass of a molecule}}} \quad (3.11)$$

At 0°C, the **speed of sound in air** is 332-m/s. To obtain the approximate speed for a temperature between ~ -50°C and $\sim +50$°C, simply add 0.61 multiplied by the temperature in °C. Thus,

$$v\left(\frac{m}{s}\right) = 332 + 0.61 \times \text{temperature (°C)} \quad (3.12)$$

For example, at a temperature of 20°C, the speed of sound is

$$v = 332 + 0.61 \times 20 = 332 + 12 = 344\text{-m/s}$$

The relative change is $12/332 = 0.04$, or 4%.

Consider the effect on a pitch pipe, which is set to produce a single musical note. We will see in the next section that the frequency of the sound produced by a pitch pipe is proportional to the wave velocity. Thus, the frequency will increase by close to 4%. In Chap. 12 we will see that this change in frequency corresponds to about 3/5 of a **semitone** interval. Thus, a musical note of *A* will change to a note close to $A^{\#}$ (A-sharp). For this reason wind instrumentalists must tune their instrument as their breath heats up the air within—if at all possible.

3.6 Standing Waves in an Air Column

Consider a long pipe, that is, one whose diameter is much less than its length ℓ, and which is open at both ends to the outside air. The pipe is thus filled with air, so that we have a column of air. It is possible to excite modes of vibration, standing waves, in a column of air.[8] These standing waves are the source of sound as a result of the variations of the pressure that the air column exerts on the air outside the column. A standing wave consists of a sinusoidal pattern of the sound pressure, whose amplitude oscillates sinusoidally in time, as in the case of a standing wave along a stretched string. In essence, air rushes in and out of the two ends of the pipe. See Fig. 3.8.

The air **displacement** has an **anti-node** at an **open end** of a pipe. A small amount of air rushes in and out of the pipe. In addition, because of the huge volume of outside air, the outside air acts like a cushion that prevents the pressure at an opening from being different from that of the outside air. The **sound pressure** has a **node** at an open end of the pipe.

In the case of the *fundamental* mode, there is no motion of air at the center. That is, the displacement has a **node** at the center. Correspondingly, the density and hence pressure are maxima there. In other words, the **sound pressure** and **density** have an anti-node at the center. The variation of the sound pressure with position in the pipe is thus the same as that for the displacement of a vibrating string with fixed ends, as seen in Fig. 3.9.

[7] More precisely,

$$v \approx \sqrt{\frac{kT}{m}} \quad (3.10)$$

where k is **Boltzmann's constant**, T is the absolute temperature (temperature in degrees Celsius + 273), and m is the mass of a molecule.

[8] We have restricted our attention to pipes with a relatively small diameter only because in this case the frequency spectrum of the modes is a harmonic series. The larger the ratio of the diameter to the length, the greater the deviation of the spectrum from a harmonic series.

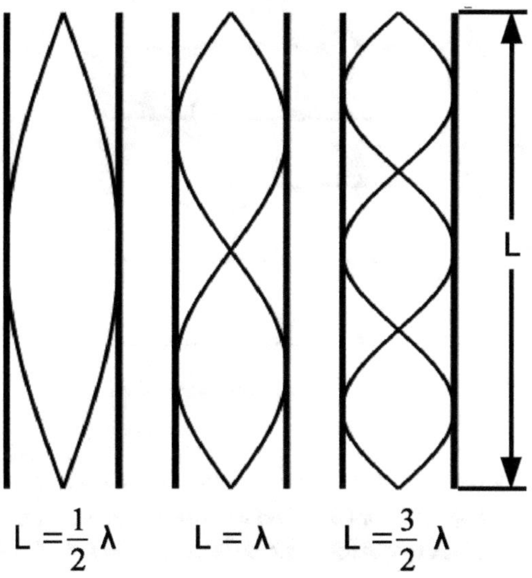

Fig. 3.9 Wave patterns of the sound pressure and sound density for the first three harmonics of an open pipe

Analysis of the standing wave vibrations of a pipe leads to the same formula for the fundamental frequency as that for the string, namely

$$f_1 = \frac{v}{2\ell} \tag{3.13}$$

where v is the wave velocity, here the speed of sound in air.

Looking beyond the fundamental mode, the mode frequency spectrum is a harmonic series, as with the string (see Eq. (2.26)):

$$f_1 = \frac{v}{2\ell}, \quad f_2 = 2\frac{v}{2\ell} = \frac{v}{\ell}, \quad f_3 = 3\frac{v}{2\ell}, \quad f_4 = 4\frac{v}{2\ell} = \frac{2v}{\ell}, \quad \ldots \tag{3.14}$$

or

$$f_n = n\frac{v}{2\ell}, \qquad n = 1, 2, 3, 4, \ldots \tag{3.15}$$

Sample Problem 3.1 Consider a person playing a flute that is open at both ends, having a length of 60-cm, with air at a temperature of 20°C. Find the fundamental frequency.

Solution Recall that we found that the speed of sound air at 20°C is 344-m/s. Hence

$$f_1 = \frac{v}{2\ell} = \frac{344}{2(0.60)} = 287\text{-Hz}$$

Now suppose that the temperature increases to 30°C by virtue of a person's warm breath. The ten degree increase in temperature leads to an increase in the speed of sound by 0.6(10) = 6-m/s, that is, to 350-m/s. The new fundamental frequency is then

$$f_1 = 350/2(0.60) = 292 \text{ Hz}$$

Alternatively, the calculation can be carried out in this illuminating way: The change in the frequency is given by Δf and can be expressed as

$$\Delta f = \frac{\Delta v}{2\ell} \tag{3.16}$$

where Δv is the change in the speed of sound. Then

$$\frac{\Delta f_1}{f_1} = \frac{\Delta v/2\ell}{v/2\ell} = \frac{\Delta v}{v} \tag{3.17}$$

This equation tells us that the **relative change in the frequency** is equal to the **relative change in the speed of sound**. Correspondingly the percent changes are equal. In our problem, the relative change in speed is 6/344 = 0.017, so that the relative change in the frequency is also 0.017. The original frequency is 287 Hz. Hence the change in frequency is 0.017(287) = 4.9-Hz and the new frequency is 287 + 4.9 = 292-Hz.

What is the benefit of going through this alternative approach? It is the following: The relative change in frequency, here 0.017, holds for any pipe, **whatever its length**, as well as for any particular mode. Thus we need not know the original length of the pipe in order to determine the effect of a temperature change. Furthermore, as we will see in Chap. 12, the relative change in frequency is directly related to the change in pitch. For example, a change of 6% corresponds to a half tone.

An aside: An increase in temperature will also lead to an increase in the length of the pipe. The actual increase depends upon the material out of which the pipe is made. That increase is minuscule. For steel, it amounts to a relative change of only one part in 100,000-pe−°C, or one part in 10, 000 (0.01%) for the temperature increase of 10 degrees in our above problem. This **increase** in length would by itself lead to a **decrease** in the frequency because the frequency is **inversely** proportional to the length. However, that decrease is imperceptibly small.

Sample Problem 3.2 What should the length of a pipe be in order to obtain a fundamental frequency of 20 Hz, with air at 20°C?

Solution We first note that

$$f_1 = \frac{v}{2\ell} \qquad \longrightarrow \qquad \ell = \frac{v}{2f_1} \tag{3.18}$$

Thus, given a speed of 344 m/s, we obtain

$$\ell = 344/2(20) = 8.6 \text{ m} \sim 28 \ ft!$$

Standing Waves in a Closed Pipe

One could use a 28 *foot* organ pipe to produce a 20 Hz note. However, often such a length is quite unwieldy. How can one avoid it? The solution lies in using a pipe which is **open** at one end but **closed** at the other. We will refer to such a pipe as a **closed pipe**, even though one end of such a pipe is open. We will refer to a pipe that is open at both ends as an **open pipe**.

At the closed end of a *closed pipe*, the air cannot move. Therefore the displacement of air flow has a **node**. On the other hand, the density has a maximum variation and therefore has an **anti-node**. Analysis leads to the result that the fundamental frequency of a **closed pipe** is half that for an open pipe of the same length.

Below we present a simple way of understanding this result. It is based upon the way in which a pulse of sound is reflected off an open end and a closed end of a pipe. The fundamental frequency is the same as the frequency of a pulse that moves back and forth down the length of a pipe.

We start by considering a pulse along a taut string, that is fixed at both ends. See Fig. 3.10. The figure depicts the displacement of the string at various times. We note that the direction of the displacement is reversed upon reflection off an end. A cycle requires but two traversals of the pulse along the length of the string.

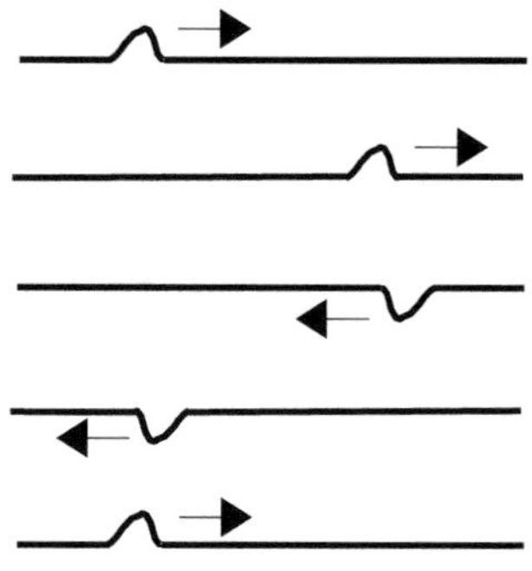

Fig. 3.10 A pulse along a string fixed at both ends

Examine the figure very carefully. You will note that the reflected pulse has a pattern that differs from the incident pulse in two ways: First the pulse is flipped upside down. Second, the steeper side of the incident pulse is on the right, while the steeper side of the reflected pulse is on the left. Thus, the front of the incident pulse remains the front of the reflected pulse.

The distance traveled by the pulse in one cycle is 2ℓ. The corresponding frequency is then

$$f = \frac{v}{2\ell} \tag{3.19}$$

Now we return to the behavior of a pulse in a pipe. When a condensation travels down a pipe and reaches an open end, the air rushes out and the reflected wave is a rarefaction. If the condensation reaches a closed end, the reflected wave is a condensation. On the other hand, when a rarefaction travels down a pipe and reaches an open end, the air rushes in, so that the reflected wave becomes a compression. If the rarefaction reaches a closed end, the reflected wave is a rarefaction. [See Fig. 3.11.] Thus, the spectra of a taut string, fixed at both ends, and an open pipe are both harmonic series.

The distance traveled by the pulse in one cycle is 2ℓ so that the frequency is

$$f = \frac{v}{2\ell} \tag{3.20}$$

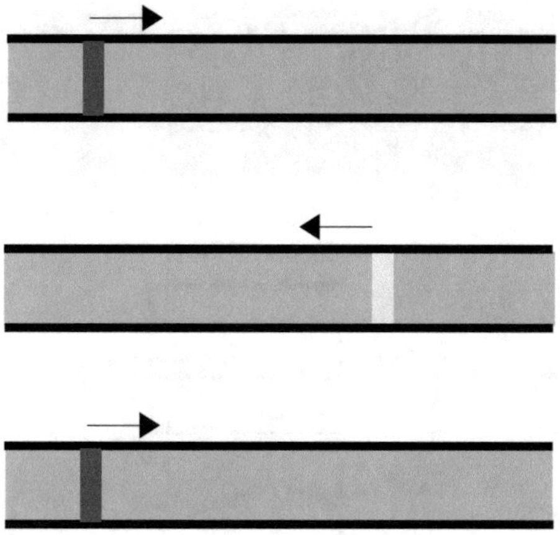

Fig. 3.11 A pulse along a pipe open at both ends

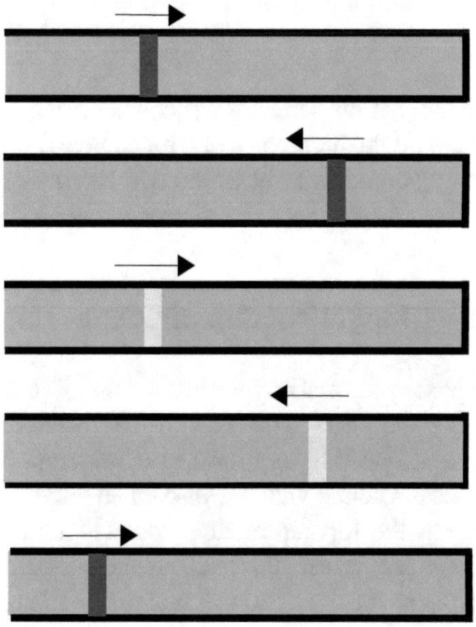

Fig. 3.12 A pulse along a pipe open at one end but closed at the other

On the other hand, when a sound pressure pulse is reflected off a closed end, its character is not changed. A condensation is reflected as a condensation and a rarefaction is reflected as a rarefaction. The result is that a cycle requires four traversals and the fundamental frequency is half that of the open pipe. See Fig. 3.12.

Correspondingly, for a given frequency, we would need only half the length of pipe, or 14-feet for our problem.

The distance traveled by the pulse in one cycle is 4ℓ so that the frequency is

$$f = \frac{v}{4\ell} \tag{3.21}$$

The frequency spectrum of a closed pipe can be shown to consist of all the **odd harmonics** of the fundamental frequency, that is

$$f_1 = \frac{v}{4\ell}, \qquad f_2 = 3\frac{v}{4\ell},$$
$$f_3 = 5\frac{v}{4\ell}, \qquad f_4 = 7\frac{v}{4\ell}, \qquad \ldots \tag{3.22}$$

Thus, for a fourteen foot pipe and sound velocity of 345-m/s, the fundamental frequency will be 20-Hz. The overtones will be

$$3 \times 20 = 60\text{-Hz}, \quad 5 \times 20 = 100\text{-Hz},$$
$$7 \times 20 = 140\text{-Hz}, \ldots$$

In Table 3.1 below, we summarize the conditions that hold at the boundaries of a vibrating air column for the three parameters: *sound pressure*, *sound density*, and *displacement* of fluid.

In Fig. 3.13 we display two graphs for each of the following: the fundamental and the first overtone, for both an open pipe and a closed pipe. One graph displays the variation of sound pressure p_s and change in density ρ_s due to the sound wave; the other graph displays the displacement y of the gas.

End Correction for Modes in a Pipe

The formula for the frequencies of the modes of a pipe is based upon the assumption that the sound pressure has a node at an open end. An open pipe has two open ends, while a so-called closed pipe has but one. At a closed end, the sound pressure has an anti-node.

In 1870, Lord Rayleigh showed that the sound pressure extends beyond an open end. His estimate was that the amount is 0.61 times the

Table 3.1 Summary of pipe properties and their behavior

Parameter	Open end	Closed end
Sound pressure	Node	Anti-node
Sound density	Node	Anti-node
Displacement	Anti-node	Node

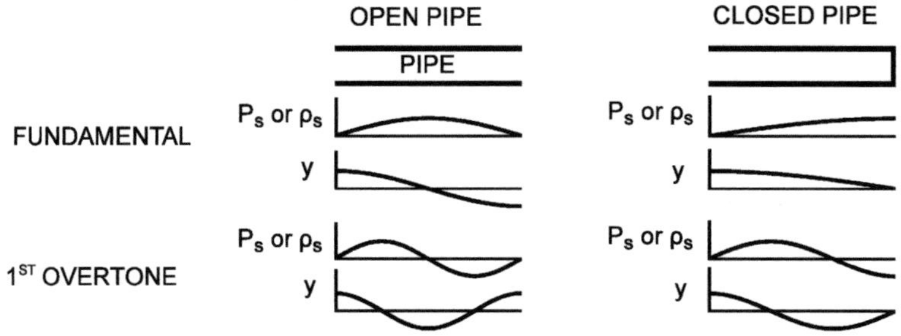

Fig. 3.13 Wave patterns of the sound pressure, sound density, and displacement for the fundamental and the first overtone for both the open pipe and the closed pipe

radius R. H. Levine and J. Schwinger[9] derived an exact expression, which rounds off to 0.61 R. These results depend upon the condition that the wavelength is much greater than the radius. That is, $\lambda \gg R$.[10]

The consequence is that the mode frequencies for a closed pipe are given by

$$f_1 = \frac{v}{4(\ell + 0.61R)},$$
$$f_2 = 3\frac{v}{4(\ell + 0.61R)}, \quad \ldots \tag{3.23}$$

Effectively, the length is increased by what is referred to as an **end correction** of 0.61R.

For an open pipe, there are two open ends so that there is an end correction of 1.22 R. The frequencies are given by

$$f_1 = \frac{v}{2(\ell + 1.22R)},$$
$$f_2 = 2\frac{v}{2(\ell + 1.22R)}, \quad \ldots \tag{3.24}$$

[9]Physical review, vol. 73, p. 383, 1948.

[10]Some organ pipes have a diameter so large and such a high frequency and therefore small wavelength that this condition can be violated. Levine and Schwinger calculated the error for a great range of frequencies.

Note that the series of frequencies remains as a set of all the harmonics of a fundamental for an open pipe and as a set of the odd harmonics of the fundamental for a closed pipe.

3.7 Magic in a Cup of Cocoa

Suppose you make yourself a cup of cocoa, mixing the cocoa powder with hot water. You stir well and then tap the top of the cup. You will hear a sound with a well-defined pitch. Now you tap the top repeatedly. You will find that as time progresses, the pitch of the sound rises steadily so much so that the final maximum pitch corresponds to more than double the original frequency. If you stir the cocoa you can repeat the above procedure with the same consequences. The question is why does this happen?

We can quickly guess that the sound corresponds to a mode of vibration of sound within the volume of cocoa. This can be checked by noting that the empty cup produces no sound in the range of the sound above. The frequency depends upon the shape of the volume—which doesn't change—and the velocity of sound. Therefore, we surmise that the effect of stirring the cocoa lowers the velocity of sound. Two possibilities present themselves: Either the stirring mixes up microscopic particles of undissolved cocoa or the stirring produces microscopic bubbles of air within.

Let us focus first on the cocoa particles as a possible explanation. We recall that the sound velocity depends upon the ratio of the bulk modulus and the mass density. This ratio must therefore change by factor of four to account for a doubling of the frequency! The particles would increase the average density of liquid, thus lowering the sound velocity; that's a good sign. However, the concentration of particles is small. [Remember that you typically put into the cup about two tablespoons in 8 oz of water and the cocoa powder is fluffy so that it takes up much less than two tablespoons.] Therefore the effect of increased density is much too small to account for the observations. Moreover, the relatively higher bulk modulus compared to that of water would *increase* the sound velocity. We therefore conclude that the stirring up of cocoa particles is not the explanation.

We next focus on the possibility that stirring produces microscopic air bubbles, which eventually rise and leave the cup. Their presence would *lower* the average density and therefore raise the frequency—but only by a small amount because of their low concentration. However, they have a much lower bulk modulus, by a factor of about one-hundred thousand! Furthermore, when a sound wave is compressing the cocoa over a typically volume of about half a wavelength, the fractional *change* in volume is extremely small. As a consequence, all of this reduction in volume can be taken up by the small volume of bubbles even if the bubbles take up a small fraction of the original volume. The effective bulk modulus of the cocoa is then dominated by the bubbles. It can be shown that all we need is a volume of bubbles equal to one part in one thousand to change the frequency by a factor of two!

With all this descriptive and semi-quantitative analysis, how can we be confident that we have arrived at the correct explanation? The answer is that a detailed theory leads to quantitative agreement with experimental observations. This is the final test to our hypothesis.

3.8 The Helmholtz Resonator

Have you ever blown over the top of an open soda bottle? The sound is like that of a hoarse human voice. The bottle is acting like a **Helmholtz resonator**, named after **Hermann Helmholtz** (Fig. 3.14).[11] who first studied this device. A real bottle with air within has many modes of vibration associated with the motion of air—standing waves that are analogous to waves in a pipe. On the other hand, the so-called Helmholtz resonator refers to a very specific mode wherein the air that is concentrated in the mouth of the bottle moves into and out of the bottle. It is analogous to the so-called **air resonance** of a violin that involves air moving into and out of the interior of the body of a violin through its **f-holes**.

We see in Fig. 3.15 an actual Helmholtz resonator, followed by a schematic drawing of a simple model of a Helmholtz resonator in Fig. 3.16. The actual shape of the bottle is not relevant; all that matters is that there be a narrow mouth whose volume is much less than the volume V of the bulk of the bottle. The mass of air that oscillates is highlighted in a dark gray shade. The air in the narrow mouth of the bottle acts as the mass of the oscillator, while the bulk of the air in the bottle acts as the spring constant. Note that the body of air within a stringed instrument such as a violin acts like a Helmholtz resonator. The corresponding mode of vibration is called the **main air resonance** of the stringed instrument.

3.8.1 The Physical Basis for the Helmholtz Resonator

We focus here as to why a mass of air moves in and out of the resonator. Here is why air moves in

[11]We will see Helmholtz's name in Chap. 11, in connection with his seminal studies of hearing, which are represented in his grand treatise *On the Sensations of Tone*. He is famous for his major contribution in the development of the **Principle of Conservation of Energy**, to be discussed in Chap. 4. And finally, he is also the inventor of the **ophthalmoscope**, which is used to examine the interior of the eye.

Fig. 3.14 Hermann von Helmholtz (photo credit: http://en.wikipedia.org/wiki/Hermann_von_Helmholtz)

and out of the chamber, leading to resonant mode of oscillation. [See Sects. 2.8.1 for a comparison with a simple harmonic oscillator.]

We will describe a single cycle:

1. Before you blow over the mouth of the chamber, the pressure in the chamber (the interior pressure) is equal to the pressure of the exterior air. When you begin to blow over its mouth, the pressure of the exterior air at the mouth is reduced, according to **Bernoulli's principle** . [This same principle is the explanation for the upwards lift on an airplane wing when the airplane is flying horizontally.] As a consequence of the lowered exterior pressure, air rushes out of the chamber. This part of the cycle is analogous to what happens when you pull a harmonic oscillator down from its equilibrium position and release it from rest, leading to the upward motion of the mass. See Fig. 2.13.

Fig. 3.15 An actual Helmholtz resonator (photo credit: http://en.wikipedia.org/wiki/Hermann_von_Helmholtz)

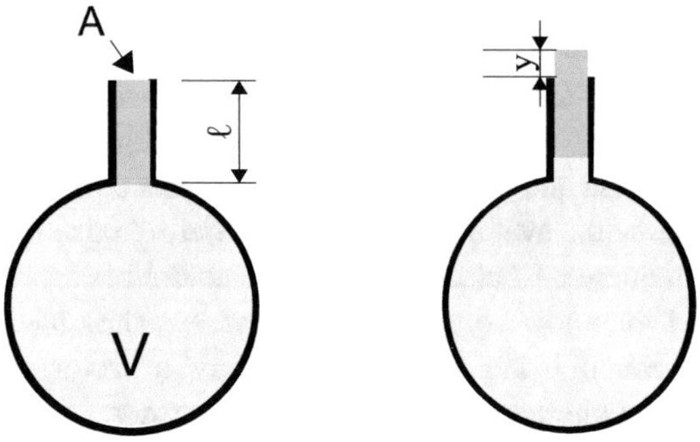

Fig. 3.16 A schematic of a model of a Helmholtz resonator

2. When air leaves the chamber, the interior pressure drops. At some point, the interior and exterior pressures are equal. However, the air rushing out cannot come to a sudden halt. Instead, air continues to leave the chamber and the interior pressure drops below the exterior pressure. As a result, air rushes back into the chamber, past the point wherein the pressures are equal and the exterior pressure again is lower than the interior pressure. This completes a single cycle.

Formula for the Frequency of Helmholtz Resonator

Three geometrical parameters determine the frequency: the volume V of the body of air; the area A of the mouth; and the length ℓ. The fourth parameter is the speed of sound in air, v. If the radius of the mouth is much less than the length ℓ of the mouth, the frequency of a Helmholtz resonator is given by

$$f = \frac{v}{2\pi}\sqrt{\frac{A}{V\ell}} \tag{3.25}$$

Derivation of the Helmholtz Formula

The mass m that oscillates is given by

$$m = \rho A\ell \tag{3.26}$$

where ρ is the mass density of the air. We show in Fig. 3.16 this mass having moved upward by a small amount y.

In equilibrium, the pressure inside the bottle is equal to the pressure outside the bottle. The force downwards on this mass is balanced by the force upwards by the air inside the bottle. As a result of the motion of the mass upwards, the pressure in the bottle is reduced, resulting in a net force downwards—or a restoring force for the displaced mass. Since this force will be found to be proportional to the displacement, the system obeys Hooke's Law and the mass oscillates like an SHO. Further details follow below.

The volume of the air in the bottle proper increases by $V = Ay$. This increased volume leads to a decreased pressure, given by Eq. (3.6):

$$\Delta p = -B\frac{\Delta V}{V} = -B\sqrt{\frac{Ay}{V}} \tag{3.27}$$

As a consequence there is a net force on the mass m given by

$$F = \Delta p A = -\frac{BA^2}{V}y \tag{3.28}$$

Comparison with Hooke's Law, $F = -ky$, where k is the spring constant, shows that the oscillator of the Helmholtz resonator has a spring constant given by

$$k = \frac{BA^2}{V} \tag{3.29}$$

Using the equation for the frequency of an SHO

$$f = \frac{1}{2\pi}\sqrt{\frac{k}{m}} \tag{3.30}$$

we obtain a frequency

$$f = \frac{1}{2\pi}\sqrt{\frac{k}{m}} = \frac{1}{2\pi}\sqrt{\frac{BA^2/V}{\rho A\ell}} = \frac{1}{2\pi}\sqrt{\frac{B}{\rho}}\sqrt{\frac{A}{V\ell}} \tag{3.31}$$

Equation (3.8), $v = \sqrt{B/\rho}$, leads to

$$f = \frac{v}{2\pi}\sqrt{\frac{A}{V\ell}} \tag{3.32}$$

3.8.2 Flutter of Air Through the Window of a Speeding Automobile

If you drive an automobile with an open window, you will create a Helmholtz resonator out of the interior of the automobile. Given the above discussion, we expect the loudness of the sound to increase as we increase the speed of the automobile, since the result will be a greater reduction of the external pressure just outside the window of the car.

Interestingly, while it is true that the sound is quite full of noise, there is a sense of **pitch**. Furthermore, the more wide open the window is, the higher the pitch, in accordance with the corresponding increase of the frequency predicted by Eq. (3.32). See Chap. 12 for details on the subject of pitch and its relation to frequency.

3.9 Terms

- Adiabatic change
- Ångstrom
- Atmospheric pressure
- Boundary condition
- Bulk modulus
- End correction
- Isothermal change
- Open-closed pipe
- Open-open pipe
- Pressure [force per area]
- Sound density change
- Sound pressure
- Standard temperature and pressure (STP)
- Helmholtz resonator

3.10 Important Equations

Pressure defined:

$$p = \frac{F}{A} \tag{3.33}$$

Fundamental relation between wavelength, frequency, and wave velocity:

$$\lambda = \frac{v}{f} \tag{3.34}$$

Bulk modulus defined:

$$\frac{\Delta V}{V} = -\frac{\Delta p}{B} \tag{3.35}$$

Speed of sound in an isotropic material:

$$v = \sqrt{\frac{B}{\rho}} \tag{3.36}$$

Speed of sound in air:

$$v\left(\frac{m}{s}\right) = 332+0.61\times\text{temperature (°C)} \tag{3.37}$$

Frequency spectrum for a pipe that is open at both ends:

$$f_1 = \frac{v}{2\ell},\ f_2 = 2\frac{v}{2\ell} = \frac{v}{\ell},\ f_3 = 3\frac{v}{2\ell}, \ldots \tag{3.38}$$

or

$$f_n = \frac{nv}{2\ell}, \qquad n = 1, 2, 3, 4, \ldots \tag{3.39}$$

Frequency spectrum for a pipe that is closed at one end:

$$f_1 = 1\frac{v}{4\ell}, \qquad f_2 = 3\frac{v}{4\ell},$$
$$f_3 = 5\frac{v}{4\ell}, \qquad f_4 = 7\frac{v}{4\ell}, \qquad \ldots \tag{3.40}$$

3.11 Problems for Chap. 3

1. Find the force acting on an area of dimensions 3″ × 4″ at the bottom of the deepest part of the Pacific Ocean, at 10,000-m (=10-km), where the water pressure is 1,000-atm.
2. What is the "sound pressure"?
3. What is the effective restoring force with respect to the velocity of a sound wave?
4. Find the mass and weight (in lbs.) of the volume of air in a room with the dimensions 4mx5mx6m.
5. The ear is most sensitive to sinusoidal sounds ("pure tones") having a frequency of about 3000-Hz. Calculate the wavelength of a sound wave with this frequency if the wave is traveling in air with a sound velocity of 340-m/s. Calculate the wavelength if the sound wave is traveling in water with a speed of 1400-m/s.
6. Calculate the sound velocity in steel, whose bulk modulus is 1.4 × 10^6-atm and mass density is 7,900-kg/m^3.
7. An open cylindrical pipe is 0.05 m long. If you ignore the end correction factor, what would be the frequency of the fundamental and the first overtone? Assume a temperature of 15°C. Repeat the above for a closed (half-open, half-closed) pipe.
8. A pipe is sounded at room temperature (20°C). It is observed that in the range 1,000-Hz to 2,000-Hz, the pipe can be made

to oscillate only at the frequencies 1,000-Hz, 1,400-Hz, and 1,800-Hz.

(a) Is the pipe an open or a closed pipe?

(b) Based upon your answer, what is the length of the pipe?

9. This problem focuses on a Helmholtz resonator.

(a) Suppose that a bottle has a volume of 1-liter = 1000-cm^3, a mouth area of 2.4-cm^2, and a mouth length of 3.0-cm. Given a speed of sound 340 m/s, find the frequency.

(b) The **mouth** for a violin are its f-holes. Suppose that their combined area is 5.0 cm^2 and that the thickness of body ℓ of the wood is 3.0-mm. Finally, suppose that the volume of the body is 1800-cm^3. Given a speed of sound 340-m/s, find the frequency. [Note that the Helmholtz formula assumes rigid walls; the flexibility of the walls of a violin reduces the actual frequency from the value calculated from the formula.]

4 Energy

We are all aware that electricity is needed to operate an audio amplifier. We casually say that we cannot get something for nothing. We pay the electric company an amount that is based upon the number of kilowatt-hours of electricity used. In fundamental physics terms, electricity is a form of **energy** that is needed to power and operate an amplifier. The expenses of the electric company include the production of electrical energy from other forms of energy and the transmission of this form of energy from the electrical generator plants to your home.

Amplifiers are rated by the number of watts output per channel. The **Watt** is a unit of **power**, which expresses the rate of energy exchange. A certain fraction of this output of energy is associated with the sound waves emitted by the loudspeakers of the audio system. We can make similar observations regarding light bulbs, which are powered by electricity.

The output from a loudspeaker is not the only factor that determines how loud a sound we hear. Our distance from the loudspeaker matters too. The closer we are to a speaker, the louder the sound. This statement is merely a qualitative one: **Loudness** is a **subjective** parameter; not only does it depend upon the individual, it cannot be given a numerical value. Nevertheless, there exists an **objective** physical parameter, **intensity**, that can be used as a reproducible reference standard and is directly associated with loudness. Intensity reflects the **concentration of power over space**.

In describing vibrations of strings and pipes, we assumed for simplicity that once excited, vibrations would last forever. We recognize that in real systems, vibrations will die out unless they are sustained by excitation from without. This phenomenon of **attenuation** is fortunate: For example, in its absence our ears would be overwhelmed by all the sounds that have been produced in the past! Nevertheless, it is reasonable to ask where all the "action" connected with sound has gone. It couldn't merely disappear.

We sometimes ask this question in the context of money flow. Money changes hands. Money is often converted from one currency to another. Were it not for the printing of money by governments (and counterfeiters) the amount of money would remain constant. Similarly, we observe changes of various sorts in physical systems. Is there something about these systems that is nevertheless constant and would allow us to make a check on our measurements so that we can discover possible errors in the measurements. The answer is yes. Generally, the physical quantity known as **energy** is a measuring stick for keeping our accounts straight, as we will see. In the context of sound that dies out, the energy associated with the sound is replaced by what is known as **thermal energy**. In this chapter, we will study the physical parameter called *energy*,

L. Gunther, *The Physics of Music and Color*,
https://doi.org/10.1007/978-3-030-19219-8_4

along with its related parameters, *power*, *intensity*, and *attenuation*. These parameters are necessary for providing us with an objective means of characterizing sound and light, so that we can better understand and appreciate our subjective experience.

4.1 Forms of Energy and Energy Conservation

Over the past 15 years or so, there has been much talk of an energy crisis. People talk about the need to conserve energy. And yet, there is a fundamental principle of physics known as the **Principle of Conservation of Energy**—that is, energy is automatically conserved, whatever we do! How can we reconcile these two contradictory claims?

The answer is simple: There are many **different forms of energy**. One can assign a numerical value to the amount of energy present among the various forms. The Conservation of Energy Principle states that whereas the amounts of each form of energy can change, the sum total of all forms of energy is constant. In other words, any amount of a given form of energy that is lost is replaced by a net increase in amounts of the other forms. Two systems can exchange energy; in the exchange process, the forms of energy may change. Nevertheless, whatever energy one system gains the other system must lose. The call to conserve energy is really a call to conserve those forms which are important to us.

We will be using the **Joule** as the preferred unit of energy in order to make direct use of the fundamental equations of physics. The "Joule" is named after **James Prescott Joule**, who identified heat as a process whereby thermal energy is transferred between one material and another. There are a number of units of energy in common use, depending upon the context. Some are listed in Table 4.1 along with their conversion to Joules. [Recall our comments earlier in the text on the need to use a consistent set of units.]

Table 4.1 Units of energy

Units of energy	Number of joules
Joule (J)	1
KiloWatt-hour (kWh)	$3.6 \times 10^6 = 3,600,000$
Food calorie = kilocalorie ($kcal$ or Cal)	4, 200
British thermal unit (	4.0
erg (erg)	1.0×10^{-7}
Electron-volt (eV)	1.6×10^{-19}

4.1.1 Fundamental Forms of Energy

1. **Kinetic Energy (KE)**—energy associated with the motion of a massive object

$$\boxed{\mathrm{KE} = \frac{1}{2}mv^2} \tag{4.1}$$

Here m is the mass of the object and v is the speed.

Sample Problem 4.1 A 10-**tonne** (1-tonne = 1000-kg $\sim$ 2200-lbs) truck is moving at a speed of 10-m/s (1-m/s = 3.6-km/hr $\sim$ 2.2-mph). What is its KE?

Solution

$$\mathrm{KE} = \frac{1}{2}(10 \times 1000)(10)^2 = 500{,}000\text{-Joules}$$
$$= 500{,}000\text{-J}$$

Sample Problem 4.2 What if the speed of the truck is **doubled** to 20-m/s?

Solution Because KE is proportional to the square of the speed, the KE is quadrupled to

$$\frac{1}{2}(10 \times 1000)(20)^2 = 2{,}000{,}000\text{-J}$$

Or, we could realize that multiplying the speed by a factor of two should multiply the KE by a factor of $2^2 = 4$. Thus

$$\mathrm{KE} = 4 \times 500{,}000\text{-J} = 2{,}000{,}000\text{-J} \tag{4.2}$$

Sample Problem 4.3 Find the KE of a 100,000-tonne oil tanker that is moving at a speed of 2m/s.

Solution

$$\text{KE} = \frac{1}{2}(100,000 \times 1,000)(2)^2 = 2 \times 10^8\text{-J}$$
$$= 200\text{-milllion Joules} \quad (4.3)$$

2. **Potential Energy (PE)**—energy associated with changing configurations of objects. This definition is bound to be obscure. Examples will help.

(a) *Potential energy of a stretched or compressed spring*
The amount is given by

$$\boxed{\text{PE} = \frac{1}{2}ky^2} \quad (4.4)$$

where k is the spring constant and y is the displacement. Note that the PE is positive whether the spring is stretched ($y > 0$) or compressed ($y < 0$).

Sample Problem 4.4 Find the PE of a spring which has a spring constant 25-N/m and a displacement of 2-cm.

Solution

$$\text{PE} = \frac{1}{2}(25)(0.02)^2 = 0.005\text{-J}$$

Sample Problem 4.5 Find the PE of the spring of the previous problem if the displacement is doubled to 4-cm.

Solution Because the PE is proportional to the square of the displacement, the PE is quadrupled to 4× 0.005 = 0.020-J.

(b) **Gravitational Potential Energy** Suppose I lift an object and then release it from rest, allowing it to fall. It will accelerate, picking up kinetic energy. We regard the increased elevation as increasing the potential energy of the object in connection with the force of gravity of the earth. This potential energy is replaced by kinetic energy as the object falls. A change in elevation results in a change in PE, which we symbolize by ΔPE. Theory leads to the relation

$$\Delta\text{PE} = \text{weight} \times \text{change of elevation} \quad (4.5)$$

If we let w be the weight and h the change in elevation we have

$$\Delta\text{PE} = wh \quad (4.6)$$

For example, if an object of weight 200-Newtons (~ 44-lbs) is raised by 3-m, its PE increases by 200× 3 = 600-Joules.
Now suppose that a person weighing 150-lbs climbs a mountain, with an increase of elevation of 4000-m. See Fig. 4.1 to give you a feeling for this activity.
What is his increase in the gravitational PE? The stress on his body—especially in rock climbing—would lead us to believe that the PE change must be enormous. To the contrary, we will find that the total energy expended in climbing is many times greater than the change in PE. The process is incredibly inefficient.
To solve this problem, we need to relate a weight in lbs to a weight in Newtons. Our starting point will be the fact that a mass of one kilogram is equivalent to a weight of 2.2-lbs. Why are mass and weight related? What is the difference between the two?
Mass is a measure of the quantity of matter. It is independent of where an object is located. On the other hand, the **weight** of an object is the force of gravity by the earth on the object. You may have heard reference to the weight of an object when situated on the surface of the moon or another planet. We would then be

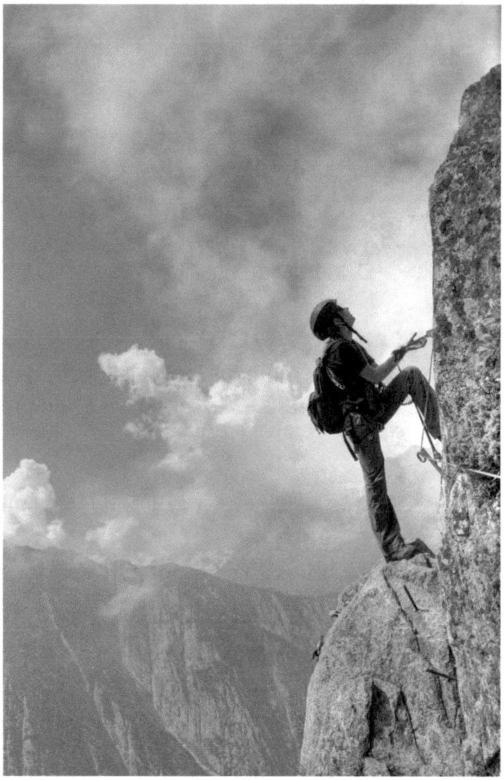

Fig. 4.1 Rock climbing (photo credit: http://vceoes.wikispaces.com/file/view/rock%2520climbing.jpg/41719235/rock%2520climbing.jpg)

referring to the force of gravity of these bodies on the object. For example, a 60-pound object on earth weighs 10-lbs on the moon. Thus the weight of an object depends upon the body exerting the force of gravity. **Isaac Newton** showed that the weight of an object is proportional to its mass m. The proportionality constant is the acceleration that an object would experience in falling in the absence of air resistance. Such a fall is referred to as "free fall." The acceleration under free fall is referred to as the "gravitational acceleration constant" and is given the symbol g. Its value is

$$g = 9.8\frac{m}{s^2} = 32\text{-}\frac{ft}{s^2} \tag{4.7}$$

Then we write

$$\text{weight} = w = mg \tag{4.8}$$

The change in PE due to a change in elevation is then given by

$$\boxed{\Delta \text{PE} = mgh} \tag{4.9}$$

Sample Problem 4.6 Find the change in PE if a 10-kg object is raised by an elevation of 2-m.

Solution

$$\Delta \text{PE} = mgh = 9.8(10)(2) = 196\text{-J}$$

Back to the Mountain Climbing Problem

We need to express the weight 150-lbs in Newtons. To obtain the conversion we can use the two corresponding values m=1-kg and w=2.2-lbs.: A mass of 1-kg has a weight of

$$w = mg = 1 \times 9.8$$
$$= 9.8\text{-Newtons} = 2.2\text{-lbs}$$

Therefore

$$1\text{-Newton} = 22/9.8 = 0.22\text{-lbs}$$

or

$$1\text{-lb} = 4.5\text{-N}$$

Therefore

$$150\text{-lbs} = (150/0.22)\text{-N} = 4.5 \times 150$$
$$= 680\text{-N}$$

The change in PE of the mountain climber is then

$$\Delta\text{PE} = wh = 680 \times 4000$$
$$= 2{,}700{,}000\text{-J}$$

Thus, we have come up with the amount of PE needed for the climb but most readers have little sense as to what this number 2,700,000-Joules means. We need to express the PE in units that are more familiar. We will do so later on in the chapter when we discuss the food calorie.

(c) Later on in this text, we will discuss the force between electric charges. There is an **electrostatic potential energy** associated with this force that depends upon the distance between the charges.

3. **Electromagnetic Radiation**—This form of energy will be discussed further in Chap. 5. For the time being, we should note that it is this form of energy that the SUN transmits to us on earth and that is our utterly basic source of energy.

4.1.2 "Derived" Forms of Energy

The reader might wonder why electrical energy or nuclear energy was not included in the list of fundamental forms of energy. The reason is that these terms refer to energy that constitutes a mixture of what physicists regard as fundamental forms of energy. We list some examples below:

- **Chemical Energy**—energy associated with the KE and PE that is directly connected with the interaction of atoms in their binding together to form molecules.
- **Electrical Energy**—energy associated with the KE of electrical charge in electric currents or chemical energy stored in electric batteries. By means of chemical reactions, the chemical energy in a battery can be harnessed so as to produce electrical energy.
- **Nuclear Energy**—energy associated with the KE and PE of protons, neutrons, and lesser known particles that is directly connected to their binding together to form the nuclei of atoms.
- **Thermal Energy**—energy associated with the random positions and velocities of atoms and molecules in a macroscopic body. Both kinetic energy and potential energy contribute to thermal energy.

Objects of all sizes can have kinetic energy and/or potential energy. Objects that are visible are referred to as **macroscopic bodies**. Taken together, the kinetic energy and potential energy of a macroscopic body is referred to as **Mechanical Energy**. That is,

$$\text{Mechanical Energy} = \text{Kinetic Energy} + \text{Potential Energy}$$

A macroscopic body consists of a huge number of atoms that move at velocities with great speeds and in random directions. In addition, the atoms interact with each other. As a consequence, there is kinetic energy and potential energy at the microscopic (here, atomic) level. This energy is called **thermal energy**. An increase in thermal energy is accompanied by an increase in the temperature. The common term for thermal energy is **heat**.

Consider a standing wave in a pitch pipe. As time progresses, the sound emitted from the open ends carries away energy. At the same time, the standing wave within the pipe loses amplitude and wave energy due to attenuation—the wave energy is replaced by thermal energy.

4.1.3 The Energy of Cheerios

Cheerios cereal has 110-Cal per oz. This means that when one oz. of Cheerios is digested, the chemical changes provide the body with 110-Cal of energy in a form that can be used by muscles to provide the body with mechanical energy. In fact, most of the chemical energy is replaced by thermal energy.

The chemical changes that take place in association with this process occur at the molecular level. Studies reveal that the typical change in the energy of a molecule is on the order of a few electron volts (eV). From this fact it is possible to estimate the number of molecules of Cheerios in an ounce of Cheerios. See problem 4.2, at the end of the chapter.

Sample Problem 4.7 Remember the mountain climber whose PE rises by 2,700,000-J in climbing an elevation of 4,000-m. What is the weight of Cheerios that she has to eat to supply this PE, assuming 100% efficiency?

Solution

$$\text{weight of Cheerios} = \frac{2{,}700{,}000\text{-J}}{(110\text{-Cal/oz})(4200\text{-J/Cal})} = 5.8\text{-oz} \qquad (4.10)$$

Only a few bowls of Cheerios are necessary! It is amazing how little Cheerios would be needed if the conversion from chemical to PE were 100% efficient. We see that mountain climbing is incredibly low in efficiency. Most of the additional food one must consume is wasted and goes into thermal energy.

Diet-conscious people have a tendency to read carefully the number of Calories per gm for every food they encounter. The fact is that the variation is not very great once we take into account the fractions that are fat, protein, or carbohydrate. This is so because there is very little variation in the Cal/gm for pure samples of these foods:

- pure oils: 9-Cal/gm $\sim$ 270-Cal/oz
- pure protein or carbohydrate: 4-Cal/g$\sim$120-Cal/oz

Thus, digesting a gallon (=128-fluid oz $\sim$128-dry oz) of oil provides 128$\times$ 270 = 35,000-Cal of thermal energy plus mechanical energy. Since digestion produces essentially the same chemical changes in oil as does **burning**—which refers to a chemical reaction of a substance with oxygen (called **oxidation**)—burning one gallon of oil produces 35,000-Cal of thermal energy, which can be used to produce mechanical energy and/or electrical energy.

Question: How many gallons of oil would provide the energy needed to make that mountain climb?

A heat engine is a device that "extracts" mechanical energy or electrical energy from thermal energy. The **Second Law of Thermodynamics** informs us that a complete conversion from thermal energy to mechanical energy and/or electrical energy is impossible. Current heat engines don't even provide us with the maximum that this law provides. To get an idea about what is currently realized, consider that typical electric power plant is about 40% efficient. Suppose it were to burn oil to run the heat engine. Note that from one gallon of oil we obtain 35,000-Cal = 1.5 $\times 10^8$-Joules = 40-kWh of thermal energy. This means that burning one gallon of oil can produce about 0.40 $\times$ 40-kWh = 16-kWh of electrical energy.

4.2 The Principle of Conservation of Energy, Work, and Heat

Suppose that the above 10-kg object is released from rest at its elevation of 2-m from ground level. According to the Principle of Conservation of Energy, the object will accelerate downward, acquiring KE as it loses PE. Its loss in PE is compensated for by its increase in KE. When it reaches ground level, its KE must equal 196-J. We can therefore calculate its speed just before it hits the ground as follows:

We set

$$\text{KE} = \frac{1}{2}mv^2 = 196\text{-J}$$

Then

$$v^2 = 2\text{KE}/m = 2\left(\frac{196}{10}\right) = 39.2$$

Thus

$$v = \sqrt{39.2} = 6.3\text{-m/s}$$

It is because one can retrieve KE in this way that the word "potential" is used to refer to the latent nature of PE.

Alternatively, if a person throws an object straight upwards with a speed of 6.3-m/s, the object will reach a maximum height of 2-m when it will be instantaneously at rest. The object will then fall downward, hitting the ground at a speed of 6.3-m/s.

Generally, the initial KE is equal to the maximum PE.

Thus,

$$\text{initial KE} = \text{max PE} \tag{4.11}$$

Symbolically, we have

$$\frac{1}{2}mv^2 = mgh \tag{4.12}$$

We can eliminate the mass and solve for the maximum height

$$h = \frac{v^2}{2g} \tag{4.13}$$

Let us repeat the above calculation so that we can see how this final expression works:

$$h = \frac{6.3^2}{2 \times 10} = 2\text{-m} \tag{4.14}$$

Sample Problem 4.8 In May 2018, the volcano Kilauea in Hawaii spewed lava into the atmosphere, reaching a maximum height of about 10,000-m.

Find the speed of the lava when it leaves the mouth of the volcano.

Solution We can solve for the speed using Eq. (4.13). We have

$$v^2 = 2gh \tag{4.15}$$

Thus,

$$v = \sqrt{2gh} = \sqrt{2(10)(10^4)} = \sqrt{200,000} \tag{4.16}$$

Or

$$v = 450\text{-m/s}$$

4.2.1 Series of Changes of Forms of Energy

We have indicated that the changes in the chemical energy of the food, known as metabolism, are the origin of the increase in PE associated with a person's raising an object. *We* say that the person does **work** in raising the object. In the framework of physics, ***work*** *is a process whereby energy is transferred from one body to another* through the application of a force and the motion of the body. The changes in chemical energy allow the person to exert that force and move the body.

On the other hand, when the temperature of a person's body rises above the temperature of the surroundings, thermal energy will be transferred from the person to the surroundings. This **heat transfer** is a second *process whereby energy is transferred from one body to another*. Overall, for the most part, chemical energy of food that is eaten by a person is replaced by thermal energy, kinetic energy, and potential energy of objects moved by the person, and chemical energy of human waste products.

4.3 Energy of Vibrating Systems

4.3.1 The Simple Harmonic Oscillator

A vibrating SHO has both KE and PE. Generally, its total vibrational energy E is given by

$$\boxed{\mathrm{E} = \mathrm{KE} + \mathrm{PE} = \frac{1}{2}mv^2 + \frac{1}{2}ky^2} \tag{4.17}$$

As the SHO vibrates, the displacement and speed are constantly changing, so that the respective amounts of KE and PE are constantly changing. Nevertheless, the total energy is constant. This result is a simple confirmation of the Conservation of Energy Principle. Let us apply the equation to an SHO which is vibrating with an amplitude A and a period T. In Fig. 4.2, we display the displacement as well as the velocity, as they vary with time over one cycle.

The displacement y varies from $-A$ to $+A$. The velocity ranges from $-v_m$ to $+v_m$. We will refer to v_m as the **velocity amplitude**; it is the maximum speed.

In Fig. 4.3, we display the PE and KE versus time over a cycle. Note that when the *SHO is in its* ***equilibrium*** *configuration, the* ***kinetic energy*** *is a* ***maximum****, while the* ***potential energy*** *is* ***zero***. On the other hand, when the SHO is *farthest* from its equilibrium configuration, the kinetic energy is ***zero***, while potential energy is a ***maximum***. Also, we see that the PE and KE both vary from zero to E, in just such a way that their sum is the constant E.

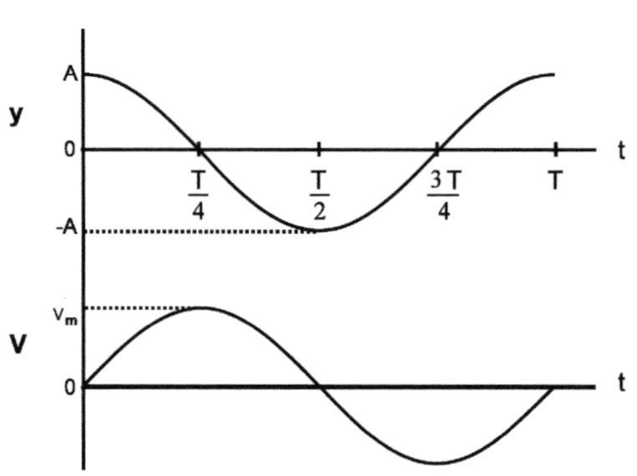

Fig. 4.2 Displacement and velocity of an SHO vs. time

Now we will see how we can relate the displacement amplitude A to the velocity amplitude v_m: Initially, the object is at rest, so that there is no KE. All the energy resides in PE:

$$\mathrm{E} = \mathrm{PE} = \frac{1}{2}kA^2 \tag{4.18}$$

On the other hand, when the object is at the equilibrium position ($y = 0$), one quarter of a cycle later, there is no PE. All the energy resides in KE:

$$\mathrm{E} = \mathrm{KE} = \frac{1}{2}m{v_m}^2 \tag{4.19}$$

The two expressions for the total energy E, Eqs. (4.8) and (4.9), are equal, so that

$$\begin{aligned} \mathrm{E} &= \frac{1}{2}m{v_m}^2 = \frac{1}{2}kA^2 \\ m{v_m}^2 &= kA^2 \end{aligned} \tag{4.20}$$

Thus,

$$v_m = \sqrt{\frac{k}{m}}A = 2\pi f A = 2\pi\frac{A}{T} \tag{4.21}$$

Now recall that in Sect. 2.8 we *estimated* the velocity amplitude as $4A/T$; this is the average speed during a cycle. The ratio of the two expressions is $2\pi/4 \sim 1.6$. We expect the maximum speed to exceed the average speed!

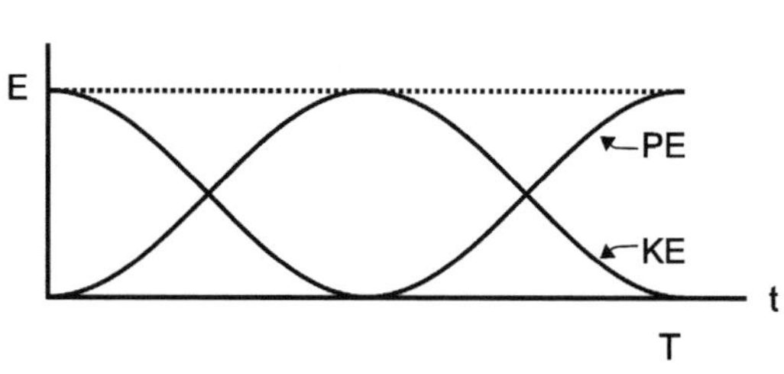

Fig. 4.3 Kinetic energy and potential energy of an SHO vs. time

Sample Problem 4.9 Find the maximum speed (velocity amplitude) and total energy of an SHO which has a spring constant of 100-N/m, a mass of 250-g, and an amplitude (displacement amplitude) of 20-cm.

Solution

$$v_m = \sqrt{\frac{k}{m}}A = \sqrt{\frac{100}{0.25}}(0.2) = 4.0\text{-m/s}$$

$$\mathrm{E} = \frac{1}{2}kA^2 = \frac{1}{2}(100)(0.2)^2 = 2\text{-J}$$

4.3.2 Energy in a Vibrating String

Let us consider all of the processes occurring when a **string is plucked**. In doing so, we will keep track of the various forms of energy involved:

- Your body stores energy obtained from the food that you have eaten. In plucking the string, the energy in your body is used to do **work** in pulling the string aside. That work leads to a storage of PE in the string.
- You then release the string. As it whips back towards the equilibrium, straight configuration, it picks up speed, hence KE, and necessarily loses PE. As the string vibrates, the amount of KE and PE oscillates, with the sum being a constant—as long as we can neglect attenuation.
- As the string moves through the air, it does work on the air in setting the air in motion. Sound waves are produced! The sound waves carry away energy in the form of KE and PE. Thus, the energy of the vibrating string must decrease. Its vibration is said to **attenuate**. Attenuation takes place also because of internal friction forces within the string, as its shape keeps on changing due to the vibration. This attenuation gives rise directly to an increase in the **thermal energy** of the string. The sound waves that are produced also attenuate, being replaced by increased **thermal energy** of the air.
- At any stage, the sum total of all forms of energy—your body's chemical and thermal energy, the string's vibrational and thermal energy, the sound wave energy, and thermal energy of the air—is a constant.

4.3.3 Energy in a Sound Wave

A sound wave has the same two forms of energy as an SHO or a vibrating string: kinetic energy—associated with the motion of the medium, such as air—and potential energy—associated with condensation or rarefaction of the medium. To understand the origin of its potential energy, it is useful to note that it takes work to compress a volume of gas, so that its potential energy is thereby increased. Correspondingly, a compressed gas has the potential to do work on its surroundings. Conversely, in rarefying a volume of gas, the potential energy of the gas is *decreased.* Work would have to be done on the gas to compress it, so as to remove the rarefaction.

While the KE and PE of an SHO is localized in space, energy is distributed continuously in the space occupied by the sound wave. For a sound wave of a specific wavelength (and hence frequency), moving in a specific direction, the distribution of the two forms of energy is characterized by the graph in Fig. 4.3, with the time axis t replaced by the spatial axis x and the period T replaced by the wavelength λ.

4.4 Power

Consider a traveling wave which is infinite in extent. There is no interest in the total energy in the wave since the total energy is infinite. We can instead deal with the amount of energy in a given length of traveling wave. The *energy per unit length* of a traveling wave is proportional to the square of the amplitude,[1] the same expression

[1] We can estimate the energy per unit length as follows: A unit length has a mass $m = \mu(1) = \mu$. Its average speed should be a bit less than the maximum velocity v_m. Thus, its KE should be a bit less than $(1/2)m{v_m}^2 = (1/2)\mu{v_m}^2$. This turns out to be the exact answer; it includes the PE too. Of course, KE=PE=constant.

Recall that $v_m = 2\pi f A$ (see Eq. (4.10)). Thus, the energy per unit length is proportional to the square of the

holds for the energy per unit length of a *standing wave* of a vibrating string or vibrating pipe.

Generally, energy is proportional to the square of the amplitude:

$$\boxed{\text{energy} \propto (\text{amplitude})^2} \qquad (4.22)$$

It is often more interesting to focus on the **rate at which energy is carried by the wave past a given point along the string**: This rate is equal to the rate at which energy would be given to the string by your hand (as in Fig. 2.6) in sending the traveling wave down the string—that is, the POWER input to the string. Why? Because whatever energy you pump into the string with your hand must pass the given point along the string.

We define **power** as

$$\boxed{\text{power} = \frac{\text{energy}}{\text{time interval}} \quad \text{or} \quad P = \frac{E}{\Delta t}} \qquad (4.23)$$

What energy E is used in this expression for power? We list some examples below:

- energy transferred in the case of WORK
- energy delivered in the form of electrical energy
- energy passing a given point along the string
- energy lost due to **dissipation**, which involves the production of thermal energy.

As we noted at the beginning of this chapter, a basic unit of power is the "Watt", which is named after **James Watt**, who contributed important improvements in the design of the steam engine:

$$\begin{aligned} \text{one Watt} &= \text{one Joule per Second} \\ 1\text{-W} &= 1\text{-J/s} \end{aligned} \qquad (4.24)$$

Another common unit of power is the **horsepower** (hp):

$$\boxed{1\text{-hp} \simeq 746\text{-W}} \qquad (4.25)$$

It is important for us to be clear as to what "power" represents in physics. When we talk about how "powerful" a person is, we are most often referring loosely to the force that a person can exert; for example, how heavy an object that the person can carry. "Force" is different from "power." The rate at which the person can increase the potential energy by lifting them would express the person's power in physics terms.

Example 4.1 The human heart does quite a bit of work over the course of a 75 year lifetime in pushing blood through a person's body—enough to raise a typical battleship 10-m upwards, if the energy could be harnessed! The energy transferred to the blood amounts to about 20 billion Joules. However, this work is done over such a long period of time that the equivalent power is quite small:

Since one year = 3×10^7-s,

$$75\text{-years} = 75 \times 3 \times 10^7\text{-s} = 2 \times 10^9\text{-s}.$$

Thus, the power of the human heart is given by

$$\frac{E}{\Delta t} = \frac{20 \times 10^9}{2 \times 10^9} = 10\text{-W}$$

Example 4.2 A 100-W light bulb consumes 100-J/s of electrical power. The electrical energy is converted to light energy, as well as thermal energy. In fact, only a few percent of the total is light energy.

Example 4.3 Only a small fraction of the electrical power fed into a loudspeaker is converted to sound wave power. Most of the electrical energy is lost to thermal energy: According to Cambridge Sound Waves Inc., their loudspeaker had an **efficiency** of a mere 0.4%. That is, 1-W of electrical power produces only 0.004-W of sound power.

Example 4.4 The horsepower unit of power is based upon a study of the performance of a real

---

amplitude A, as in the case of the energy of a standing wave.

Table 4.2 Power of sound sources

Source of sound	Power in watts
Orchestra of 75 instruments	70
Bass Drum	25
Trombone	6
Piano	0.4
Average sound power of an orchestra of 75	0.09
Flute	0.06
Clarinet	0.05
French Horn	0.05
Average Speech	0.000024
Softest Violin passage	0.0000038

horse. In fact, horses can generate **many** horsepower. It is not difficult for a person to generate one hp of work: This would be accomplished by a person weighing 155 lbs by running up a hill and increasing his/her elevation at a rate of 3.6 ft/s. It is understood that the entire 1 hp is going into an increase of gravitational potential energy.

Power of Various Sources of Sound

In the table below, we present the sound power produced by a number of musical instruments. Except for the cases so indicated, the maximum power is given. The range of powers is quite extraordinary, the ratio of the largest to the smallest value being $70/0.0000038 \sim 2 \times 10^7$, or 20 million to one! It is no small wonder that our auditory system can be responsive to such a large range (Table 4.2).

4.5 Intensity

Loudness and **brightness** are subjective, psychological experiences. They cannot be quantified and are not scientific parameters. They both *reflect* the physical parameter called **intensity** that is applied to both sound and light. For a given frequency spectrum, we can generally expect loudness and brightness to increase with increasing intensity. **Intensity** characterizes how concentrated the flow of energy is in space. Specifically, intensity is the rate at which energy passes through a unit area. We write

$$\boxed{\begin{aligned} \text{Intensity} &= \frac{\text{Power}}{\text{Area}} \\ I &= \frac{P}{A} \end{aligned}} \tag{4.26}$$

which has the common units W/m^2 and W/cm^2. We have used the letter A to represent the area.

The problems below should make the definition of intensity clear.

Sample Problem 4.10 Find the intensity of my voice, with a power of 10^{-3}W, which is traveling down a pipe of radius 2-cm.

Solution

$$A = \pi r^2 = \pi(2 \times 10^{-2})^2 = 1.3 \times 10^{-3} \text{m}^2$$

$$I = \frac{P}{A} = \frac{10^{-3}}{1.3 \times 10^{-3}} = 0.8\text{-W/m}^2$$

Sample Problem 4.11 Find the intensity of a straight laser beam, with a power of 20-mW = 20×10^{-3}-W and a beam diameter of 1-mm (= 10^{-3}-m).

Solution We have

$$A = \pi(\text{diameter})^2/4 = \pi(10^{-3})^2/4$$
$$= 7.9 \times 10^{-7}\text{-m}^2$$

so that

$$I = \frac{P}{A} = \frac{20 \times 10^{-3}}{7.9 \times 10^{-7}} = 2.5 \times 10^4\text{-W/m}^2$$

Definition The intensity of sunlight just above our atmosphere is known as the **solar constant** and is given by 1400-W/m^2.

4.6 Intensity of a Point Source

Let us turn our attention back to the *sun*. We can measure the solar constant by placing a light detector in a satellite just above our atmosphere.

An interesting question is: What is the rate at which light energy is emitted by the sun, that is, the **luminosity** of the sun? The *solar constant* is related to the luminosity and the distance from the earth to the sun. We expect the intensity of sunlight to increase with increasing luminosity and to decrease with increasing distance from the sun. We now discuss the precise relationship between the three parameters.

To an excellent approximation, the sun emits light in all directions with the same intensity. We say that it is an **isotropic source**.[2] An isotropic source has the following important characteristics:

1. The intensity *outside* the source does not depend upon the *radius* of the source. In fact, because the radius can shrink to any size whatever without a change in the exterior intensity, an isotropic source is often referred to as a **point source**.
2. The intensity of a point source depends only upon the distance r from the center of the source and its power P, and not upon the direction from the source. Specifically, the intensity is inversely proportional to the square of the distance d.

$$\boxed{I = \frac{P}{4\pi r^2}} \tag{4.27}$$

We now provide a proof of Eq. (4.27): Suppose that the source has a power P. Now consider a spherical surface of radius r, centered at the point source. The rate P at which energy leaves the source equals the rate at which energy flows through the sphere. Finally, the energy flows radially through this surface.

In Fig. 4.4, we have a point source at the center of a spherical surface. The power P is the rate at which energy is emitted from the source. The rate at which energy crosses the sphere must be P also since we have a steady state situation. The area A of the spherical surface is $4\pi r^2$.

[2]The mere existence of sunspots tells us that the emission of light from the SUN cannot be perfectly isotropic.

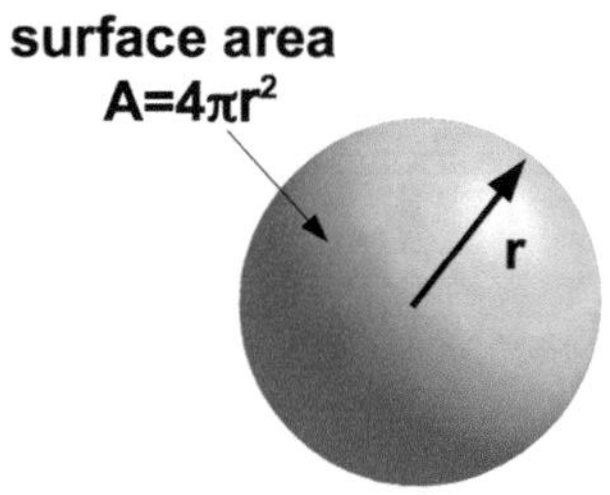

Fig. 4.4 Point source

Equation (4.27) follows by direct substitution from Eq. (4.26).

Sample Problem 4.12 Estimate the light intensity of a 100-W light bulb having an efficiency of 3% at a distance of 1-m from the bulb by treating it as a point source.

Solution light power = $P = 0.03 \times 100 = 3\text{-W}$ with r = 1-m,

$$I = \frac{P}{4\pi r^2} = \frac{3}{4\pi (1)^2} = 0.24\text{-W/m}^2$$

Sample Problem 4.13 The sun is 150 million-km away from the earth. Find the light power emitted by the sun. This is the rate at which energy is emitted by the sun.

Solution Recall that the sun's light intensity at the earth is 1400-W/m^2. We set $r = 150$ million-km. Thus, we find

$$P = 4\pi r^2 \times I = 4\pi (150 \times 10^6 \times 10^3)^2 \times 1400$$
$$= 4.0 \times 10^{26}\text{-W}$$

For comparison, in 2016 the total electric power consumption in the world was 2.3×10^{14}-W. The source is Wikipedia, https://en.wikipedia.org/wiki/List_of_countries_by_electricity_consumption.
Thus the sun can supply the electric power needs of about $10^{26}/10^{14} = 10^{12}$ = one-trillion worlds.

Sample Problem 4.14 Suppose a loudspeaker emits sound power at 0.004-W, isotropically in

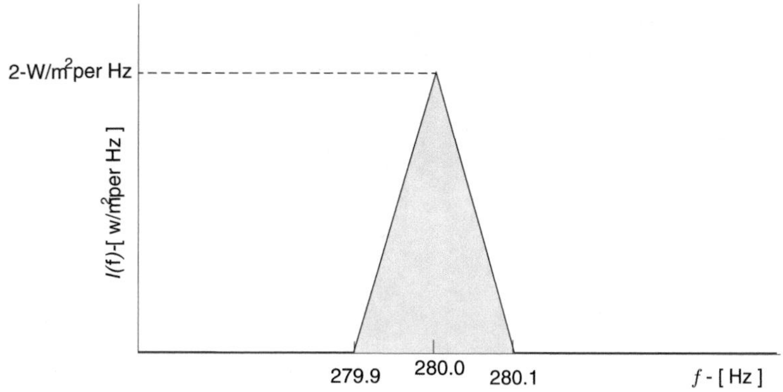

Fig. 4.5 Simplified spectral intensity of the fundamental of a violin at 280-Hz

the forward direction alone. Estimate the sound intensity at a distance of 2-m.

Solution As an approximation, we can replace the area $4\pi r^2$ by $2\pi r^2$. Then

$$I = \frac{P}{2\pi r^2} = \frac{0.004}{2\pi (2)^2} = 1.6 \times 10^{-4}\ \mathrm{W/m^2}$$

4.7 Spectral Intensity with Respect to Frequency

In Chap. 2 we displayed the spectrum of the sound of a violin in Fig. 2.29 and the spectrum of the sound of a flute in Fig. 2.30. These spectra are obtained from a **Fourier analysis** of the respective waves. It is assumed that the sound spectrum is constant for all time. There is a mathematical method for carrying out a Fourier analysis from the waveform.[3] There also exists an electronic instrument that can give us the spectrum if the sound is picked up by a microphone that is connected to the instrument. We should note that we cannot encounter a wave that lasts forever. Even if the sound wave were to exist over a long period of time, we would want to obtain a spectrum by inputting the sound into the device over the course of a finite interval of time.

Note that we didn't indicate on the vertical axis what parameter is being plotted. To understand what this parameter is, we will begin by introducing the **spectral intensity with respect to the frequency**, which is symbolized by $\mathbf{I}(f)$.[4,5]

The **spectral intensity** must be distinguished from the parameter we introduced earlier, referred to as the **intensity**. The spectral intensity can be plotted as a function of frequency. As an example that I will use in order to facilitate our discussion, I will replace the actual complex peak of the violin's spectrum about its fundamental frequency at 280-Hz by the greatly simplified graph shown in Fig. 4.5. Here are its features: The spectral intensity vanishes for frequencies less than 279.9-Hz or greater than 280.1-Hz. It increases linearly from 279.9-Hz to 280.0-Hz and decreases linearly from 280.0-Hz to 280.1-Hz, reaching a peak value of 2-W/m^2 per Hz. Note that the vertical axis is labeled with units of "W/m^2 per Hz."

We cannot define the intensity at a specific frequency. If we want to measure the intensity *at a frequency*, the device used to make this measurement will always measure the intensity over a range of frequencies. For example, if the device's dial is set at 440-Hz, the output might

[3]The mathematical method is known as a **Fourier Transform**.

[4]Beware: We will see later that Figs. 2.29 and 2.30 are **not** plots of the spectral intensity.

[5]In Chap. 14 we will introduce a different spectral intensity $\mathrm{I}(\lambda)$: the **spectral intensity with respect to the wavelength**. The two spectral intensities are related. In fact there is an equation that expresses this relationship. With v being the wave velocity, we have

$$\mathbf{I}(f) = \frac{\lambda^2}{\mathrm{v}}\mathrm{I}(\lambda). \tag{4.28}$$

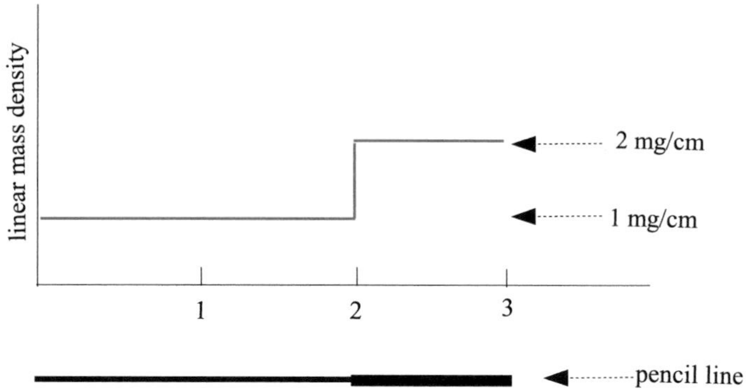

Fig. 4.6 Linear mass density of a pencil line

be the total intensity over a range of frequencies from 439.99-Hz to 440.01-Hz.

For comparison sake, suppose you were to use two pencils to draw two segments of lines on a piece of paper, with different thicknesses. The lines contain microscopic particles of graphite that stick to the paper. The line segments therefore have **mass**. Suppose that the first segment has a low thickness, while the second line segment is thicker, as shown in Fig. 4.6. In the figure, we display both the appearance of the segments as well as a graph of the **linear mass densities** expressed in units of mg/cm.[6]

There is **no mass at all at a given point** along the line segments. However, we can determine the total mass along the lines as follows. Along the first segment we have 1-mg/cm over a length one cm. This leads to a mass 1-mg/cm × 2-cm = 2-mg. This number is the area of the rectangle under the horizontal line segment at a vertical position, 1-mg/cm. In the second segment we have a mass 2-mg/cm × 1-cm = 2-mg. This number is the area of the rectangle under the horizontal line segment at a vertical position, 2-mg/cm. The total mass is 2+2=4-mg. Generally, if we plot the linear mass density of a line along a piece of paper, the total mass between two points is the **area under the curve**.

Now we return to the spectral intensity displayed in Fig. 4.5. The **intensity** contributed by the fundamental is given by the **area under the curve** determined as follows. The area under the curve of the spectral intensity is the area of the **triangle** in the figure. The base of the triangle is 0.2-Hz. The height of the triangle is 2-W/m^2 per Hz. Given that the area of a triangle is (1/2) the base times the height, we obtain for the **intensity**

$$\begin{aligned} I &= (1/2)[0.2\text{-Hz}] \times [2\text{-W/m}^2 \text{ per Hz}] \\ &= 0.2\text{-W/m}^2 \end{aligned} \tag{4.29}$$

Note Now let us return to Figs. 2.29 and 2.30. They were produced by the superb software **Amadeus Pro**, which analyzes audio files such as mp3s, thereby providing us with an excellent sense of their frequency spectra. However, the spectrum that is displayed is *not* the spectral intensity **I**(f). Instead, Amadeus Pro displays the absolute value[1] of the frequency spectrum of the **amplitude**, which is symbolized by $|A(f)|$. The spectral intensity **I**(f) is the square of this function.[2]

$$\mathbf{I}(f) = |A(f|^2 \tag{4.30}$$

[1] The absolute value of a number is the magnitude of the number; thus, for example, $|2| = 2$ and $|-2| = 2$.

[2] I am indebted to Martin Hairer for providing me with this information. He is not only the creator of Amadeus Pro but is a world renowned mathematician, being a winner of the prestigious **Fields Medal** in mathematics. See https://news.softpedia.com/news/Martin-Hairer-a-Mac-App-Developer-Has-Received-the-Highest-Honor-A-Mathematician-Can-Get-454768.shtml.

[6]Compare this linear mass density with that of a vibrating string.

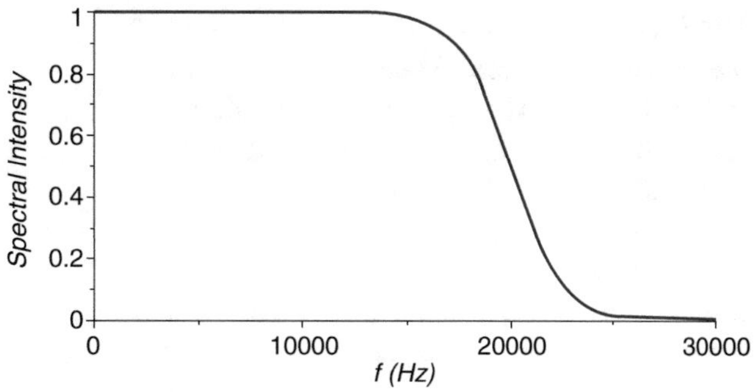

Fig. 4.7 An example of a spectral intensity of white noise

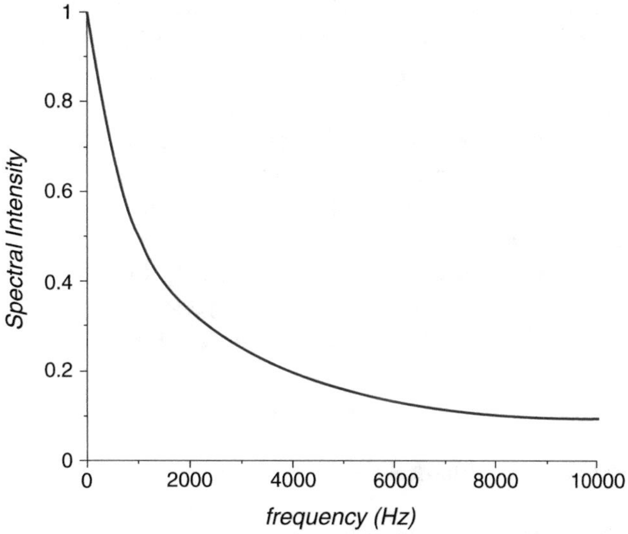

Fig. 4.8 Spectral intensity of pink noise

4.7.1 White Noise and Pink Noise

A very important spectrum of sound and of electronic signals is referred to as **white noise**.[7] It is so designated because the spectral intensity with respect to frequency is constant. Actually realizable noise that is referred as being white has a constant spectrum up to a high frequency beyond that which is relevant for the signal. For example, it might be constant up to a frequency just beyond the audible range. We see an example of such a spectral intensity in Fig. 4.7.

It is referred to as being "white" because the spectral intensity with respect to **wavelength** for *light* is constant. However, it can be shown that if $\mathbf{I}(f)$=constant, the corresponding spectral intensity $I(\lambda)$ with respect to wavelength is inversely proportional to the square of the wavelength and therefore not constant. See Eq. (4.28).

Another important type of noise of a sound or of electronic equipment is known as **pink noise**. Its spectral intensity is inversely proportional to the frequency and is shown in Fig. 4.8. We have

$$\mathbf{I}(f) \propto \frac{1}{f} \tag{4.31}$$

In electronics the noise is often referred to as "**one-over-f noise**."

4.8 Sound Level and the Decibel System

The range of intensities of sound that a person would perceive as sound—that is, can detect and yet feel no pain—is enormous: From ~

[7]Noisy sound is often used in the waiting room of a medical doctor or a psychotherapist to maintain the privacy of the patients. The noise is *not* typically white noise.

10^{-12} W/m^2 to $\sim$ 1-W/m^2. Thus, the ratio of the highest to lowest intensities is 10^{12} or a trillion to one.[8] The symbol for the minimum audible intensity is I_0. This intensity corresponds to the minimum audible sound pressure p_0=2 $\times$ 10^{-10}N/m^2 mentioned in Chap. 3.

Note that the loudest sound does not feel a trillion times louder than the quietest perceivable sound. Furthermore, doubling the sound intensity produces a change in loudness which most people would describe as being much less than a doubling in the loudness. Thus sound intensity does not give us a qualitative sense of loudness. Because of the above facts, a more meaningful and useful way of reflecting the sound intensity was devised called the **sound level** abbreviated by "*SL*," or alternatively the **sound pressure level**. The unit for the sound level is the **decibel**, abbreviated **dB**, in honor of **Alexander Graham Bell**, who is credited with inventing the telephone. In order to appreciate this system we need to first review the important properties of logarithms.

4.8.1 Logarithms

The **logarithm** is a mapping of numbers in that the logarithm replaces any number by another number. (In mathematics, a mapping is often referred to as a **function**.) The function depends upon the choice of **base**, which we choose to be ten. Some simple examples follow:

$$1/1000 \to \log(1/1000) = \log 10^{-3} = -3$$

$$1/100 \to \log(1/100) = \log 10^{-2} = -2$$

$$1/10 \to \log(1/10) = \log 10^{-1} = -1$$

$$1 \to \log 1 = 0$$

$$10 \to \log 10 = 1$$

$$100 \to \log 100 = 2$$

$$1000 \to \log 1000 = 3$$

Generally, $10^n \to \log 10^n = n$. So much for the logarithm of powers of ten.

For *integers* between 1 and 10 we have:

$$\log 2 = 0.30$$

$$\log 3 = 0.48$$

$$\log 4 = 0.60$$

$$\log 5 = 0.70$$

$$\log 6 = 0.78$$

$$\log 7 = 0.85$$

$$\log 8 = 0.90$$

$$\log 9 = 0.95$$

The logarithm of a number between one and ten can be calculated mathematically and is available in tables and from pocket calculators.

General mathematical properties of logarithms:

$$\log xy = \log x + \log y$$

$$\log x/y = \log x - \log y \qquad \text{Note:} \log(1/y) = \log 1 - \log y = 0 - \log y = -\log y$$

$$\log x^y = y \log x$$

$$\text{If } z = \log x, \qquad x = 10^z$$

[8]Coincidentally, the range of sensitivity and tolerance of vision to light intensities is also understood to be about twelve orders of magnitude, from $\sim$ 10^{-10} W/m^2 to $\sim$ 100 W/m^2.

Examples of the logarithm of some numbers that illustrate the use of these mathematical properties:

$$\log 40 = \log(4 \times 10) = \log 4 + \log 10 = 0.60 + 1 = 1.60$$

$$\log(3 \times 10^{11}) = \log 3 + \log(10^{11}) = 0.48 + 11 = 11.48$$

4.8.2 Sound Level

We will now define a quantity that is an objective complement of the sound intensity. We define the **sound level**, SL, as follows:

$$\boxed{SL = 10\log\frac{I}{10^{-12}\text{-W/m}^2}} \quad (4.32)$$

Note that $10^{-12} - W/m^2 = I_0$ is the lowest audible intensity. I_0 is referred to as the **reference level of the sound level**. We have

$$\boxed{SL = 10\log\frac{I}{I_0}} \quad (4.33)$$

Note It can be shown that the sound intensity is proportional to the square of the sound pressure [1].

$$I \propto p_s^2 \quad (4.34)$$

In parallel, $I_0 \propto p_0^2$, where p_0 is the sound pressure corresponding to I_0, that is, $2 \times 10^{-10}\text{N/m}^2$.

As a result the sound level can be expressed as

$$\boxed{SL = 10\log\left(\frac{p_s}{p_0}\right)^2 = 20\log\frac{p_s}{p_0}} \quad (4.35)$$

[1] The relation is akin to the potential energy of an SHO being proportional to the square of the displacement: PE $\propto y^2$. See Chap. 2.

Let us see how we calculate the sound level.

Example 4.5 $I = I_0 = 10^{-12}\ W/m^2$:

$$SL = 10\log\frac{I}{I_0} = 10\log 1 = 0\text{-dB}$$

The intensity 10^{-12} W/m^2 is called the **reference level**, corresponding to a SL of 0-dB, and is approximately the lowest intensity that can be heard.

Example 4.6 $I = 0.1I_0 = 10^{-13}$ W/m^2, so that $I/I_0 = 0.1$:

$$SL = 10\log(0.1I_0) = 10\log(0.1) = 10(-1)$$
$$= -10\text{-dB}$$

A sound level can be negative!

Example 4.7 $I = 1$ W/m^2,

$$SL = 10\log\left(1/10^{-12}\right) = 10\log\left(10^{12}\right)$$
$$= 10 \times 12 = 120 \text{ dB}$$

We note that the range of audible sound is from $\sim$ 0 dB to $\sim$ 120 dB. This is not to imply that there aren't people who can hear a sound that has a negative sound level.

4.8.3 From Sound Level to Intensity

We know how to calculate the sound level from the intensity. Suppose that we know the sound level and we want to calculate the corresponding intensity. We can invert Eq. (4.32) so that:

$$I = 10^{-12+SL/10} \quad (4.36)$$

Sample Problem 4.15 Suppose that the sound level is, say, 85-dB. What is the corresponding intensity?

Solution

$$I = 10^{-12+SL/10} = 10^{-12+85/10} = 10^{-12+8.5} \quad (4.37)$$

Thus,

$$I = 10^{-3.5} = 3.2 \times 10^{-4}\ \mathrm{W/m^2} \quad (4.38)$$

We next describe how changes in intensities are related to **changes** in sound level.[9]

Sample Problem 4.16 A sound intensity is doubled, from I to $2I$. Find the change in SL.

Solution The change in SL is given by

$$\begin{aligned} SL \equiv \Delta SL &= 10\log\left(\frac{2I}{I_0}\right) - 10\log\left(\frac{I}{I_0}\right) \\ &= 10[\log(2I) - \log I_0] \\ &\quad - 10[\log I - \log I_0] \\ &= 10\log\left(\frac{2I}{I}\right) = 10\log 2 \\ &= 10(0.30) = 3\text{-dB} \end{aligned}$$

Note This is a very important characteristic of the SL: The **change** in SL is independent of the initial intensity I; it depends only upon the ratio of intensities. Here is the proof:

$$\begin{aligned} \Delta SL &\equiv SL_2 - SL_1 \\ &= 10\log\frac{I_2}{10^{-12}} - 10\log\frac{I_1}{10^{-12}} \end{aligned} \quad (4.39)$$

or

$$\boxed{\Delta SL = 10\log\frac{I_2}{I_1}} \quad (4.40)$$

Equation (4.40) can be "inverted": Suppose that we know the change in sound level ΔSL and want to know the corresponding ratio of intensities. Then we have

$$\boxed{\frac{I_2}{I_1} = 10^{\Delta(SL/10)}} \quad (4.41)$$

Sample Problem 4.17 The sound level is increased by 10-dB. By what factor does the intensity increase? What if ΔSL= 25-dB?

Solution Let I_1 = initial intensity and I_2 = final intensity. We seek I_2/I_1, given that ΔSL = 10-dB. According to Eq. (4.41)

$$I_2/I_1 = 10^{\Delta(SL/10)} = 10^{\Delta(10/10)} = 10 \quad (4.42)$$

If $\Delta SL = 25$ dB,

$$\frac{I_2}{I_1} = 10^{\Delta SL/10} = 10^{25/10} = 10^{2.5} = 316 \quad (4.43)$$

We close this section by pointing out that we have assumed that the wave has a specific frequency, and hence wavelength. What happens if we have a more complex sound wave. An example is a wave that is produced by more than one source. The source might produce waves that have a definite fixed phase relation. In this case, we say the wave is **coherent**. Otherwise the wave is said to be **incoherent**. We will discuss these situations in Chap. 7.

4.9 Attenuation

Up to this point, we have assumed for simplicity that once a wave is established it will last forever. In fact, we know from experience that waves die out spontaneously. The technical term for this process is **attenuation** or **damping**. In the case of a string, attenuation is mainly due to the force of the surrounding air on the string. In order to compensate for attenuation so as to keep a string vibrating, an external excitation force on the string must be maintained. Attenuation of sound is due to the very same intermolecular forces that sustain the wave itself. Energy is conserved in the process of attenuation through

[9] An excellent website for appreciating *changes* in sound levels is http://www.phys.unsw.edu.au/~jw/dB.html.

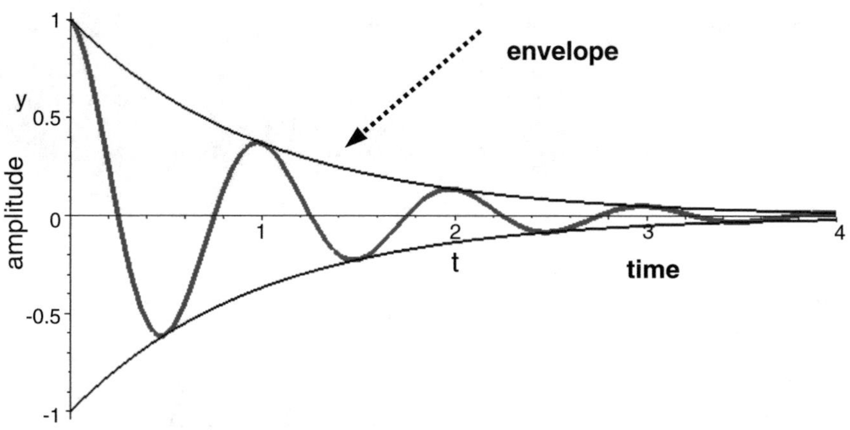

Fig. 4.9 Attenuated sine wave

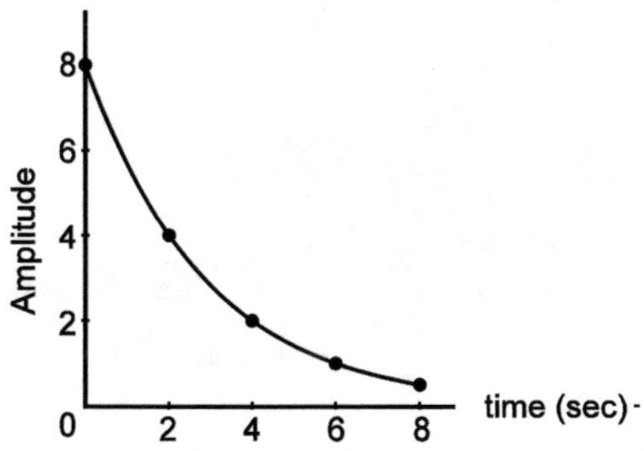

Fig. 4.10 Amplitude vs. time due to attenuation

the production of thermal energy. This aspect of attenuation is referred to as **dissipation**. In this section, we will discuss how we characterize attenuation numerically.

4.9.1 Attenuation in Time

We will first deal with *attenuation in time* of a mode of vibration. We will focus on the *amplitude* of the vibration. Its attenuation is exponential and is depicted in Fig. 4.9.

The black curves above and below the sinusoidal blue curve together comprise the **envelope**. The envelope above represents the attenuated amplitude and is shown as a specific quantitative curve in Fig. 4.10.

Let us define the **attenuation time** T as the **time it takes for the amplitude to be reduced in half**. In the figure, we have an initial amplitude of 8 units and we have an attenuation time of 2-s. We note that after 2-s, the initial amplitude has been cut in half—to 8/2=4. After an additional 2-s, amounting to a total of 4-s elapsed, the amplitude is reduced by additional factor of 2—to 4/2=2, and so on.

Question: What will be the amplitude in the above example after 6-s? After 8-s? After 10-s?

The attenuation time is different for each mode. Generally, the higher the frequency is, the stronger will be the attenuation, so that the *shorter* will be the attenuation time. This fact explains why when a string is excited in a haphazard manner and the vibrations are allowed to attenuate, eventually one sees a pattern of vibration very close to that of the fundamental mode, which has the lowest frequency. Similarly, when striking a tuning fork, one often hears the fundamental frequency, which is usually the desired frequency, masked by an overtone frequency. Eventually, only the fundamental frequency is heard, because the fundamental mode attenuates slowest.

We can make this point clearer from the following mathematical digression: Suppose that only the first and second modes are excited, with initial amplitudes $A_1 = 16$ units and $A_2 = 64$

Table 4.3 Attenuation vs. time

Time period—s	A_1—number of units	A_2—number of units
0	16	64
2	$16/2 = 8$	$64/2^2 = 64/4 = 16$
4	$16/2^2 = 16/4 = 4$	$64/2^4 = 64/16 = 4$
6	$16/2^3 = 16/8 = 2$	$64/2^6 = 64/64 = 1$
10	$16/2^5 = 16/32 = 1/2$	$64/2^{10} = 64/1,024 = 1/16$

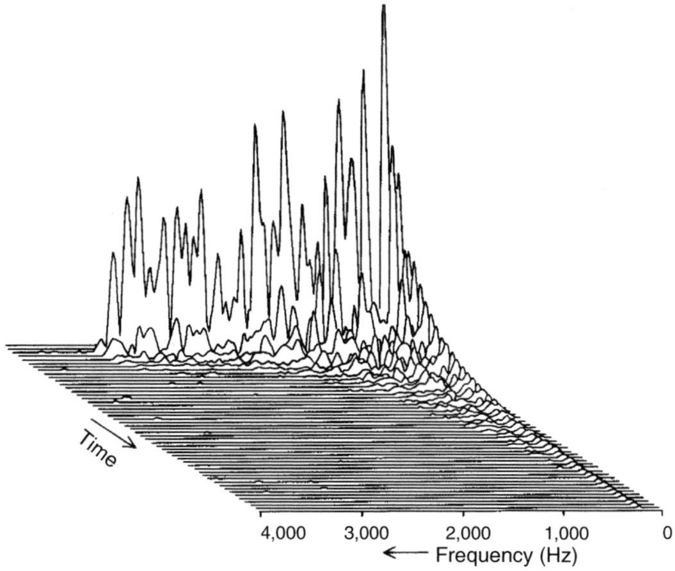

Fig. 4.11 The change in the frequency spectrum of a tom-tom with time (photo source: *The Science of Musical Sound*, by John R. Pierce (Scientific American Books, New York, 1983))

units, respectively. Suppose further that the respective attenuation times are $T_1 = 2$s and $T_2 = 1$s. In the table below, we provide the two amplitudes, A_1 and A_2, at a number of times after the initial time (Table 4.3).

We see that while the ratio of the initial amplitudes A_1/A_2 was $1 : 4$, after 10-s it is $8 : 1$. The fundamental dominates.

The behavior summarized above can be seen clearly in Fig. 4.11, wherein we see a depiction of how the spectrum of the sound of a tom-tom changes with time after the tom-tom is struck.

What we see is a set of many sound spectra taken one after another. Each spectrum runs from right to left. The earliest spectrum, at the rear of the set, has a series of many peaks, each peak reflecting the excitation of a particular mode of the tom-tom. As time progresses, each peak decreases in amplitude. However, we see that the higher the frequency of a mode, the faster the mode dies out. Towards the end, only the peak of the fundamental is observable, albeit with quite a small amplitude.

Note We should be aware that it takes some time interval over which the sound is sampled to obtain a single spectrum. The time interval must be well chosen: If it is not much longer than the period of the fundamental, the peak of the fundamental would be washed out and not be clearly discernible. If the interval is comparable to and longer than the attenuation times, the above spectra would not accurately reflect the attenuation at a given point in time.

The ear must also be carrying out an analysis that takes into account this dilemma: To produce a sense of pitch, a pure tone must last long enough to contain many oscillations. Its duration must be much longer than the period. To hear a sense of pitch, the ear must be sampling the sound over such a long duration of time. [That is one reason why it is difficult to have a sense of pitch of a percussive sound having a very low frequency and therefore a very long period.] On the other hand, since we do sense changing pitches, the sampling time must be shorter than the interval of time over which the pitch is changing in order to discern that changing pitch.

4.9.2 Response and Resonance in the Presence of Attenuation

Suppose that you want to excite a system by interacting with it in a periodic way with a frequency f. In Chap. 2 we noted that we have **resonance** when we have system A disturbed by another system B in a periodic way at a frequency equal to the frequency of one of the modes of system A, that is, we have $f_A = f_B$. Resonance reflects the ability of system A to respond to system B. The simple discussion in Sect. 2.14 seems to imply that a **resonant response** requires an **exact equality of the two frequencies**. By *exact* I mean equal to an infinite number of significant figures. The fact is that it is impossible to have an exact equality. However, in practice, we don't need to have an exact equality in order to have a response. Attenuation reduces the degree of equality necessary to have a response, as we will now see.

In Fig. 4.12 we see the **response** of an SHO of frequency f_{SHO} to an external force having a frequency f. We see a **peak response** around $f \approx f_{SHO}$ that is sharper, the lower the attenuation. In this case, if you want a significant response, the frequency f must be relatively very close to the frequency of the SHO. At a higher attenuation, you can get a significant response even with a greater mismatch. Generally, it can be shown that the width of the **resonance peak** is on the order of the inverse of the attenuation time.[10]

There are situations when you want to drive an SHO and *avoid the resonance* so that you

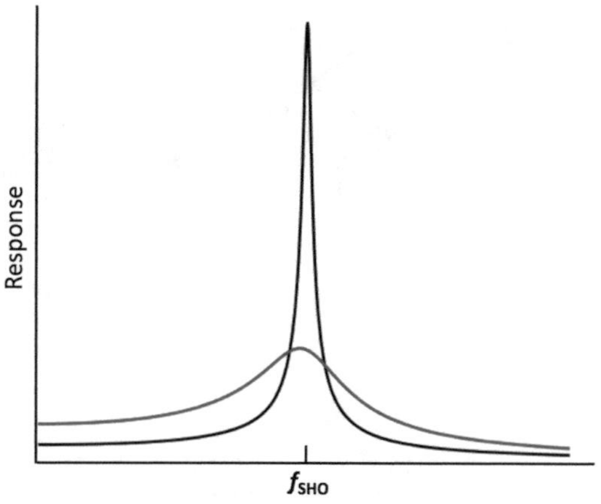

Fig. 4.12 Response of an SHO vs. frequency of applied force Black=low attenuation; Red=high attenuation

can have a uniform response with respect to a specific range of frequencies. This is the case for a **loudspeaker**, which has an SHO connected to the **loudspeaker cone** that vibrates so as to produce sound waves. Note that with low attenuation, as we see in the black curve, the response is nearly constant for low frequencies. Thus, for a loudspeaker, we want the frequency of the SHO to be considerably above the range of audible sound, that is, above 20-kHz, and an attenuation time of the SHO that is much greater than the inverse of the frequency of the SHO, so that you can avoid the resonance and have a flat response.

4.9.3 Attenuation of Traveling Waves—Attenuation in Space

Suppose a steady source produces a sinusoidal wave traveling in *one direction*,[11] such as a wave along a string or a sound wave down a very long pipe. In the absence of attenuation, the amplitude of the wave will be constant along the length of the wave. However, because of attenuation, the waveform will have an ever de-

[10] If you examine the graphs carefully you will note that the peak of the graph lies at a lower frequency for a higher attenuation. This change is not a mistake but reflects the actual behavior.

[11] It is **essential** to keep in mind that even in the absence of attenuation, the intensity of a wave that is emitted by a *point source* will decrease according to the "inverse square law" of Eq. (4.27). Attenuation will produce an additional contribution to the decrease of the intensity with increasing distance from the point source.

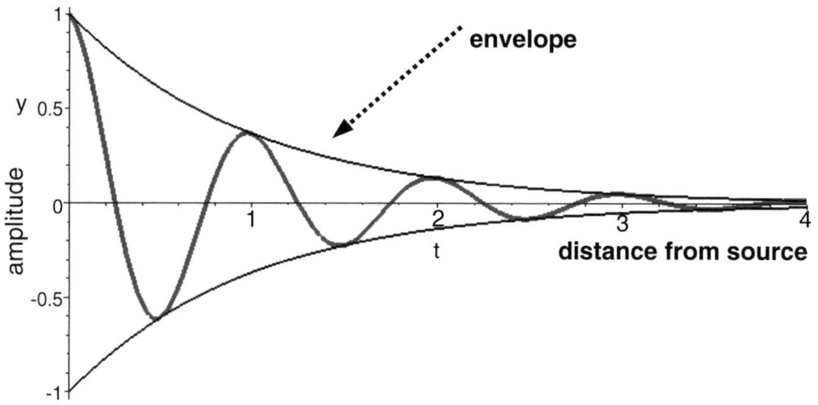

Fig. 4.13 Amplitude vs. distance from source due to attenuation

creasing amplitude, as shown in Fig. 4.13. The **envelope** of the waveform describes an attenuation of the amplitude *in space*. Numerically, the envelope is characterized by the **attenuation length**, which is the *distance* along the wave over which the amplitude decreases by a factor of two. The stronger the attenuation, the *shorter* the attenuation length. In the case of Fig. 4.13, the attenuation length is about 0.6 units. The **attenuation constant** α_L is an alternative way to characterize the attenuation of sound or light through a medium.[12] In precise terms it is the decrease in the decibel level (ΔSL) per distance x, so that

$$\boxed{\Delta SL = -\alpha_L x} \tag{4.44}$$

[We can show that the attenuation constant is inversely proportional to the attenuation length.[13]] Suppose that the attenuation constant is α_L = 20-dB/km for a sound wave and that the sound travels through a distance of 5-km. Then the change in sound level will be

$$\Delta SL = -\alpha_L x = -20(5) = -100 \text{ dB} \tag{4.45}$$

Note Recall that the decibel is used for light too: The symbol ΔSL is used to characterize a change in the intensity of light.

In Fig. 4.14,[14] we see graphs of the attenuation constant of sound in air versus **relative humidity**, for various frequencies of sinusoidal sound waves.

Here are the highlights of the curves.

First, for a fixed humidity, the higher the frequency, the greater the attenuation. We noted a similar behavior with respect to the modes of vibration of a system: The higher the mode frequency, the greater the attenuation.

We next note that for a given frequency, as the relative humidity varies from zero to 100%, the attenuation first increases and then decreases. The peaks in the individual curves indicate that for a given frequency, adding moisture to the dry air first leads to an increase in the degree of attenuation (hence, a decrease in the attenuation length) and that further increases of moisture lead to a decrease in the degree of attenuation.

The problems below will explain how we use the graph.

[12]It is also often referred to as the **attenuation coefficient**.

[13]Here is the proof: The attenuation length is the distance over which the amplitude decreases by a factor of two. The intensity will drop by a factor of four, corresponding to $\Delta SL = -10 \log 4 \approx -6$-dB. We must then have $\alpha_L \times$ attenuation length ≈ 6, so that $\alpha_L \approx 6$/attenuation length.

[14]The graph was produced using the standard **ISO 9613-1:1993**. See http://www.iso.org/iso/catalogue_detail.htm?csnumber=17426. I obtained the formula from the website (2-2-2011): http://www.sengpielaudio.com/AirdampingFormula.htm.

I am grateful to Eberhard Sengpiel for his help. Sadly, he passed away in 2014.

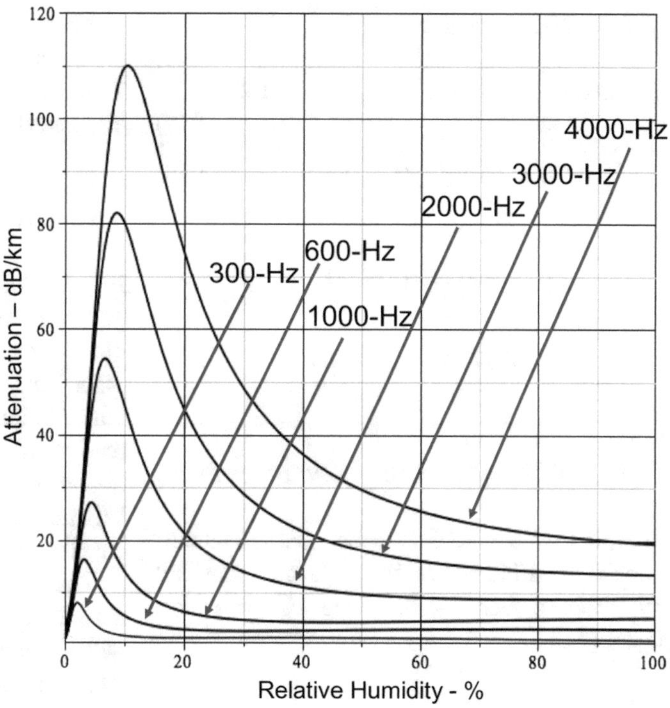

Fig. 4.14 Attenuation constant vs. relative humidity and frequency

Sample Problem 4.18 Suppose we have a sound wave of frequency 600-Hz and the relative humidity is 20%. Suppose further that the initial intensity I_1 is 10-W/m^2. Find the intensity I_2 after a distance of 100-m, after 1-km, and after 20-km.

Solution According to Fig. 4.14, the curve for a frequency of 5-kHz gives an attenuation of about 5-dB/km.

We begin with a distance of 100-m, which equals 0.1-km. Then the reduction in SL is

$$\Delta SL = -0.1\text{-km} \times 5\text{-dB/km} = -0.5\text{-dB} \tag{4.46}$$

Thus,

$$\Delta SL \equiv 10\log(I_2/I_1) = -0.1 \tag{4.47}$$

According to Eq. (4.41),

$$\frac{I_2}{I_1} = 10^{-\Delta SL/10} = 10^{-0.01} = 0.98 \tag{4.48}$$

so that

$$I_2 = 0.98I_1 = 0.98 \times 10 = 9.8\ W/m^2 \tag{4.49}$$

After 1-km, we have

$$\Delta SL = -1\text{-km} \times 5\text{-dB/km} = -5\text{-dB} \tag{4.50}$$

so that

$$\Delta SL \equiv 10\log(I_2/I_1) = -1 \tag{4.51}$$

and

$$\log(I_2/I_1) = -0.1 \tag{4.52}$$

Thus,

$$\frac{I_2}{I_1} = 10^{-0.1} = 0.79 \tag{4.53}$$

and

$$I_2 = 0.79I_1 = 0.79 \times 10 = 7.9\ \text{W/m}^2 \tag{4.54}$$

After 20-km, we have

$$\Delta SL = -20\text{-km} \times 5\text{-dB/km} = -100\text{-dB} \quad (4.55)$$

$$\Delta SL \equiv 10\log(I_2/I_1) = -10 \quad (4.56)$$

or

$$log(I_2/I_1) = -1 \quad (4.57)$$

Thus,

$$I_2/I_1 = 10^{-1} = 0.1 \quad (4.58)$$

and

$$I_2 = 0.1 I_1 = 0.1 \times 10 = 1 \text{ W/m}^2 \quad (4.59)$$

Sample Problem 4.19 Suppose that the attenuation is 10-dB/km. Find the distance x traveled by the wave such that the intensity drops by a factor of two.

Solution We set

$$\Delta SL = 10\log(I_2/I_1)$$
$$= -(10 \text{ dB/km}) \times x \text{ [in km]}$$

where $I_2/I_1 = 1/2$. Since log(1/2)= -log 2, we have $-10\log 2 = -10x$ or

$$x = \log 2 = 0.3 \text{ km} = 300 \text{ m}$$

4.9.4 Attenuation of Light in a Transparent Medium

The attenuation of light in common glass is not negligible, amounting to on the order of 10,000-dB/km or 10-dB/m. Thus, in passing through a one meter thickness of such glass, the light intensity will be reduced by a factor of ten. Fiber optics communication has depended upon the development around the year 1970 of glasses having an extraordinarily low attenuation of about 0.1 to 1-dB/km! The transition from wired connection to optical fiber connection resulted in an increase in the speed of communication by a factor of about ten.

Note that a 1-dB drop corresponds to a decrease in the intensity by a factor of $10^{0.1} = 1.26$. That is, $I_2/I_1 \approx 0.8$ over a distance of one-kilometer!

4.10 Reverberation Time

If you produce a sound in a room, the sound you hear might resound like an echo. Such is often the case for a large empty room. On the other hand, if the room is small and/or is loaded with furniture, the sound dies out so fast that we tend to label the room as being **dead** to sound. Sound produced in an open field is quite dead. What is the process that determines the resulting character of the sound?

The sound produced by a source bounces off the surfaces of the room. Eventually the sound will die out—mostly because it passes on into these surfaces. How long a sound lasts is characterized by the **reverberation time** ($\equiv RT$).[15] It is defined as the time it takes for the sound intensity in the room to drop by a factor of one-million. The corresponding **drop** in the sound level is

$$\Delta SL = 10\log(10^6) = 60\text{-dB}$$

Over the course of an interval of two reverberation times, the intensity will drop by a factor of (one-million)^2 = one-trillion.

Very long reverberation times produce an echo effect; very short reverberation times produce a dead space. In the design of auditoria or music rooms, one will design the space so as to produce a reverberation time that suits one's preference. Typically, a reverberation time on the order of one second is desired.

Sabine's Law

When a sound wave is incident upon a surface, a certain fraction of the intensity is reflected, while

[15]For more information on the reverberation time see the website: http://www.yrbe.edu.on.ca/~mdhs/music/oac_proj97/music/reverb.html. You will be able to calculate the reverberation time of a room given the volume of the room and the area and absorption constant of each surface within the room.

the remainder is transmitted into the surface. If the surface is very thick, all of the transmitted sound will be absorbed. On the other hand, a wall might be so thin that some of the sound intensity passes through on into the air or other material on the other side of the wall. Often, this second fraction alone is referred to as the transmitted sound. For walls in a room, one often refers to the sum of both absorption and transmission as **absorption**.

Suppose now that a sound is produced in an empty room. The room has walls, floor, and ceiling. Each of these surfaces will reflect a certain fraction of sound. The fraction that is not reflected is called the **absorption coefficient**—that is represented by the symbol α. The absorption coefficient depends upon the frequency of the sound wave and the surface material.

Note The symbol α should not be confused with the symbol α_L used for the attenuation constant.

Suppose that all the surfaces have the same absorption coefficient and that the total surface area is A. It can be shown that the reverberation time RT is approximately given by **Sabine's Law**:

$$\boxed{RT = 55.2\frac{V}{v\alpha A}} \tag{4.60}$$

Here v is the sound velocity in air, V is the volume of the room expressed in m^3, A is expressed in m^2, and the reverberation time is expressed in seconds. Note that the greater the volume of a room, with fixed area, the longer the reverberation time. Consider a room in the shape of a cube of side L. It is easy to see that reverberation time is given by

$$RT = 9.2\frac{L}{v\alpha} \tag{4.61}$$

Because the volume increases faster than the total area, the reverberation time increases with increasing size of the room.

If $v = 340$ m/s,

$$RT = 0.16\frac{V}{\alpha} \tag{4.62}$$

Usually a room has surfaces with different absorption coefficients. Then, with the surfaces having areas $A_1, A_2, A_3, \ldots$, and absorption coefficients $\alpha_1, \alpha_2, \alpha_3, \ldots$, respectively, the factor αA in Eqs. (4.60) and (4.62) is replaced by a sum:

$$\alpha A \Rightarrow \alpha_1 A_1 + \alpha_2 A_2 + \alpha_3 A_3 + \ldots \tag{4.63}$$

Acoustic tile, suspended from the ceiling, has an absorption coefficient close to unity:

Suspended acoustic tile	125 Hz	250 Hz	500 Hz	1000 Hz	2000 Hz	4000 Hz
	0.76	0.93	0.83	0.99	0.99	0.94

Below is a chart with absorption coefficients for a variety of materials and for various frequencies (Table 4.4).
Source:http://www.sfu.ca/sonic-studio/handbook/Absorption_Coefficient.html

Note Suppose that the absorption coefficient were unity ($\alpha = 1$). Then any sound wave incident on a wall would be completely removed from the room. The reverberation time would have no meaning since any sound created in the room would be completely lost in a time at most equal to the time it takes for a sound wave to cross the room. The sound intensity would drop suddenly from its initial value to zero in this time. And yet, the Sabine's formula, Eq. (4.60), leads to a finite reverberation time. Obviously, the formula is in error and can only be an approximation. It can be shown that Sabine's formula holds only when $\alpha \ll 1$.
In Eq. (4.64) we exhibit a more accurate formula that is identical to Sabine's formula except that α is replaced by $-\ln(1-\alpha)$. Here, **ln** is the natural log [or log to the base "e"]. Thus, Eq. (4.60) is replaced by

$$\boxed{RT = 55.2\frac{V}{-\ln(1-\alpha)\cdot A}} \tag{4.64}$$

Table 4.4 Chart of absorption coefficients

Material	128 Hz	256 Hz	512 Hz	1,024 Hz	2,048 Hz	4,096 Hz
Draperies hung straight, in contact with wall, cotton fabric, 10 *oz*. per square yard	0.04	0.05	0.11	0.18	0.30	0.44
Rock wool (1 inch)	0.35	0.49	0.63	0.80	0.83	–
Carpet on concrete (0.4 inch)	0.09	0.08	0.21	0.26	0.27	0.37
Carpet, on 1/8″ felt, on concrete (0.4″)	0.11	0.14	0.37	0.43	0.27	0.27
Concrete, unpainted	0.010	0.012	0.016	0.019	0.023	0.035
Wood sheeting, pine (0.8 inch)	0.10	0.11	0.10	0.08	0.08	0.11
Brick wall, painted	0.012	0.013	0.017	0.020	0.023	0.025
Plaster, lime on wood studs, rough finish (1/2 inch)	0.039	0.056	0.061	0.089	0.054	0.070

Note that for very small α, $-\ln(1-\alpha) \approx \alpha$, *so that we would recover Sabine's Law—Eq.* (4.60). *Note also that if* $\alpha = 1$, *corresponding to total immediate absorption, this formula leads correctly to* $RT = 0$, *since* $\ln(0) = -\infty$. *That is, according to this formula, any sound that is produced dies out immediately, as expected.*

4.11 Terms

- attenuation constant
- attenuation length
- attenuation time
- brightness
- British Thermal Unit (BTU)
- Calorie or food calorie (Cal)
- chemical energy
- decibel scale
- dissipation
- electrical energy
- electromagnetic energy
- electron-volt (eV)
- energy
- envelope of an attenuated wave
- exponential behavior
- gravitational potential energy
- heat transfer intensity
- Joule (J)
- kilowatt-hour (kWh)
- kinetic energy
- loudness
- nuclear energy
- point source [=isotropic source]
- potential energy
- power
- principle of conservation of energy
- reference level reverberation time
- sound level
- stroboscope
- thermal energy
- weight
- work
- Watt (W)

4.12 Important Equations

Kinetic energy:

$$\mathrm{KE} = \frac{1}{2}mv^2 \tag{4.65}$$

Potential energy of a simple harmonic oscillator:

$$\mathrm{PE} = \frac{1}{2}ky^2 \tag{4.66}$$

Change of gravitational potential energy with change of elevation:

$$\Delta\mathrm{PE} = \text{weight} \times \text{change of elevation} = \Delta\mathrm{PE}$$
$$= wh = mgh \tag{4.67}$$

Intensity defined:

$$\text{Intensity} = \frac{\text{Power}}{\text{Area}} \qquad I = \frac{P}{A} \tag{4.68}$$

Intensity of a point source:

$$I = \frac{P}{4\pi r^2} \quad (4.69)$$

Sound level defined:

$$SL = 10\log\frac{I}{10^{-12}} \quad (4.70)$$

$$\Delta SL \equiv SL_2 - SL_1 = 10\log\frac{I_2}{I_1} \quad (4.71)$$

Change in sound level with change in intensity re-expressed:

$$\frac{I_2}{I_1} = 10^{(\Delta SL/10)} \quad (4.72)$$

Change in sound level with distance due to attenuation:

$$\Delta SL = -\alpha_L x \quad (4.73)$$

where α_L is in dB/km and x is in km.

4.13 Problems for Chap. 4

1. Estimate the number of Calories, Joules, and kiloWatt-hours (kW-hrs) in a gallon of oil.
2. One ounce of Cheerios provides about 100 Calories of food energy through chemical changes. On the average, one *molecule* of Cheerios provides about 5 eV of food energy. From this information, estimate the number of molecules in an ounce of Cheerios. [You will need to make use of the number of eV there are in a Calorie.]
3. How many *lbs*. of Cheerios would one have to eat to provide the mechanical energy output of about 25-billion J of a human heart in a 75 year lifetime?
4. What are the two basic modes of transfer of energy from one system to another?
5. When a person is said to be "powerful enough to lift a 200 *lb*. object," is one referring to the person's "power" according to the way that term is used in physics? **Explain.**
6. Calculate the power delivered by a jet engine that increases the kinetic energy of a jet plane from zero to 40-billion Joules in five minutes. [40-billion Joules of KE corresponds to a mass of 1000 tonnes (1 tonne = 1,000 kg ≅ 2,200 lbs) moving at 300 m/s (≅ 1000 km/h).]
7. A loudspeaker delivers 0.002 W of sound power down a tube of diameter 4 cm. Calculate the sound intensity traveling down the tube.
8. A **stroboscope** is a device that produces a series of flashes of light having an extremely short duration in time (μs) but very large energy, with a variable frequency, that can range from about one flash per second to tens of thousands of flashes per second. Consider a particular stroboscope—model MVC-4100—manufactured by the *Electromatic Equipment Corporation*. The corporation provides the following specifications for this model on its website:
(http://www.inspectionstroboscope.com/prods/MVC-4000?PHPSESSID\=7933e3da79f0401c1e54c9ea5d0e8367)
Input energy per flash: 5.41 Joules; Duration of a pulse of electrical: 30 μs = 30 microseconds; radiometric light output (light energy per flash): 0.210 Joules.
   (a) Calculate the average electric power delivered in a single flash over the duration of a pulse of electric energy.
   (b) Calculate the efficiency in the production of light energy from electric energy.
9. Calculate the light intensity at the earth from a star that is 100 light years away (1 light year = 9.5×10^{15} m) and emits light with a power equal to that of our sun (4.0×10^{26} W).
10. (a) If the sound intensity from an isotropic source is 0.2 W/m^2 at a distance of 1 m, what will the sound intensity be at a distance of 2 m?
   (b) Calculate the sound level (in dB) corresponding to each of the two intensities of part (a). What is the corresponding change in sound level?

11. (a) What is the "solar constant"?
    (b) The atmosphere absorbs sunlight. Suppose that as a result during some period the sunlight intensity on the surface of the earth is one-half the solar constant. Suppose further that we have an array of solar cells on a roof with dimensions 10 m by 6 m facing directly towards the sun. Finally assume that for every 100 Joules of incident sunlight energy the cells produce 15 Joules of electrical energy. [Their efficiency is then 15%.] Calculate the electrical power output of the array.
12. Describe in detail, using all the physical principles we've covered so far, how the sound emitted by a tuning fork varies in time after it has been struck hard.
13. (a) What is the "attenuation time"?
    (b) What is the "attenuation length"?
14. Suppose that a ball is released from rest at a meter above a hard floor. It bounces back up and reaches a height of one-half meter. Now suppose that it keeps bouncing, with each bounce leading to a maximum height that is one-half the previous height.
    After how many bounces will the maximum height be one Å above the floor? [One Å $= 10^{-10}$m.]
    The significance of 1Å is that this elevation is about equal to the size of a molecule, so that a bounce disappears within the typical motion of the molecules. For all intents and purposes, the ball has stopped bouncing.
    Hint: Let $n = number\ of\ bounces$. Find an expression for the height after n-bounces and take the log of that expression.
15. If the amplitude of vibration of an SHO decreases from 8 cm to 4 cm in one minute, what will be its amplitude after two minutes? After three minutes?
16. Suppose that the intensity of the fundamental of an organ pipe is initially 8-units and decays to 7-units after one second.
    How many units of intensity does the fundamental have after a total of 3 s?
17. The following is a fascinating application of attenuation time. We will consider a pipe that is excited in a particular mode. On the one hand, we recall that the assumption is that the ends are nodes for the sound pressure. In the section on standing waves, it was assumed that pulses are totally reflected from the ends. If this were exactly so, the pipe wouldn't produce a sound wave outside the pipe, so that we wouldn't hear the excited pipe!
    In fact, the ends are not nodes precisely and sound is emitted from an open end any time a pulse reaches that end. In order to produce a sound associated with a particular mode, we need a compromise, a balance between reflection and emission of sound into the air outside. The pertinent transmission coefficient T is defined as the fraction of sound energy that is emitted when a wave reaches an open end.
    (a) Now imagine a wave moving back and forth in the pipe. With each incidence on an open end, a fraction T of sound intensity is lost. In order for the mode to be well defined and be heard with a clear frequency, it is necessary that there be many oscillations with negligible attenuation.
        What condition does this place on the transmission coefficient?
    (b) Suppose that the wavelength is much larger than the diameter of the pipe. It can be shown that the transmission coefficient is approximately given by the ratio of the square of the circumference to the wavelength—that is,

$$T = \left(\frac{2\pi R}{\lambda}\right)^2 \tag{4.74}$$

        Does the transmission coefficient increase or decrease with higher frequency?
    (c) Now consider a flute of radius one centimeter played at a frequency of 440 Hz. Calculate the corresponding transmission coefficient.

18. If the frequency of sound increases, the attenuation length will:

    decrease/increase/remain the same.

19. If the humidity of air increases, the attenuation length will:

    decrease/increase/remain the same/

    all three of the previous choices are possible.

20. What range of intensity is associated with a **negative** sound level?
21. Suppose the light intensity drops by a factor of four in passing through one meter of a certain glass.
    (a) What is the corresponding drop in dB?
    (b) What drop would there be in passing through *four* meters of glass?
22. (a) Use Fig. 4.14 to determine the attenuation of sound in dB/km of 2-kHz sound in air at 20% humidity.
    (b) If a sound wave of frequency 2-kHz is traveling in a straight line through the air, what would the change in sound level be and by what factor would the intensity drop in traversing a distance of 2-km?
23. Choose a room with which you are familiar. You will have to know a bit about the materials out of which the walls, floor, and ceiling are made. If you are not sure, make a guess. [Most walls in homes are made of gypsum. Ceilings are usually made of gypsum. Sometimes they are made of acoustic tile, which have a higher absorption coefficient.] Then go to one of the websites below and estimate the RT [reverberation time] for that room. Discuss your result in relation to the sound quality of the room.
    (a) A simple website to use to calculate the reverberation time (RT) for a room. But you obtain the RT for each of a set of frequencies, one at a time http://www.saecollege.de/reference_material/pages/Reverberation%20Time%20Calculator.htm
    (b) This website calculator is a bit more cumbersome for inserting data but gives you RT for a number of frequencies after one click. http://www.atsacoustics.com/cgi-bin/cp-app.cgi
    (c) This site has audio files so that you can hear the difference among various reverberation times: http://www.armstrong.com/reverb/main.jsp
24. Suppose that a room is a cube with sides of length 3 m. Compare the RT with a larger cubic room with sides of length 4 m. That is, find the ratio of the RTs. Note that both the volume V and the total area A increase. Would you have expected RT to increase or decrease? Explain.
25. Suppose that a sound wave were traveling across a room the shape of a cube of side L.
    (a) How long would the wave take to travel directly from one side to the other and back? Compare this time to the RT as expressed in Eq. (4.60).
    (b) Note that for each "collision" of a sound wave against the wall, the intensity drops by a factor equal to (1-α). For n round trips the intensity will drop by a factor equal to $(1-\alpha)^{2n}$.
    What is the corresponding change in decibel level, ΔSL?
    (c) How many collisions will there be in a time t?
    (d) RT is the time it takes for $\Delta SL = 60$-dB. Find an expression for RT. Compare your expression with Eq. (4.64), when the room is a cube of side L, as in part (a) above. Make use of the relation $\log x \approx 0.43\, ln\, x$.
26. **Reverberation Time with Wall Absorption and Attenuation in Air**

    In Sect. 4.10, we studied the Reverberation Time **RT** in a room associated with the decrease in the sound level due to absorption on surfaces in the room. Remember that the RT is the time it takes for a sound to decay by 60-dB. In Sect. 4.8.3 we studied attenuation of a wave while traversing a medium; such attenuation would be experienced by sound in air. Yet the latter attenuation was not taken into account with respect to the RT. One might think that this neglect is reasonable since attenuation of sound in air is typically on the order of 10 to 100-dB/km. We will now see that this neglect is not always warranted.

Suppose that RT = one second due to absorption of sound by the walls and that the attenuation in air is 100-dB/km.

(a) Imagine a sound wave starting out at some point in the room and bouncing off the walls specularly as shown below. The red circle is the source. Continue the rays in the figure so as to exhibit three more reflections off the walls. The rays are assumed to be propagating in a horizontal plane. Draw a few more rays of sound in Fig. 4.15.

(b) What is the total distance traveled by the sound wave in one second, assuming a sound velocity of 340 m/s.

(c) What would be the corresponding reduction in sound level?

(d) Note that we shouldn't add the two contributions to the RT—one from absorption by objects in the room and one from attenuation in the air. We would expect the attenuation in the air to reduce the RT! In fact, it can be shown that the resulting RT is given adding inverses:

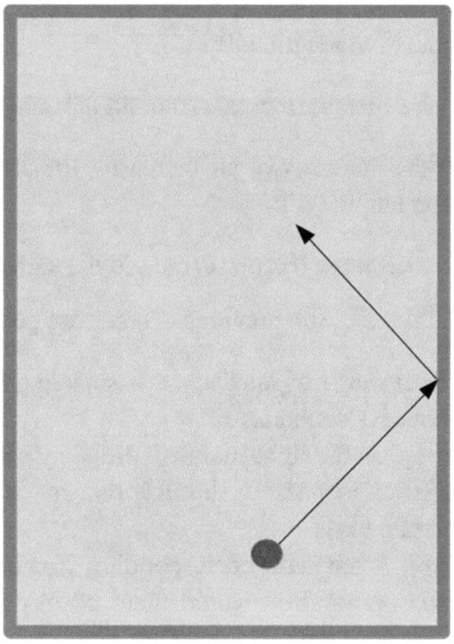

Fig. 4.15 Sound reflecting off walls with absorption by the wall and by the air

$$\frac{1}{RT_{\text{total}}} = \frac{1}{RT_{\text{absorption}}} + \frac{1}{RT_{\text{air}}} \tag{4.75}$$

Thus, if $\frac{1}{RT_{\text{absorption}}}$ =1 s and $\frac{1}{RT_{\text{air}}}$ =1 s, the total RT is one-half second.

Confirm this result.

Electricity, Magnetism, and Electromagnetic Waves

5

We have mentioned that **light** is an **electromagnetic wave** with a frequency that lies in a particular range: $\sim 4 \times 10^{14}$ to $\sim 7 \times 10^{14}$ Hz. But what is an electromagnetic wave? To answer this question we will need to study **electricity** and **magnetism**. The principles of this branch of physics are the basis of operation of the various electronic instruments used in sound reproduction, such as radio transmitters and receivers, amplifiers, microphones, and speakers. There is an interesting further relevance: The **atom** often serves as a primary *source* of light and is the *receiver* of light in our eyes. How the atom performs these functions, as well as how the atom is held together, depends upon the laws of electricity and magnetism. We experience numerous manifestations of **electricity** which are not dependent upon technological developments: They were known by mankind before the age of science. Most powerful and majestic are thunder and lightning, which involve enormous currents of **electric charge** flowing from thousands of feet above, down to the earth's surface. **Static electricity** is another manifestation. It is said that **Thales of Miletus** was the first to note around 600 BC that amber rubbed with fur attracts straw. In the *magnetism arena*, many of us are familiar with "**lodestone**," which is the rock mineral "magnetite" in a **magnetized** state. Lodestone is capable of attracting iron and, for centuries, served as a **compass** for guiding sailors out at sea.

5.1 The Fundamental Forces of Nature

We note that electricity and magnetism are manifested in an obvious way by *forces*, the so-called **electric force** and **magnetic force**. These two **fundamental forces** in nature were originally incorrectly believed to be unrelated to one another. Subsequently, as we shall see later in this chapter, they were found to be so interrelated that they are together referred to as one force, the "**electromagnetic force**." Other fundamental forces include the "**gravitational force**," the "**nuclear force**," and the "**weak force**." [Interestingly, physicists have a goal to see whether *all* these forces are manifestations of only *one* all-encompassing and more fundamental force.]

Until this century, the force of gravity was the best understood of all forces, thanks to the "Universal Theory of Gravitation" that was expounded by **Isaac Newton** (1642–1727) (Fig. 5.1).

Below we highlight the main features of Newton's Theory of Gravitation.

1. Matter has a quantifiable attribute called **mass**. [Common units of mass are the **gram** (gm) and the **kilogram** (kg).] Mass gives a body **inertia** (resistance to having a change in velocity) and is also the source of the gravitational force.

L. Gunther, *The Physics of Music and Color*,
https://doi.org/10.1007/978-3-030-19219-8_5

Fig. 5.1 Isaac Newton—Portrait by Sir Godfrey Kneller in 1702 (source: http://en.wikipedia.org/wiki/File:SirIsaacNewtonbySirGodfreyKneller,Bt.jpg)

2. The gravitational force between two bodies can be determined in terms of the force between two **point masses**. These are idealized bodies which take up no space. Given the force between two point masses, one can determine the force between any two real bodies which take up space. The most important aspects of this force are that it is always attractive, with each body attracting the other towards its respective self with a force of the same magnitude and that its magnitude decreases with increasing separation between the bodies.
3. Isaac Newton successfully used his theory of gravitation to account for the motion of the planets and the Earth about the Sun, the motion of the Moon about the Earth, as well as the motion of projectiles such as cannon balls flying through the air near the earth's surface. However, he was deeply perplexed about how one object could exert a force on another with only empty space between them. We are referring here to what is called the **action-at-a-distance** enigma.

It is important to realize that this issue must be regarded as a philosophical one, being outside the realm of science. Ultimately, the goal and the measure of success of a scientific theory is found in terms of its ability to provide relationships among measured physical quantities, rather than its ability to explain why the phenomena take place. "Why" questions essentially seek underlying, more fundamental principles that can be applied to the specific phenomenon at hand. Thus, we might ask why the sky is bright with the answer being that the molecules of the air scatter sunlight. But why do the atoms scatter light? Well, because they are comprised of electric charges, which experience a force from the light and so on. The sequence of questions must ultimately end when we reach the most fundamental level of theory that encompasses all phenomena below it. At that point, "why" questions cease to have meaning. Of course, physicists are never content to conclude that they have definitely reached the most fundamental level. They are always open to revelations of the new.

We will see in this chapter that the action-at-a-distance question did lead to the introduction of the concept of the "force-field," which, while it may not have satisfyingly addressed the action-at-a-distance question, has been extremely useful in the development of physics.

5.2 The Electric Force

In the late 1700s, experimental studies of the force between electrically charged bodies led to a theory for the electric force which is similar to the theory for gravitational forces. It is embodied in "**Coulomb's Law**." We summarize the theory below:

1. Bodies can have an attribute called **electric charge**.

   There are two types of charge—referred to as **positive charge** and **negative charge**. Letting q_1 and q_2, respectively, be the numerical values of the charges of two bodies, we find that the force is:

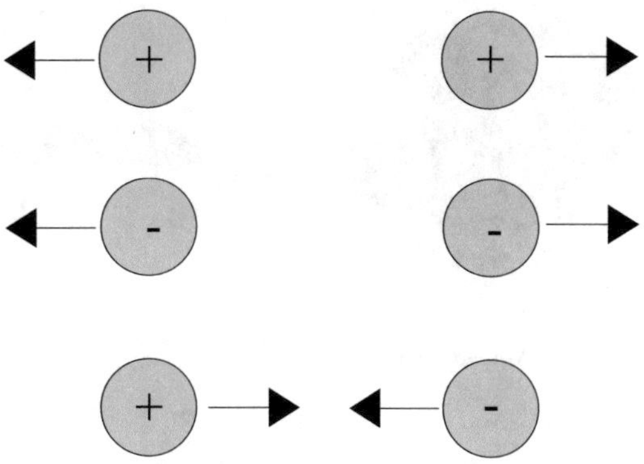

Fig. 5.2 Forces between electric charges—two positive, two negative, and one positive with one negative

- **repulsive** if q_1 and q_2 have the same sign (hence either both positive or both negative);
- **attractive** if q_1 and q_2 have opposite signs.

This summary is exhibited in Fig. 5.2:

2. The **action-at-a-distance** issue is manifest with electric forces too. How can two electric charges affect each other when they are not, so to speak, *touching*? In fact, a study of the force on your body by another object along with the structure of atoms and the electric forces therein that are in fact responsible for this force reveals that the force on your body does not involve contact between the charges therein.

5.3 Electric Currents in Metal Wires

We begin this section by summarizing some important information about all matter on earth. Matter consists of atoms. These in turn consist of **nucleus** and **electrons**. The electron has a negative electric charge. The nucleus consists of a tightly bound collection of **protons** and **neutrons**. Protons have a positive charge. Neutrons, while being a charged structure, have no net electric charge—they are "electrically neutral." An **ion** is a neutral atom which has gained or lost electrons and thus is electrically charged.

Roughly, materials can be divided into two categories with respect to electrical behavior: The

Fig. 5.3 Random motion of electrons in a wire

first category consists of the **insulators**. While they have electric charge, the charges are bound and can't move freely within the material. In a sample of the second category—consisting of **conductors**—there are charges that are free to move throughout the material. **Metals** are a specific type of conductor. The atoms of a metal are **ionized**: Atoms have lost some electrons, which are then free to move about a solid array of positive ions. The former, the so-called **free electrons** (also called **conduction electrons**), move about randomly at speeds averaging about 1000 km/s! An **electric current** in a metal wire represents a *net* flow of free electrons. We might refer to this flow on a large scale as a **macroflow** of free electrons, commonly referred to as an electric current. For comparison sake, the so-called flow of a river is a **macroflow** of water: Water molecules in a river are in random motion, with speeds on the order of a km/s, even when the river as a whole is still. The flow of a river reflects an average motion in some specific direction that is added to the background random motion. See Fig. 5.3.

One can compare the motion of the charges to that of a pinball in a pinball machine; the pinball moves helter-skelter, nevertheless making overall progress down the board.

The direction of an electric current is the same as the direction of motion of the positive charges, but opposite to the direction of the negative charges. If both signs of charge are present and moving, we must subtract the contribution of the negative charges from that of the positive charges. [See Fig. 5.4.] In the case of a current in a metal wire, there is no contribution from the positive ions. That is, only the conduction electrons contribute to the current and the current is opposite to the direction of the conduction electrons.

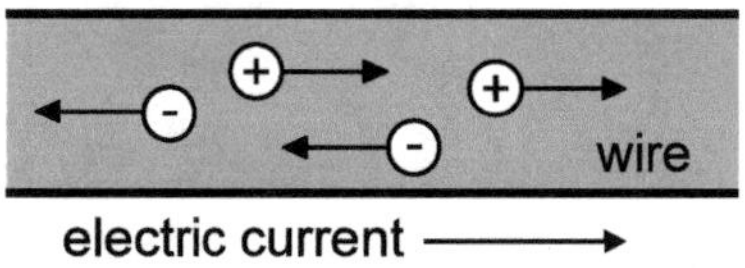

Fig. 5.4 Electric current in a wire

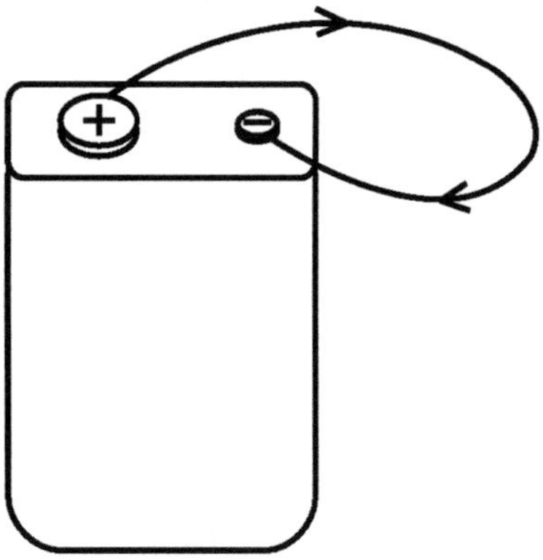

Fig. 5.5 Electrons flowing from the negative terminal to the positive terminal of a battery

The common **electric battery** drives current from the **positive terminal** through the wire and back into the **negative terminal** of the battery, as shown in Fig. 5.5. Thus, electrons flow through the wire from the negative terminal to the positive terminal.

A battery is characterized by a physical parameter referred to as its **voltage**, expressed by the unit of a **volt**. For example, small batteries usually have voltages of 1.5-volts or 9-volts. Automobile batteries usually have a voltage of 12-volts. The voltage is analogous to the **force** needed to tug a boat through the water. In addition, a battery stores electric charge. This charge is used to provide the voltage. As a battery is used, the stored charge is dissipated.

5.4 The Magnetic Force

Originally, magnetic forces were confined to magnetized bodies, called **magnets**. We note that a sample of magnetic material need not be magnetized. Magnetite is a naturally occurring magnetic material which when magnetized

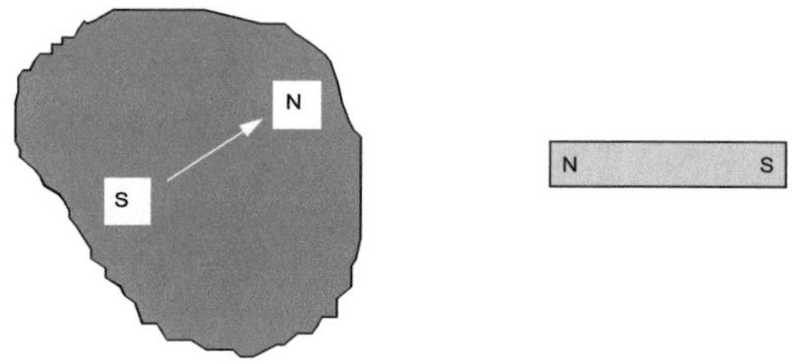

Fig. 5.6 Magnetite and a bar magnet

is called **lodestone**. A sample of iron can be magnetized by a piece of magnetite. The earth as a whole acts as if it has a huge mass of magnetized material. Typically, a pair of magnetized bodies exert forces on each other which tend to orient the bodies in a certain direction with respect to each other.

To the eye, a magnetized body appears **isotropic**. In order to appreciate what this implies, consider the rotation of a magnet having a spherically shaped body. If the magnet were rotated, you would not notice any change in its appearance. In fact, at an invisible level, the body is actually **anisotropic**. Its magnetic behavior can be characterized by picturing the body as having an axis with a **north pole** and a **south pole**, as seen below in the sample of magnetite. Thus, as far as its magnetic properties are concerned, one could easily detect that the magnet had been rotated. We indicate the north and south poles of a sample of magnetite and a **bar magnet** in the figure below. *N* represents the north pole, while *S* represents the south pole. See Fig. 5.6.

The magnetic south pole of the earth is shown in Fig. 5.7 to be in *Northern* Canada.[1]

Finally, a bar-like magnet can be bent into a **horseshoe magnet** so as to produce strong magnetic forces between its poles (see Fig. 5.8).

[1]The *north* magnetic pole is diametrically opposite, in the *southern* hemisphere, close to New Zealand. During this century it has moved on average 10 km per year. See the Wikipedia article (12-26-2010): http://en.wikipedia.org/wiki/Magneticdeclination. Also: http://obsfur.geophysik.uni-muenchen.de/mag/news/e_nmpole.htm.

Fig. 5.7 The Earth's magnetic south pole indicated by the small white circle in Northern Canada

Fig. 5.8 Horseshoe magnet

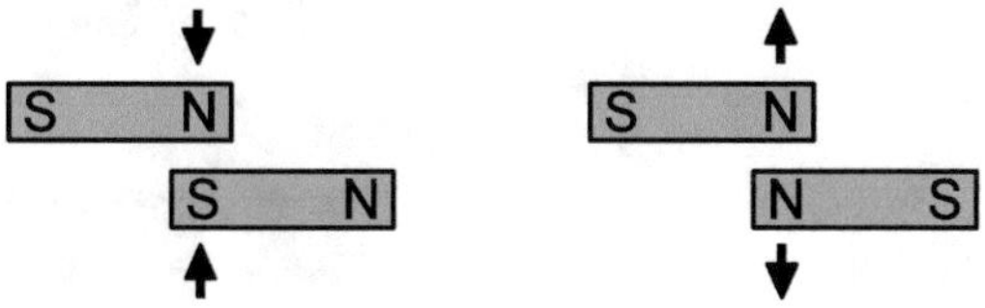

Fig. 5.9 Repelling poles

5.5 Characterization of Magnetic Forces

1. As with electric forces, action-at-a-distance is manifest.
2. Like poles repel; unlike poles attract. See Fig. 5.9.
3. Two freely suspended bar magnets tend to rotate so as to line up with parallel axes (Fig. 5.10).
4. A non-magnetized material can be **magnetized** through the presence of an already magnetized body. When the bar magnet is removed, the degree of magnetization may diminish, sometimes to an undetectable level (Fig. 5.11).

What happens to a bar magnet when it is cut in half? One might be inclined to guess that we end up with one-half being a north pole and the other a south pole. To the contrary, we end up with two shorter bar magnets, each with a north pole and a south pole, as seen in the figure below (Fig. 5.12).

This result should be contrasted with what happens when a bar of electrically polarized material is cut in half. In Fig. 5.13, we see what happens to a bar of metal that is polarized by a neighboring point electric charge. Each half of the bar is charged, one positively, the other negatively. If the point charge is removed, the two halves remain charged. On the other hand, if a bar of insulating material is so polarized and halved, the two halves are not charged. In the presence of the point charge, each half will be polarized; while, if the point charge is not present when the bar is halved, neither half is charged or polarized.

5.6 Is There a Connection Between Electricity and Magnetism?

In 1820, **Hans Oersted** (Fig. 5.14) is reported to have been lecturing his class on this issue, exhibited in Fig. 5.15.

A compass needle was placed close to an electric wire that was connected to a switch and a battery. In advance of closing the switch and thus turning on a current through the wire, he told his class that it was obvious that there would be no effect on the wire. Alas, he was wrong! The needle tended to be oriented in a particular way, as depicted in Fig. 5.16.

We depict the wire oriented perpendicular to this page. A number of compass needles are suspended along a circle with the wire at its center. Current is in a direction out of the page.

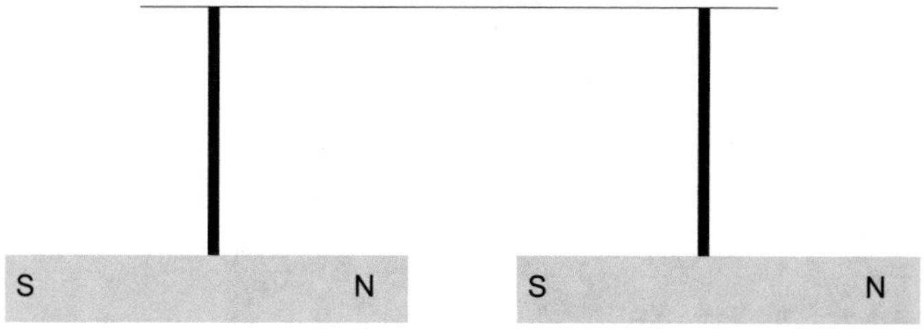

Fig. 5.10 Suspended magnets rotate so that axes align

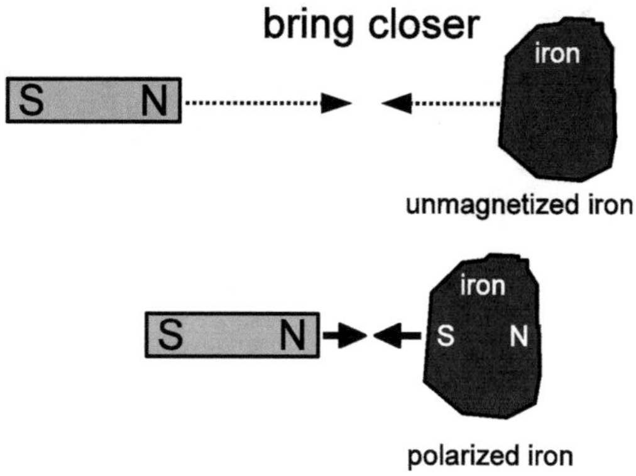

Fig. 5.11 Polarizing an unmagnetized piece of iron

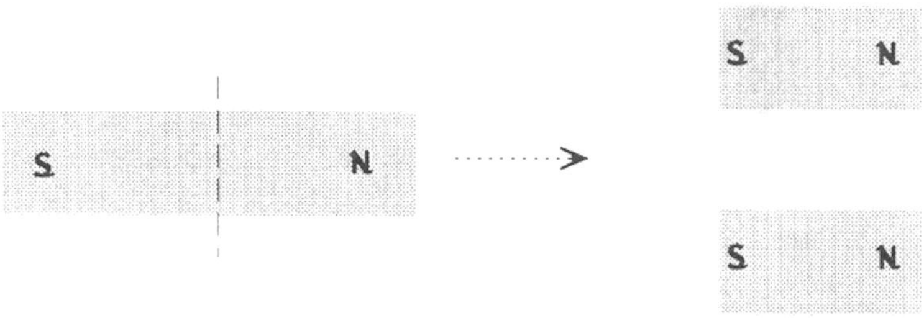

Fig. 5.12 Splitting a bar magnet creates two bar magnets

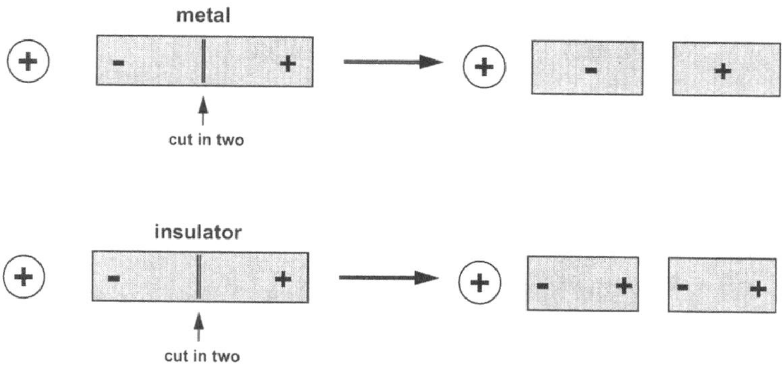

Fig. 5.13 Results of splitting polarized materials: metals versus insulators

Fig. 5.14 Hans Oersted (source: http://en.wikipedia.org/wiki/Hans_Christian_\T1\Orsted)

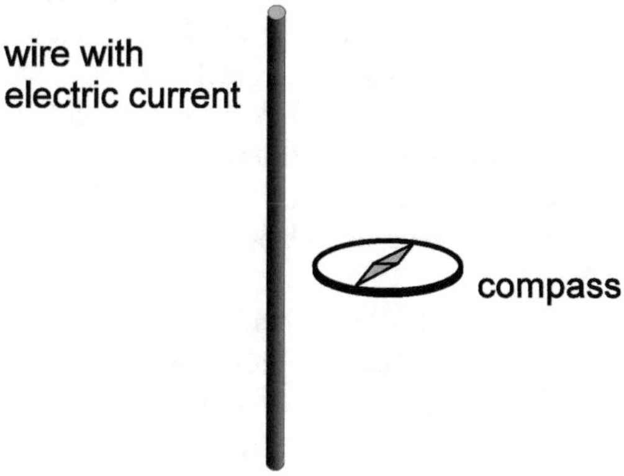

Fig. 5.15 Electric current exerting a force on a magnet

In equilibrium, the axis of each compass needle is tangent to the circle.

Note We will use the symbol ⊙ to indicate a direction *out of the page*. It represents the head of an arrow. We will use the symbol ⊗ to indicate a direction *into the page*. It represents the tail of an arrow.

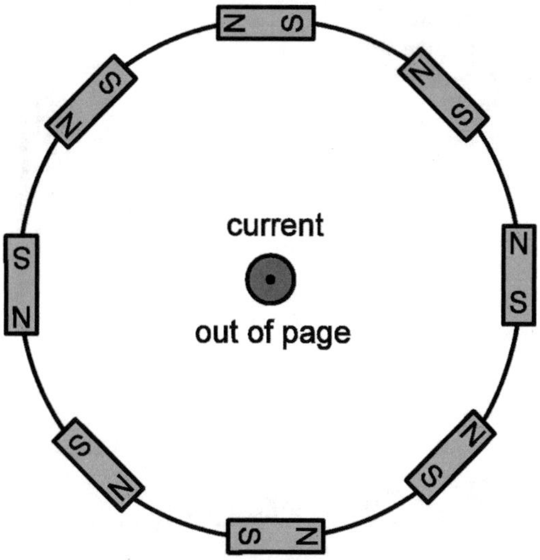

Fig. 5.16 Orientation of compasses around a current-carrying wire

We are inclined to conclude that in being able to orient a bar magnet, an electric current acts like a magnet.

The ability of a wire carrying an electric current to orient a magnet can be exhibited very nicely using iron filings. (These are elongated bits of iron, around a millimeter or less long.) In the figure below, iron filings have been distributed on the surface of a piece of cardboard. The cardboard is held horizontal (so that the filings will not fall off) and a wire carrying an electric current is passed through its center. The electric current magnetizes the filings, so that they provide us with a distribution of a huge number of bar magnets on the piece of cardboard. The alignment of the filings shows up clearly in Fig. 5.17.

The details of how specific shapes of wires carrying a current affect magnets were determined experimentally by André Ampère. See Fig. 5.18.

Another significant discovery by Ampere was that two wires carrying a current exert a force on each other. No magnet is required. We can appreciate this result better by examining further

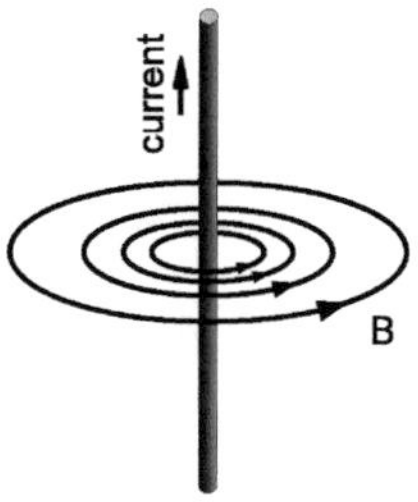

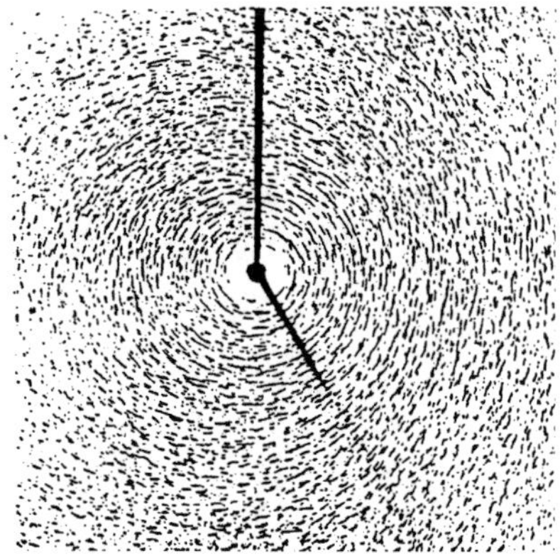

Fig. 5.17 Iron fillings aligning in the direction of the magnetic field

Fig. 5.18 André Marie Ampère (source: http://en.wikipedia.org/wiki/Andrel_Ampere)

Oersted's discovery that a current-carrying wire exerts a force on a magnet. We will see below that a magnet exerts a force on a current-carrying wire.

5.6.1 Action–Reaction Law and Force of Magnet on Current-Carrying Wire

We can learn something entirely new about magnets and electric currents by using **Newton's** 3rd**Law** of Dynamics. It is often referred to as the **Law of Action and Reaction** and states that

> a force of one body on a second body is always automatically accompanied by a force of the second body on the first.

This law is often difficult to comprehend; it is baffling and seems to contradict our intuition about forces. For example, if I push on a wall with a force of, say, ten lbs, the wall must be concurrently pushing on me with a force of ten lbs. Thus, as a further example, we arrive at the remarkable fact that it is impossible for a horse to push on me with a force that is greater than the force I exert on it! It is then reasonable to wonder how a live healthy horse can win a tug-of-war. To solve this dilemma, we turn to Fig. 5.19, wherein the forces on each of the participants are exhibited.

We note that both the horse and I experience forces by the ground: The ground pushes upward on the horse and on me, forces that overcome the downward force of gravity. The ground supports our weights. However, in addition, the ground provides a tangential force that tends to prevent us from slipping. Yet there is a limit to how much tangential force the ground can exert before we slip. While the force of the horse on me is exactly equal to the force I exert on the horse, the horse is capable of experiencing a much greater force of the ground on it than the force of the ground on me. As a result, I will slip before the horse does and I will lose the competition.

What do we learn from the above? According to **Newton's Third Law**,

> a magnet must also exert a force on a wire carrying a current!

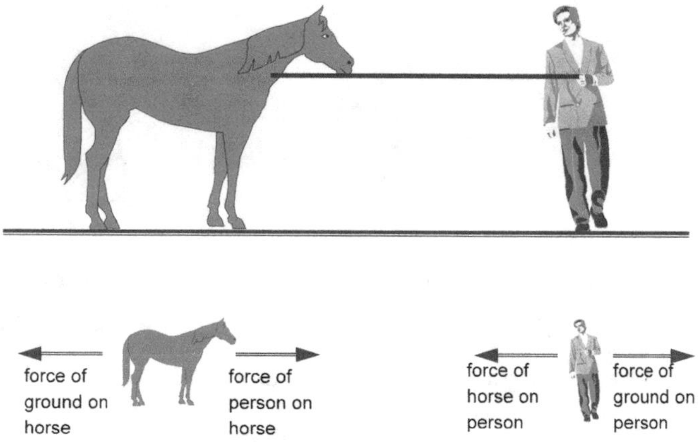

Fig. 5.19 Tug-of-war between a man and a horse. How does the horse win?

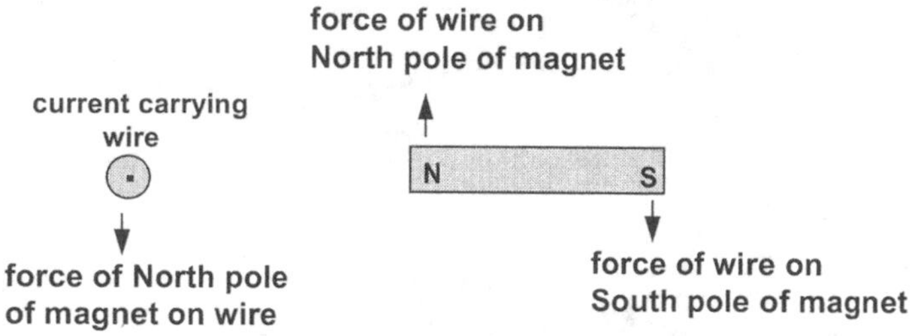

Fig. 5.20 Forces acting on a compass near a wire

A clarifying way to exhibit this second force is with the configuration in Fig. 5.20.

Note The force is perpendicular to both the direction of the current and the North–South axis of the bar magnet.

Now suppose that the direction of the current in the wire is reversed. Then:

1. The compass needles in Fig. 5.16 will tend to be oriented in the reverse direction as seen in Fig. 5.21.
2. The force of the wire on the poles of a bar magnet is reversed in direction (Fig. 5.22).

A very important configuration of electric current in a wire is the **solenoid**. This consists of a wire wound like a helix (Fig. 5.23).

Note that reversing the direction of the current *or* the orientation of the helix (*clockwise* or *counter clockwise*) exchanges the poles of the

Fig. 5.21 A compass aligns in the opposite direction when the current is flowing into the page

solenoid. For example, we might have a solenoid as seen in Fig. 5.24.

Henceforth, we will represent the solenoid by a set of parallel line segments as illustrated in Fig. 5.25.

Experiments show that a long solenoid behaves very much like a bar magnet, with the poles as indicated in the above figure. As such, it is often referred to as an **electromagnet**—a device that owes its magnetization to an imposed electric current: When the current is turned off, the magnetization vanishes. In contrast, a piece of lodestone or a piece of iron that has been magnetized by a magnet and has retained its magnetization even after the magnet is removed are called **permanent magnets**.

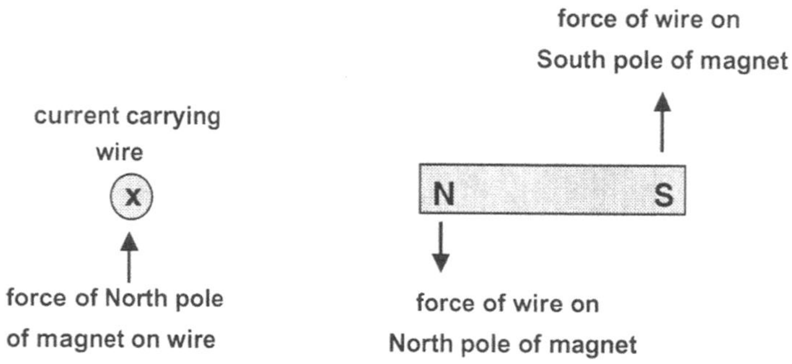

Fig. 5.22 Forces by a current-carrying wire on a bar magnet

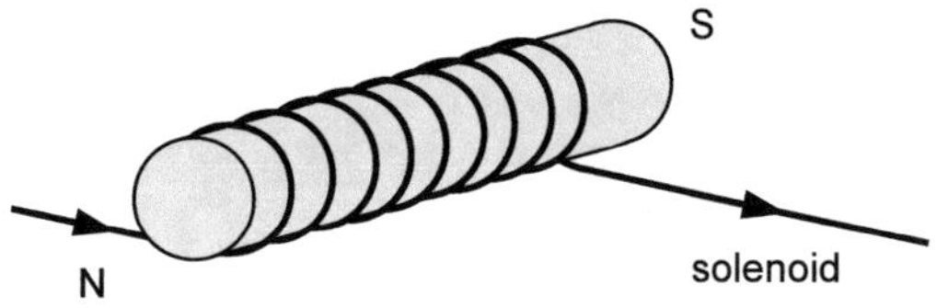

Fig. 5.23 A solenoid

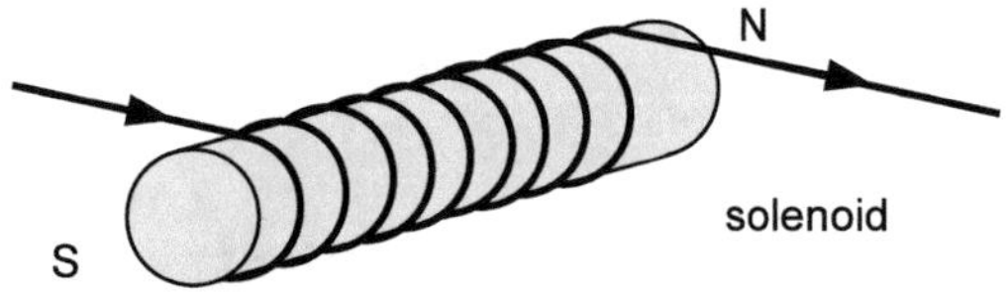

Fig. 5.24 A solenoid with reversed winding

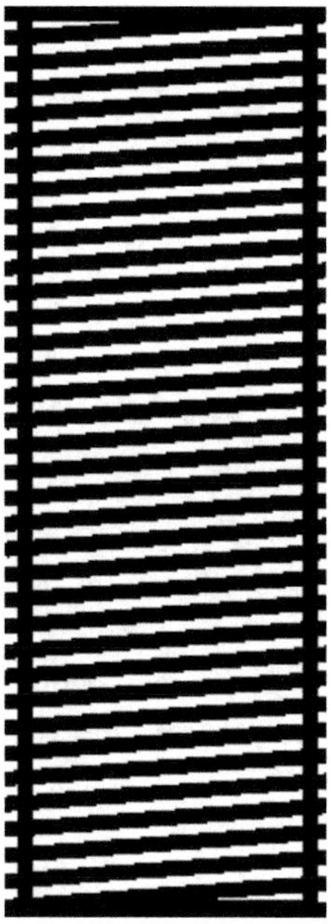

Fig. 5.25 Simplified figure of a solenoid

5.7 The Loudspeaker

The solenoid provides the basis for the operation of a **loudspeaker** (represented in Fig. 5.26) as follows: A solenoid is attached to the cone of the speaker. The sound pattern is fed into the solenoid by an electric signal from the amplifier. The amount of displacement of the solenoid and cone depends upon a balance between the force of the neighboring permanent magnet on the solenoid and the force of a spring (not shown in the figure) which tend to pull the solenoid towards an equilibrium position.

The strength of the magnetic force on a solenoid can be increased greatly by inserting a cylinder of iron into the core of the solenoid. Such a unit is used in buzzers, bells, and the telegraph apparatus.

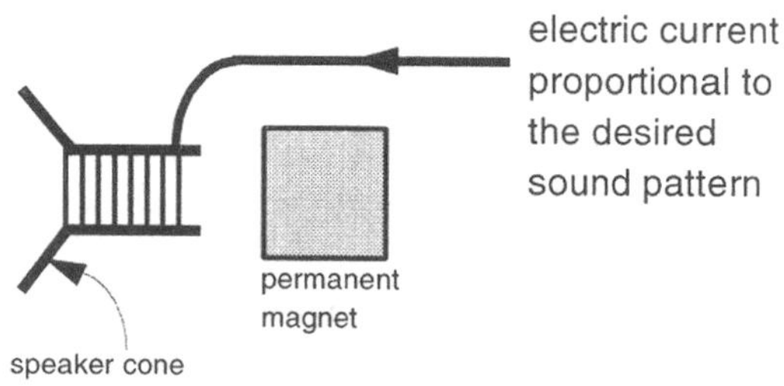

Fig. 5.26 A schematic of a loudspeaker

5.8 The Buzzer

In Fig. 5.27 we depict a **buzzer**. When the switch is closed, an electric current flows through the solenoid and attracts the iron plate.

Once the iron plate loses contact with the pointer, electric current ceases to flow through the solenoid and the iron plate drops back into its original position, thus allowing current to once again flow through the solenoid. This cycle is repeated at a high frequency, thus producing the sound of the buzzer. If the plate is attached to a hammer that can strike a bell shaped piece of metal. The buzzing sound is replaced by the sound of a ringing bell.

5.9 The Electric Motor

The "**electric motor**" also makes use of the interaction between a permanent magnet and a solenoid electromagnet (Fig. 5.28).

A coil of wire is wound around a cylinder and placed between the poles of a permanent magnet. An electric current is caused to flow through the coil, which acts like a solenoid. The magnetic force on the coil causes the cylinder to rotate so that the "poles" of the solenoid line up with those of the permanent magnet. But by the time the poles line up, the cylinder has "rotational inertia," so that it cannot come to a dead halt. Instead, it moves on. As so far described, there would be a force which would tend to reverse the sense of rotation of the solenoid. Instead, the so-called brushes of the motor cause the current in the coil to reverse its direction, so that the poles are oppositely aligned and the cylinder is caused to rotate further in the same sense. This sequence is repeated again and again as the motor rotates.

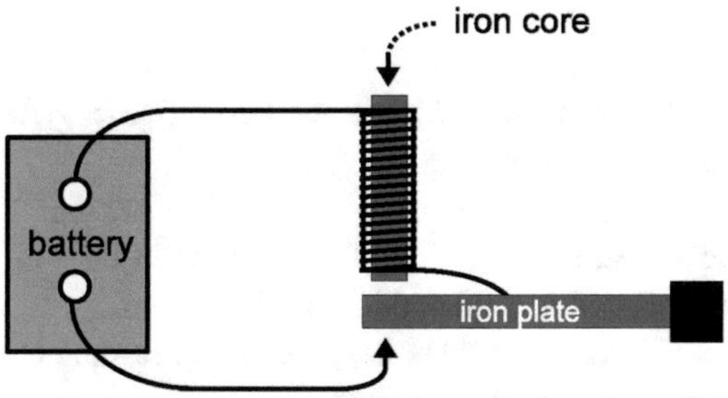

Fig. 5.27 A schematic of a buzzer

5.10 Force Between Two Wires Carrying an Electric Current

What happens if two wires carrying an electric current are in each other's neighborhood?

They exert forces on each other too! The details were explored by Ampère in 1820, within months of Oersted's discovery. The experiment is depicted in Fig. 5.29. We see that if two parallel wires carry respective currents in the same direction, the wires **attract** each other. [Like attract, in contrast to electric charges!] If the currents are anti-parallel, the wires repel each other, as you would expect.

The force between the two wires is not an electric force since the wires are electrically neutral. Instead, physicists concluded that electric currents behave magnetically, not only with respect to their interaction with permanent magnets, but also with respect to their interaction with each other. We do not need permanent

Fig. 5.28 Electric motor

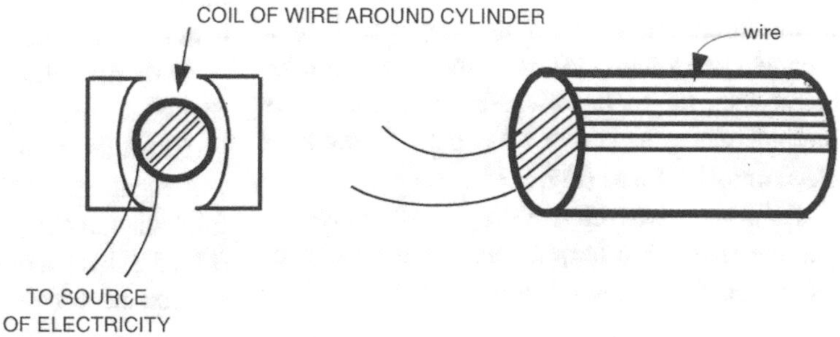

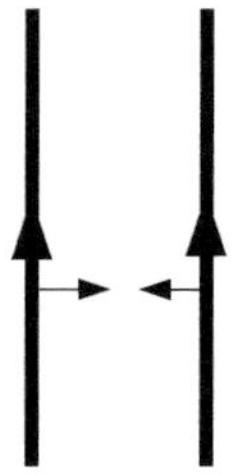
PARALLEL WIRES WITH
CURRENTS IN THE SAME
DIRECTION ATTRACT

Fig. 5.29 Force between two current-carrying wires

magnets to observe magnetic forces. Eventually, experimental and theoretical studies revealed that even permanent magnets owe their magnetism to small, microscopic electric current loops made by electrons within the material. The conclusion was that

All magnetic phenomena are due to electric currents.

5.11 The Electromagnetic Force and Michael Faraday

The next major advance in electricity and magnetism was made by **Michael Faraday**; see Fig. 5.30. In the 1830s through his discovery of the **induced electromotive force** (EMF). Generally speaking, an **electromotive force** refers to a means whereby electric charges are given an electric force which enables them to move through a material against the presence of internal friction, called **electrical resistance**. Faraday's discovery led to the technologically revolutionary source of EMF called the "**electric generator**."

Prior to Faraday's discovery, electric currents were produced by attaching wires to an electric **battery** (called a "pile" in England and France, albeit with different pronunciations). A battery was made by putting together a pile of discs of dissimilar metals such as copper and zinc, arranged like a sandwich, with sheets of insulator in between the metal discs. In driving currents through wires, a battery is said to produce an **electromotive force** (Fig. 5.31).

Faraday discovered that if one moves a magnet through a loop of metal wire, an electric current will flow around the loop.

Fig. 5.30 Michael Faraday (source: http://en.wikipedia.org/wiki/Michael_Faraday)

An **induced EMF** is said to drive current around the loop (Fig. 5.32).

If the direction of motion of the magnet is reversed, the direction of the current in the loop is reversed as depicted in Fig. 5.33.

What do you think happens if you reverse the orientation of the magnet?

Suppose now that the *magnet is stationary and the loop is moved to the left* (Fig. 5.34). The relative motion is the same as in Fig. 5.32. The result is the same.

Only the relative motion of the magnet with respect to the loop is relevant.

Given the above observations, electric current was eventually identified as constituting moving fundamental charges, such as electrons in a metal (Fig. 5.35b). Oersted's observation that a wire with an electric current experiences a force in the

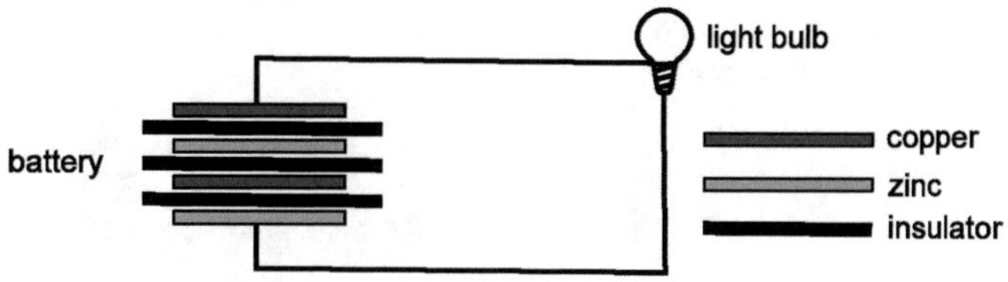

Fig. 5.31 A battery lighting up a light bulb

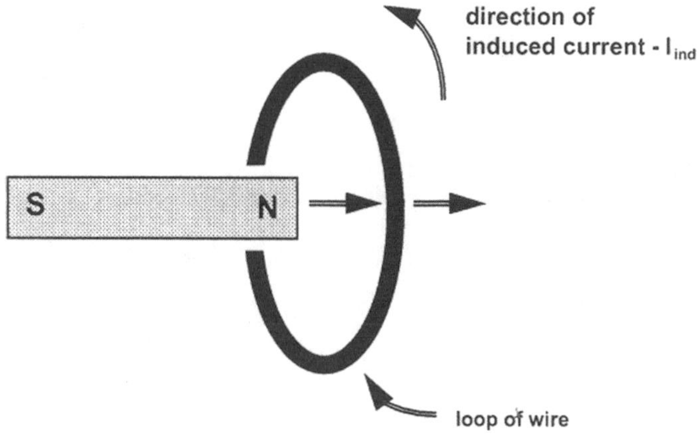

Fig. 5.32 Faraday induction of a current in a metal loop by a moving magnet

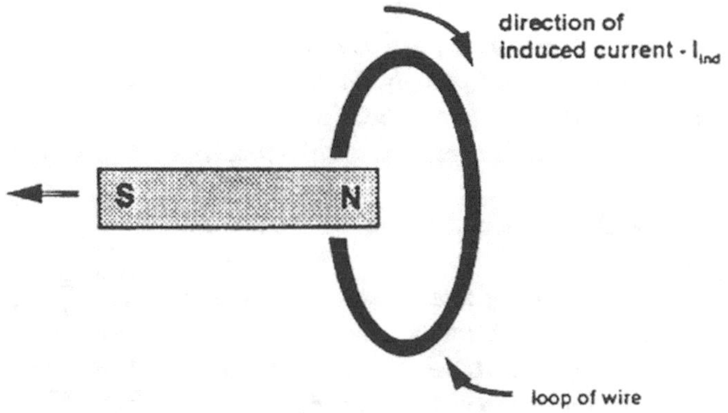

Fig. 5.33 Reversing the direction of motion of the magnet reverses the current direction

presence of a magnet can then be understood on the basis of the principle that:

A moving electric charge experiences a magnetic force in the presence of a magnet.

The fact that an electric current is induced to flow in a wire that is moving in the presence of a magnet can be understood on the basis of this same principle since the wire has electric charges that experience a magnetic force. This observation accounts for the current induced in the loop of Fig. 5.34.

Consider a straight metal wire that is being pulled to the right so as to move between the poles of a magnet as shown in Fig. 5.35a. There will be an induced current flowing upward. We see in the microscopic view presented in Fig. 5.35b that the free electrons experience a **downward** force, perpendicular to the direction of motion of the wire, and move downward so as to produce the upward current. On the other hand, the positive ions experience an upward force but remain close to their equilibrium positions and do not contribute to the current.

However: *From the viewpoint of a second observer who sees the wire stationary and the magnet moving, there is no moving charge and hence no magnetic force. This second observer describes the situation in terms of a changing mag-*

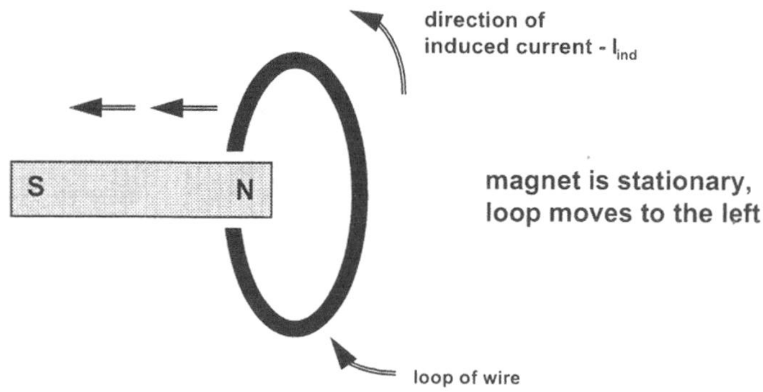

Fig. 5.34 A moving loop with a stationary magnet

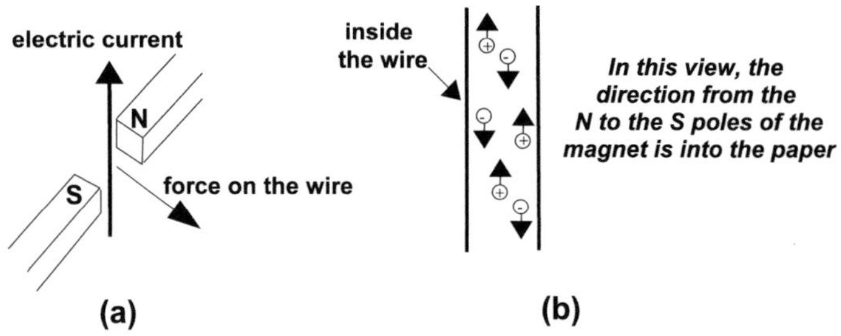

Fig. 5.35 Force on a current-carrying wire by a magnet

netic field because the second observer observes a moving magnet. Accordingly, the charges in the wire experience an electric field, and therefore an ***electric force****, in accordance with Faraday's Law.*[2]

In the above example, the basic physical principle that accounts for the EMF depends upon the state of motion of the observer with respect to the wire and the magnet. In one case the basis is an electric force, while in the other case the basis is a magnetic force. (For an observer with respect to whom both the wire and the magnet are moving, both a magnetic force and an electric force must be used as a basis!)

Stated differently, the question as to whether a magnetic force and/or an electric force is present is meaningless in the absence of a specification of an observer along with a specification of the state of motion of that observer. We will refer to this aspect of observation as the **Relativity of Description**.[3]

[2]The reader who is extremely probing of this process will note that even for the second observer, the electrons in the wire are moving in association with the induced current. As a consequence, the second observer (as well as the first observer) predicts that there will be an additional force—a magnetic force for both observers. This force on the electrons turns out to be directed to the left. As a result, an external force must be applied on the wire in order to prevent it from acquiring a motion to the left.

[3]**This is an application of the Principle of Relativity.** We experience one of its consequences when we sit in a subway train and watch a second train moving relative to us while having poor visibility of any other objects such as a train station. We wonder whether it is our train or the second train that is moving with respect to the tracks. For another common example of this principle, imagine yourself in a car stopped at a red traffic light on an upgrade. You see beside yourself a second car that is slowly moving forward. You then check whether your brake pedal is securely pressed because you worry whether it is in fact your car that is slowly moving backward. In this situation, until you discern the state of motion of the road or some other objects beside the road *relative to yourself*, you are finding it difficult to decide which of the two cars is actually moving with respect to the road. Now imagine yourself in a spaceship in outer space. You look out the window and see a second spaceship moving past you. Which of the two spaceships is moving, you might ask. Such a question has no answer. You can say that the second spaceship is moving with respect to yours or vice versa. Or, you might decide to investigate the state of motion of both spaceships with respect to the earth and find that both are moving with respect to the earth! It is clear that the state of motion, that is, the velocity of an object is a relative one; it depends upon the observer.

We see then that Faraday's Induced EMF follows from Oersted's observations.

We should note another interesting aspect of the above system: The electrons must overcome electrical resistance as they flow through the wire. This resistance produces thermal energy. Where does this energy come from? Analysis reveals that in order to maintain the motion of the wire, a force must be exerted on the wire. This force provides the necessary energy.

5.12 Applications of Faraday's EMF

The **microphone** is essentially a loudspeaker in reverse (Fig. 5.36).

A diaphragm is attached to a solenoid. Sound waves impinge on the diaphragm, setting the diaphragm and solenoid in motion. A neighboring magnet induces an EMF in the solenoid which is passed on to an amplifier. An electric generator signal from the amplifier can be used to drive a speaker or make a recording.

An **electric generator** is essentially a motor in reverse (Fig. 5.37).

A cylinder has a coil of wire wound around it. The cylinder is rotated by an external force—say a waterfall or steam engine. The presence of the poles of a permanent magnet produces an EMF in the coil, which is used to run electrical devices. Electric companies use huge generators to "produce electricity." "Producing electricity" refers, in fact, to providing electric power needed to maintain the EMF used by devices hooked up to the companies electric lines.

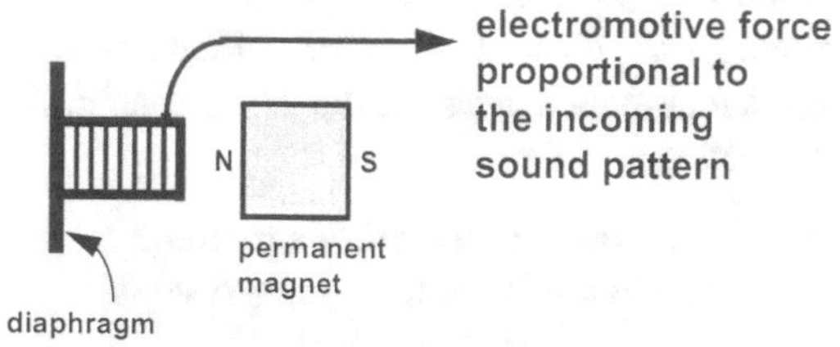

Fig. 5.36 A microphone

5.13 A Final "Twist"

We have realized that permanent magnets can be replaced by wires carrying a current; notably, a solenoid with a current behaves like a magnet. Thus, we can produce an EMF in a coil by moving a solenoid relative to the coil. Basically, the source of the EMF can be regarded as simply a magnetic force.

Consider the following alternative: Instead of moving the solenoid, *change the current in the solenoid*. It turns out that this change will also produce an EMF in the coil!

In the figure below, we depict two wire loops close to each other. The left loop is connected to a switch and a battery. The second loop is connected to a light bulb. When the switch is closed the light bulb lights up briefly (Fig. 5.38).

What has happened is that initially there was no current in the first loop. After the switch is closed, the current in the first loop changes to some stationary value, albeit usually over a very short period of time, perhaps a hundredth of a second. *It is only during this short period of changing current that the second coil experiences an EMF*. In the graphs, we depict the variation in time of both the current in the solenoid and the induced EMF in the adjacent loop. The current does not rise instantaneously once the switch is closed. In this example, it is seen to take about 4 ms (milli-seconds) to reach its final value. Also, notice that there is an induced EMF only when the current in the solenoid is changing in time,

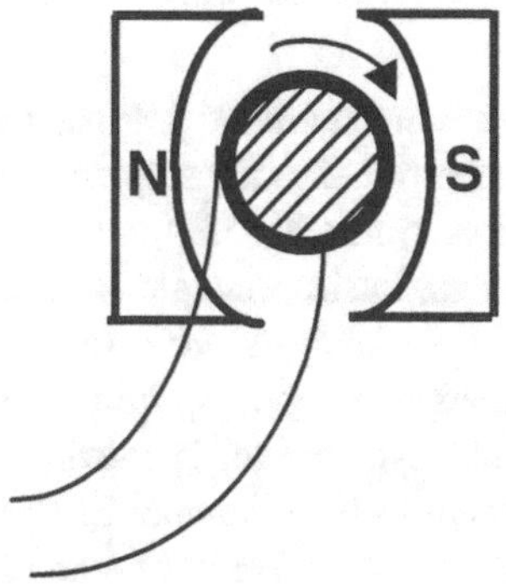

Fig. 5.37 An electric generator using an electromotive force

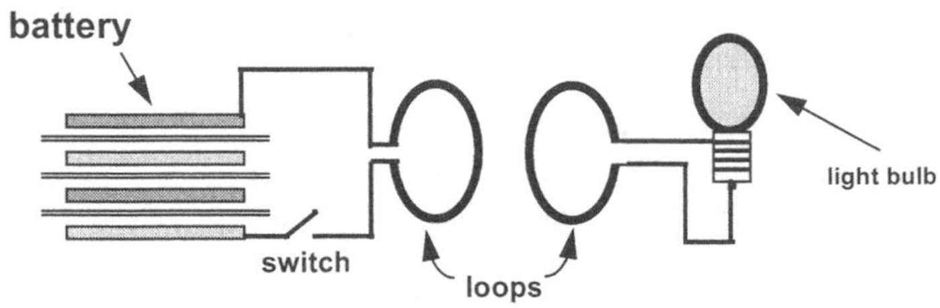

Fig. 5.38 Induction by a wire loop on another wire loop

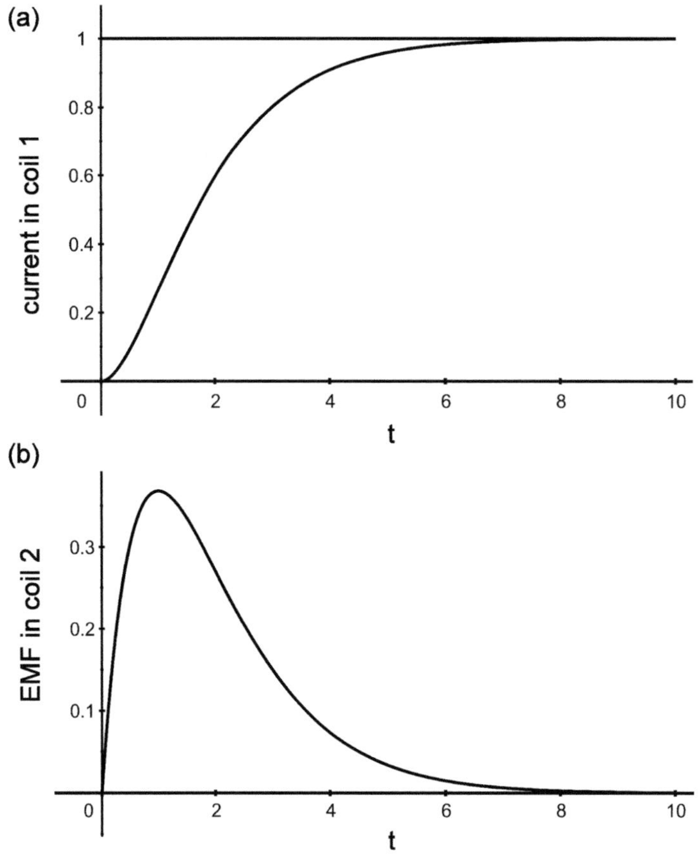

Fig. 5.39 Changing current in one coil induces an EMF in the second coil

with a peak value at around 3/4 ms. The induced EMF is proportional to the slope of the graph of the current vs. time (Fig. 5.39).

What is the basic principle behind this phenomenon? It certainly is *not* a **magnetic force**, since no charges are moving initially in the second solenoid. No magnet is moving, so that it does not seem to be a Faraday EMF of the sort we introduced earlier. We have noted that, in the presence of a solenoid carrying a current, we can produce an EMF in a second solenoid *either* by moving one solenoid with respect to the other *or* by changing the current in the first solenoid. An interesting question is:

Are these two methods related?
If so, what is the unifying principle behind them?

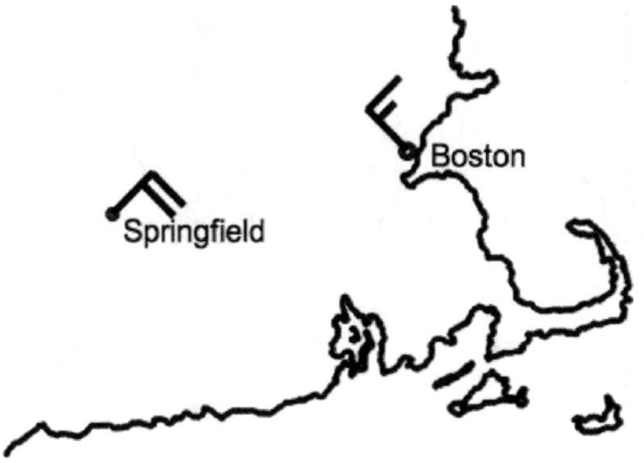

Fig. 5.40 Wind velocity map

5.14 Action-at-a-Distance and Faraday's Fields

In an effort to explain how electric and magnetic forces can act at a distance, Faraday proposed the existence of an "**electric field**" and a "**magnetic field**."

What is a **field**? To answer this question we will cite some familiar examples. Weather reports provide us with the value of temperature, pressure, and wind velocity at various points on a map. All *three* parameters are "fields." The first two are specified by numbers alone, such as 20°C or 30°C for temperature, and 29 inches mercury or 31 inches mercury for pressure. Wind velocity is specified by a *direction* as well as a *number*, and is an example of a "**vector field**." As we will shortly see, both electric and magnetic fields are vector fields.

Below is an example of the symbol used on weather maps to provide both direction and magnitude for wind velocity (Fig. 5.40).

Meaning of the symbols: The wind velocity at Boston's Logan Airport is 15 knots≈17 mph NW. The wind velocity in Springfield Center is 20 knots≈23 mph NE.

5.15 The Electric Field

According to Faraday, an electric charge is accompanied by an **electric field** that is present throughout space. Below we depict a *positive point charge* with its electric field. See Fig. 5.41.

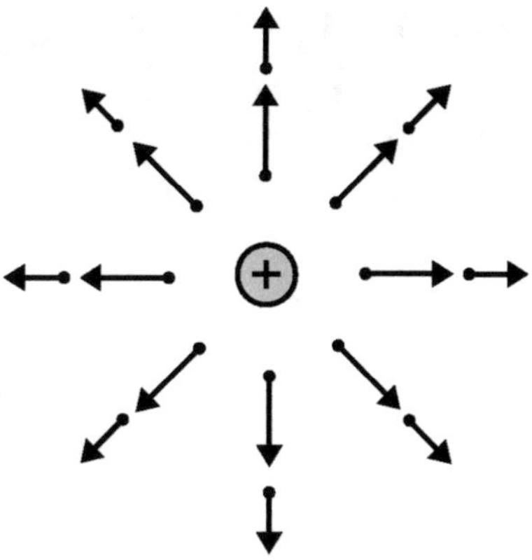

Fig. 5.41 Electric field of positive charge

The direction of the field at the base (the heavy dot) of the arrow is indicated by the direction of the arrow. All arrows point directly away from the point charge. The *length* of an arrow is proportional to the *magnitude* of the field at the corresponding point. We note that the magnitude decreases with increasing distance from the point charge and is the same for equal distances. For a *negative* point charge, the arrowspoint *towards* the charge.

We will represent the electric field by the symbol E. An alternative way to represent the electric field on a map is to use **continuous electric field lines**. See Fig. 5.42.

Here, it is the relative closeness of the field lines that indicate the magnitude of the electric field at various locations in space. Remember that in fact a charge exists in three-dimensional space, so that the figure above shows the field lines in a plane running through three-dimensional space.

While the *magnitude* of the electric field in some region of space is proportional to the density of the field lines (how close together the lines are), the *direction* of the field at a point on a field line is along the *tangent* to the field line at that point. See Fig. 5.43, where the field E at point P is indicated.

If there are many charges present, the total electric field in space is a superposition, that is, sum, of the contributions of each charge taken separately.

Below are examples of the electric fields of interesting charge configurations.

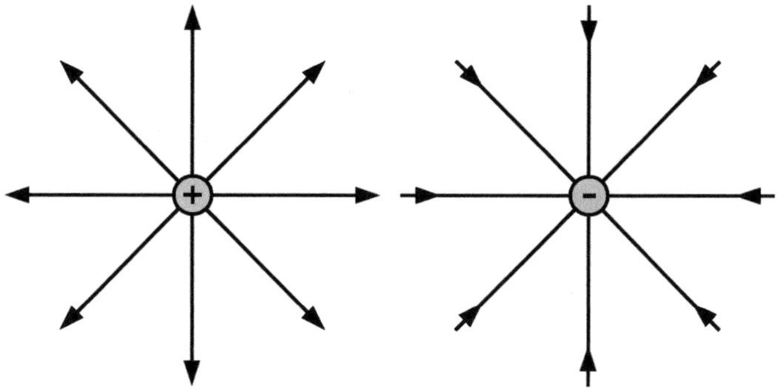

Fig. 5.42 Electric field lines of point charges—a positive charge on the left and a negative charge on the right

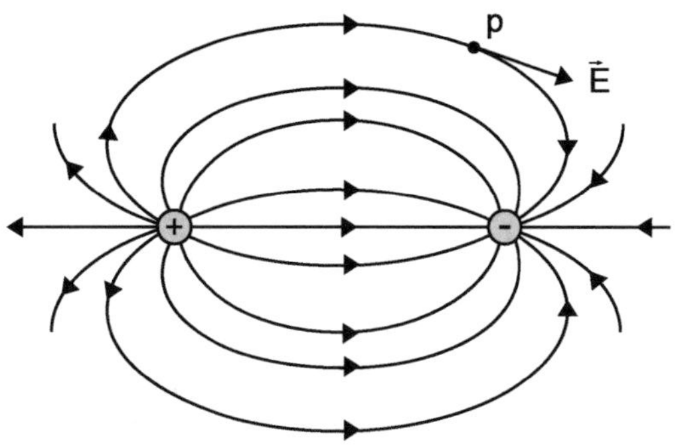

Fig. 5.43 Electric dipole field

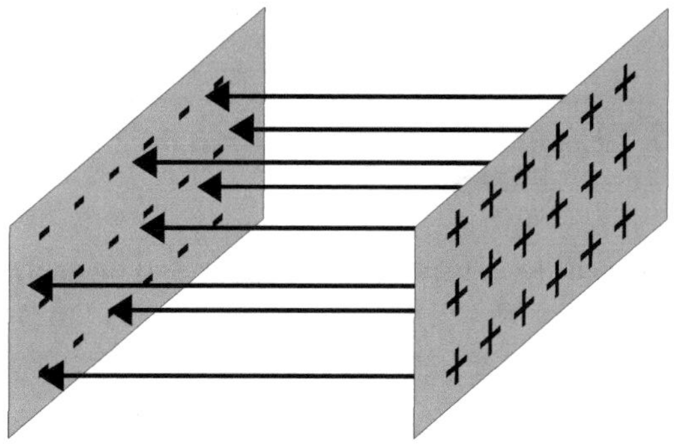

Fig. 5.45 Electric field of two charged sheets

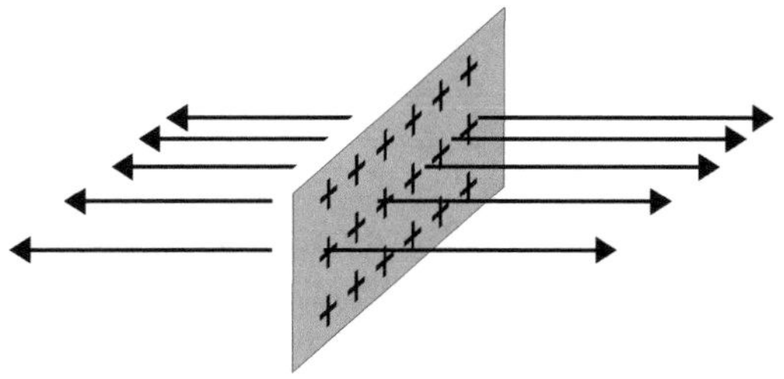

Fig. 5.44 Electric field of a sheet of charge

1. A positive and a negative point charge, side by side—called a "**electric dipole**." See Fig. 5.43.
2. An infinite sheet with a uniform distribution of positive charge. See Fig. 5.44.
3. Two infinite sheets with positive and negative uniform charge distributions, respectively. The electric field is confined between the sheets. See Fig. 5.45.

What is the significance of the electric field? The field represents the *potential* of the charges associated with it to exert a force on an additional charge, say q, placed in the field. Thus, the force $\vec{F}$ (note the arrow on the symbol, since force is a vector) on q due to the changes producing the field $\vec{E}$ is given by:

$$\vec{F} = q\vec{E} \tag{5.1}$$

where $\vec{E}$ is the electric field at the location of the charge q.

Comment: The electric field of a charge may be likened to the halo drawn by artists above a saintly person or a hero in order to indicate the potential of the person to influence others in a spiritual or holy way.

The electric field can be regarded as a modification of the space between two charges, and thus deals with the philosophical action-at-a-distance issue: The electric force between two charges is mediated by the electric field. But is the electric field a *real* thing? What properties would give it reality? In fact, physics properly cannot answer such a question:

The essential goal of physics is to establish a theoretical framework for describing in a quantitative way what we decide to and are able to measure. That framework makes use of models, concepts, and images. However, its ultimate content is a set of mathematical equations, which we call laws. The laws are as simple and all-encompassing as possible, and provide relationships among measurable quantities.

5.16 The Magnetic Field

Now we turn to **magnetic phenomena**: A magnet or a current-carrying wire is understood to fill space with a magnetic field, which has a specific magnitude and direction at every point in space. We will represent the magnetic field by the symbol $\vec{B}$. There exists a prescription for determining the magnetic field for a given permanent magnet or a current-carrying wire—a prescription which is beyond this course. Below, we have sketched the magnetic field for a number of cases.

1. Bar magnet as seen in Fig. 5.46.

   Note that the magnetic field lines pass through the bar magnet itself. Also note the similarity of the pattern with the electric field lines of an electric dipole.

2. An infinite straight wire with current. Here current is directed out of the paper. See Fig. 5.47.

   Note The relation between the magnetic field and the current has a complex form and is referred to as **Ampère's Law** in honor of its discoverer, who was mentioned earlier in this chapter. Essentially, **the magnetic field $\vec{B}$ is proportional to the current I that produces it**. Thus

$$\vec{B} \propto I \tag{5.2}$$

3. A long solenoid with tightly wound coils. Compare the field line configuration with that of a bar magnet. See Fig. 5.48.
4. A horseshoe magnet. See Fig. 5.49. Here we have drawn only the field outside the magnet. Noteworthy is the closeness of the magnetic lines between the poles: The field is most intense in this region. Also, between the poles the field lines tend to be straight.

Note: Generally, in contrast with the electric field lines described so far, magnetic field lines are "closed"—that is, they have no beginning or end. This fact can be shown to be connected with their being produced by moving electric

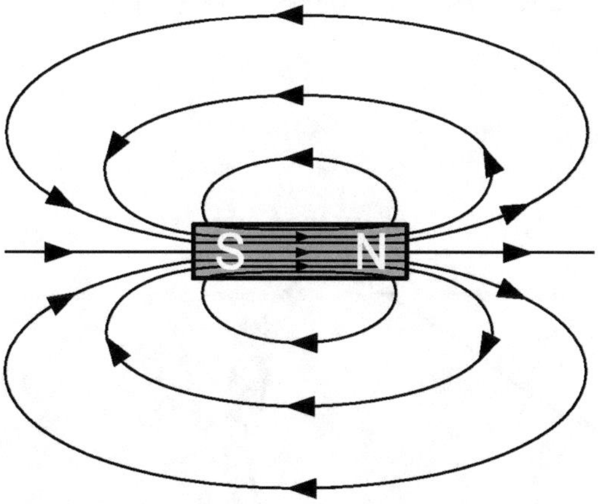

Fig. 5.46 Bar magnet

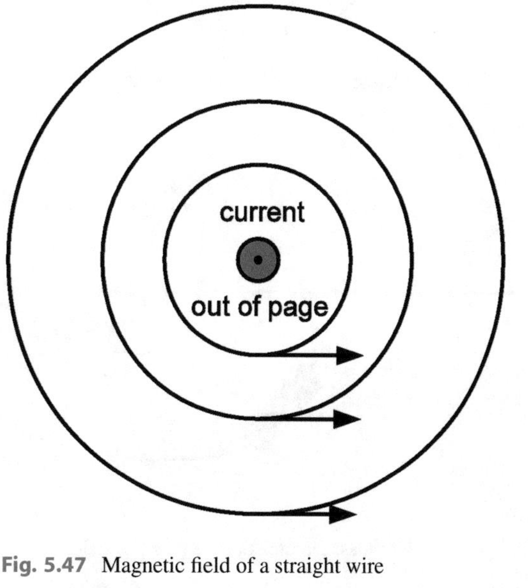

Fig. 5.47 Magnetic field of a straight wire

charge, rather than by what physicists refer to as "magnetic charges." It is also connected with the fact that when a permanent magnet is split in two, we end up with two whole magnets, both with north and south poles. (See Fig. 5.12.) Later we will discuss electric fields that have closed field lines.

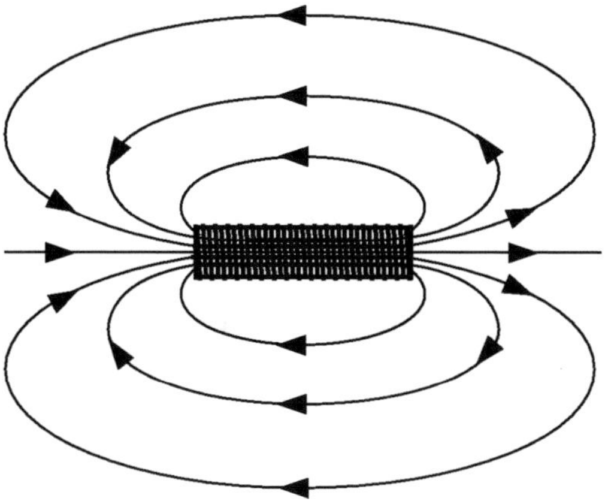

Fig. 5.48 Magnetic field of a solenoid

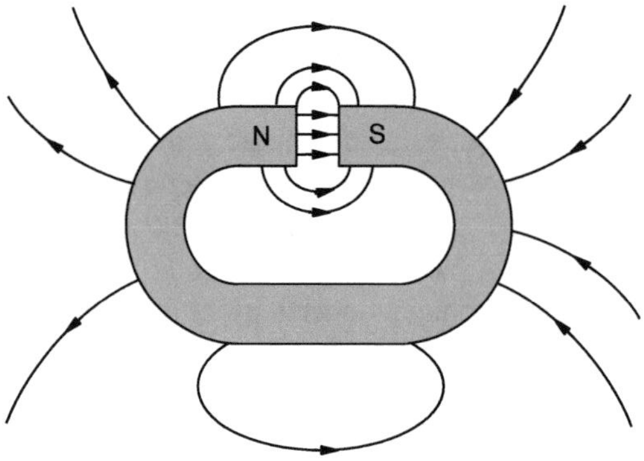

Fig. 5.49 Magnetic field of a horseshoe magnet

How can we determine the direction of the magnetic field at some point in space? The answer is simple. A bar magnet experiences a torque (twisting force) in the presence of a magnetic field which lends it to line up with its *South to North direction* parallel with the direction of the magnetic field. Thus, the bar magnet tends to line up tangent with a magnetic field line. We can therefore use compass needles to determine the direction of the magnetic field. Below, in Fig. 5.50, we depict the orientation of several compass needles in a given magnetic field.

5.17 Magnetic Force on a Moving Charge

Point charge: Generally, the force is perpendicular to both the velocity $\vec{v}$ of the charge and the magnetic field B at the location of the charge. Also, **the force vanishes if the velocity is parallel to the magnetic field $\vec{B}$.**

Below, in Fig. 5.51, we depict the force on a positive charge with various velocity directions with respect to a magnetic field. (If the charge is negative, the direction of the force is reversed.) Note that the force $\vec{F}$ is perpendicular to the plane determined by $\vec{v}$ and $\vec{B}$.

For the simple case that the velocity is perpendicular to the magnetic field, that is, $\vec{v} \perp \vec{B}$:

$$F = qvB \tag{5.3}$$

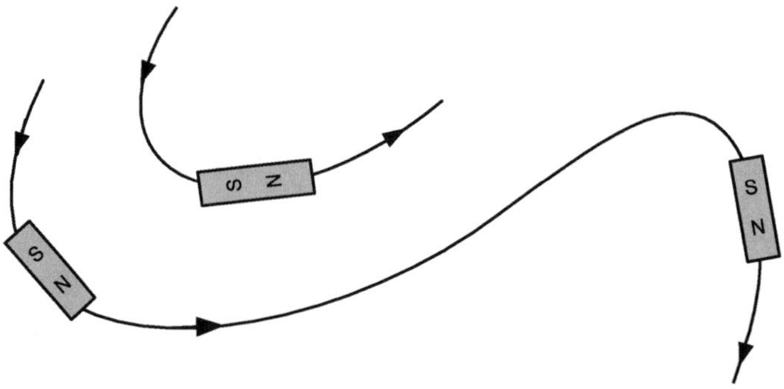

Fig. 5.50 Compass needles line up with the field

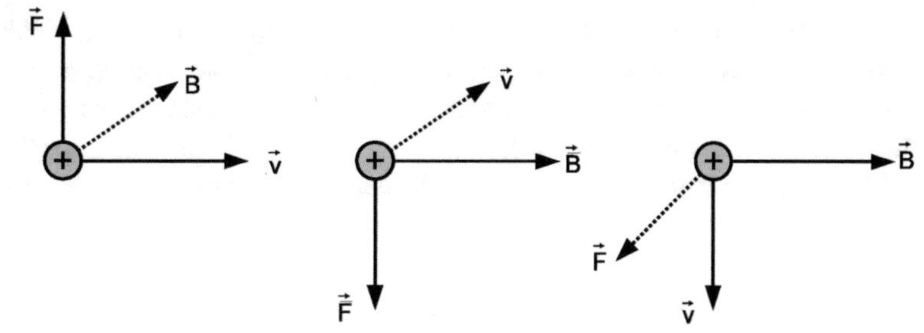

Fig. 5.51 Magnetic force on a point charge

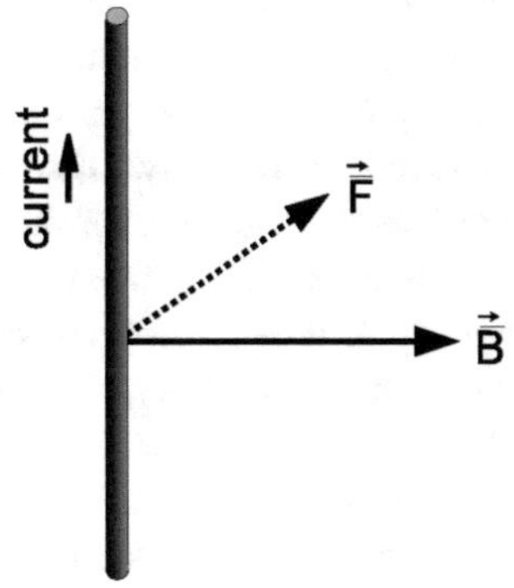

Fig. 5.52 Magnetic force on a wire

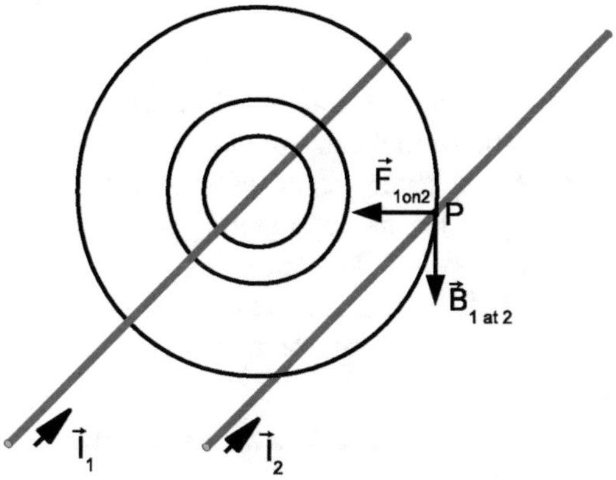

Fig. 5.53 Magnetic force between two long current-carrying wires

In Fig. 5.52 we depict a wire that is carrying a current in the presence of a magnetic field that is perpendicular to the wire. (Note that a current in the upward direction can be produced by positive charges that are moving upward **and/or** negative charges moving downward.)

5.18 Force Between Two Parallel Wires Carrying Currents

Each wire produces a magnetic field which, in turn, accounts for a magnetic force on the other wire. With some rather painful analysis it can be shown that

- the wires **attract** each other if the currents are in the **same direction**;
- the wires **repel** each other if the currents are in **opposite directions**.

The Analysis: In Fig. 5.53 we depict two wires that are carrying currents in the same direction. Let us label the wires #1 and #2, respectively. To find the force of wire #1 on wire #2, we need to find the magnetic field $\vec{B}_1$ due to wire #1 at the position of wire #2. We will label this field $\vec{B}_{1 \text{ at } 2}$. Now let us focus our attention on point P on wire #2.

From the direction of $B_{1\text{at}2}$ and of I_2 we can determine the force $F_{1 \text{ on } 2}$ of wire #1 on wire #2 as being to the left. Thus we see that wire #2 is **attracted** by wire #1.

5.19 Generalized Faraday's Law

Recall that an electromotive force is produced when a wire is moving in the presence of a magnet—whether it be permanent or otherwise. This result can be understood as being associated with a magnetic force on an electric charge which is moving in the presence of a magnetic field. However, the EMF produced in a stationary wire, in the presence of a moving magnet **or** in the presence of a second wire which carries a current which changes in time, cannot be attributed to a magnetic force. In this case, a new principle is needed. This principle is embodied in **Faraday's Law**, which can expressed as follows:

Any change in a magnetic field with respect to time is accompanied by an induced electric field.

In more concrete mathematical terms:

$$\vec{E} \propto -\frac{\Delta B}{\Delta t} \tag{5.4}$$

That is, there is an induced electric field $\vec{E}$ that is proportional to the rate of change (symbolized by the Greek letters Δ) of the magnetic field $\vec{B}$ with respect to time. *The reason for inserting a minus sign will be clear later.*

This induced electric field can drive electric charges through a wire that may be present. Thus, in these cases,

*the EMF is attributed to an **electric** force!*

A number of very important observations are in order:

1. We have seen that a current-carrying wire produces a magnetic field. Thus, it is absolutely clear that changing the current will change the magnetic field and hence produce an electric field.

   Now, suppose a magnet is moving relative to me at some constant velocity v. The magnetic field at some fixed location will change, so I will observe the presence of an electric field. However, suppose you move *with* the magnet, at the same velocity v with respect to me, so that it is *motionless relative to you*. You will therefore observe no change in the magnetic field at any of your fixed locations and so *you will not observe an electric field*!

   Thus, the observation of an induced electric field—even the very question of its existence—depends upon the state of motion of the observer. It is not appropriate to ask whether there is an induced electric field. All we can say is that there is an EMF as far as all observers are concerned. How this EMF is accounted for depends upon the state of motion of the observer.

2. Consider again the situation when a wire loop and magnet are moving with respect to each other. There will be a current induced in the loop with a magnitude which is usually very close to being independent of the state of motion of the observer.[4]

   Now consider two specific observers: One is at rest with respect to the magnet and observes a moving loop, while the second observer is at rest with respect to the loop and observes a moving magnet. Both observe an induced current. However, they account for the induced current in two different ways: The **first observer** accounts for the current in terms of a **magnetic force** due to the charges in the wire loop moving in a **constant magnetic field**. For this observer, there is **no electric field**. On the other hand, the **second observer** accounts for the current in terms of an **electric force** due to an electric field that is brought about by the moving loop leading to a **changing magnetic field**.

3. We see that the question of the existence of an induced electric field depends upon the state of motion of the observer. So it is with a magnetic field. Consider, for example, that someone who observes an electric charge moving with a constant velocity v will perceive the presence of a magnetic field. Someone else, moving at the same velocity v as the charge, does not observe a moving charge and hence must account for the consequent observations without the presence of a magnetic field from the charge.

   You might feel upset that the question of the existence of an electric or a magnetic field can depend upon the state of the observer. However, we all have to deal with this apparent dilemma in asking the question as to whether an object at rest on the surface of the earth has kinetic energy. With respect to an observer at rest on the earth, the object has no kinetic energy. However, a person at rest with

[4]Einstein' Theory of Special Relativity does predict a dependence of the current on the state of motion of the observer. The dependence is small when velocities are much less than the speed of light in vacuum (3×10^8 m/s).

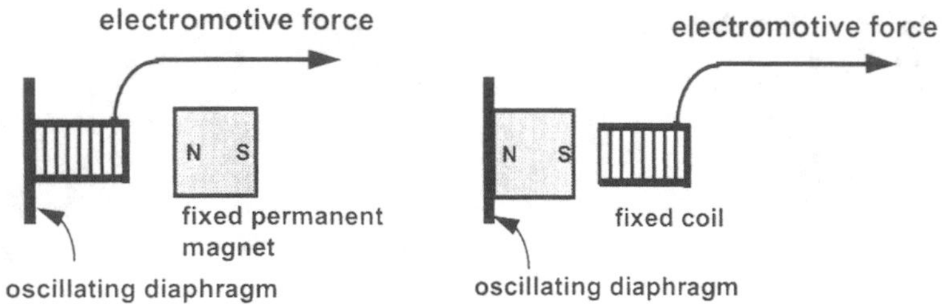

Fig. 5.54 Two designs for a microphone

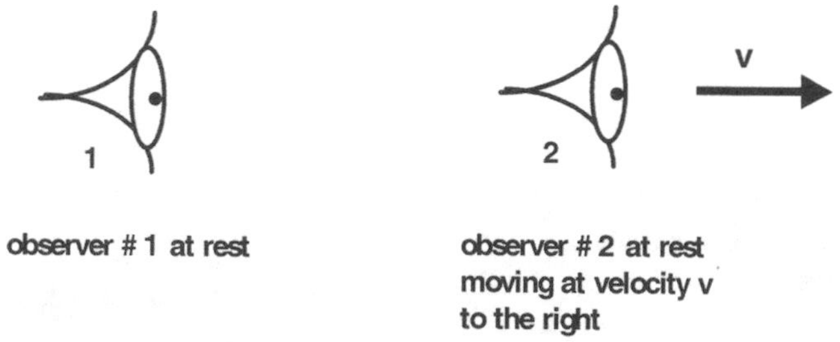

Fig. 5.55 Symbols denoting the state of motion of an observer

respect to the sun would say that the object is moving around the sun along with the earth and has kinetic energy. Thus, the amount of kinetic energy that an object has depends upon the observer.

4. The two ways of producing an EMF are used in two designs for a microphone depicted below. Both designs work! Microphone (1) works on the basis of a *magnetic* force on charges in the solenoid. Microphone (2) works on the basis of a *electric* force on charges in the solenoid (Fig. 5.54).
5. Let the schematic drawings of an eyeball in Fig. 5.55 represent the state of motion of an observer.

In Figs. 5.56 and 5.57 we represent two different situations that we have just discussed in pictorial form.

Case I: A permanent magnet is in the presence of various observers (Fig. 5.56). How these observers account for their observations depends upon the relative velocity of the observer and the magnet.

Observer #1 accounts for his/her observations in terms of both a magnetic field and an electric field $\vec{E}_1$. The magnet produces a magnetic field by virtue of its being a magnet. It produces an electric field because of its motion with respect to the observer.

Observer #2 accounts for his/her observations in terms of a magnetic field alone since the observer and the magnet are moving at the same velocity with respect to the paper and hence have no relative velocity.

Observer #3 accounts for his/her observations in terms of both a magnetic field and an electric field $\vec{E}_3$ which is different from $\vec{E}_1$. Note that the velocity v' is at an angle with respect to the velocity v.

Case II: An electric charge is in the presence of various observers. How these observers account for their observations depends upon the relative velocity of the observer and the charge (Fig. 5.57).

Observer #1 accounts for his/her observations in terms of an electric field alone: Both charges are at rest with respect to this observer. Thus, according to this observer, they exert only an electric force upon each other.

Observer #2 accounts for his observations in terms of both an electric field and a magnetic field. There are both an electric and a magnetic field due to the upper charge. Therefore, the lower moving charge experiences both an electric force and a magnetic force. One can reverse the roles of the two charges.

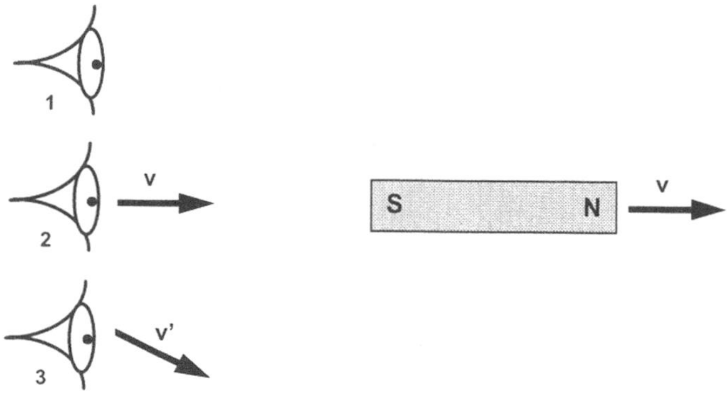

Fig. 5.56 Three observers of a magnet

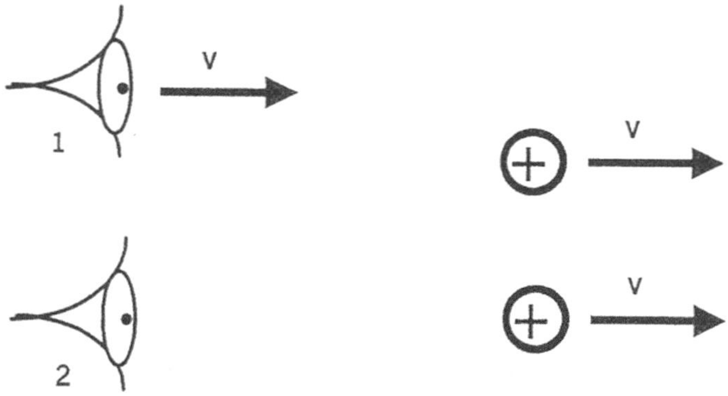

Fig. 5.57 Two different observers of two moving charges

SUMMARY: Moving permanent magnets, moving electromagnets (wires with electric current), and stationary electromagnets with a changing electric current—all produce a change of the magnetic field and therefore an induced electric field. If a loop of wire is present, this electric field can drive an electric current and people refer to the presence of an "induced EMF." It is important to realize that *the induced electric field is present whether or not a loop of wire is present*!

5.20 What Do Induced Electric Field Lines Look Like?

Below we show some examples.

1. Moving bar magnet or solenoid.

   In Fig. 5.58, we see a bar magnet moving to the right. The magnetic field of the bar magnet points to the right of the magnet. As a result of the motion, the magnetic field is increasing to the right: We say that there is a change $\Delta\vec{B}$ in the "B-field" to the right. There is a consequent electric field with electric field lines shown to be clockwise on the surface of the curved rectangular surface.

   In Fig. 5.59 we see the magnet moving to the left. To the right of the magnet, the field is still pointing to the right but is decreasing in magnitude. As a result, the electric field lines are seen to be counter-clockwise on the rectangular surface to the right of the bar magnet. To the left of the magnet, the magnetic field is also pointing to the right, but is **increasing** in magnitude. As a result, the electric field lines on the surface to the left of the bar magnet are clockwise.

   We can represent the above schematically in a very much simplified Fig. 5.60.

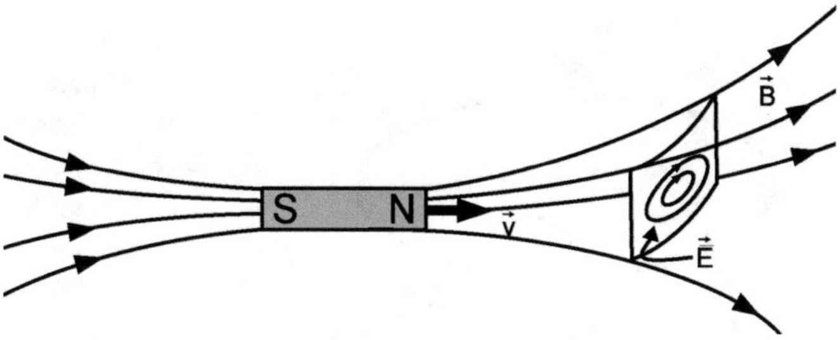

Fig. 5.58 Induced electric field from a moving magnet

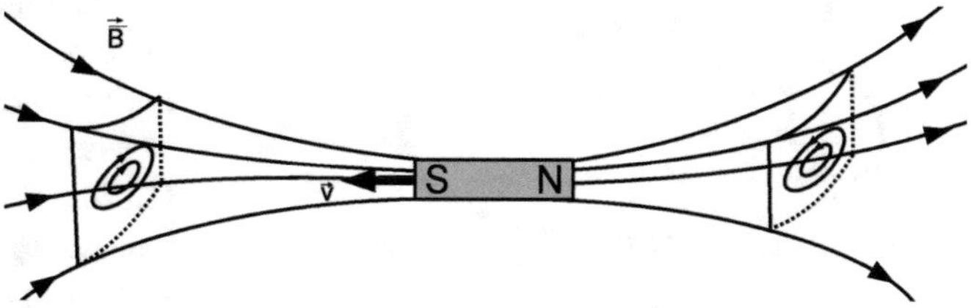

Fig. 5.59 Induced electric field from moving magnet—reversed direction

Fig. 5.60 Changing magnetic field leads to an electric field

Note that if $\vec{B}$ is pointing to the *right* and its magnitude is decreasing, $\vec{B}$ is pointing to the *left*.

2. Stationary solenoid with a changing current.

   Home exercise: Describe $\vec{E}_{\text{ind}}$ to the right and left of the solenoid.

3. Long straight wire with a changing current.

   Note that the change in current ΔI produces a change in the magnetic field $\Delta\vec{B}$, which is said to induce the presence of an induced electric field $\vec{E}_{\text{ind}}$.

4. Short segment of wire with a changing current.

   Because the wire is infinite in length in case (3), the lines of E_{ind} are parallel straight lines. In the present case, $\Delta\vec{B}$ is concentrated around the wire segment so that $\vec{E}_{\text{ind}}$ is also concentrated. We can also see how $\vec{E}_{\text{ind}}$ has closed.

5.21 Lenz's Law

Notice that in both figures—Figs. 5.61 and 5.62—the induced **electric** field produced by a changing current is in a direction opposite to that of the change in current, as far as positions along the wire are concerned. Thus, E_{ind} opposed the change in current. The result is that in order to change the current in a wire, work has to be done to overcome the consequence induced electric field. That is why extra power is needed while an electric motor is being started up.

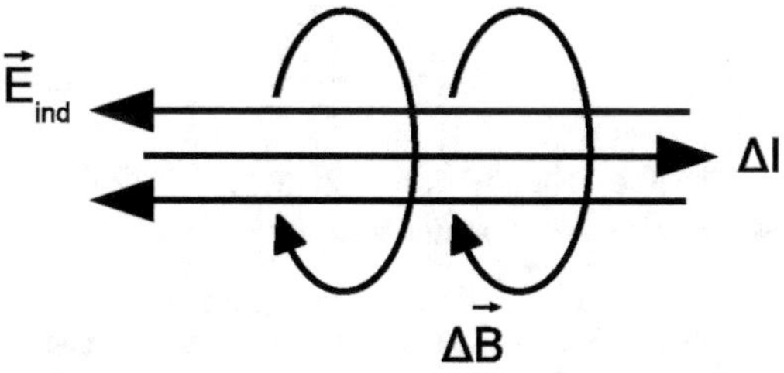

Fig. 5.61 Changing current leads to an induced electric field

Notice too, how the sense of rotation of loops is related to the direction of straight arrows: The relation between E_{ind} associated with $\Delta\vec{B}$ (determined by Faraday's Law) is different from the relation between $\vec{B}$ and the current I which produces that magnetic field. [Here a changing current ΔI is producing a changing $\Delta\vec{B}$.] This

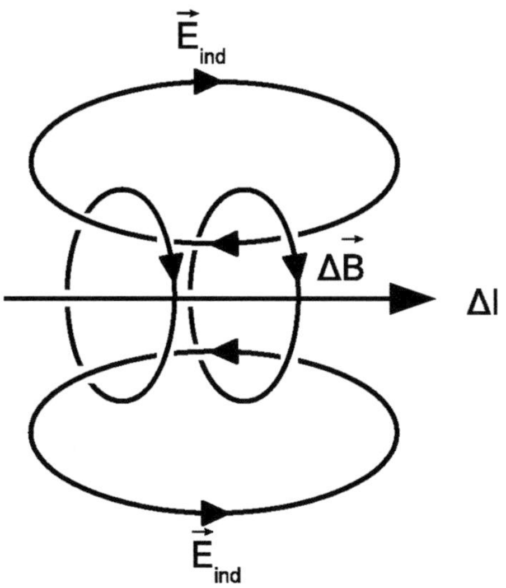

Fig. 5.62 Changing current in a short wire leads to an induced electric field

difference is reflected by the minus sign in Faraday's Law:

$$\vec{E}_{\text{ind}} \propto -\frac{\Delta \vec{B}}{\Delta t} \tag{5.5}$$

The above behavior reflects what is referred to as **Lenz's Law**. It states that

> The current induced by a changing magnetic field produces a magnetic field that opposes the change in magnetic field.

Let us examine a few experiments to see how Lenz's Law applies.

Figure 5.32: Here we see a magnet moving towards a loop of wire. The magnetic field is thereby increased to the right. The induced current has a direction that produces a magnetic field to the left along the axis. This figure corresponds to Fig. 5.58. To the right of the latter figure, we see the induced electric field lines labeled $\vec{E}$. If a loop of wire were present along an electric field line loop, there would be an induced current around the loop, as seen in Fig. 5.32. This induced current will produce a magnetic field to the left, thus opposing the increased magnetic field to the right.

Figure 5.61: Here, the change in current, ΔI produces a changing magnetic field ΔB. In turn, this changing magnetic field produces the induced electric field $\vec{E}_{\text{ind}}$. This induced electric field produces an induced current (over and above the original current ΔI). This additional contribution to the current is in the same direction as the electric field and thus in a direction **opposite** to the changing current ΔI.

NOTE: Lenz's Law is not an independent, new law. It merely accentuates the significance of the minus sign in Faraday's Law. It is an extremely useful tool for predicting the net qualitative result of Faraday's Law without having to carry out complete mathematical calculation. Furthermore, it has extremely important consequence regarding the stability of electromagnetic systems: Let us suppose that the sign in Faraday's Law were a PLUS sign. Let us consider a ring of metallic wire. It has a huge number of free (mobile) electrons that are moving at incredible speeds in random directions. On the average, in equilibrium, their currents essentially cancel and we end up with an essentially vanishing net current. However, these individuals do not exactly cancel. As a consequence, there are always fluctuating changing net currents. With a PLUS sign, a changing current would produce an electric field that will increase that current in the same direction. We would have a **positive feedback**. The current can be shown to increase exponentially! The system would be unstable. The actual NEGATIVE sign provides a **negative feedback** and always brings the system back towards its equilibrium vanishing net current.

5.22 The Guitar Pickup

We will now discuss the design of a **guitar pickup** that makes use of **magnetic polarization**. It is based upon the following phenomenon: Materials such iron or steel or nickel can be magnetized so as to form a permanent magnet, as described above. These materials are said to be **magnetic materials**. [Aluminum or copper cannot be magnetized and are said to be **non-magnetic materials**.] Magnetic materials also exist in **non-magnetized states**—witness a steel paper clip that you buy from the store. However,

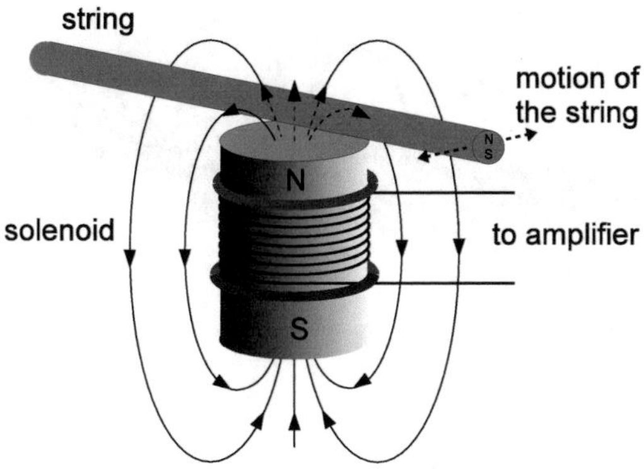

Fig. 5.63 Guitar pickup using magnetic polarization and a Faraday EMF

if a piece of magnetic material is placed in the vicinity of a magnet, the piece can become temporarily magnetic—the technical term is **magnetically polarized**. Often, when the magnet is removed, the piece will return to its non-magnetic state, so that its magnetic state is dependent upon the presence of the other magnet. A guitar pickup depends upon this residual induced magnetism being negligible.

In Fig. 5.63 we depict details of a practical design for a common guitar pickup. The permanent magnet polarizes the steel string. The solenoid experiences the magnetic fields of both the permanent magnet and the magnetically polarized vibrating string. When the string vibrates, the magnetic field of the polarized string changes in time, resulting in an induced EMF in the solenoid pickup coil. The induced EMF that is passed on to the amplifier will have a pattern in time that mirrors that of the velocity of the string.

5.23 Maxwell's Displacement Current

Faraday's Law describes how a change in the magnetic field with respect to time is accompanied by an induced electric field. Around 1860, **James Clark Maxwell** (Fig. 5.64) discovered that

Fig. 5.64 James Clerk Maxwell (source: https://www.findagrave.com/memorial/16871396/james-clerk-maxwell)

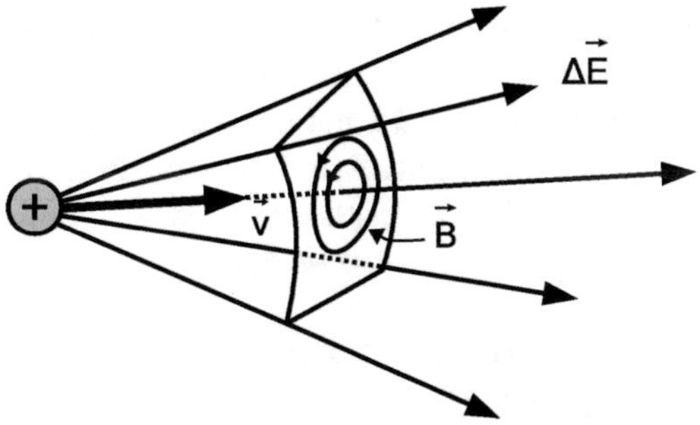

Fig. 5.65 Electric and magnetic fields of a moving charge

> a change in electric field with respect to time must be accompanied by an induced magnetic field.

The rate of change of $\vec{E}$ with respect to time is known as the **displacement current**. In simplified form, we have

$$\vec{B}_{\text{ind}} \propto +\frac{\Delta \vec{E}}{\Delta t} \tag{5.6}$$

There is a situation where such a relation would not be surprising: A moving point charge will be associated with a magnetic field and an electric field which changes with respect to time as shown in Fig. 5.65.

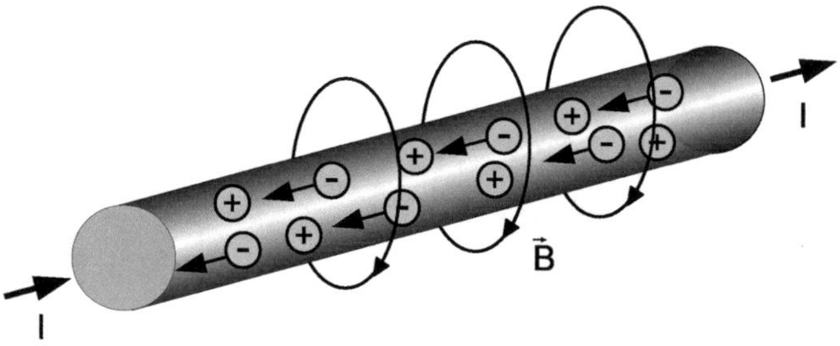

Fig. 5.66 Magnetic field from an electric current

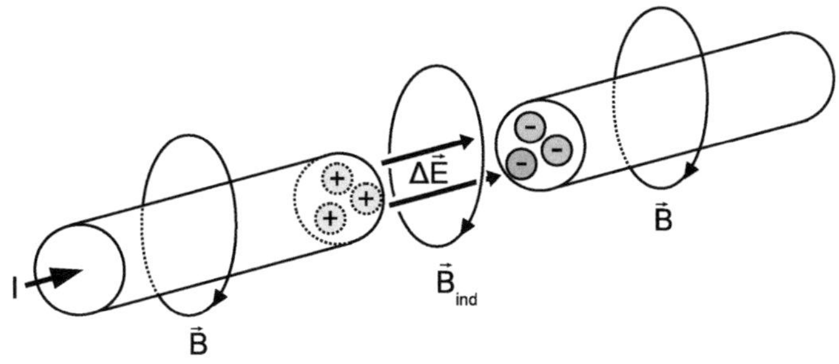

Fig. 5.67 Induced magnetic field due to changing electric field—the **Displacement Current**

However, this latter magnetic field is the one usually associated with Ampère's Law. In order to appreciate the difference between an "Amperian" magnetic field and one associated with the displacement current (and therefore to appreciate Maxwell's contribution), we will consider instead the following two situations:

1. Let us study more closely a metal wire carrying a current. Positive ions are stationary, electrons account for current, and the metal remains electrically neutral. Outside the wire, all we have is a magnetic field (Fig. 5.66).
2. Now suppose that a segment of wire is removed while the EMF which drives current through the wire continues to "pump" electrons in from the right and draw them off from the left. The ends of the wire will accumulate respective positive and negative charge and we will have a changing electric field in the space between the two ends (Fig. 5.67).

   We observe a magnetic field $\vec{B}_{\text{ind}}$ in the gap region not due to a current, but rather due to a changing electric field, the **displacement current** $\Delta\vec{E}/\Delta t$. Maxwell's great contribution was that more generally, a changing electric field must be accompanied by a magnetic field. The resulting relation led to his deduction that there exists an electromagnetic disturbance which we call **electromagnetic waves** and that light is an example of such a wave.

5.24 Electromagnetic Waves

Now we are ready to see how Faraday's Law and Maxwell's Displacement Current can generate an electromagnetic wave pulse and thereby describe the nature of an **electromagnetic wave**. Electromagnetic waves are often referred to as **electromagnetic radiation**.[5]

Consider first the propagation of a pulse on a long string under tension. The equilibrium state is a straight string. Figure 5.68 shows the string at some early stages after its being plucked at its center.

Next we turn to the propagation of an electromagnetic wave pulse. Here the equilibrium state is absence of an electromagnetic wave. For a wave in vacuum, there is nothing. We can start the pulse by having a localized electric field $\vec{E}_1$.

[5]The term **radiation** is used to refer to beams of tenuous material. Examples besides electromagnetic waves are beams of protons or electrons.

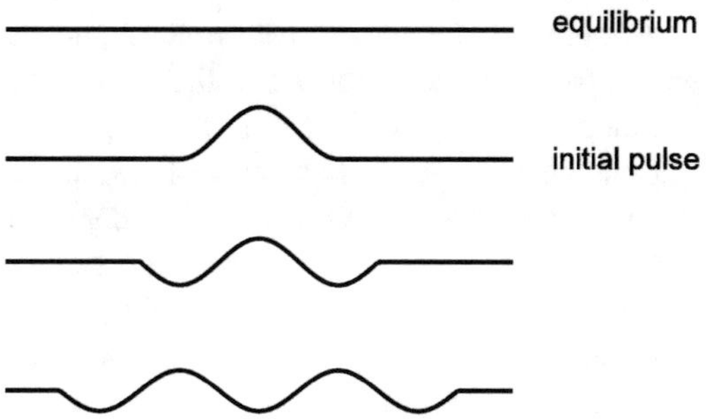

Fig. 5.68 Propagation of a pulse along a stretched string

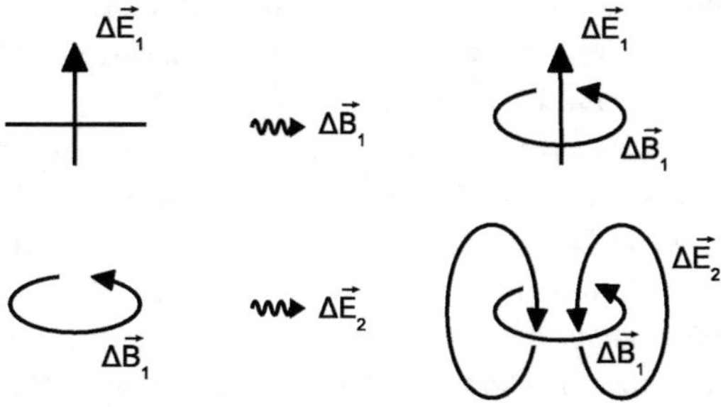

Fig. 5.69 Generating an electromagnetic field—stage one

(One way to do this is by giving an electric charge a sudden jerk—in our example, a downward jerk on a positive charge.) This represents a change in the electric field, so that a magnetic field $\vec{B}_1$ is generated (via the mechanism of Maxwell's Displacement Current). See Fig. 5.69, where the change in field is observed at a position just **below** the charge. In turn, $\vec{B}_1$ generates a second contribution to the electric field, $\vec{E}_2$ (via Faraday's Law). Then, in turn, $\vec{E}_2$ generates $\vec{B}_2$. And so on.

If we add $\Delta\vec{E}_1$ and $\Delta\vec{E}_2$, we obtain an electric field which, along the horizontal axis, has a direction and qualitative magnitude shown in Fig. 5.70.

This process actually takes place continuously in time and space. The result is that an electromagnetic wave pulse propagates outwardly from the source.

Maxwell himself derived specific mathematical equations to express the laws of electricity

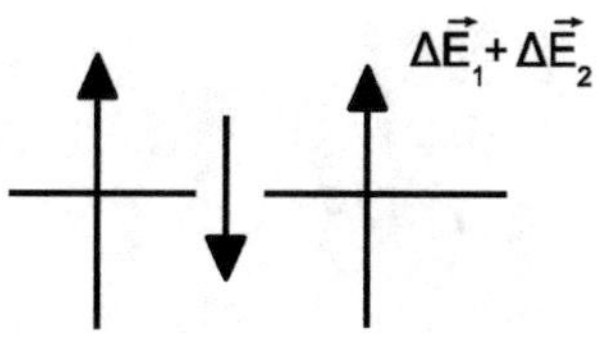

Fig. 5.70 Generating an electromagnetic field—stage two

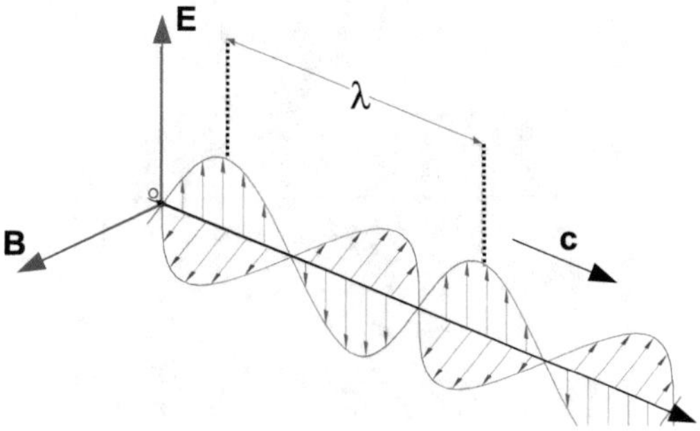

Fig. 5.71 Electromagnetic field—detail

and magnetism. These equations have an infinite number of solutions which describe combinations of motion of electric charge and electric and magnetic fields. Most significantly, Maxwell showed that there exists a class of solutions which describe the propagation through space of electric and magnetic fields as a so-called electromagnetic wave. Maxwell's analysis provides mathematical rigor behind our description above. In Fig. 5.71 we depict an electromagnetic wave traveling to the right.

The direction of both the electric and magnetic fields is perpendicular to the direction of propagation. Hence, an **electromagnetic wave is said to be transverse**. In addition, the electric field and the magnetic field are perpendicular to each other.

What is the wave velocity for these EM waves?

Remember the two relevant equations:

$$\vec{E}_{\text{ind}} \propto -\frac{\Delta\vec{B}}{\Delta t}$$
$$\vec{B}_{\text{ind}} \propto +\frac{\Delta\vec{E}}{\Delta t} \qquad (5.7)$$

Fig. 5.72 Heinrich Hertz (source: http://horowhenua.kete.net.nz/en/site/images/show/450-heinrich-hertz-b1857-d1894?view_size=large photo credit: Horowhenua Historical Society, Inc)

In these equations there appear two constants of proportionality involving two constants:

The **permeability of free space** and the **permittivity of free space**.

Maxwell found that the wave velocity of electromagnetic waves—*c*—is given by:

$$c = \frac{1}{(\text{permeability of free space} \times \text{permittivity of free space})^{1/2}} \quad (5.8)$$

The **permittivity of free space** is analogous to the mass density and the **permeability of free space** is analogous to the inverse of the effective force in Eq. (2.27).

When Maxwell evaluated this expression, he obtained

$$c = 3.0 \times 10^8 \text{ m/s}$$

This value is just the measured value for the speed of light in vacuum and confirmed his identification of a light wave as an electromagnetic wave. Can you imagine how Maxwell might have felt at this discovery!

Comment: The fact that Maxwell's equations have EM waves as solutions without the need for electric charge means that the waves are **self-sustaining**: We do not know of EM waves that were not originally produced by charge. But once the EM waves are produced, the charges may be removed afar. This self-sustaining feature is dependent upon the existence of Maxwell's Displacement Current.

A dramatic example of this self-sustaining property occurs in the phenomenon of **electron–positron pair annihilation**. When these two fundamental particles are close to each other, they have a high probability of both disappearing and being entirely replaced by EM radiation.

Heinrich Hertz (Fig. 5.72) is given credit for being the first to detect EM waves and being able to identify them as such in the framework of the then commonly accepted set of electromagnetic phenomena. A schematic of his experimental apparatus is shown in Fig. 5.73, while the actual apparatus is shown in Fig. 5.74. In the experiment, a high voltage source produced a spark across a gap in a circuit. Across the room was a metal ring with a gap. A spark jumped across the ring's gap in response to the original spark.[6]

The first spark involves a changing electric current which produces a changing magnetic field. Here it is a $\Delta\vec{B}$ which is the initial source of the EM pulse. Only an EM wave could account for the great distance traveled by the EM disturbance with such small attenuation.

In time, a wide variety of disturbances, produced under different circumstances, have been identified as EM waves. *The only difference among them is the range of frequencies.* See Fig. 5.75.

[6]Figure 5.74 of Hertz's apparatus was generously provided by John Jenkins, who directs the Spark Museum. For more details about the museum, see its website at: http://www.sparkmuseum.com.

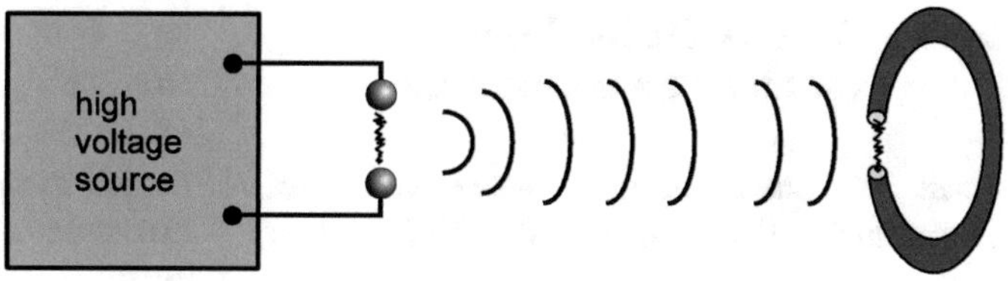

Fig. 5.73 Generating an electromagnetic field from an electric spark

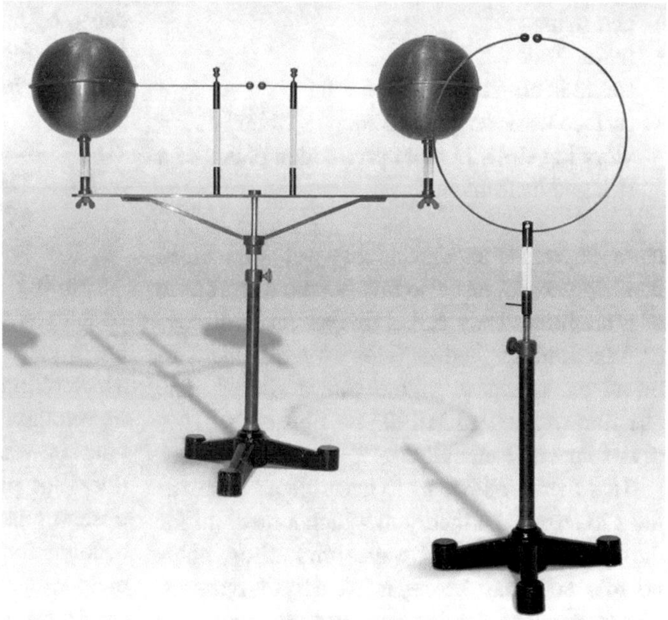

Fig. 5.74 Heinrich Hertz's apparatus for the detection of electromagnetic waves (photo provided by John Jenkins)

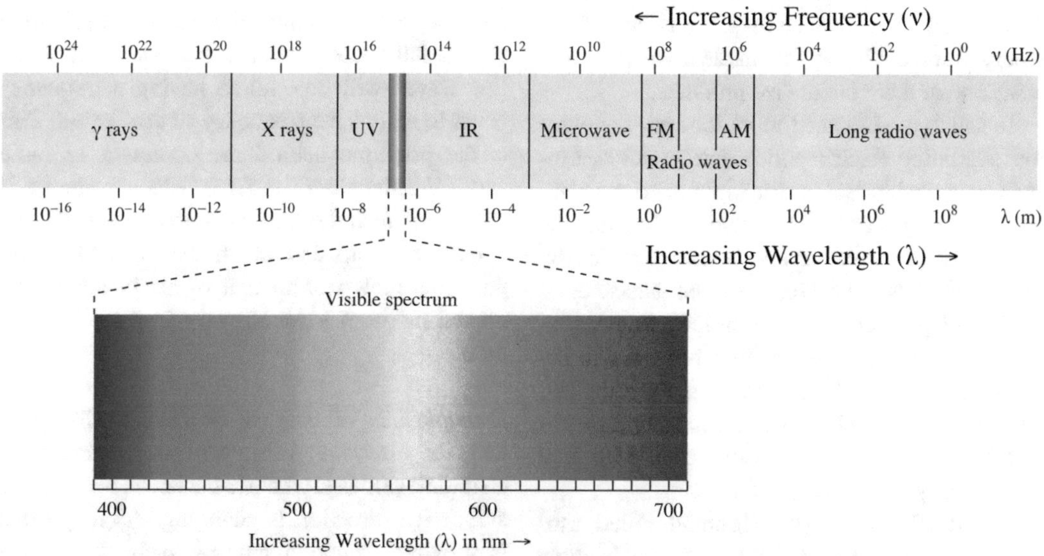

Fig. 5.75 Names of electromagnetic waves for various ranges of frequencies (source: http://en.wikipedia.org/wiki/Electromagnetic_radiation)

5.25 What Is the Medium for Electromagnetic Waves?

Aside from EM waves, all the waves we have discussed so far in the course involve the disturbance of a **medium**:

- taut string
- tuning fork
- sound in air gas, liquid, or solid
- surface wave on a liquid such as water
- vibrating Chladni plate or wooden plates of a stringed instrument.

Note Physicists understandably were convinced that EM waves must also involve the disturbance of a **medium**. They called this as yet undiscovered medium the "ether." However, all efforts to reveal the existence of the "ether" failed. The question was settled in 1905 by Einstein's "Theory of Special Relativity."

This theory allows us to describe all observations in terms of a theory in which a medium for the propagation of EM waves, the **ether**, plays no role and may be regarded as non-existent. The stupendous details, ramifications, and consequences of the Theory of Special Relativity are beyond the scope of this course. You are encouraged to read a layman's account of the theory. Suffice it to say that nuclear weapons and energy were a couple of "by-products."

In satisfying his need to understand the enigmas regarding electric and magnetic fields, Einstein presumably had no foresight as to the awesome consequences of his studies. In fact, the possibility of a nuclear bomb was regarded by most physicists as absurd, even into the 1930s.

Should physicists stop thinking? Some, perhaps many people, feel so. I do not want to belabor this issue much here. I would, nevertheless, encourage you to think about an entirely different area for the sake of comparison because the area is seemingly more benign to us.

About 100 years ago, Sigmund Freud ushered in the modern age of psychology. People have benefitted greatly from this development. However, psychiatry gave birth to the age of manipulation of the masses with highly sophisticated methods of propaganda and advertising. Most glaringly, the Nazi horrors including tens of millions dead directly due to WWII and about twelve million exterminated victims of the holocaust, by far exceeds the losses to humankind, at least, so far, due to nuclear weapons. *Should we stop studying psychiatry and psychology?*

5.26 The Sources of Electromagnetic Waves

According to the laws of electromagnetism which were formulated by Maxwell, EM waves are produced by **accelerating charge**. Recall that acceleration is the measure of the rate of change in velocity. Thus, a charge that is moving at a constant velocity and therefore is not accelerating does not produce EM waves, even though an observer uses both electric and magnetic fields to account for the forces produced by the moving charge.

We say that an accelerating charge "radiates" EM waves or emits EM radiation.

Example 5.1 If a charged particle is oscillating sinusoidally like a simple harmonic oscillator, EM waves will be emitted having a frequency equal to that of the frequency of oscillation. This is the principle behind the operation of radio and TV antennas (Fig. 5.76). Electric charge is "pumped" in and out of the antenna. The electric current runs up and down the antenna. Compare this motion with that of air in the fundamental mode of a sound wave in a semi-closed tube!

Example 5.2 **Velocity** is specified by its speed and its direction. **Acceleration** represents a changing velocity, whether the speed is changing and/or the direction is changing. A charge that is moving around a circular path at constant

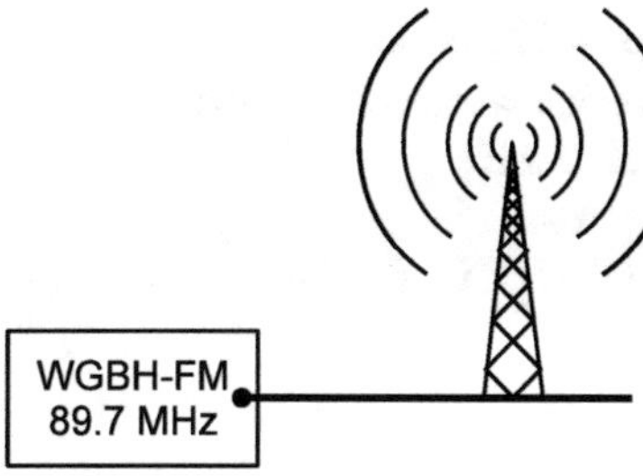

Fig. 5.76 Generating a radio wave

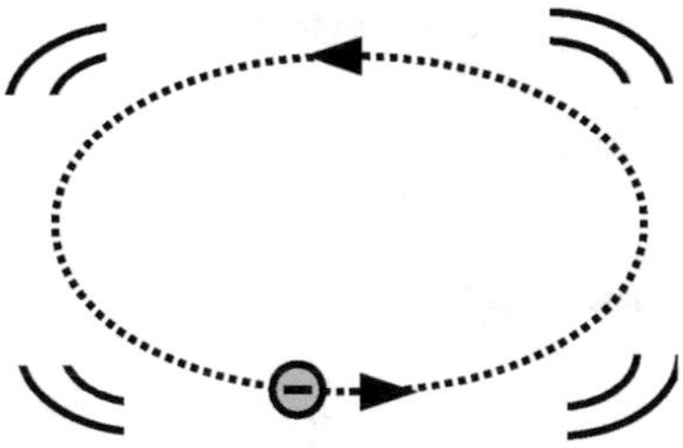

Fig. 5.77 EM field from a circulating charge

speed still has an ever-changing direction. Therefore, it is accelerating. As a consequence, it will emit EM waves. Not surprisingly, perhaps, the waves have a frequency equal to the frequency of revolution of the charge. See Fig. 5.77.

Example 5.3 Resonance between two electrically charged simple harmonic oscillators.

In Fig. 5.78, we have depicted two SHOs whose masses have electric charge and whose frequencies of oscillation are identical. Suppose that the left SHO is set into oscillation. It will emit EM waves which reach the SHO to the right. At the position of the right SHO we have an electric field (associated with the EM waves) which oscillates up and down sinusoidally. It therefore produces a sinusoidal force on the right charge at a frequency which equals the "natural" frequency of oscillation of the SHO and so causes the right charge to oscillate with a relatively large amplitude. We have a resonant response.

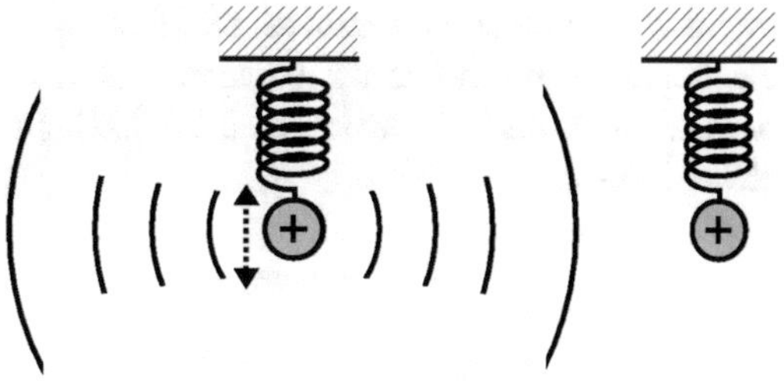

Fig. 5.78 Resonance between two charged SHOs

SUMMARY of Electricity and Magnetism

Electromagnetic phenomena are ultimately manifested by forces between electric charges. These forces can be determined, in principle, by the state of the charges alone. Thus,

1. two charges exert an electric force on each other that is determined by their relative position.
2. two charges that are both moving exert a magnetic force on each other that depends upon both their position and their relative velocity.

These forces can also be understood in terms of electric and magnetic fields, a concept introduced by Michael Faraday. Thus,

1. a charge produces an electric field;
2. a moving charge produces a magnetic field;
3. a charge experiences an electric force in the presence of an electric field;
4. a moving charge experiences a magnetic force in the presence of a magnetic field;
5. an accelerating charge produces a combination of electric and magnetic fields referred to as an electromagnetic wave. Light is an example of an electromagnetic wave.

Added note: The question of whether electromagnetic waves are real is not a question within the domain of Physics; it is a philosophical question. However you choose to think of an electromagnetic wave, it is a mathematical quantity that can be used to determine the behavior of electric charges.

5.27 Terms

- action–reaction
- γ-ray
- *Ampère's Law*
  (relating current to magnetic field)
- *compass*
- *conductor*
- *electrically polarized*
- *electric battery*
- *electric charge*
- *electric force*
- *electric current*
- *electric field*
- *electric generator*
- *electric motor*
- *electromagnetic energy*
- *electromagnetic field*
- *electromagnetic force*
- *electromagnetic wave*
- *electromotive force ("EMF")*
- *electron-volt (eV)*
- *electroscope*
- *Faraday's Law of Induction*
- *free electrons*
- *Galilean Principle of Relativity*
- *gravitational force*
- *induced EMF*
- *insulator*
- *Lenz's Law*
- *loudspeaker*
- *magnet*
- *magnetic field*
- *positive and negative charge*
- *magnetic force*
- *power*
- *magnetite (or "lodestone")*
- *Maxwell displacement*
- *metal*
- *microphone*
- *microwave*
- *neutral (electrically)*
- *Newton's Third Law*
- *north and south pole of a magnet*
- *nuclear force*
- *optical fiber*
- *Relativity of Description*
- *solenoid*
- *weak force*
- *x-rays*

5.28 Important Equations

Electric force:

$$F \propto qE \tag{5.9}$$

Magnetic force:

$$F \propto qvB \tag{5.10}$$

Ampère's Law:

$$B \propto I \tag{5.11}$$

Faraday's Law

$$E \propto -\frac{\Delta B}{\Delta t} \tag{5.12}$$

Maxwell's displacement:

$$B \propto +\frac{\Delta E}{\Delta t} \tag{5.13}$$

5.29 Problems for Chap. 5

1. What is an "electromagnet"?
2. Where is the magnetic N-pole of the earth? (Approximately?)
3. What was Oersted's discovery?
4. Draw the electric field lines of an electric dipole.
5. Draw the magnetic field lines of a bar magnet.
6. In Fig. 5.79 we see a single Neodymium bar magnet of length one-inch, with its north and south poles at opposite ends.
   (a) In Fig. 5.80 we see two such magnets in stable positions, parallel to each

Fig. 5.79 A single neodymium bar magnet

Fig. 5.80 A pair of neodymium bar magnets

Fig. 5.81 Three neodymium bar magnets

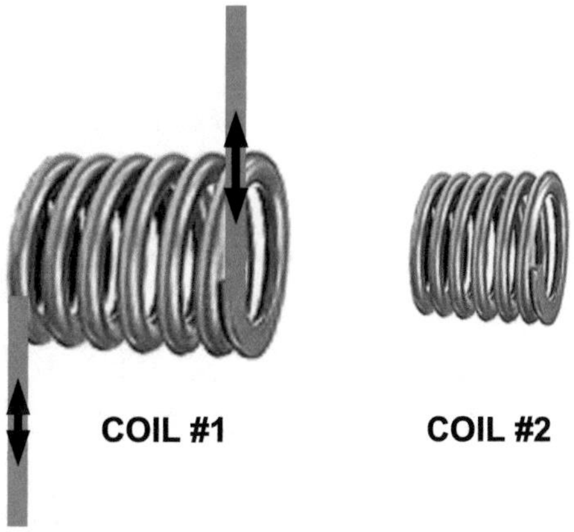

Fig. 5.82 Two coils in close proximity

other. Indicate on the figure, or express in words, a possible set of **polarities** of the four ends of the two magnets.

(b) Describe another stable configuration of the two magnets.

(c) In Fig. 5.81 we see three magnets together side by side in a triangular configuration, with a tie around them. If the tie is removed, explain why the three magnets cannot remain stable in a triangular configuration.

(d) Describe the likely stable configuration of the three magnets that start out in a triangular configuration.

(e) Describe another stable configuration of the three magnets.

7. Magnetic field lines are: always closed/may be closed/never closed.
8. Electric field lines are: always closed/may be closed/never closed.
9. State Faraday's Law.
10. In Fig. 5.82 we see two coils of wire. An electrical wave generator produces a sinusoidal current in the first coil, as indicated.
    (a) **DESCRIBE** fully what will happen in the second coil as a result of the current in the first coil.
    (b) **EXPLAIN** in detail how this second current is produced. Make sure that you mention **all** of the physical principles and/or laws that are involved!
11. How do the following work? **motor**; **generator**; **loudspeaker**.
12. Describe the operation of a **microphone** with two different designs, using fundamental physics principles.

13. Describe a situation which illustrates the Principle of Relativity in electricity and magnetism.
14. Describe a situation wherein the presence of an electric field depends upon the motion of the observer.
15. Describe a situation wherein the presence of a magnetic field depends upon the motion of the observer.
16. What is Maxwell's Displacement Current?
17. Electromagnetic radiation is produced when an electric charge is behaving in what way?
18. Calculate the wavelength of microwaves in a microwave oven having a frequency of 2500-MHz. On the basis of your answer discuss the effect on cooking of the nodal lines of a standing wave that could set up in the microwave oven.

6 The Atom as a Source of Light

We have noted that according to Maxwell's theory of electromagnetism, light is nothing but a visible electromagnetic (**EM**) wave that has a frequency in the narrow range $\sim 4 \times 10^{14}$ Hz to $\sim 7 \times 10^{14}$ Hz. The corresponding range of wavelengths is 4000 $\AA$ to 7000 $\AA$. Furthermore, EM waves are produced and emitted by accelerating electric charge. There are two interesting questions that immediately confront us: (1) We accept the premise that animal eyes evolved so as to be sensitive to sunlight. Still, what characteristics of animal eyes make them sensitive to this particular narrow range of frequencies? (2) What are the physical characteristics of the sun that cause sunlight to be concentrated in a particular range of frequencies?

The answer to the first question is connected with the fact that the eye uses conglomerates of atoms, that is, molecules, as detectors of EM waves. This chapter therefore focuses on the atom as a source and receiver of EM waves. The answer to the second question has to do with the fact that the sun is a body that is in equilibrium, with a surface temperature of about 5, 800 degrees above absolute zero (that is, 5, 800 Kelvin/5,800 K = 6,073°C). That there is a connection between temperature and frequency must be surprising to the beginning student of physics; as we will see later on this chapter, the connection stems from their common link with energy.

6.1 Atomic Spectra

Matter is constantly emitting EM radiation and absorbing EM radiation which strikes it. Ordinarily this EM radiation is invisible and of low intensity. However, when matter is heated up from room temperature, it emits EM radiation which becomes more intense and eventually, in a certain range of temperatures, becomes visible. When dilute gases of atoms are heated up sufficiently, as they are in a star, we can see EM radiation which comes directly from the star's atoms. That radiation reflects the behavior of a single atom. The frequency spectrum of radiation around the visible range can be determined by using a **diffraction grating** or a **prism**. We call this process **spectral analysis**. (How these devices work will be discussed in Chaps. 7 and 9, respectively.)

In Fig. 6.1, we see a narrow beam of white light incident upon a prism from the lower left corner. A fraction of the light is reflected off the surface of the prism, with reduced intensity. The other fraction of the beam passes through the prism. A fraction of this beam exits down towards the bottom right corner in a fan of light of varying frequency. The higher the frequency, or the lower the wavelength, of a Fourier component of the incoming beam, the more is the component deflected away from the incoming

L. Gunther, *The Physics of Music and Color*,
https://doi.org/10.1007/978-3-030-19219-8_6

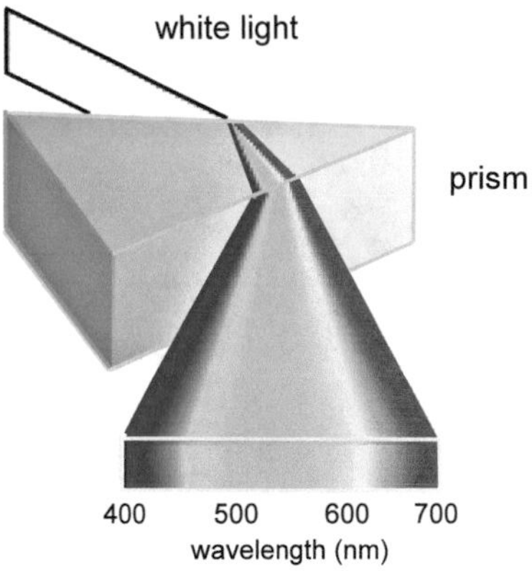

Fig. 6.1 Prism analyzing white light (source: http://en.wikipedia.org/wiki/File:Dispersive_Prism_Illustration_by_Spigget.jpg)

direction. Thus, the blue end is deflected most. [See Sects. 9.8 and 9.9, which show how this effect is a result of **dispersion** and **refraction**.] This outgoing beam can be viewed and analyzed, revealing a rainbow of colors.

On the other hand, if a gas of atoms is heated up to a very high temperature (1000s of Kelvins) or has been subjected to a very high electric voltage, it emits light with a discrete frequency spectrum, akin to the spectrum of frequencies of a vibrating system such as a violin string or drum head.

When the beam is restricted so as to produce a narrow column of light, each frequency component shows up as a bright colored line. For this reason, scientists make reference to the **line spectrum** of an atom. See Fig. 6.2 for the line spectra of hydrogen and of iron.

The frequency spectrum light emitted by an atom is **discrete**, as opposed to the spectrum of sunlight, which is **continuous**. Each atom has its own unique spectrum, which can serve as its "fingerprint" for identification purposes. This fact allows us to determine which atoms and molecules are present in outer space, such as in stars, quasars, and interstellar gases.[1]

[1]The following website (12-29-2010) enables you to see the emission spectrum of elements shown in the periodic table. http://chemistry.bd.psu.edu/jircitano/periodic4.html.

6.2 The Hydrogen Spectrum of Visible Lines

The following are the frequencies of the visible spectral lines of hydrogen, which is the simplest element:

$$f_1 = 4.57 \times 10^{14}\ \text{Hz}$$

$$f_2 = 6.17 \times 10^{14}\ \text{Hz}$$

$$f_3 = 6.91 \times 10^{14}\ \text{Hz}$$

$$f_4 = 7.32 \times 10^{14}\ \text{Hz}$$

The fundamental question is: What is the theoretical basis for these spectra?

To appreciate the meaning of the above question, let us review the case of the vibrating string with fixed ends. The frequency spectrum is a harmonic series:

$$f_1,\ f_2 = 2f_1,\ f_3 = 3f_1,\ \ldots$$

What counts essentially are the ratios of the frequencies of the overtones to the fundamental frequency, $1:2:3:4:\ldots$. The fundamental frequency sets the scale for the whole spectrum. Recall that we can express the frequency spectrum as follows:

$$f_n = nf_1 \quad \text{where } n = 1, 2, 3, \ldots \tag{6.1}$$

We showed that this spectrum follows from the fact that the modes of vibration of a string have periods which are an integral *fraction* of the time that it takes for a pulse to make a round trip along the length of the string. Since according to Maxwell's theory, the source of EM radiation is accelerating charge, a reasonable hypothesis to account for atomic spectra would be that:

1. The charges in an atom have modes of vibration associated with atomic forces. In order to determine the modes an as yet unknown model of the atom would be needed.
2. Each particular frequency in the frequency spectrum of an atom is associated with a particular mode of vibration. In order to understand this idea, recall the charged SHO of Sect. 5.26 and exhibited in Fig. 6.16.

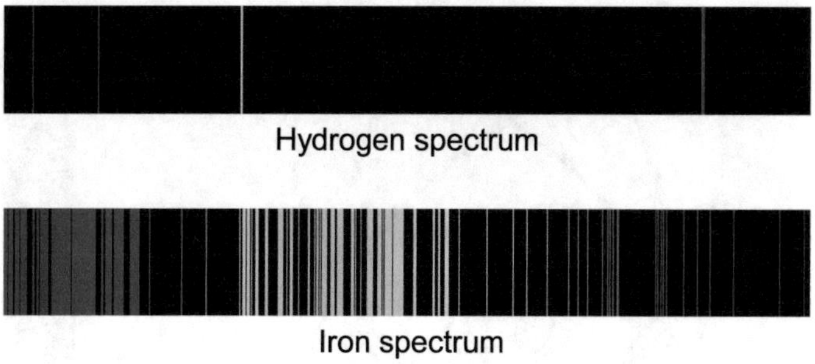

Fig. 6.2 Spectra of hydrogen and iron (source: http://en.wikipedia.org/wiki/Emission_spectrum)

Unfortunately, no one could find a model for an atom that accounted for the observed atomic spectra. A hint was provided in 1884 by Johann Balmer, who discovered that the observed visible spectrum of hydrogen, that is, the four frequencies given above, were fitted well with the following complicated formula:

$$f_n = 3.29 \times 10^{15}\ \text{Hz} \cdot \left[\frac{1}{4} - \frac{1}{(n+2)^2}\right] \tag{6.2}$$

Thus, for example,

$$\begin{aligned} f_1 &= 3.29 \times 10^{15}\ \text{Hz} \cdot \left[\frac{1}{4} - \frac{1}{(1+2)^2}\right] \\ &= 4.57 \times 10^{14}\ \text{Hz} \end{aligned} \tag{6.3}$$

Two points must be stressed: *First*, it can be proved that any finite set of numbers (here four numbers) can be fitted precisely with any one of an infinite number of formulas such as Eq. (6.2).[2] *Second*, Balmer's formula had no theory to give it physical significance when it was first presented. It was purely **empirical**.

Formulas similar to Balmer's were later found to fit the observed spectral lines of hydrogen that lie both in the ultraviolet region (i.e., just *above* the visible frequency range) and in the infrared region (i.e., just *below* the visible frequency range). Later, Neils Bohr gave Balmer's formula a theoretical basis. As we will shortly see, the formulas did provide Bohr with a clue to his theory in that it involved a difference between two numbers (the two terms between the brackets of Eq. (6.2)).

It is interesting to note that after Maxwell showed that there must exist a displacement current contribution to the magnetic field, and identified light as an electromagnetic wave, it was believed that the existing set of fundamental laws of physics was complete; that is, the laws could in principle account for all observations thus far and henceforth to be made.[3] This set of laws, along with the accompanying concepts, is referred to as **classical physics**. Unfortunately, as far as atomic spectra were concerned, no one succeeded in finding a model for an atom that accounted for the observed spectra. In fact, it ultimately became clear that classical theory was inadequate and would have to be modified or improved so as to account for phenomena at the atomic level. Thus, limitations of mathematical solvability were not the issue here.

In 1911, on the basis of experiments of alpha-particles scattered by atoms of gold, **Ernest Rutherford** proposed a model for the atom in which a collection of negatively charged electrons revolve in planetary-like orbits about a positively charged nucleus. The radius of the electron orbits is on the order of Ångstroms. (1 $\mathring{A}$ = 10^{-10} m.) The nucleus is relatively utterly minuscule, with

[2]As an example, the set of numbers, $\{1, 2, 4\}$, can be represented by 2^{n-1} <u>or</u> by $[n(n-1)/2+1]$, where $n = 1, 2,$ and 3, respectively.

[3]The term **"in principle"** means that one merely had to solve the mathematical equations and one would find that the theory would be confirmed by experiment. In practice, there are many phenomena that require the solution of equations that are too difficult and complicated to solve, so that the theory cannot be tested. However, unsolvability does not imply that the equations and the theory that they represent are incorrect.

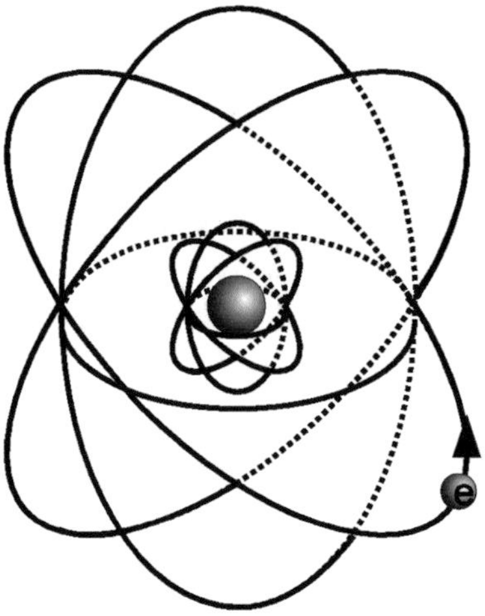

Fig. 6.3 Electron orbits in the Rutherford model

a diameter that is about $1/100,000$ that of the atom as whole and therefore occupying only about one part in $(1/100,000)^3 = (1/1,000,000,000,000,000)$ = one-thousand-trillionth of the volume of the atom as a whole. Since the electrons are regarded as being much smaller than the nucleus, most of an atom's volume is empty space! We see a reflection of Rutherford's model, expanded upon by Bohr, in the common symbol for an atom (Fig. 6.3).

There was a major difficulty with the Rutherford model in the context of Classical Physics in that the atom is unstable.

Suppose an electron is orbiting a nucleus at some radius. Because of its acceleration, the electron will emit EM radiation having a frequency equal to the frequency of revolution. Also, the radiation has energy, so that the orbiting electron must lose energy, which here is a sum of kinetic energy (KE) and potential energy (PE). Since the radius of the orbit decreases with decreasing energy, the electron will spiral into the nucleus. Furthermore still, since the frequency of revolution decreases with decreasing radius, the radiation will have an ever continuously decreasing frequency. And finally, classical theory predicts that an electron that starts out at a radius of $1\AA$ would spiral into the nucleus in about one-billionth of a second! Thus, an atom would collapse and thus be quite unstable.

Fig. 6.4 Neils Bohr (source: http://en.wikipedia.org/wiki/Niels_Bohr)

Two responses were reasonable at this point in the search for a theory of the atom:

1. One could search for a new model while keeping the classical laws.
2. One could keep the basic Rutherford model but find new laws.

6.3 The Bohr Theory of the Hydrogen Atom

When there is disagreement with experiment, indicating a need for revised laws, physicists try hard to preserve as much of the essence of existing laws. Such was the case when, in 1913, **Neils Bohr** (Fig. 6.4) proposed a theory of the hydrogen atom that incorporated the Rutherford model but combined classical laws with a modification that restricted the orbits.

We can summarize the theory as follows:

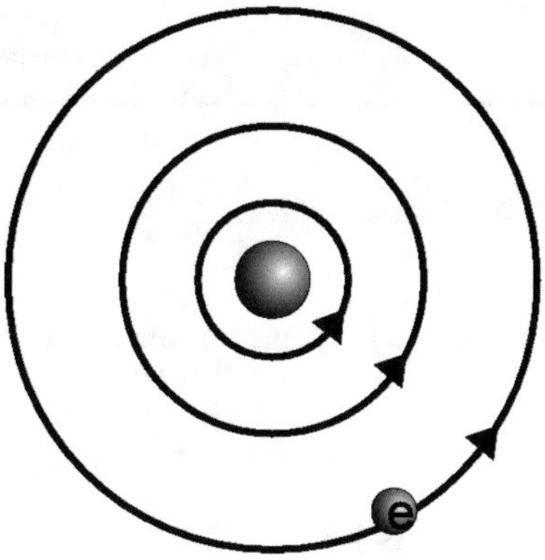

Fig. 6.5 Bohr orbits of the hydrogen atom

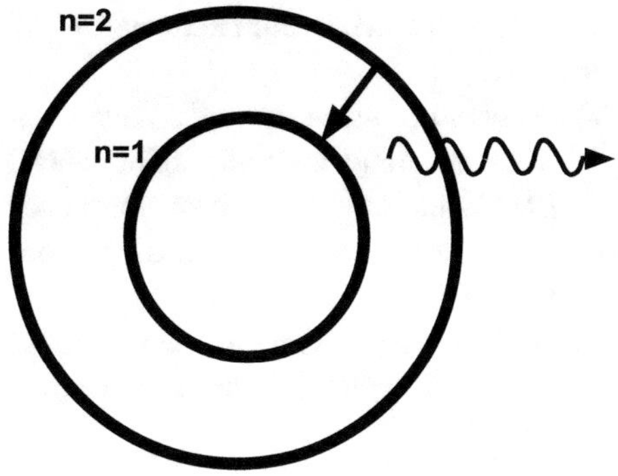

Fig. 6.6 Transitions in the Bohr model that lead to emission of EM radiation

1. According to classical theory, the electron orbits are ellipses, with any **size** or degree of flatness (referred to as "eccentricity") being possible. Circular orbits can have any radius. Bohr proposed that only certain discrete orbits are possible. We exhibit Bohr's discrete concentric circular orbits in Fig. 6.5.

   In Bohr's theory, the radii are equal to the following multiples of the so-called **Bohr radius**, which is equal to about 0.53Å, and to which we will give the symbol a_0:

$$r = a_0,\ 4a_0,\ 9a_0,\ 16a_0,\ \ldots$$

$$\text{generally, } r = n^2 a_0,\ \text{ where } n = 1, 2, 3, \ldots \tag{6.4}$$

   The allowed orbits correspond to certain allowed energies:

$$E = -13.6\text{ eV}, -3.4\text{ eV}, -1.5\text{ eV}, -0.85\text{ eV}, \ldots$$

$$\text{or generally, } E = -13.6\text{ eV}/n^2, \text{ where } n=1, 2, 3, \ldots \tag{6.5}$$

   Note that while the energies are **negative**, the energies of the orbits increase (are less negative) with increasing radius.

   We will see that only the **difference** between pairs of energies is important. Thus, if E' is less negative than E, $E' > E$ and $E' - E$ is positive.
2. Next we consider Bohr's theory of emission of EM radiation. According to Bohr, this takes place not because of the acceleration of charge, but rather in association with *transitions* of the electron from one orbit to another of lower energy. A transition from the $n = 2$ orbit to the $n = 1$ orbit is depicted in Fig. 6.6.

   A transition from one orbit to another of lower energy is accompanied by the emission of a specific discrete amount of EM radiation. This unit of EM radiation is called a **photon**, which is represented by the wiggly arrow in the figure. By the Principle of Conservation of Energy, the photon must have an energy equal to that lost by the atom.

   Thus, photon energy equals energy lost by an atom or

$$\boxed{E_{\text{ph}} = E_i - E_f} \tag{6.6}$$

   which is also called the photon emission condition.

   The emission takes place without the need of an interaction of the atom with an external system. The process is therefore referred to as **spontaneous emission**.

   In Eq. (6.6), E_i and E_f are the initial and final energy of the atom, respectively. For example, if the atom makes a transition from the $n = 2$ orbit to the $n = 1$ orbit, the emitted photon has an energy

$$E_{\text{ph}} = (-3.4\text{ eV}) - (-13.6\text{ eV}) = 10.2\text{ eV}$$

What are the characteristics of a photon?

(a) A photon has a *specific frequency.* In fact, monochromatic (sine wave) EM radiation of a given frequency f consists of a collection of photons having the same frequency.

(b) A photon has a *specific energy* related to its frequency via the **Planck Relation**:

$$\boxed{\text{energy of photon} = E_{\text{ph}} = hf} \tag{6.7}$$

In Eq. (6.7), h is a universal fundamental constant of nature known as **Planck's Constant**—named after **Max Planck**. It has the value

$$\boxed{h = 4.1 \times 10^{-15}\ \text{eV per Hz}} \tag{6.8}$$

Thus, suppose that the radiation has a frequency $f = 5.0 \times 10^{14}$ Hz. Then,

$$E_{\text{ph}} = 4.1 \times 10^{-15} \cdot 5.0 \times 10^{14} = 2.0\ \text{eV}$$

The condition of Eq. (6.7) for photon emission then reads

$$\boxed{hf = E_i - E_f} \tag{6.9}$$

This is known as the **photon emission condition**. Therefore, the emitted photon has a frequency given by

$$f = \frac{E_i - E_f}{h} \tag{6.10}$$

Sample Problem 6.1 Find the frequency of the photon emitted by a hydrogen atom that makes a transition from the $n = 2$ orbit to the $n = 1$ orbit.

Solution The energy of the photon is

$$E_{\text{ph}} = E_2 - E_1 = (-3.4) - (-13.6)$$
$$= 13.6 - 3.4 = 10.2\ \text{eV}$$

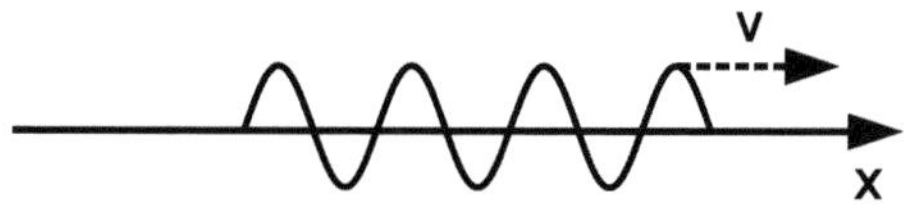

Fig. 6.7 A photon wave packet

so that its frequency of the photon is

$$f = E_{\text{ph}}/h = \frac{10.2}{4.1 \times 10^{-15}}$$
$$= 2.5 \times 10^{15}\ \text{Hz}$$

This frequency is in the invisible ultraviolet region.

(c) A photon also has an associated *wavelength*, which is related in the usual way to the frequency:

$$\lambda = \frac{\text{v}}{f} = \frac{c}{f} \tag{6.11}$$

(d) A photon usually is *localized in space*, both longitudinally, in the direction of its motion, and transversely. We will describe only the longitudinal extent, which we will refer to as the **length of the photon**. A photon that is emitted by an atom is about ten-million oscillations in extent. Such a finite segment of a sine wave is generally called a **wave packet** and is depicted in Fig. 6.7.

Sample Problem 6.2 Find the wavelength and the length of a photon of frequency 5.0×10^{14} Hz and ten-million (10^7) oscillations in extent?

Solution From Eq. (6.11) we have a wavelength

$$\lambda = \frac{c}{f} = \frac{3.0 \times 10^8}{5.0 \times 10^{14}} = 6.0 \times 10^{-7}\ \text{m}$$

so that the length is given by

$$\text{photon length} = \text{number of oscillations} \times \text{wavelength} = 10^7 \cdot 6.0 \times 10^{-7} = 6\ \text{m}$$

Note The spatial extent of a photon is connected with the time it takes for the atom to emit the photon, which is typically on the order of 10^{-8} s: Thus, since the period is the inverse of the frequency,

$$\text{photon emission time} = \text{number of oscillations} \times \text{period} = 10^7 \times \frac{1}{f} = \frac{10^7}{5 \times 10^{14}} = 2 \times 10^{-8}\ \text{s}$$

3. The spiraling of an electron into the nucleus according to Classical Theory is replaced by consecutive **spontaneous transitions** of the atom from one orbit to another orbit of lower energy, as shown in the figure below. The process is usually referred to as **spontaneous emission**, in contrast to stimulated emission, to be discussed later in this chapter (Fig. 6.8).
4. The stability of the atom against total collapse is provided by the existence of the orbit having the lowest energy, the $n = 1$ orbit. This orbit is referred to as the **ground state**.
5. It is observed that an atom will absorb monochromatic EM radiation only if its frequency is equal to one of the frequencies in the *emission spectrum*. That is, the *absorption spectrum* is the same as the emission spectrum. According to Classical Theory, **absorption** of EM radiation is understood to result from the force of the electric field of EM radiation that is incident upon the electrons of an atom and thus transfers its energy to the electrons. Preferential absorption of certain frequencies is due to resonance (see Sect. 2.16): The incoming EM wave has a frequency equal to one of the modes of vibration of the collection of interacting charges in the atom.

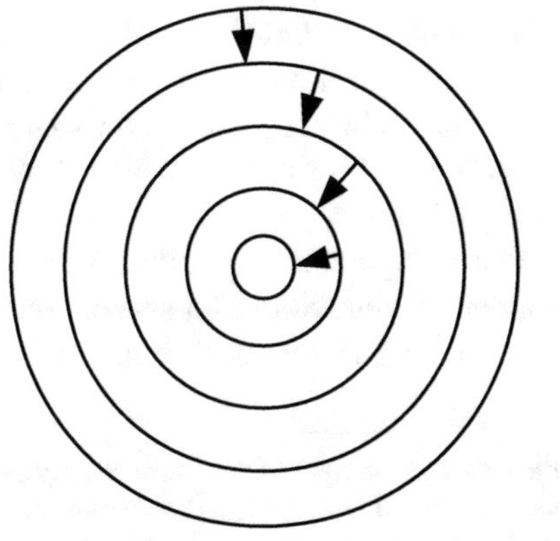

Fig. 6.8 A sequence of quantum transitions representing the collapse of the atom

According to the Bohr Theory, *absorption* occurs when a photon is incident upon an atom and has an energy equal to the difference between the energy of the initial orbit and the energy of an orbit of higher energy. See Fig. 6.9:

According to the Principle of Conservation of Energy, we must have

$$\boxed{E_{\text{ph}} \equiv hf = E_f - E_i} \tag{6.12}$$

which is also known as the **photon absorption condition**.

Thus, for absorption of a photon of frequency f to take place,

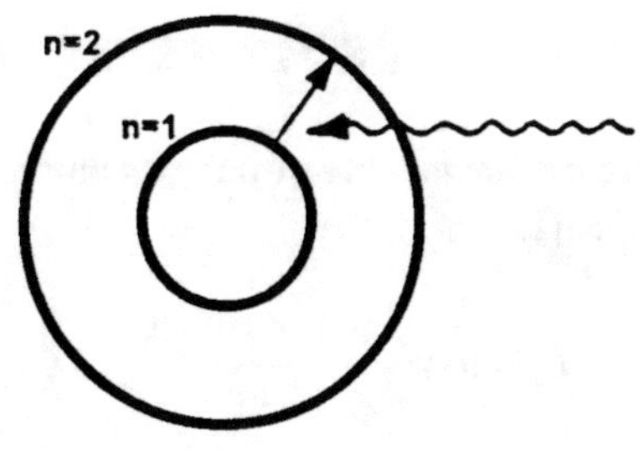

Fig. 6.9 Absorption of a photon by an atom

$$f = \frac{E_f - E_i}{h} \tag{6.13}$$

Sample Problem 6.3 What must the frequency of a photon be for it to be absorbed by a hydrogen atom that starts out in the $n = 2$ orbit and is to make a transition to its $n = 3$ orbit?

Solution

$$E_{\text{ph}} = E_3 - E_2 = (-1.5) - (-3.4) = 1.9 \text{ eV}$$

$$f = \frac{E_{\text{ph}}}{h} = \frac{1.9}{4.1 \times 10^{-15}} = 4.6 \times 10^{14} \text{ Hz}$$

Note An atom in the $n = 3$ orbit can make a spontaneous transition to the $n = 2$ orbit and simultaneously emit a photon of the same energy and frequency. It is clear from Eqs. (6.10) and (6.13) why, in the framework of the Bohr Theory, the absorption spectrum is the same as the emission spectrum.

Sample Problem 6.4 Given that the light flash of a camera has a light power of 100-W lasting 1/1000 s with light having an average frequency of 5.5×10^{14} Hz, calculate the number of photons emitted in the flash.

Solution The energy of the flash is given by

$$E_{\text{flash}} = \text{power} \times \text{time} = 100 \times (1/1000) = 0.1\text{J} \tag{6.14}$$

Since 1-eV = 1.6×10^{-19} J, the energy of a single flash is $0.1J/(1.6 \times 10^{-19} \text{ J/eV}) = 6.3 \times 10^{17}$ eV.

The energy of a single photon is given by

$$E_{\text{photon}} = hf = 4.1 \times 10^{-15} \text{ eV/Hz} \times 5.5 \times 10^{14} \text{ Hz} = 2.1 \text{ eV} \tag{6.15}$$

Therefore the number of photons emitted in a single flash is

$$\text{Number of photons} = \frac{6.3 \times 10^{17}\text{eV}}{2.1 \text{ eV}} = 3.0 \times 10^{17} \tag{6.16}$$

6.4 Quantum Theory

The Bohr Theory accounted for the atomic spectrum of hydrogen very well. But it failed to quantitatively account for the spectrum of any other atom except those having only one electron, such as a Helium atom that has lost one of its two electrons. Ultimately the Bohr Theory was supplemented by a more comprehensive theory called **quantum theory**.

Recall that according to the Bohr Theory, an electron orbits a nucleus as do the planets about the Sun. However, this picture is not substantiated by experiments. Then what is the path of an electron? We will be describing the behavior of electrons in an atom that defies our understanding of the way particles should behave. Quantum Theory is precise in accounting for our observations. Most significantly, it predicts that

> No experiments can allow us to describe how electrons move about. This statement is likely to strike the uninitiated reader as being preposterous since it implies that Physics cannot answer questions that are fundamental to its own ultimate purpose. See Appendix K for a detailed description of the mysterious behavior of photons that leads to the same conclusion about photons.

Instead, we account for experimental observations in terms of the atoms being in one of a set of the so-called **quantum states**, whose significance will be elucidated below:

Suppose an atom is known to be in a certain quantum state. Rather than knowing the precise orbit of the electrons, quantum theory provides us with the **relative probability** for finding an electron at various locations, called the **probability density**. The probability density for some of the quantum states of the hydrogen atom is depicted in Fig. 6.10 below.[4]

It is very important to note that: While a classical mode of vibration has a specific *frequency* of vibration, a quantum probability mode has a

[4]The numbers to the right of the figure correspond to the quantum numbers $n = 1, 2$, and 3. You see one probability density for $n = 1$, two for $n = 2$, and three for $n = 3$. Absent are two others for $n = 2$ and five others of $n = 3$. Thus, in place of Bohr's one state for each n, there is one state for $n = 1$, four for $n = 2$, and nine for $n = 3$.

specific *energy*. At the simple level of description provided by the Bohr theory, the Quantum Theory of absorption and emission of EM radiation is essentially the same as that of the Bohr Theory except that quantum states are not associated with well-defined orbits of electrons. We have merely to replace the word *orbits* with the term *quantum state* in our text.

In Fig. 6.10, the brighter the region, the more likely is it to find an electron in that region. Notice that the patterns of these probability densities bear a remarkable resemblance to the patterns of the modes of vibration of a brass plate. For this reason, we will also refer to these quantum states as **probability modes**.

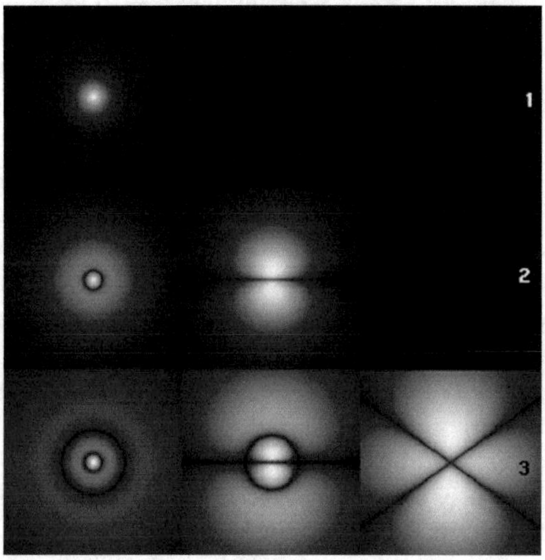

Fig. 6.10 Electron probability clouds (source: http://en.wikipedia.org/wiki/File:HAtomOrbitals.png)

At the top of the figure is the probability cloud for the ground state. The Bohr theory describes the ground state as a classical orbiting electron at the Bohr radius. In its place, quantum theory and experiment describe the electron as not having a well-defined orbit. To the incredible contrary, the electron is most likely to be found at the origin. The second and third rows display probably clouds corresponding to $n = 2$ and $n = 3$, respectively.

Currently, it is possible to lay down a number of atoms on a substrate of made of other atoms, arranged in a pattern of choice. One such example is shown in Fig. 6.11.

We see 48 atoms of iron arranged in a circle—known as a **quantum corral**. Each peak represents the probability distribution of electrons on a single iron atom. An added delight is the presence of concentric crested rings, which represent a standing wave of probability density! Note the similarity with a surface wave of water that is produced by placing a mug of water in a sink and turning on the garbage disposal. If you count the crests, you can determine which circularly symmetric mode is being excited.

Quantum Theory provides us with a mathematical means (that is far beyond the scope of this text to present) of correctly determining the quantum states of all atoms, molecules, and indeed, macroscopic samples of matter containing huge numbers of atoms, that is, solids, liquids, and gases. Quantum theory can account for the

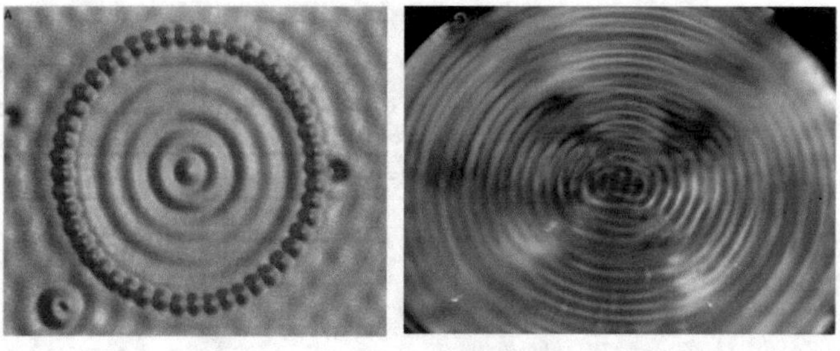

Fig. 6.11 The "Iron Corral": 48 iron atoms on a copper surface, with a standing wave of probability density; Standing surface wave in a mug of water (sources: Corral: From M. F. Crommie, et al., *Science* **262**, 218 (1993); Reprinted with permission from AAAS. photo of water surface by Konstantinos Metallinos)

properties of all materials.[5] Ultimately, quantum theory enables us to explain such questions as to why, at room temperatures, copper is solid, is opaque with a shiny orange appearance, is pliable, and conducts electricity well, while water is liquid, is transparent and colorless, and conducts electricity very poorly.

The quantum theory of the emission of EM radiation can be summarized as follows:

1. Any system has a set of Quantum States (Probability Modes), each with a specific energy.
2. **Emission of EM radiation** occurs via a transition from one quantum state to another having a lower energy, accompanied by the emission of a photon. One might regard a quantum state as a **probability mode**.
3. **Absorption of EM radiation** is the reverse process: A photon of frequency f is absorbed by an atom making a transition from one quantum state to another of higher energy.

   For both emission and absorption, the photon energy and hence frequency is accounted for by the change in energy of the system, as in the Bohr Theory of the hydrogen atom, as given by Eqs. (6.9), (6.10), (6.12), and (6.13).
4. Among all the states of any system, there exists a *state having the lowest energy*. This state is called the **ground state**. The states with higher energy are called **excited states**.
5. Sometimes more than one quantum state has the same energy. We then refer to this set of states as a **quantum energy level**, or simply, **energy level**. (In particular, even the so-called ground state can consist of more than one state.) The set of energies of an atom is referred to as the system's **energy spectrum**, in analogy with the frequency spectrum of a vibrating system.

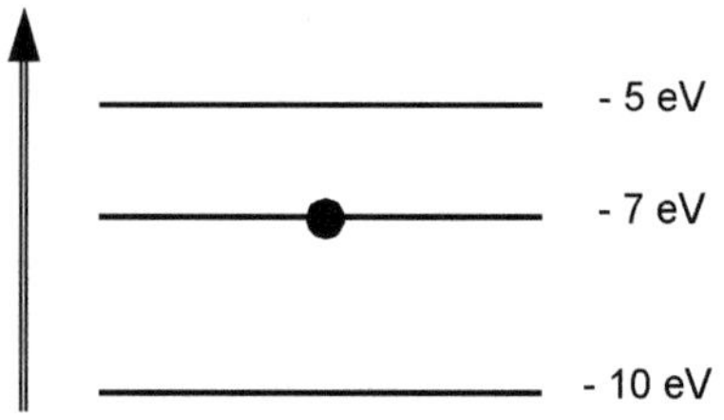

Fig. 6.12 Energy level diagram

It is useful to summarize an energy spectrum in the form of an **energy level diagram**. See Fig. 6.12. Such a diagram is depicted below for a *fictitious* energy spectrum that has, in order to simplify our discussion, integer values of energy when expressed in electron volts. The system is in its first excited level, as indicated by the large dot.

Normally, at room temperature that is, essentially all atoms are in their ground states. An atom can make a transition to an excited state by absorbing energy through a collision with another atom. Transitions from the ground state to one or the other of two excited states are shown in Fig. 6.13. Here, E_1 is the ground state energy.

Once in an excited sate, an atom can spontaneously make a transition to a state having a lower energy. Each such transition is accompanied by the emission of a photon. Thus, referring to the fictitious energy level diagram of Fig. 6.12, with the atom initially in the 2nd excited state, three transitions can take place, with (a) and (b) being consecutive ones. See Fig. 6.14.

Photons having three different energies and frequencies can be emitted, corresponding to the three possible quantum transitions (a)–(c), as follows:

1.

$$hf_a = E_3 - E_2 = -5 - (-7) = 2\,\text{eV}$$

$$f_a = \frac{2}{h} = \frac{2}{4.1 \times 10^{-15}} = 5 \times 10^{15}\ \text{Hz}$$

[5] The theory provides us with equations that need to be solved in order to calculate the material properties. In practice these equations are so complex that they can be solved only approximately. However, any discrepancy between calculated values and observed values is accountable by the approximation of the calculation and not any shortcomings of the theory.

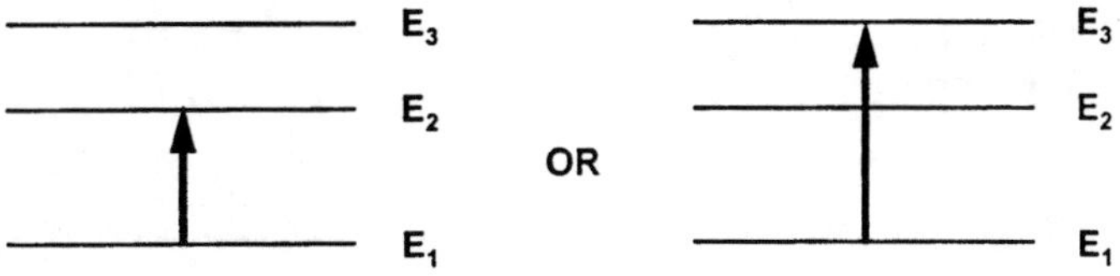

Fig. 6.13 Transitions from the ground state

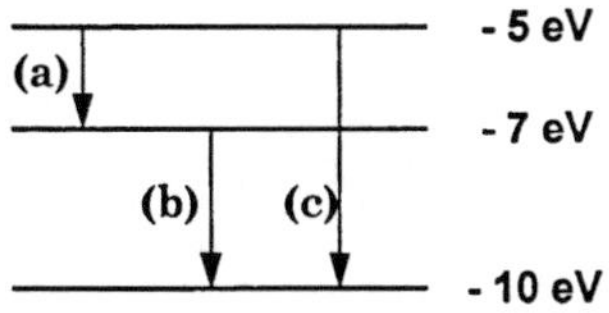

Fig. 6.14 Transitions to lower levels for a three level system

2.

$$hf_b = E_2 - E_1 = -7 - (-10) = 3 \text{ eV}$$

$$f_b = \frac{3}{h} = \frac{3}{4.1 \times 10^{-15}} = 7 \times 10^{14} \text{ Hz}$$

3.

$$hf_c = E_3 - E_1 = -5 - (-10) = 5 \text{ eV}$$

$$f_c = \frac{5}{4.1 \times 10^{-5}} = 1.2 \times 10^{15} \text{ Hz}$$

Photons (a) and (c) are in the ultraviolet regime. Photon (b), being at the boundary between the visible and ultraviolet regimes, is barely visible.

Question: Suppose there are four quantum states. How many transitions are possible from the 3rd excited state? Beware; there are more than four!

Resonance can occur between two identical atoms via the emission of a photon by one atom and the subsequent absorption of that photon by a second atom. This process is depicted on the following pair of energy level diagrams, depicted in Fig. 6.15.

Compare this process with that of the two charged SHOs shown in Fig. 6.16. The force of one atom on another, as described classically, is due to electromagnetic waves. We reproduce the figure here:

In the classical case, resonance has been described as requiring the equality of the frequency of the respective modes of the two oscillators. For quantum systems, resonance requires the equality of the difference in energy between a pair of energy levels.

Note Recall now that we mentioned in our introductory remarks that animal eyes are sensitive to EM radiation because of the behavior of atoms. We need to expand a bit on this remark: The eye is sensitive to a broad band of wavelengths. If individual atoms were the visual sensors in the eye, the absorption spectrum would consist of a small number of spectral lines in the visible regime and the absorption spectrum would be far from continuous. Instead, molecules are used that have a band of a great many excited states, all with energies above the ground state that correspond to visible photons. Absorption with an energy band is quite continuous.[6] We can surmise that evolution produced visual sensors that are sensitive to light from the Sun. As we will see below, the spectrum of sunlight is centered in the visual region of animal eyes.

6.5 Line Width

We will now refine our description of transitions between quantum states of an atom and the resulting atomic spectral lines. In the previous section we have seen the parallels between resonance between two charged oscillators and two atoms. We will now refine our understanding of resonance.

To appreciate line width, it will help to point out the parallels between a **charged harmonic oscillator** and an atom that makes a transition between two states.

For a given spectral line, we have stated that there should be a match between the difference of

[6]See Sect. 15.13 for more details.

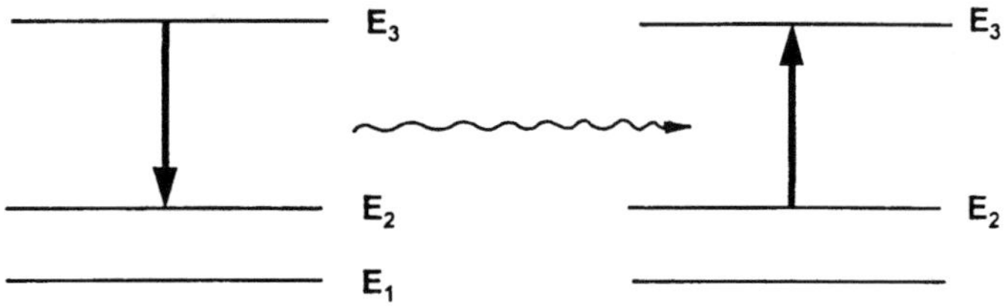

Fig. 6.15 Resonance between two atoms

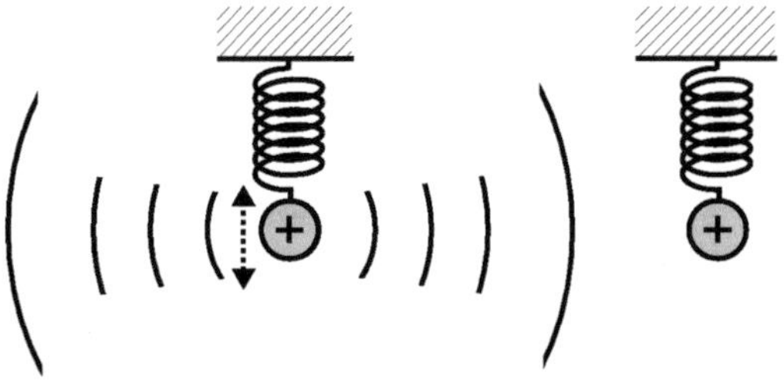

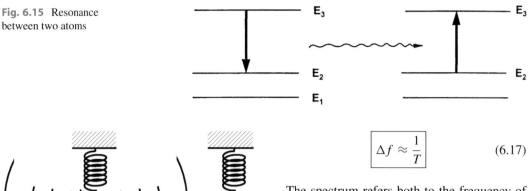

Fig. 6.16 Resonance between two charged SHOs

energies of the atomic states and the energy of the corresponding photon. This should hold true for both the **emission** of a photon—see Eq. (6.9)—and the **absorption** of a photon—see Eq. (6.12). However, the equality is only approximate. At a more precise level, there are factors that reduce the need for precise equality.

If the amplitude of the charged oscillator were constant, and thus not attenuating, the frequency spectrum of the electromagnetic wave emitted would be a spike—infinitesimal in width. Every real oscillator has some degree of attenuation.[7] The resulting attenuated sine wave of an oscillator is shown in Fig. 4.10. As a result of the attenuation, Fourier analysis can be used to determine the Fourier spectrum, which looks like the **response function** of an attenuating oscillator, as shown in Fig. 4.12.The spectrum of the electromagnetic waves emitted by an oscillating charged harmonic oscillator is displayed in Fig. 6.17.

The width of the spectrum of the charged harmonic oscillator is on the order of the inverse of the **attenuation time** T of the oscillator.

Symbolically we write

$$\boxed{\Delta f \approx \frac{1}{T}} \tag{6.17}$$

The spectrum refers both to the frequency of the electromagnetic waves emitted and to the response of the charged oscillator to incident electromagnetic waves.

As we mentioned above, a transition between two states of an **atom** corresponds—but is not identical—to a charged harmonic oscillator. The important question for us now is this: What corresponds for an atom to the attenuation time of an oscillator? The answer is the following: When an atom is in an excited state, it takes time for it to make a transition to a state of lower energy. The actual time that the transition takes place cannot be predicted. However, there is an **average time**—referred to as the **lifetime of the excited state**, with respect to the transition. We will use the symbol τ to denote the lifetime. Analogous to the attenuated harmonic oscillator, the width Δf of a spectral line is on the order of the inverse of the lifetime. Thus

$$\boxed{line - width \approx \frac{1}{lifetime}} \tag{6.18}$$

or

$$\Delta f \approx \frac{1}{\tau} \tag{6.19}$$

Excited states of atoms range over many decades of times, from a small fraction of a second to many years. Suppose that an excited state has a lifetime τ of 10^{-10}-s. Correspondingly, the width Δf is 10^{10}-Hz. Let us note that the highest frequencies of the spectral lines of hydrogen are on the order of 10^{14}-Hz to 10^{15}-Hz. Thus, we see that in this case the width is extremely small compared to the frequency

[7]If this were not so, we would be observing oscillators oscillating today that were set into motion millennia ago. *Attenuation or dissipation is necessary for us even to be able to grab hold of a door knob! In its absence, we might grab the knob and our hands would bounce back..*

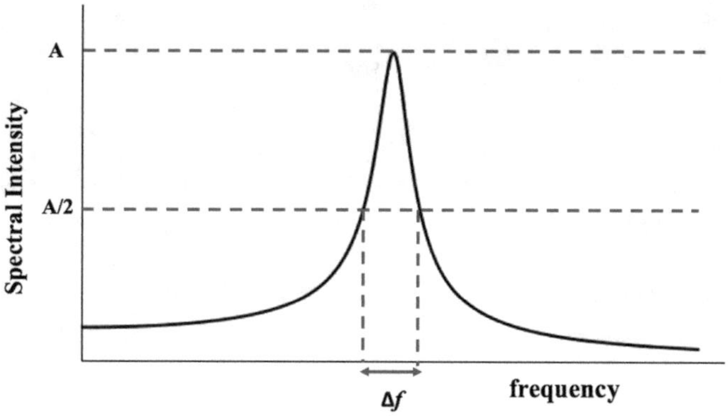

Fig. 6.17 Spectrum of electromagnetic waves emitted by a charged attenuated harmonic oscillator. Δf is the width of the peak at half the maximum A. The curve also describes the spectrum of an atomic spectral line—applying to the **emission spectrum** as well as the **absorption spectrum**

by a factor on the order of one part in 10,000 to 100,000 and intrinsic spectral line is extremely narrow. This is the typical case.

Note The spectral intensity shown in Fig. 6.17 applies to both the photons emitted by an atom in an excited state and the probability that an atom will absorb an incident photon and be excited to a state with higher energy.

Note In practice, there are factors that contribute to the observed line-width in addition to intrinsic properties of an atom. A very significant factor is the *motion* of atoms. In Chap. 9 we will discuss the **Doppler effect**. The effect describes a shift—Δf—in the frequency of a wave that results from either motion of the source of the wave or the motion of the observer of the wave. The Doppler effect applies also to the frequency of photons. Thus, the frequency of the photon emitted by a moving atom is shifted upward if the atom is moving in the same direction as the emitted photon or downward if the atom is moving in a direction opposite to the direction of emission. If a photon is incident on a moving atom, the observed frequency is shifted upwards or downwards if the atom is moving towards or away from the photon. The motion of this atom will affect its absorption. The relative shift of the frequency is on the order of the ratio of the speed v of the atom to the speed of light. We have

$$\frac{\Delta f}{f} \approx \frac{v}{c} \tag{6.20}$$

For an atom in a *gas* at room temperature, the average atomic speed could be about 300-m/s.[1] In this case $v/c \approx 10^{-6}$.

[1] This speed can be shown to be about the speed of sound in the gas.

6.6 Complex Scenarios of Absorption and Emission

We have discussed absorption and emission of photons as two independent processes. We now describe a number of scenarios that are more complex, involving both absorption and emission.

6.6.1 Rayleigh Scattering

The simplest obvious complex scenario is the absorption of a photon by an atom (or a general quantum system such as a molecule) followed by the emission of a photon with the same energy. The atom makes a transition from an initial state i to an excited state f and then returns to the initial state i. Thus we have i to f and back to i. We see this process depicted in figure

The photon emitted is coherent (has a definite phase relation) with the incident photon. When we have a huge rate of photons incident on many such atoms, the overall process is observed as the scattering of a light beam—hence the insert of the term "scattering" in the name **Rayleigh scattering**. Since the outgoing photon has the same energy as the incoming photon, the scat-

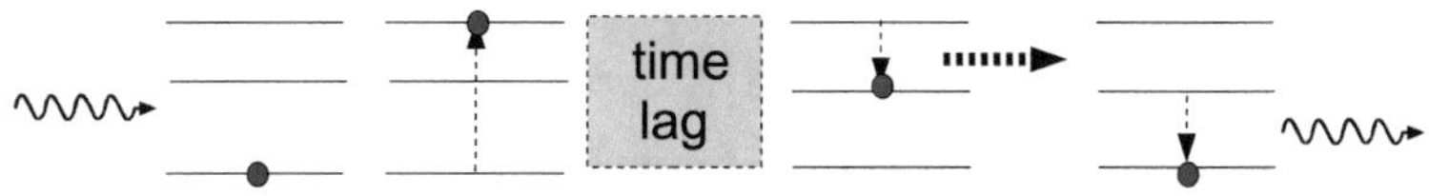

Fig. 6.18 General fluorescence in an energy diagram

tering is regarded as being elastic. It was **Lloyd Rayleigh** who identified this type of scattering of sunlight as being responsible for our blue skylight—at a time when quantum theory was not yet formulated.[8]

6.6.2 Resonant Fluorescence

If a gas of atoms is illuminated with monochromatic radiation having a frequency equal to that of one of its spectral lines, the atoms will continuously emit radiation of the frequency of that spectral line. Simply put, the atoms are *cycled*—repeatedly being excited from the lower state to the corresponding higher state, emitting a photon upon returning to the lower state. The intensity of the emitted light is relatively high. Furthermore, while the external incident radiation can be uni-directional, the resulting emitted radiation is multi-directional.

6.6.3 General Fluorescence

The last scenario we discuss here is **general fluorescence**. In this case, we have absorption of a photon by a molecule from an initial state to an excited state. After an unpredictable time, the molecule gives up its internal energy to the environment and makes a transition to a state of lower energy. Concurrently, or shortly thereafter, the molecule makes a transition to a state of lower energy along with emission of a photon. See Fig. 6.18.

The thick dashed arrow represents the process whereby a part of the excitation energy is replaced by forms of energy such as vibrational or thermal energy. If the time for emission of a photon is relatively long—as long as hours—the phenomenon is referred to as **phosphorescence**.

[8]See Sect. 9.1.2.

This process occurs when the intermediate state from which the final transition to the ground state occurs has a long lifetime.[9] A familiar example is the result of shining ultraviolet radiation on a rock and observing visible radiation (light) afterwards.

The energy given up to the environment can simply be an increase in thermal (heat) energy. In the case of the cones of the retina of an eye some or all of the energy given up goes into the nerve impulse that travels down the optic nerve. See Chap. 13 on The Eye.

6.6.4 Stimulated Emission

Consider an atom that is in an excited state. We know that it can make a transition to a state of lower energy along with the emission of a photon having an energy given by the difference of the energies of the two states. We mentioned in Sect. 6.3 that the process is called **spontaneous emission**. Now imagine that initially the atom is in the above excited state and that a photon of the above energy is incident on the atom. The incident photon can then induce the transition to the lower state, with the emission of a photon. We will then have two photons of the same energy. See Fig. 6.19 in which we see depicted an incident photon of energy hf equal to the energy difference ΔE between two atomic states.

At first you might think that nothing has been gained: We could view the process as a simple combination of spontaneous emission that occurs coincidently with the passage of a photon of the same energy past the atom. The actuality is that the two photons are now correlated: they are in phase and therefore said to be **coherent**.

[9]Note that typically the average time for an atomic transition is on the order of 10-nanoseconds (10^{-8}-s).

Fig. 6.19 Stimulated emission

Stimulated emission is at the heart of the **Laser**. A beam of photons is created by excitation via electric discharge and subsequent deexcitation with photon emission. The photons travel down the length of the laser and contribute to photon production via stimulated emission from excited atoms that have not yet been deexcited spontaneously. Many round trips produce the clusters of coherent photons.

Lasers have very special characteristics. First, the spectrum of the laser beam is extremely sharply peaked, as we see in Fig. 6.20 for a Helium-Neon laser. Second, the laser beam consists of many clusters of photons in each of which all photons are in phase (coherent) with respect to each other. Photons of different clusters aren't in phase. We say that the laser beam is highly coherent.[10] On the other hand, light from most common light sources, such as an incandescent bulb or a fluorescent bulb, is quite **incoherent**. The phases of the photons are randomly distributed because they are emitted in an uncorrelated way. For example, fluorescent light is a result of a huge number of **independent** atomic transitions from many atoms.

6.7 Is Light a Stream of Photons or a Wave?

Until we discussed the Bohr Theory of the atom, we treated light as a wave—which is a continuous disturbance. However, in the context of the Bohr Theory or Quantum Theory of absorption or emission of radiation by an atom, we stated that the radiation can only be absorbed in discrete, indivisible "quanta" of radiation called **photons**. These are the so-called particles of light. Which picture is correct? How can they both be correct?

Controversy as to whether light is a beam of discrete particles or a wave was significant in Newton's day ~ 300 years ago. Newton himself believed in the particle theory. His contemporary, Christiaan Huygens professed the wave theory of light and is responsible for much of what we know about wave propagation today.

The argument regarding the nature of light propagation seemed close to being settled in the 1830s, when Young performed the first wave "interference" experiments[11] which clearly demonstrated the wave nature of light. Further support for the wave theory came from the demonstration that the speed of light in an "optically dense medium" (for example, glass) is smaller than that in vacuum: The particle theory predicted the contrary. Finally, Maxwell had produced a theory of light as an electromagnetic wave and Hertz confirmed some of its predictions. It became accepted that light propagated as a wave and not as a stream of particles.

However, the controversy was not yet settled. In the last decades of the nineteenth century, results of studies of the spectrum of EM radiation emitted by ovens heated to high temperatures—so-called **Black Body radiation** experiments—disagreed with the predictions of classical laws. This led Max Planck, around 1901, to propose that EM radiation is emitted and absorbed in discrete multiples of hf—first called **quanta** and then later called **photons**. Nevertheless, Planck believed that EM itself was not quantized. Rather, only the process of emission or absorption was quantized. Einstein emphasized that since all means of detecting radiation involve absorption, and the absorption process is **quantized**, we might as well regard the radiation itself as being quantized, that is, consisting of a beam of quanta. We have already seen how, at the beginning of the twentieth century, the quantum theory of the atom and EM radiation invoked particle attributes to light.

[10] See the website http://hyperphysics.phy-astr.gsu.edu/hbase/optmod/qualig.html for more details about lasers.

[11] See Chap. 7 for details.

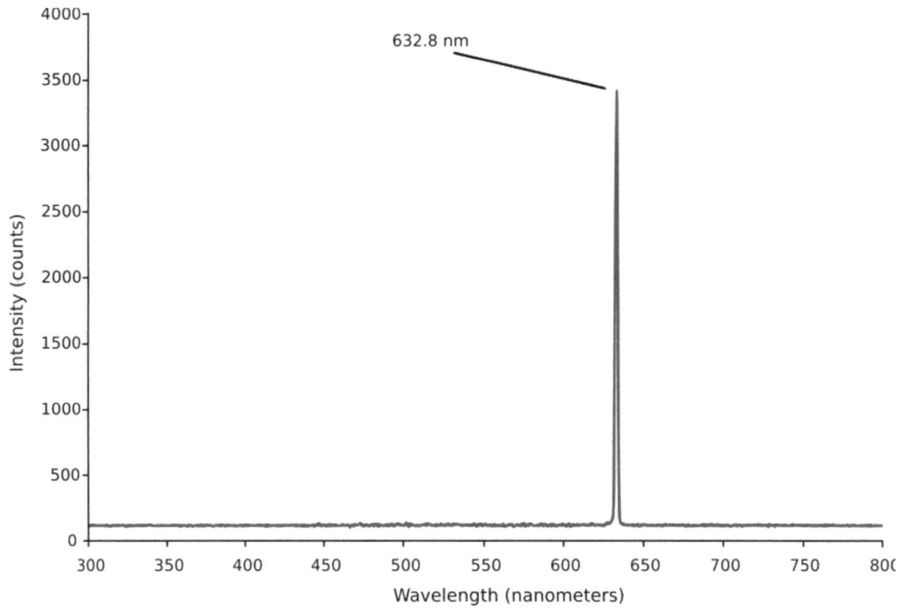

Fig. 6.20 Spectrum of a Helium-Neon laser (source: http://wapedia.mobi/en/Laser)

Thus, we see that some experiments indicate that EM radiation propagates as a wave while others indicate that light is propagated as a stream of particles. Which is the correct description, one might understandably ask? At the level of this text, the question will, unfortunately, have to remain unanswered. All we can say for now is that light, or EM radiation, exhibits *both* wave-like properties and particle-like properties. No experiment has ever, nor, most physicists believe, will ever, be able to reveal the "true" nature of EM radiation in terms of common, familiar concepts. The bewildering wave-particle nature of light is akin to that of the electron. You are invited to study Appendix K, which describes an experiment that reveals the impossibility of describing the path of a photon.

6.8 The Connection Between Temperature and Frequency

We are now prepared to explain why it is that the frequency spectrum of the Sun is concentrated in the visible region. We mentioned above the role that the study of blackbody radiation played in the development of the quantum theory of radiation. In Fig. 6.21 we see the spectrum of blackbody radiation graphed against wavelength (*not* frequency) for various temperatures (absolute temperature $T = {}^\circ C + 273$). The graph is a plot of the **spectral intensity with respect to the wavelength**. See Sect. 14.3 for a detailed discussion of this function.

Notice how the curves change as the temperature increases.[12] We note that the *height of the peaks* increases with increasing temperature. This is so because it is a general property of matter that energy increases with increasing temperature. In the case of black body radiation in particular, the average energy of the photons is proportional to the absolute temperature T. According to the Planck equation frequency is proportional to the energy—$E = hf$. As a consequence, the average frequency of the photons increases with increasing temperature.

[12]Regarding Fig. 6.21 you can see the Black Body curve for any chosen temperature as well as the corresponding color of the spectrum using the following physlet: (12-29-2010): http://ephysics.physics.ucla.edu/physlets/eblackbody.htm.

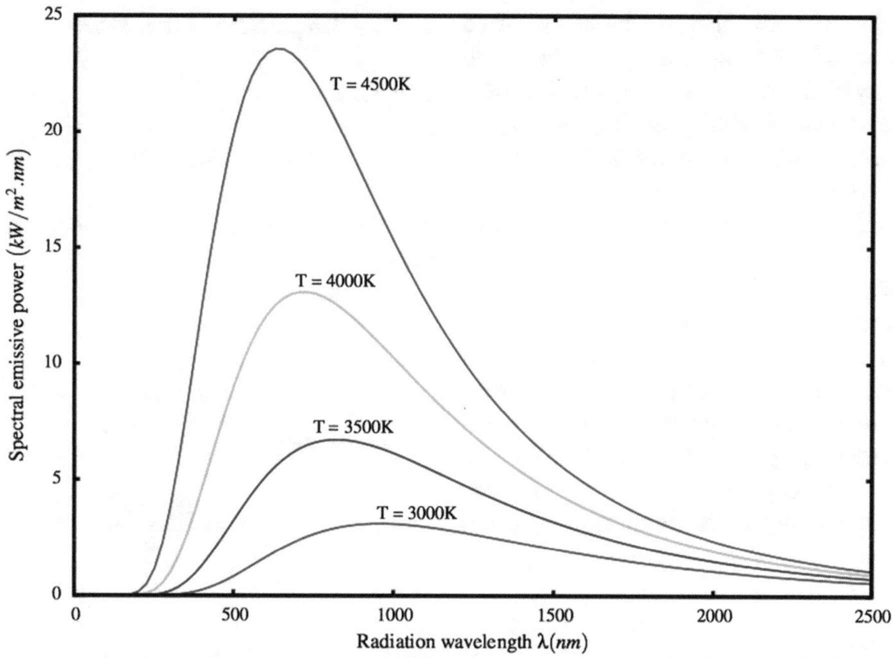

Fig. 6.21 Blackbody frequency spectrum (Source: https://neutrium.net/heat_transfer/blackbody/)

Now let us turn to the *position of the peaks*. Note that it moves towards lower wavelengths as the temperature increases.[13] Here is the explanation: As the temperature increases, the frequency at the peak increases. The wavelength at the peak decreases correspondingly because of its inverse relation with the frequency: $f = c/\lambda$.

Correspondingly, the frequency associated with the peak increases with increasing temperature, being proportional to the temperature T. That is,

$$f_{\text{peak}} \propto T \tag{6.21}$$

The decreasing peak wavelength with increasing temperature is associated with the changing color of materials that are heated up. For example, as a flame gets hotter, its color changes from red (at the low frequency end of the visible spectrum) to white (a broad spectrum centered around the middle of the visible spectrum; and finally to blue (at the high frequency end of the visible spectrum). (See Chaps. 13 and 14 for details regarding the connection between color and the frequency spectrum.) Notice that the spectrum of the Sun is very close to that of blackbody radiation at a temperature of 6, 000 K, which is the surface temperature of the Sun. This connection between color and frequency and temperature enables astronomers to determine the temperature of distant matter in outer space.

One of the most outstanding applications of our knowledge of blackbody radiation is in the connection with the 3 K blackbody radiation of the **Big Bang**. Its discoverers, Penzias and Wilson, had detected unexpected radiation that they attributed to *noise* [random radiation] from the Earth. Ultimately, this radiation was shown to be associated with radiation produced as a result of the Big Bang. The first step in this identification was their fitting the spectrum of this radiation to a blackbody curve as shown above and fitting the temperature to 3 K. Coincidentally, cosmologists, who study the origin and evolution of the Universe, had already predicted

[13]The interested reader can carefully check from the curves that the wavelength at the peaks is inversely proportional to the absolute temperature.

the existence of such radiation with an estimated temperature of 3 K. This result gives support to the validity of Einstein's theory of Gravitation—known as **General Relativity**. Recent penultimately precise measurements of the *variation of the temperature with direction in the sky* have led to further significant cosmological information.

Our study of the world of atoms and light has enabled us to see how physics weaves an intricate web of concepts and establishes quantitative relationships among various physical parameters, such as frequency, wavelength, temperature, energy, pressure, displacement, and time. We can appreciate why physicists feel that they are obtaining a knowledge of the ultimate truths about the Universe.

6.9 Terms

- absolute temperature
  (Example: 100 K = 100 Kelvins)
- absolute zero temperature
- atomic spectra
- black body radiation
- Bohr radius—a_0
- Bohr Theory of the hydrogen atom
- classical physics vs. modern physics
- diffraction grating
- discrete spectrum
- empirical formula
- energy level
- energy level diagram
- ground state
- line spectrum
- photon
- photon absorption
- photon emission
- Planck relation
- Planck's constant
- probability density
- probability mode
- quantized process
- quantum state
- quantum theory
- spectral analysis
- wave packet

6.10 Important Equations

Planck relation:

$$E_{\text{ph}} = hf \tag{6.22}$$

Note below the order of the two energies, the initial E_i and the final E_f.

photon emission:

$$f = \frac{E_i - E_f}{\hbar} \tag{6.23}$$

photon absorption:

$$f = \frac{E_f - E_i}{\hbar} \tag{6.24}$$

6.11 Problems for Chap. 6

1. What are the frequencies and wavelengths of all photons that would be ultimately emitted by an atom that has energy levels with the following energies: 3 eV, 2 eV, and 0 eV?
2. What key feature of Balmer's formula was used by Bohr in his theory of the hydrogen atom? Explain.
3. What does the term "empirical" mean? Why was Balmer's formula merely empirical?
4. What catastrophe plagued the Rutherford model within the context of classical theory? How does the Bohr Theory deal with this catastrophe?
5. Determine the frequency and wavelength of a photon having an energy of 1 J; of 1 keV.
6. Determine the energy of a microwave photon of frequency 2500 MHz.
7. Read problem (8) in Chap. 4 that deals with the **stroboscope**.

   Estimate the number of photons emitted by the stroboscope in a single flash.
8. What is the length of a photon whose frequency is 7×10^{14} Hz if it consists of 100 oscillations?
9. Describe resonance between two atoms according to quantum theory.

10. Given that the wavelength at the peak of the black body intensity curve (Fig. 6.21) is 4,300 $\AA$ at a temperature of 7,000 K, calculate the corresponding wavelength at a temperature of 5,000 K using the relation: peak photon frequency $\propto$ absolute temperature. Compare your result with that taken from Fig. 6.21.
11. Suppose that a 100 Watt yellow incandescent light bulb has an efficiency of 3% for producing visible light. (That is, only 3% of the input electric power is converted into light). Its average frequency is about 5.5 $\times 10^{14}$-Hz. An observer looks directly at the bulb at a distance of 10-m from the bulb.
    (a) Assuming that the light is isotropic, what is the intensity of the light incident on the eye?
    (b) Estimate the area of the pupil of an average human eye.
    (c) Neglecting reflection and absorption by the pupil of the eye, determine the power of the light entering the eye through the pupil.
    (d) How much visible light energy enters the eye in one second?
    (e) For simplicity, suppose that all the visible light entering the eye has a wavelength of 6000 $\AA$. What is the frequency of the EM radiation?
    (f) How much energy does a single photon of this radiation have?
    (g) How many photons enter the eye in one second?
12. Lawrence Livermore National Laboratories reported making a laser that could produce a laser pulse that had a peak power of about one petawatt [= 10^{15} watts] and that lasted for about one-half picosecond [= 0.5×10^{-12} s]. [See the website 12-26-2010: https://www.llnl.gov/str/Petawatt.html].
    (a) Calculate the total energy of one pulse.
    (b) Given an intensity of 700 Giga 10^9-Watts/cm^2 during the pulse, calculate the area of the laser beam.
    (c) How long would a 100-W light bulb have to burn to produce the same energy?
    Note how the laser concentrates light in many ways:
    - **frequency with respect to spectral intensity**,
    - **intensity with respect to area,**
    - **phase with respect to coherence,**
    - **and here, energy with respect to time**.
    (d) If the wavelength of the laser light were 530-nm, how many photons are there in a single pulse?

7 The Principle of Superposition

Suppose there are two sources of waves of a given type. For example, there may be two loudspeakers emitting sound waves or two accelerating charges emitting EM waves. What is the resultant wave? Or, suppose that two pulses are sent down a string, one after the other, so that the second pulse "collides" with the first one after the first one has been reflected from the opposite end. What happens as a result of this collision? Such questions are answered by the **principle of superposition**, which states that:

The wave that results from two independent sources—the so-called **resultant wave**—is a simple sum of the two waves that would in turn be produced by the respective sources if each were present alone. This principle will be illustrated by numerous applications in this chapter.

7.1 The Wave Produced by Colliding Pulses

In Fig. 7.1 we depict the wave pattern at five different times that results from a "collision" of two pulses that are traveling in opposite directions along a string. The solid curve represents the observed wave. Initially we see the two pulses far apart. In the next figure we see the two pulses having just begun to overlap—we might say they are "colliding." The two individual pulses are shown as dashed curves, as they would appear if they were present alone, each without the other. What is actually observed is the solid curve, which is a sum of the two dashed curves. Notice how, in the third curve of Fig. 7.1, the displacement vanishes everywhere along the string. There is no potential energy at this instant. What has happened to the energy that was in the pulses? All of the energy of the wave resides in kinetic energy. Finally, we see how the two pulses survive the collision intact, moving away from each other, as if the collision had not taken place.

Below, in Fig. 7.2, is a second example of two colliding pulses—here square pulses. They have the same amplitude and move past each other at the same speed of 1 unit/s. We see the resulting wave at four different times. At $t = 1$ s, they are about to "collide." At $t = 2$ s, we see how each of the pulses would look if each were traveling alone. However, the actual wave is obtained by adding the two graphs. There is complete cancelation in the middle segment. The resultant is shown as the orange graph. At $t = 3$ s and thereafter, the two pulses are seen retreating from each other as if neither were affected by the other.

7.2 Superposition of Two Sine Waves of the Same Frequency

By itself, a sine wave is characterized by its amplitude and either its period in time or its wavelength in space. However, when two sine

L. Gunther, *The Physics of Music and Color*,
https://doi.org/10.1007/978-3-030-19219-8_7

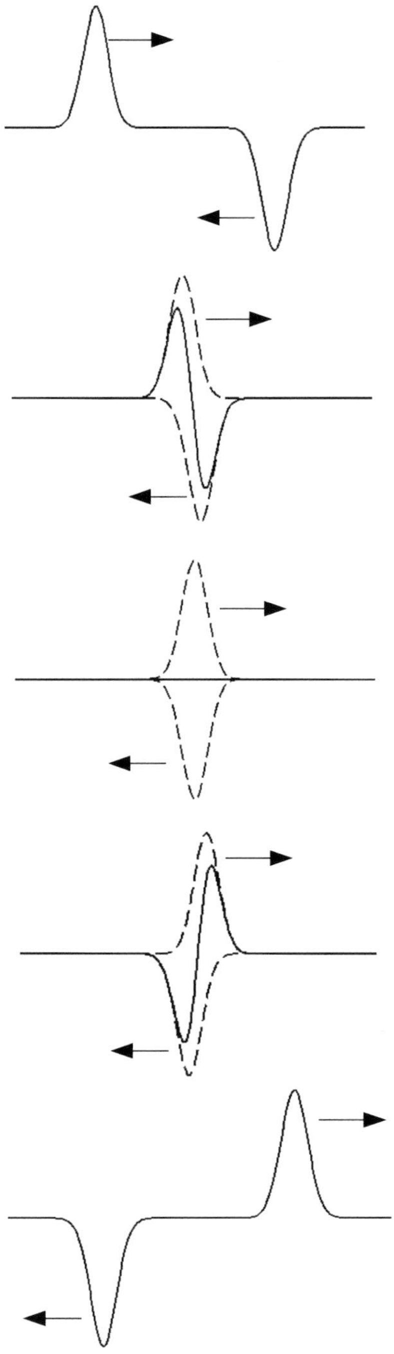

Fig. 7.1 Collision of two pulses

waves are to be added, the relative position of their peaks matters too. Recall from Chap. 2 that this characteristic is called the **phase difference**, or equivalently, the **relative phase**.

Below, in Fig. 7.3, we see three sine waves having the same period in time (one second) and the same amplitude, but a non-zero relative phase with respect to each other. We simply say that they are **out of phase** or *have different phases*.

Clearly, since the crests of wave A coincide with the troughs of wave B, the waves will cancel each other when added. Thus $A + B = 0$.

Generally, the relative positions of the waves can be expressed in fractions of a cycle or as an angle ranging from 0° to 360°, with 360° representing a full cycle.

Thus, relative to wave A:

- Wave B is $1/2$ cycle or 180° ahead or behind A.
- Wave C is $1/4$ cycle or 90° ahead of A.

NOTE: Two waves with no phase difference are said to be **in phase**.

Note Sine waves have the following amazing property that we mentioned in Chap. 2 and that we will review in greater detail here:

Two sine waves having the same period add up to a sine wave of the same period.

The amplitude of the resulting sine wave depends upon the amplitudes of the component sine waves and their relative phase.

Thus, we see in Fig. 7.3 that the sum of waves A and C is a sine wave with the same period as A and C.

Note Above, we have described the superposition of two sine waves as they vary in time. The behavior is purely mathematical. Thus, if we have two sinusoidal patterns in space, we can apply the above results by replacing the word *period* by the *wavelength*.

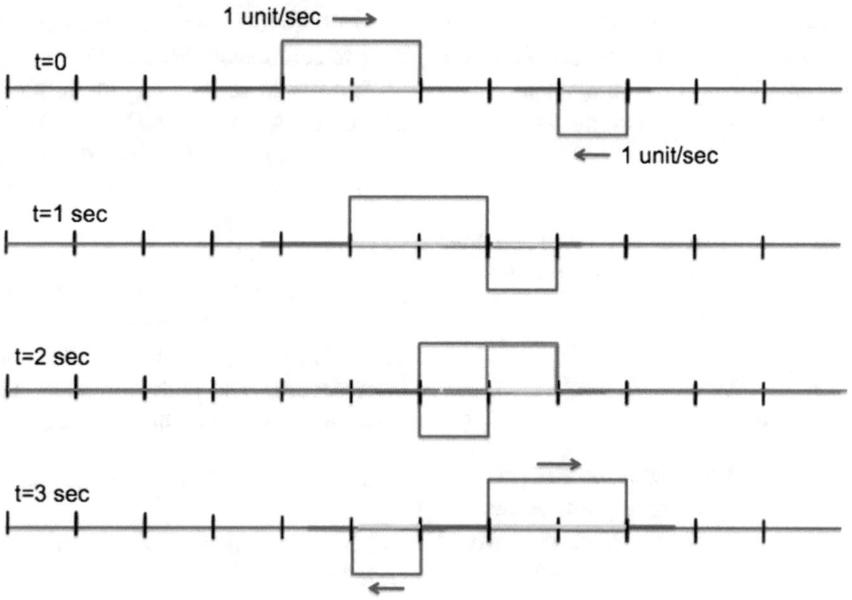

Fig. 7.2 Collision of two square pulses

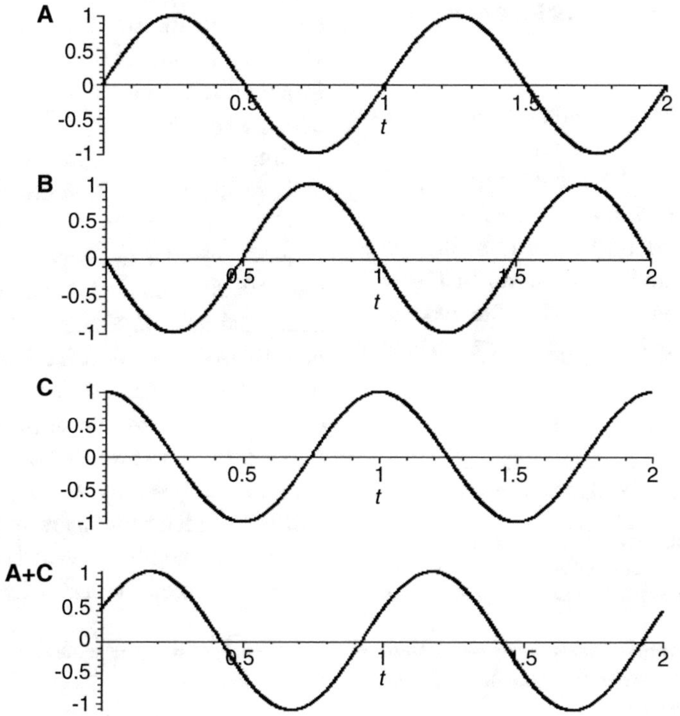

Fig. 7.3 Superposition of two sine waves

Let A_1 and A_2 be the amplitudes of the respective components and A the amplitude of the sum or **resultant** amplitude. If the components are *in phase*, the resultant amplitude is given by

$$A = A_1 + A_2 \tag{7.1}$$

Note that if the two amplitudes are equal, the resultant amplitude is double the individual amplitude.

If the components are 180° *out of phase*, the resultant amplitude is given by[1]

$$A = |A_1 - A_2| \tag{7.3}$$

Thus, referring to the figure on the previous page, waves A and C are out of phase and have the same amplitude. If added, they would therefore cancel and we would be left with no wave at all![2]

7.3 Two-Source Interference in Space

Consider the resultant traveling wave from two **point sources**,[3] S_1 and S_2, respectively, which emit sine waves of the same frequency (and hence wavelength), with amplitudes A_1 and A_2, respectively. We will assume here that the waves start out from S_1 and S_2 in phase. In Fig. 7.4, we have captured an instant when the two waves have a crest at their respective sources. The two sets of concentric circles represent the crests of the two component waves.

The key to determining the resultant wave at any particular position P in space lies in the *difference between the respective distances from the point to the two sources*—which we will refer to as the **path difference**.

In the figure, where two circles cross, the path difference is a multiple of the wavelength. At such points, the component waves arrive in phase. Then we have what is called **constructive interference**. The path difference satisfies the equation, with an infinite sequence

$$\overline{S_2P} - \overline{S_1P} = 0, \lambda, -\lambda, 2\lambda, -2\lambda, \ldots \tag{7.4}$$

which is the condition for constructive interference. Here, $\overline{SP}$ refers to the distance between two points, S and P.

Such a point P is shown in Fig. 7.5. We see two sine waves leaving the sources in phase. The distances to the point P are $\overline{S_1P} = 5 + 1/4$ wavelengths and $\overline{S_2P} = 6 + 1/4$ wavelengths. (They leave the source with a value of zero and end up at a peak.)

Since the components are sinusoidal in time, their resultant sum is sinusoidal. The resultant amplitude is given by $A = A_1 + A_2$ (Eq. 7.1).

Now consider the opposite extreme, wherein the path difference is a multiple of λ plus an additional half wavelength. (The path difference is said to be an **odd-half-integer** number of wavelengths.) The two sine waves, which started out in phase, arrive out of phase at point P. We have **destructive interference** and the amplitude at point P is given by $A = A_1 - A_2$ (Eq. 7.3). In particular, if the component amplitudes are equal, the resultant amplitude vanishes (Fig. 7.6).

The condition for destructive interference[4] is

$$\overline{S_2P} - \overline{S_1P} = \frac{\lambda}{2} - \frac{\lambda}{2}, \lambda + \frac{\lambda}{2} = 3\frac{\lambda}{2}, -3\frac{\lambda}{2}, \ldots \tag{7.5}$$

which is the condition for destructive interference.

[1]Generally, the amplitude is given by

$$A = \sqrt{A_1^2 + A_2^2 + 2A_1A_2\cos(\phi_1 - \phi_2)} \tag{7.2}$$

Note that the phase difference is $\phi_1 - \phi_2$. You can plug into this expression the two phase differences, $\phi_1 - \phi_2 = 0$ or 180^0, to reproduce the special cases—in phase or out of phase.

[2]For a general phase difference ϕ (which ranges from 0° to 360°), and when the two-source amplitudes are equal, the amplitude is given by

$$A = 2\cos^2\frac{\phi}{2}.$$

[3]A real source takes up space. A point source is a term used for a source that is so localized that we can regard it as taking up no space.

[4]When $A_1 = A_2$, and for general path differences, the resultant amplitude is given by $2A\cos^2\phi/2$, where the phase difference $\phi = 2\pi(\overline{S_2P} - \overline{S_1P})/\lambda$. Compare this expression with Eq. (7.1).

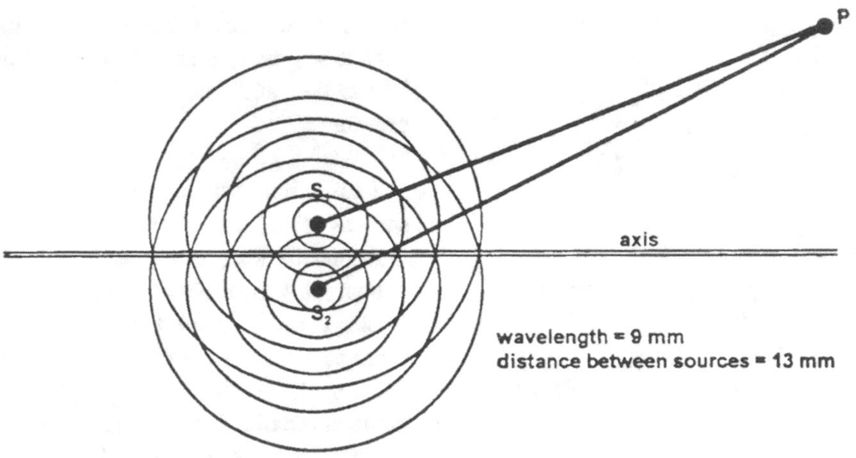

Fig. 7.4 Two point sources

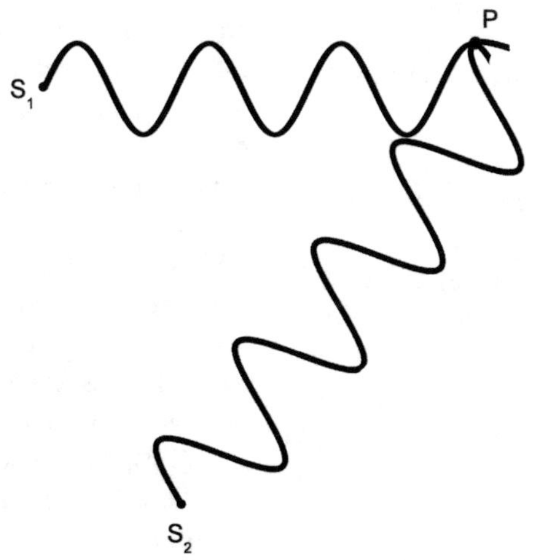

Fig. 7.5 Point where waves arrive in phase

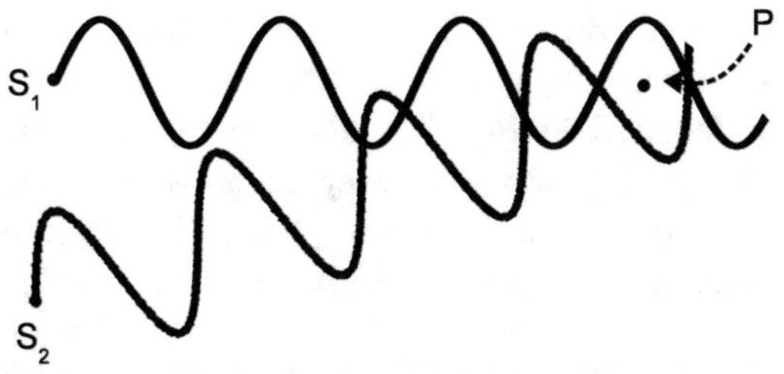

Fig. 7.6 Point where waves arrive out of phase

In Fig. 7.7 we depict the curves in space all along which there is constructive interference. These curves are called **hyperbolas** by mathematicians. The two sources are a distance d apart. Far away from the two point sources, the curves approach straight lines, indicated by the dotted lines, at various angles θ with respect to the axis. They are labeled $m = 0, 1, 2, \ldots$ corresponding to the path differences $0, \lambda, -\lambda, 2\lambda, -2\lambda, \ldots$.

The angles satisfy the equations:

$$\begin{aligned} \sin\theta_1 &= \frac{\lambda}{d} \\ \sin\theta_2 &= 2\frac{\lambda}{d} \\ \sin\theta_n &= n\frac{\lambda}{d} \end{aligned} \tag{7.6}$$

where $\sin\theta$ is the trigonometric sine function.

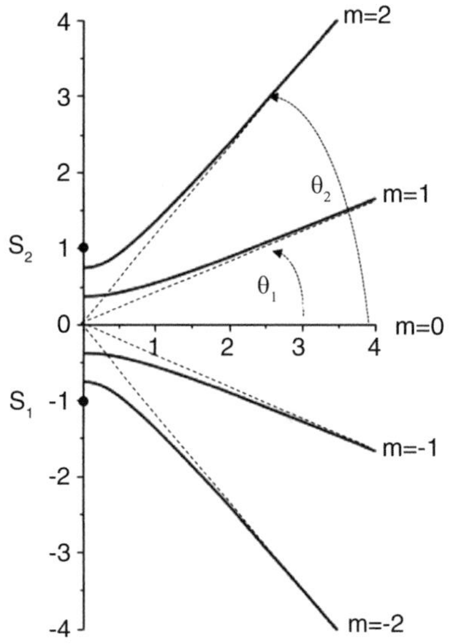

Fig. 7.7 Hyperbola curves along which there is constructive interference

Each angle specifies an **order of interference**.

Sample Problem 7.1 Suppose that $\lambda = 6 \times 10^{-7}$ m and $d = 16 \times 10^{-7}$ m. Find all of the angles of constructive interference. This ratio $\lambda/d = 3/8$ of the wavelength to the separation of sources is shown in Fig. 7.7, where d is equal to four units on the y-axis.

Solution We have

$$\sin\theta_1 = \frac{\lambda}{d} = \frac{6}{16} = 3/8,$$

so that $\theta_1 = 22.0°$, corresponding to the 1st **order**.

Also,

$$\sin\theta_2 = 2\frac{\lambda}{d} = 2 \times 3/8 = 3/4,$$

so that $\theta_2 = 48.5°$, corresponding to the 2nd **order**.

Note that $\theta_2 \neq 2\theta_1$.

Next, note that according to our Eq. (7.6),

$$\sin\theta_3 = 3\frac{\lambda}{d} = 3 \times 3/8 = 9/8,$$

which is impossible, since the sine of an angle cannot be greater than one. The equation cannot be satisfied, and thus, $m = 2$ is the highest order present.

One might ask how we can produce two sources that start out in phase since two independent sources typically do not have a known and controllable phase relation. Here is one commonly used method: We start with a single source of light, preferably light from a laser. Laser light is an intense beam of monochromatic light that is an ensemble of a many clusters; each cluster has a huge number of photons that are in phase with one another. We say that the light is extremely **coherent**. The beam of light is projected onto an opaque surface that has two parallel slits. These slits produce our two sources starting out in phase with each other, as seen in Fig. 7.8. We then view the light projected on a distant screen.

The image on the screen consists of a strip with varying brightness. Analysis predicts that the brightness will vary along the screen as shown in the figure below. We note the set of vertical lines representing bright light. These lines are referred to as **fringes**. According to Eq. (7.6), the **closer** the two sources are (the smaller d is), the **greater** the angles $\theta_{1,2,3,...}$ are and the **further** apart will be the fringes on the screen.

Sample Problem 7.2 Suppose that a sound wave of wavelength one meter is incident on a surface with two holes that are 1.5 m apart, as depicted in Fig. 7.9. You are to determine whether the waves arriving at point P are in phase or completely out of phase. The distance of point P from the surface is 2 m.

Solution We cannot use a simple formula for the path difference. We must calculate it in association with this particular problem. Thus, the shorter path (from the upper hole)—call it ℓ_1—is simply one-meter. The second path has a distance ℓ_2 equal to the hypotenuse of the right triangle, namely

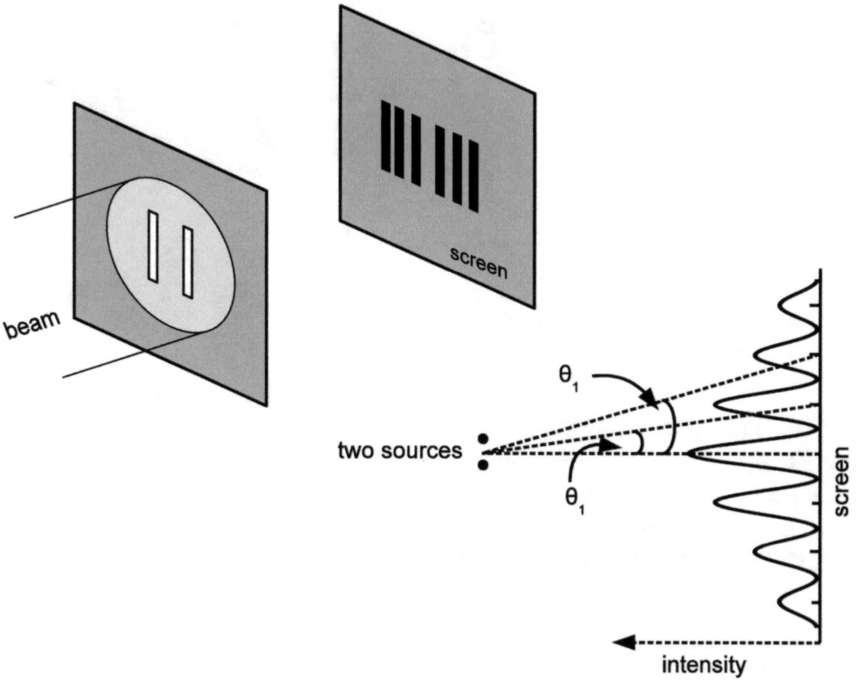

Fig. 7.8 Two-slit interference

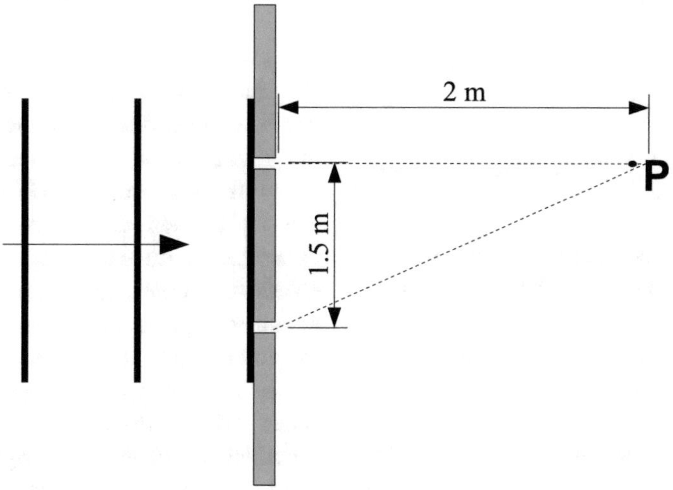

Fig. 7.9 Wave incident upon an Opaque plane with two slits

$$\ell_2 = \sqrt{1.5^2 + 2^2} = 2.5\ m \tag{7.7}$$

Since the wavelength is one-meter, the path difference $[\ell_2 - \ell_1]$ is two and one-half wavelengths. Therefore the waves arrive completely out of phase and we have destructive interference.

Sample Problem 7.3 Look at Fig. 7.10. It represents the following scenario: Waves from an FM radio station at point S with a carrier wave frequency of 100.1-MHz are received by a radio in a car A that is 5-km away. In addition, a neighboring car B that is a distance of 3-m from

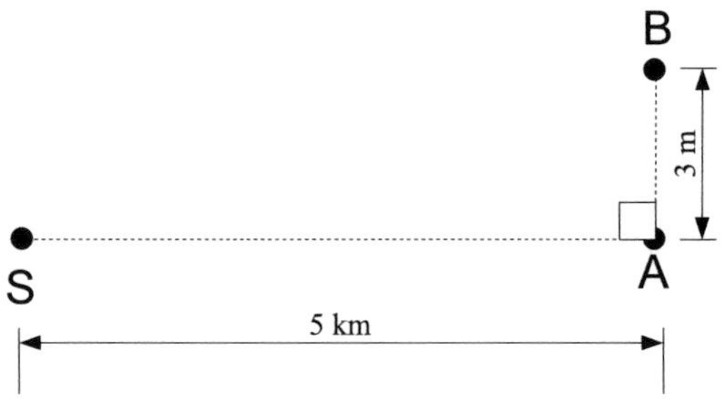

Fig. 7.10 Interference of a radio wave by a neighboring car in traffic

A car **reflects** the radio waves from the radio station towards car A, so that car A's radio receives two interfering waves. The indirect wave travels along the path S to B and then from B to A.

IMPORTANT INFORMATION: When the radio wave reflects off car B, the phase of the wave is shifted by one-half cycle. Thus, at car B, the radio wave is flipped, as shown for a wave along a taut string in Fig. 2.5.

Determine whether or not the waves, the direct wave and the indirect wave, arrive in phase.

Solution We first calculate the wavelength of the carrier wave. We have

$$\lambda = \frac{c}{f} = \frac{3.00 \times 10^8}{100.1 \times 10^6} = 3.00\text{-m} \qquad (7.8)$$

Next we calculate the path difference. We have ℓ_1=5-km=5000-m. The second path length is the **sum** (because of the reflection) of the hypotenuse and the vertical leg of the right triangle (=3-m). The hypotenuse is given by

$$\sqrt{(5000)^2 + 3^2} = 5000 + 0.0009 \sim 5000\text{-m} \qquad (7.9)$$

Therefore the second path length is $\ell_2 \sim 5003$-m. Finally, the path difference is about 3-m and equal to the wavelength. Without the flipped phase due to reflection, the waves would arrive in phase. With the flip, the waves arrive OUT of PHASE and we have destructive interference.

As the cars move along, there will be evolving relative positions that will result in the two waves having varied phase relations.

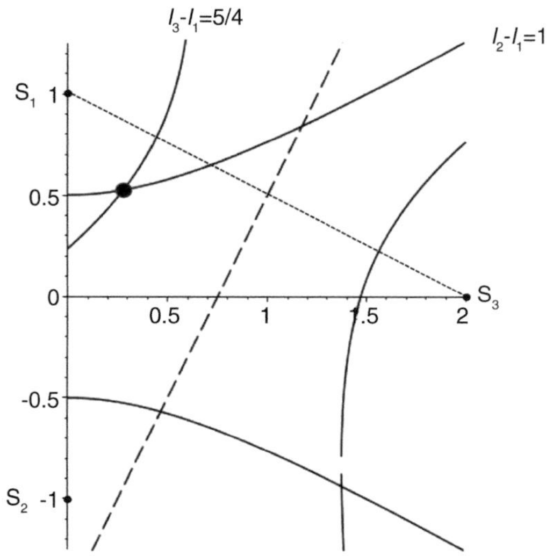

Fig. 7.11 Detecting the location of a gunshot

7.3.1 Detecting the Epicenter of an Earthquake or the Location of a Gunshot

The mathematics of **two-source interference** can be expanded to study the interference from three sources of waves, which we will refer to as **three-source interference**. Its essence can be applied to the determination of the **epicenter of an earthquake** or the **source of gunshots** in cities. The signal from the epicenter of an earthquake travels through the earth via longitudinal sound waves as well as **transverse waves** associated with the resistance of rocks to shear.[5] We will discuss the location of gunshots, for which sound waves propagate the signals.

In Fig. 7.11 we see three detectors, located at positions S_1, S_2, and S_3. Distances are expressed in units of the distance traveled by sound in one second.[6] The goal of the detectors is to determine

[5]The two types of waves travel at different speeds and can be distinguished by having different characteristics.

[6]A light year is the distance traveled by a light signal in vacuum over the course of a year. This distance is about 9.5 trillion-km. We could call our distance "one sound-second." If the speed of sound is 340-m/s, the unit is 340-m.

the position of the gun. According to the figure, detector #1 receives the sound 1-s before the sound from #2 and 5/4-s before the sound from #3. The curves in the figure are hyperbolae, analogous to the hyperbola in Fig. 7.7. The intersection of the hyperbolae locates the gun—at the large black dot.[7]

7.3.2 Sound Level with Many Sources

In Chap. 4, we saw how the sound level is defined in terms of the intensity. Suppose that the wave has **many sources**. What is the resulting sound level? What matters is the intensity. The intensity depends upon whether or not the sources are **coherent**. Two waves are said to be **coherent** if they have a definite phase relation. In this case we also say that the sources are coherent.

Consider first two coherent sources that have the same frequency and produce two individual waves that have equal amplitude and are in phase at some location in space; we have constructive interference. In this case, we know that the amplitude is doubled. Consequently, the intensity is **quadrupled**! How can we obtain four times the intensity instead of the doubling we would expect? Are we gaining energy from nowhere, thus violating the principle of conservation of energy? The resolution of this dilemma is that there are other places in space where we have destructive interference. There, the intensity vanishes. Thus, on the average, the intensity is double the intensity of one source.

The result is that the sound level is increased by 10 log(4) = 6dB at points of constructive interference. If we have n-coherent sources that produce waves that are equal in magnitude and are all in phase, the sound level is increased by $10\log\left(n^2\right) = 20\ \log(n)$ dB.

[7]"ShotSpotter" is a company that provides cities with equipment that includes a lattice of sound detectors (microphones) that are set up throughout the city in optimum locations. The detectors send signals to a central system that analyzes all sounds so as to identify gunshots and find the location of a gunshot. See https://www.shotspotter.com for details about this corporation.

Next, suppose that we have two sources that are **incoherent** and produce the same intensity at some location in space. The result is that the total intensity is doubled, so that the sound level is increased by 10 log(2)=3-dB. With n-incoherent sources, each with the same intensity at some location in space, the intensity is multiplied by a factor of n, so that the sound level is increased by 10 log(n)-dB.

As an application, suppose that we have a string orchestra with 100 instruments each producing a sound level of 1-dB at some location. The resulting sound level will be (1+10 log(100))=21-dB.

7.3.3 Photons and Two-Slit Interference

We learned in Chap. 6 that while light propagates as if it were a continuous wave, in fact the wave characteristics represent a distribution of photons in space. The lack of continuity is reflected by the fact that if we have an ultra-low intensity of light, so that there is an ultra-low density of photons, one can use a **photon detector** to look for photons. The detector always detects individual photons. It never detects part of a photon. If two detectors are in different locations, only one of the detectors will detect a photon at a time.

Let us consider a two-slit experiment with a light beam of such low intensity that only one photon at a time can be found between the two slits and the screen. The result is that the photons will impinge on the screen in a random fashion but be distributed according to the interference pattern we observe with a high intensity of light. In Fig. 7.12 the top image represents a screen; each spot represents a position where a photon struck the screen. We see that the spots aren't distributed uniformly. Instead, the distribution reflects the interference pattern shown below that would be observed with a high intensity beam.

We can place two detectors in the apparatus, one behind each of the slits. The result is that only one or the other detector registers that a photon

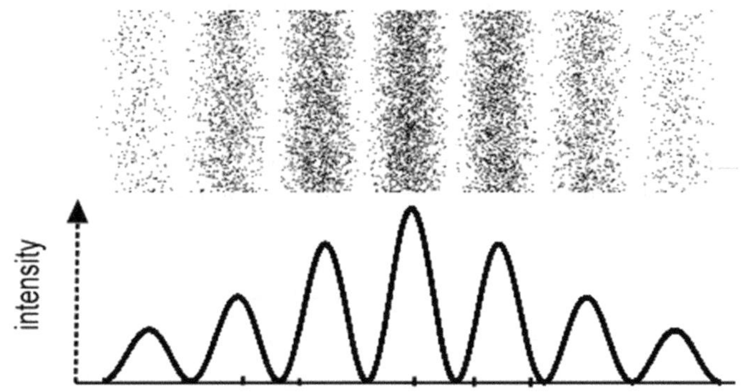

Fig. 7.12 Interference with a low intensity beam of photons

has passed through a slit. It would seem fair to say that a photon exits one slit or the other—never both slits at a time; it doesn't split into two pieces. The question is how the photons can land on the screen with a pattern of interference.

It is beyond the scope of this book to discuss the answer to this question here. My goal is only to try to entice you, to get you to think and wonder. For a detailed discussion of this experiment, I recommend that you to see the Feynman video referred to in Sect. 6.4.

7.4 Many-Source Interference

We will now discuss the interesting behavior of various ensembles, each of which has many sources that are arranged in a periodic manner in a plane or in three-dimensional space. There are cases where the sources are not point sources. Nevertheless, it can be shown that if each of the multitude of sources is spread out while the sources are identical, the result is an interference pattern that is distributed in space the same way as if the sources were point sources.

7.4.1 Gratings

Suppose now that we have a long line of point sources, all of the same wavelength and equally spaced at a distance d (Fig. 7.13, below). At any given point in space, the resulting wave in time is a superposition of many sine waves. Again the resultant wave will be sinusoidal in time.

It can be shown that *far away from the sources*, the curves of constructive interference are again straight lines satisfying the same Eq. (7.6) as with two sources. The difference is that the regions of brightness are much more sharply defined. Thus, in the case of a light wave, projection on a screen gives a brightness pattern shown in the figure below.

We can produce a similar result without using point sources by having a light beam pass through or reflect off of a **diffraction grating**: In the first type, the grating consists of a plate of transparent material, such as glass or plastic, which has a surface upon which identical and equally spaced parallel grooves have been engraved. In the second type, parallel grooves are coated with a reflecting material such as silver. Such is the case with a compact disc (CD), which has closely and equally spaced concentric circles of holes that characterize the digital storage of information, be it audio or data storage. White light that is incident on the surface of the CD is reflected off the grooves, with interference of light depending upon the wavelength—hence leading to the rainbow of colors. Whether the gratings transmit light or reflect light, we essentially have an array of identical, equally spaced sources. See Fig. 7.14 below.

For a grating, instead of having a periodic array of **point sources**, we have a periodic array of identical **extended sources**: It can be shown that in this case, we obtain the same arrangement

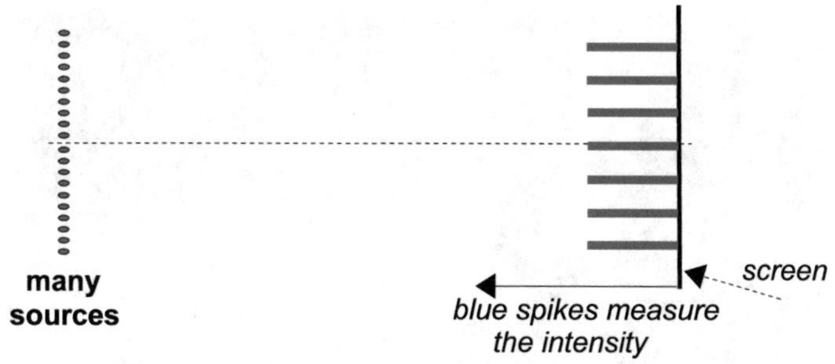

Fig. 7.13 Interference from a linear array of many sources that are equally spaced

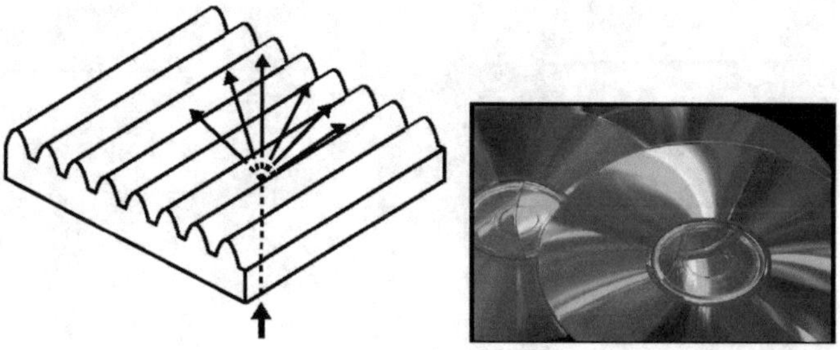

(a) (b)

Fig. 7.14 Diffraction grating (photo by Leon Gunther)

of fringes as we would from point sources except that the shape of each fringe is modified according to the wave coming from each groove. The same holds true for the interference pattern discussed below for a two-dimensional mesh and for a crystal.

7.4.2 Diffraction Through a Mesh

Let us pass a light beam through **silk screen**, which is a piece of "mesh" material that is a weave of threads arranged in two mutually perpendicular parallel arrays. The result is an interesting interference pattern shown in Fig. 7.15.

In the figure you see two silk screens above, labeled L and R; below them are two diffraction patterns, labeled (a) and (b) that were produced by the two lasers that were set side by side and can be seen through the silk screens. The number of threads per inch is greater on the right (R); correspondingly, the distance between neighboring threads is smaller. One might think that each spot is produced by a single hole in the mesh through which the laser beam passed. This is not the case at all; each spot is produced by light that has passed a number of holes in the silk screen. Without looking at the answer in the footnote, can you determine how the pairs are matched up? L with (a) R with (b) or the converse?[8]

Finally, notice that the number of spots is limited. If the holes were infinitesimal in diameter, the light coming through a single hole would exhibit strong diffraction—See Sect. 9.1 on diffraction. Then the hole would act like a point source. In this case, the spots would go on far the right, left, above, and below. With a finite size hole, the angle of spreading (in radians), due to diffraction of the outgoing wave, is on the

[8]Image (a) was produced by the screen labeled R and (b) by the silk screen labeled L. The reason is analogous to the fact that in two-source interference, the closer the sources, the further apart are the fringes on a distant screen.

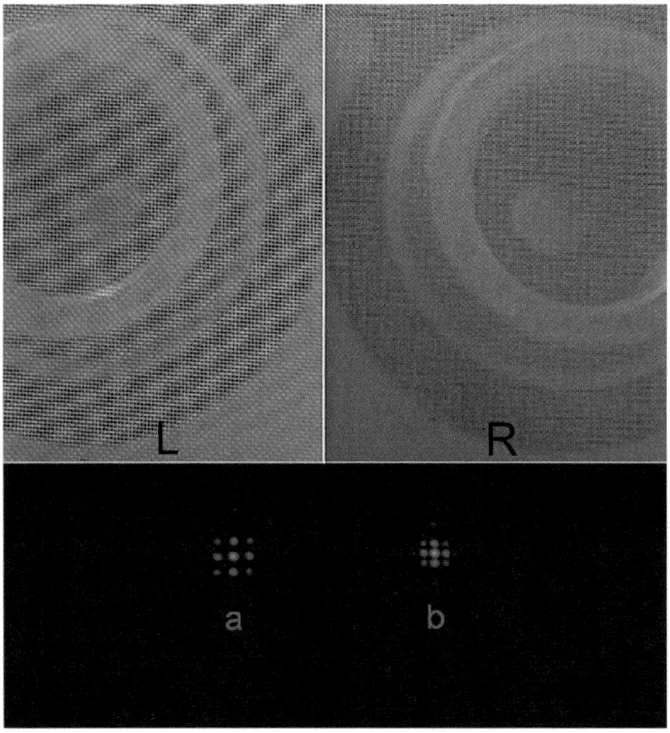

Fig. 7.15 Diffraction through a mesh (photo by Konstantinos Metallinos)

order of the wavelength divided by the diameter d of a hole. (See Eq. (9.1).) As a result, the number N of spots along a line can be shown to be given by

$$N \approx \frac{\text{spacing d between adjacent holes}}{\text{diameter of a hole}} \quad (7.10)$$

7.4.3 X-ray Diffraction Off Crystals

We have just made the transition from a linear array of sources to a two-dimensional array of sources, the mesh material. We now move on to three dimensions (3D): In a **crystal**, atoms or groups of atoms are arranged in a periodic array in 3-D space, as depicted by the model of a sodium chloride crystal in the Fig. 7.16 below.

Suppose that we cast a beam of monochromatic EM waves on the crystal. We will get an interference pattern that is determined by the crystal structure. The outgoing wave is strong only in certain directions. One requirement is that the wavelength λ be on the order of or less than or about equal to the spacing d between neighboring atoms (the distance between a large ball and a small ball). This distance is typically on the order of Ångstroms.

In the case of sodium chloride, $d = 5.63/2 = 2.82\AA$. We would then need a wavelength $\lambda < 2.82\AA = 2.82 \times 10^{-10}$ m. The corresponding frequency of the EM wave is then $f = c/\lambda = (3 \times 10^{8})/(2.8 \times 10^{-10}) = 1.1 \times 10^{18}$ Hz, a frequency in the X-ray region.

The intensity pattern produced on a screen by X-rays scattering off a crystal of sodium chloride is shown in Fig. 7.17.

7.5 Terms

- Constructive interference
- Demodulation
- Destructive interference
- Diffraction grating
- In phase

- Modulation
- Order of interference
- Out of phase
- Path difference of two waves
- Phase difference
- Principle of Superposition
- Relative phase
- Resultant amplitude
- Resultant wave

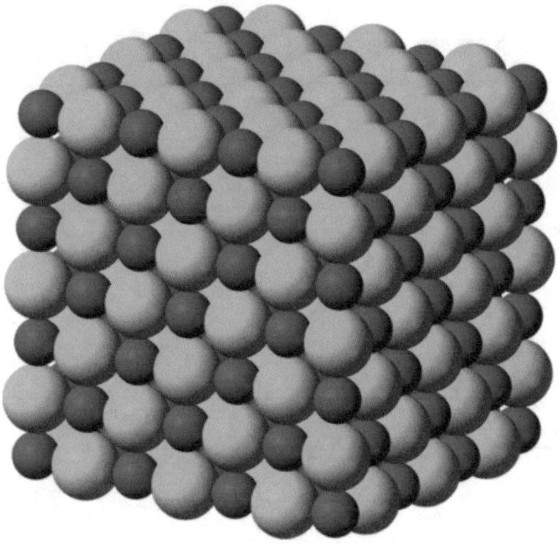

Fig. 7.16 Crystal lattice structure of sodium chloride. Sodium is purple while chlorine is green (source: https://en.wikipedia.org/wiki/Sodium_chloride)

7.6 Important Equations

For constructive interference:

$$\overline{S_2P} - \overline{S_1P} = 0,\ \lambda,\ 2\lambda,\ \ldots \tag{7.11}$$

Angles for constructive interference:

$$\sin\theta_1 = \frac{\lambda}{d},\quad \sin\theta_2 = 2\frac{\lambda}{d},\ \ldots \tag{7.12}$$

Angles for destructive interference:

$$\overline{S_2P} - \overline{S_1P} = \lambda/2,\ \lambda + \lambda/2 = 3\lambda/2,\ 2\lambda + \lambda/2 = 5\lambda/2,\ \ldots \tag{7.13}$$

7.7 Problems for Chap. 7

1. The first figure below shows two specially shaped pulses traveling along a string, represented by thick line segments. The blue pulse is moving to the left, while the red pulse is moving to the right. Eventually they pass through each other on the string. List below which of the five figures—(a) through (e)—represents the shape of the string at some future time (Fig. 7.18).

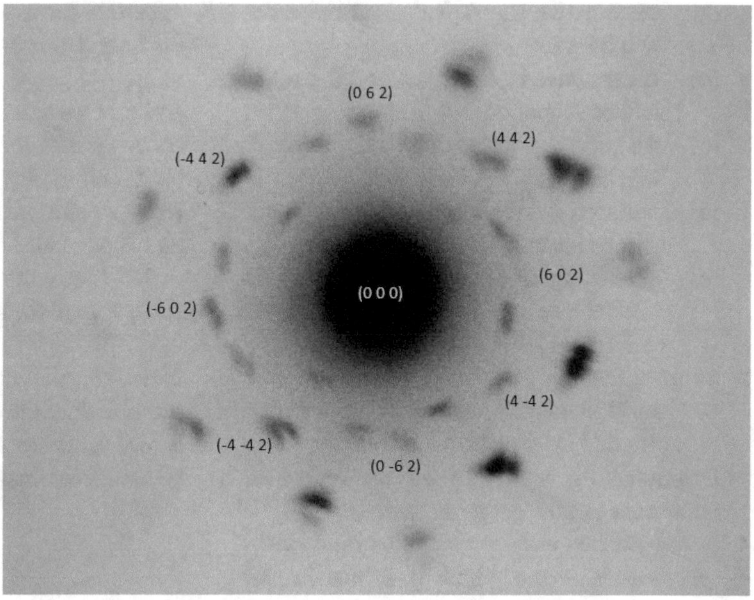

Fig. 7.17 X-ray diffraction pattern of a sodium chloride crystal (source: http://physicsopenlab.org/2018/01/22/sodium-chloride-nacl-crystal/)

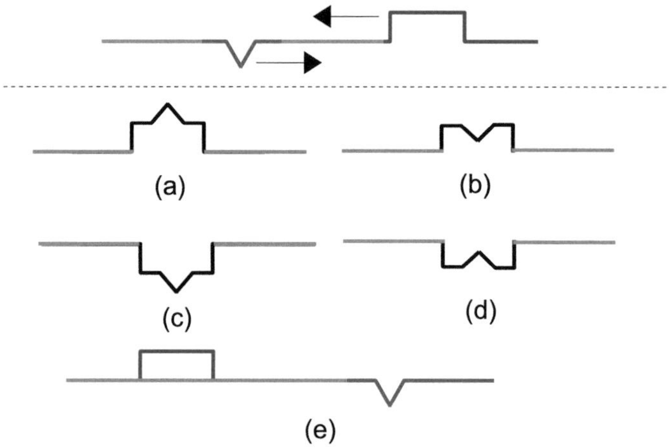

Fig. 7.18 "Colliding" waves

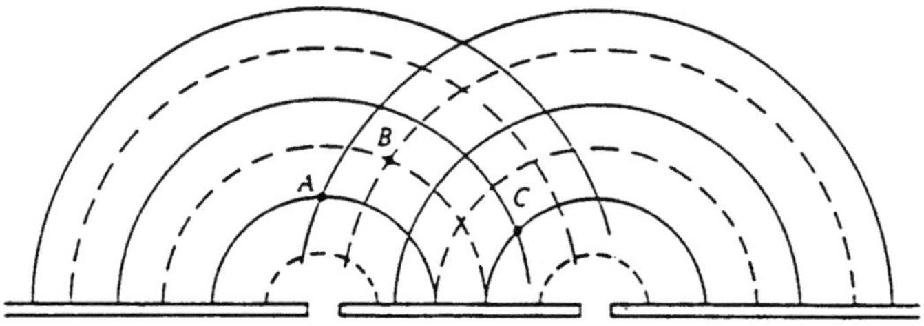

Fig. 7.19 Two-slit interference

2. The figure below shows waves that are passing through two slits in the barrier. The solid lines are crests, and the dashed lines are troughs. Therefore, at points A, B, and C, there will be
   (a) constructive interference, and the water will be still.
   (b) constructive interference, and the water will be in motion.
   (c) destructive interference, and the water will be still.
   (d) destructive interference, and the water will be motion.
   (e) alternating constructive and destructive interference, so the water will be in motion.
3. Suppose that a light wave of wavelength 4×10^{-7} m is incident upon a pair of slits that are separated by a distance of 14×10^{-7} m. Determine the angles of all the orders of interference of the outgoing wave (Fig. 7.19).
4. A two-slit interference pattern of light is observed on a screen which is at distance of 4 m from the slits. The slit separation is 0.2 mm, while the distance between neighboring fringes is 0.9 cm. Find the wavelength of the light.
5. A certain star is situated at point S, a distance 10^{14} km from the earth. An atom in the star emits a light wave of wavelength 4000 Å, and this wave is detected at two locations on earth—points A and B shown in Fig. 7.20.

   A and B are 200 km apart, and form a triangle with the star. The signals at points A and B are added together.

   Do they add constructively or destructively? Explain using a calculation.

Note You will need to use the following approximation: Given a right triangle, with legs a and b, where $a \ll b$, the hypotenuse c is given approximately by

$$c \cong b + \frac{a^2}{2b}$$

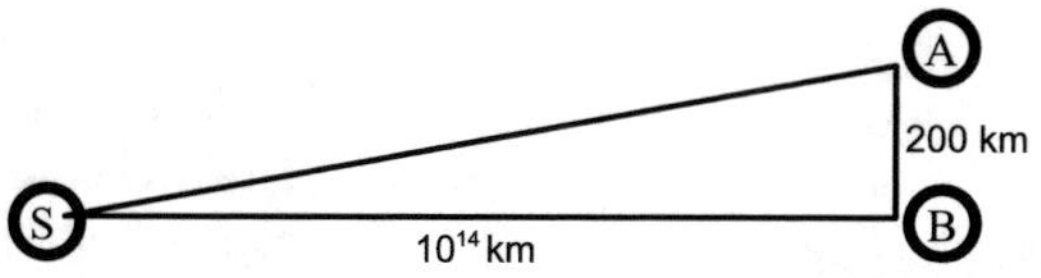

Fig. 7.20 Interference of light from a star

6. A diffraction grating has 5000 lines per cm. Determine the angles of the various orders of the interference pattern produced by a light beam having a wavelength of 4.4×10^{-7} m.
7. Add graphically the two waves in each of the two figures, (a) and (b), in Fig. 7.21 below.
8. Below, in Fig. 7.22, we exhibit a drawing of two sources that are emitting waves having the same wavelength. The two sets of concentric circles are the crests. The equal spacing between neighboring circles equals the wavelength. Note the light rays that radiate from the region between the two sources. One such ray is horizontal, being the perpendicular bisector of the line joining the two sources. Note that all of the crossings between a pair of circles emitted by the two sources lie along the bright rays. These crossings are points of**constructive interference**. The bright rays are separated by dark rays; we have **destructive interference** along these rays. Both sets of rays are straight lines far from the sources.

   You are to study the second and fourth orders of interference shown in the figure as follows. Lengths should be measured using a ruler.
   (a) Determine the distance between the two sources.
   (b) For each order, draw a line from the midpoint between the sources along the center of the corresponding bright rays (at the crossings of the circles) where the rays are straight.

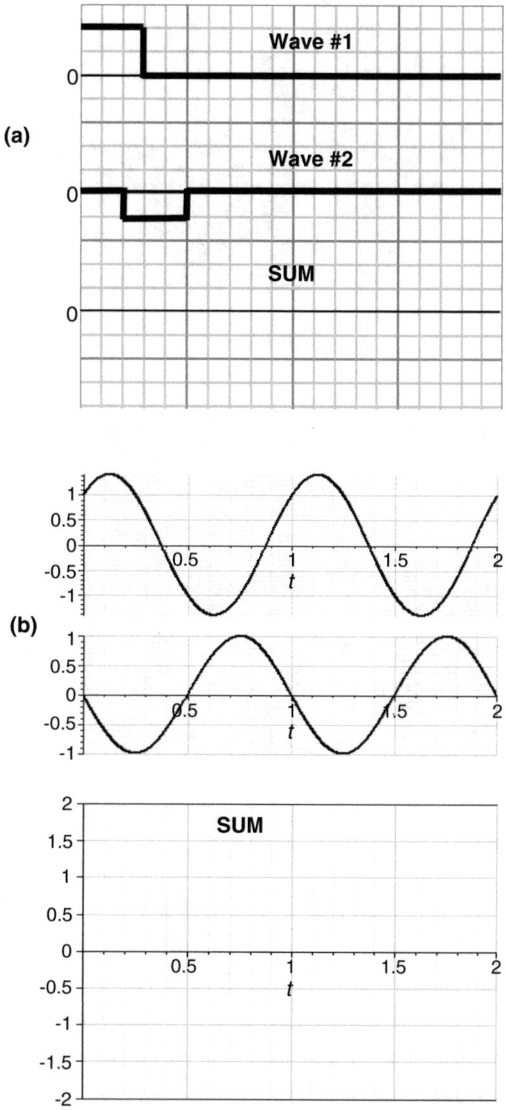

Fig. 7.21 Addition of waves

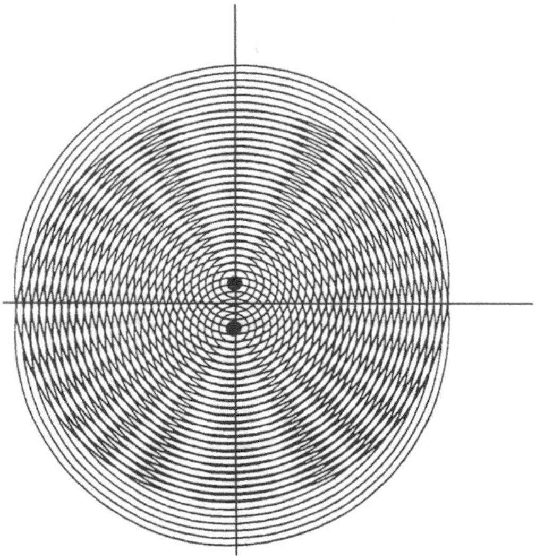

Fig. 7.22 Two spherical sources

(c) For each order, determine the angle that each ray makes with the horizontal. You might choose to measure the slope by completing a right triangle. The angle is the inverse tangent of the slope.

(d) For each order, use Eq. (7.6) to determine the wavelength. Compare the value you obtain with the value you measure directly from the figure.

Complex Waves

8

In Chap. 7 we studied the wave that results from adding two sine waves having the same frequency, typically produced by having a single sine wave incident upon a set of slits or a grating. In this chapter we will study the nature of more complex waves: Topics included are:

- **beats** that result from having two waves with nearly equal frequencies
- **modulated waves**, such as AM and FM waves, that are central for radio communication
- **spectrograms** that provide us with a visual representation of a complex wave such as is produced by the human voice, a bird, or a complex electronic signal
- **polarized light** that is associated with the fact that the electric field, which characterizes a light wave, oscillates in a direction that is transverse to the direction of propagation. This characteristic of light enables bees to navigate, enables us: to manipulate the path of a light beam, to cut down glare from sunlight, and to study the Universe soon after the Big Bang.

The original version of the chapter has been revised. Corrections to this chapter can be found at https://doi.org/10.1007/978-3-030-19219-8_16

8.1 Beats

Suppose we superimpose two sine waves having nearly the same frequency. Below we see one sine wave that has ten cycles in a one second interval while a second sine wave has nine cycles during that one second interval. Their frequencies are 9-Hz and 10-Hz, respectively. Notice how the two waves are in phase initially. Halfway, at one-half second, the waves are out of phase. The first wave has gone through five oscillations while the second has gone through but four and a half oscillations. At the end of the entire interval, the first has gone through ten oscillations and second nine oscillations. It takes one second for them to be in phase again so that one second is the beat period. See Fig. 8.1.

The wave pattern is that of a sine wave modulated by an envelope that oscillates at a frequency of 1-Hz, which is the difference between the two input frequencies. (Compare the pattern with that of the coupled SHOs, in Fig. 2.25.)

This phenomenon is called **beating**. The peaks in amplitude are called **beats**. The frequency of the envelope is called the **beat frequency**.

Generally, it can be shown that if f_1 and f_2 are the two respective input frequencies, the beat frequency is given by

$$\boxed{f_B = |f_2 - f_1|} \tag{8.1}$$

L. Gunther, *The Physics of Music and Color*,
https://doi.org/10.1007/978-3-030-19219-8_8

Fig. 8.1 Beats

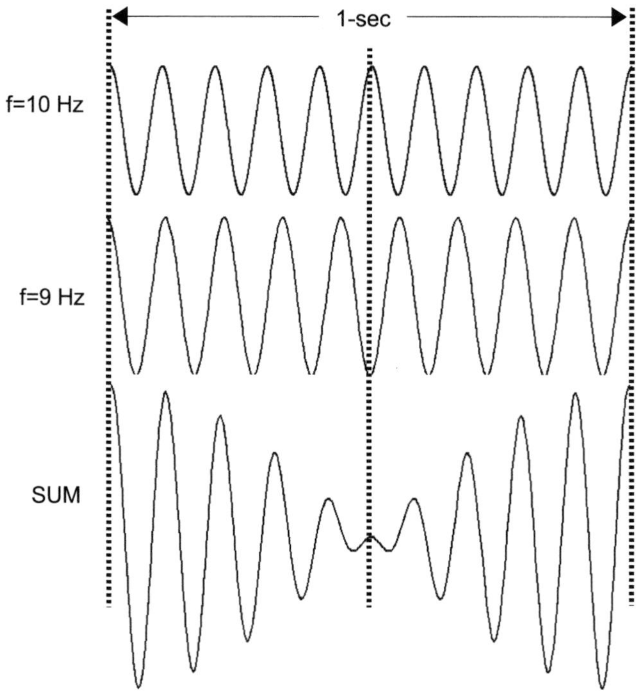

Note the absolute value sign, since the beat frequency is a positive quantity and its value doesn't depend upon which of the two frequencies is the greater one.

This result can be obtained from trigonometric identities.[1]

Let us introduce the symbol $\boxed{f}$ for a sinusoidal wave of frequency f and a unit amplitude.

A wave of frequency 200-Hz and amplitude 3.0 is written as

$$3.0\,\boxed{200}$$

Suppose that we have two sine waves, with frequencies f_1 and f_2.

Then, we obtain the following trigonometric identity for the product of two sinusoidal waves:

$$\boxed{f_2} \otimes \boxed{f_1} = \frac{1}{2}\boxed{f_2 + f_1} + \frac{1}{2}\boxed{f_2 - f_1} \quad (8.6)$$

In the context of beats, wherein we let $f_2 = f$ and $f_1 = f_B/2$, we have

$$\boxed{f_B/2} \otimes \boxed{f} = \frac{1}{2}\boxed{f + f_B/2} + \frac{1}{2}\boxed{f - f_B/2} \quad (8.7)$$

The difference between the two frequencies on the right-hand side is $(f + f_B/2) - (f - f_B/2) = f_B$. This is the **beat frequency**. Such a wave is displayed in Fig. 8.2 with $f = 10$-Hz and $f_B = 2$-Hz. We see the envelope wave of 1-Hz in red modulating the 10-Hz wave in blue. The

---

[1]

$$\cos(a + b) = \cos a \cos b - \sin a \sin b \quad (8.2)$$

Now let $\overline{f} = (f_1 + f_2)/2$. This is the average of the two frequencies. Assume also that $f_2 > f_1$. *The result doesn't depend upon which frequency is greater.* Notice that

$$f_1 = \overline{f} - f_B/2 \quad (8.3)$$

and

$$f_2 = \overline{f} + f_B/2 \quad (8.4)$$

We can then show that

$$\cos(2\pi f_1\, t) + \cos(2\pi f_1\, t) = 2\cos(\pi f_B\, t)\, \cos\big(2\pi \overline{f}\, t\big) \quad (8.5)$$

The significance of this equation is that the resulting wave is a sine wave with frequency $\overline{f}$ modulated by a sine wave having the beat frequency.

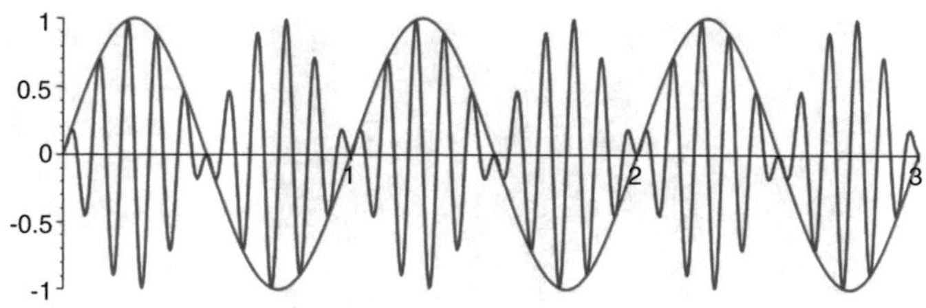

Fig. 8.2 1-Hz wave modulating a 10-Hz sine wave, with a 2-Hz beat frequency

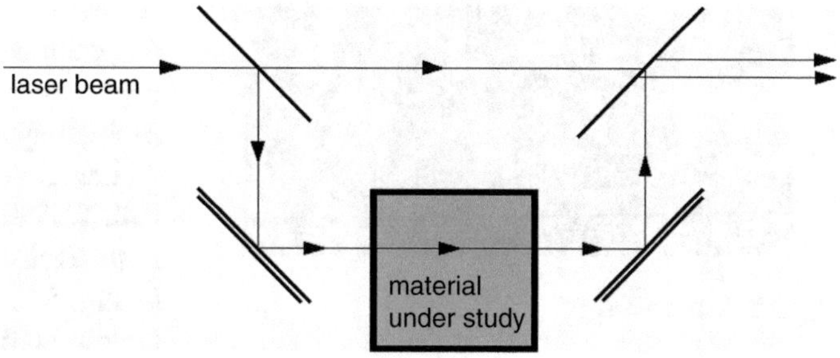

Fig. 8.3 Using beats to study materials

beat frequency is **double** the frequency of the envelope wave.

Sample Problem 8.1 What is the beat frequency if the two input frequencies are 440 Hz and 442 Hz , respectively?

Solution $f_B = 442 - 440 = 2\text{-Hz}$

Sample Problem 8.2 A tuning fork whose fundamental is 440-Hz produces a beat frequency of 5 Hz when its sound is added to the sound of a violin A-string. What can one say about the frequency of the A-string?

Solution We only know the absolute value of the difference between the two input frequencies. Thus, the violin string's frequency is either 435 Hz or 445 Hz .

The phenomenon of beating has a number of applications. We will mention three of them here:

1. The tuning of stringed instruments requires high level of accuracy in matching the frequency of a string with a standard frequency source (such as a tuning fork) and/or with the frequency of other strings. Beating provides a means to attain the required accuracy.
2. In Chap. 9 we will discuss how beating is used by radar detectors to measure the speed of an automobile.
3. If a laser beam is passed through a medium consisting of a transparent liquid that has large molecules or very small sub-micron size particles in suspension, the outgoing beam will have a frequency component that is very slightly shifted from the input frequency. That shift reflects valuable information about the properties of the medium. The shift can be determined by beating an unaffected portion of the input laser beam with a portion that has been shifted. See Fig. 8.3 above, where a typical experimental setup is depicted. We note that the material has shifted the frequency of the laser beam from f_1 to f_2.

   Four mirrors are used to direct the laser beams. Splitting and combining is accomplished by using three **half-silvered mirror**: There are two outgoing beams, one that has been reflected by the mirror and the other that has passed straight through the mirror. They each have an intensity equal to half the incident intensity. What is remarkable is that a beat frequency of about 15 Hz can be detected as against a laser beam frequency that is on the order of 6×10^{14}-Hz. This represents a sensitivity of two parts in 100-trillion!

Fig. 8.4 Tibetan bell

8.1.1 Beats of a Tibetan Bell

Often people seek perfection. Perfection in a bell, such as the Tibetan bell shown in Fig. 8.4, might be expected to be represented by a perfect circular circumference at the top. It should be clear that one cannot realize an absolutely perfect circle, having no deviations whatever. A perfect circle is regarded as a mathematical construct. In fact, the top perimeter of the Tibetan bell in the figure is an ellipse, with the inner dimensions having a **major axis** of length 107.2-mm and a **minor axis** of length 106.2-mm, corresponding to difference of a mere 0.9%![2] This small difference in the axes was intentional in producing the bell so that there would be **beats** when the bell is struck, due to the nearly equal frequencies of the two lowest modes. Figure 8.5 displays thesound wave of the bell in Fig. 8.4, produced by the software Amazon Pro, wherein the beats are quite pronounced.

We can see ten beats in a time interval of about 3.2-s, corresponding to a beat frequency of about 10-beats/3.2-s=3.1-Hz. Amazon Pro produces the frequency spectrum shown in Fig. 8.6. It also determines the frequencies of the fundamental and nearby second mode (that beat together) with great precision as 1377-Hz, as well as the frequency 2534-Hz of the third mode

Let us now examine the lowest modes of the bell. We begin with a bell that has perimeter that is **perfectly circular**, as represented to the left in Fig. 8.7. Suppose we consider a diameter of the circle at an **arbitrary angle** with respect to the horizontal. We add a second diameter that is perpendicular to the first. Next, the bell is struck at the end of one of the diameters as represented by the red arrow. The result will be an oscillation of the bell, starting with change towards an elliptical shape. The following is a complete cycle: A compression towards an elliptical shape, stopping at the red ellipse and moving back towards the original circular shape. This is half a cycle. The motion will continue on towards the second maximal shape of the green ellipse and then back towards the original circle, thus completing a full cycle.

Next we consider a bell with an elliptical perimeter, as shown at the left in Fig. 8.8. We see two dotted lines, one horizontal, the other vertical. There are four very small blue discs at their ends. We also see another pair of dotted lines at an angle to the first pair. They end in very small blue rings. When the ellipse is very close to being circular, as is the case most Tibetan bells as well as mine, the angle between the pairs can be taken to be 45^0.

The fundamental of the bell will be excited if the bell is struck at any of the four blue discs. The oscillation of the fundamental is shown at the right in Fig. 8.8.

The second mode of the bell will be excited if the bell is struck at any of the four blue rings. The oscillation of the second mode is shown at the right in Fig. 8.9.

[2]Regarding perfection: The **Enso** (https://en.wikipedia.org/wiki/Ensō) is a Zen Buddhist circle that is hand drawn and meant to help a person achieve a high level of enlightenment. There is a Zen tale that focuses on the circle: In a version I read years ago, a novice sees a monk seated next to a piece of paper on which the monk has drawn a circle. The novice asks, "Dear Master, the figure you have drawn does not appear to be a perfect circle." To which the master replies, "What I have drawn is a perfect whatever it is."

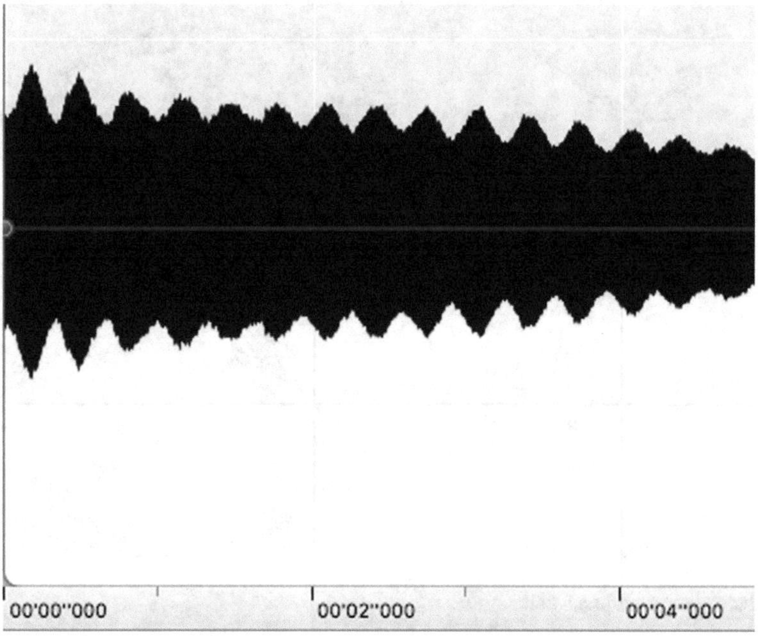

Fig. 8.5 Sound wave of Tibetan bell

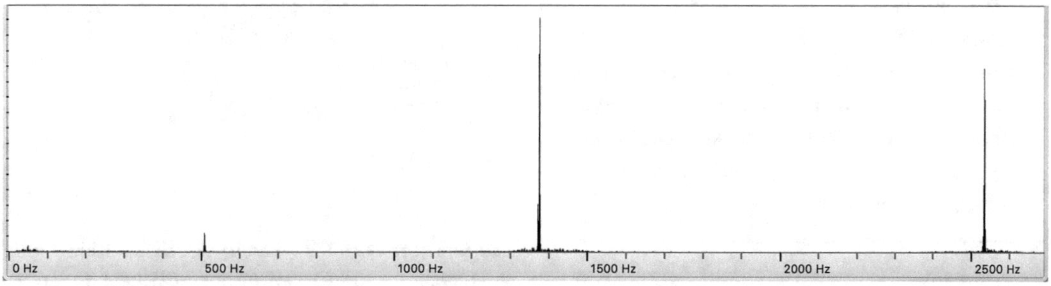

Fig. 8.6 Frequency spectrum of the sound wave of a Tibetan bell

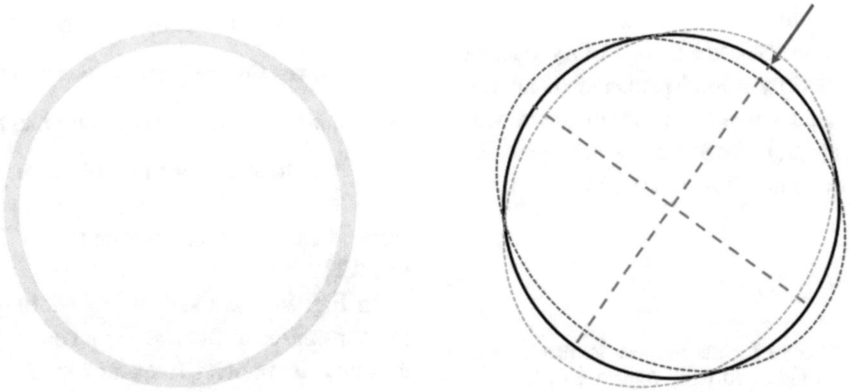

Fig. 8.7 Fundamental mode of a circular bell

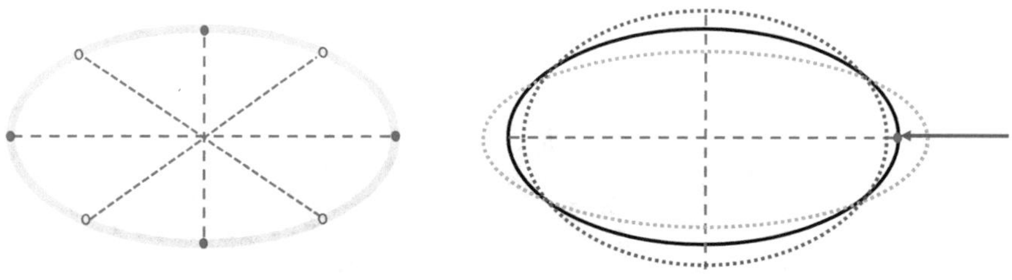

Fig. 8.8 Fundamental mode of an elliptical bell

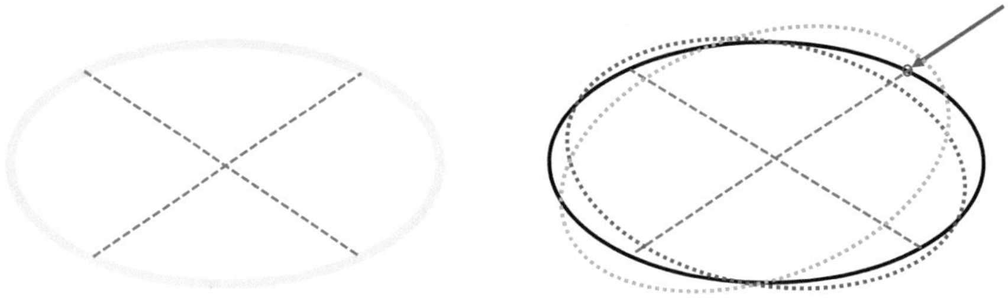

Fig. 8.9 Second mode of an elliptical bell

> **Producing Beats**[3]:
> If the bell is struck at any point around the perimeter that is NOT at one of the eight points shown in Fig. 8.8, both modes will be excited and we will hear beats. The closer the bell is struck to one of these eight points, the more will the corresponding mode frequency be dominant.

8.2 AM and FM Transmission

Radio waves are used to transmit a pattern that is ultimately reflected in the pattern of a sound wave that is emitted by a loudspeaker or headphone. However, radio waves are electromagnetic waves with a very high frequency in comparison to the range of audible sound: Audiblesound waves range from about 20-Hz to about 20-kHz while AM waves range from about 500-kHz to 1600-kHz and FM waves range from about 87.5 to 108.0-MHz. The question is

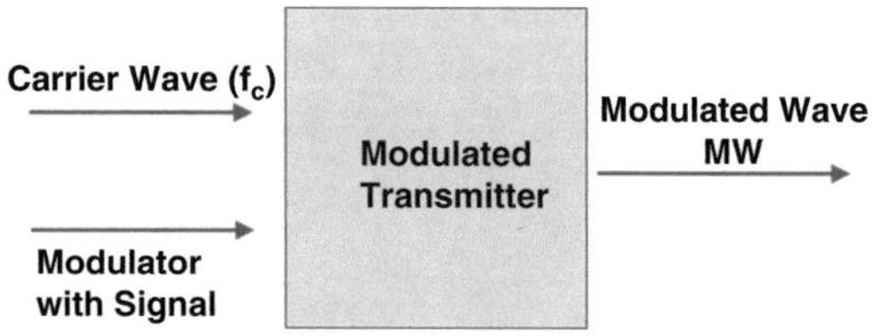

Fig. 8.10 Modulation

How can radio waves carry the pattern of sound waves that have much lower frequencies?

The solution involves a procedure that is referred to as **carrier modulation**.

Here is the scheme, represented in Figs. 8.10 and 8.11.

In Fig. 8.10 we see an input into a modulator/transmitter of two waves: a carrier wave that is a sine wave with frequency f_c along with a modulator wave. The **modulator wave** carries the signal $f_s(t)$ that we want to transmit and is combined with the carrier wave.

[3]The physical basis for the beats is attributed to Lord Rayleigh, who discussed it in Volume 1 of his *Theory of Sound*, second edition, [MacMillan and Co. London, 1894]. Rayleigh is most famous for his explanation for why the sky is blue, a subject that is discussed in Chap. 9.

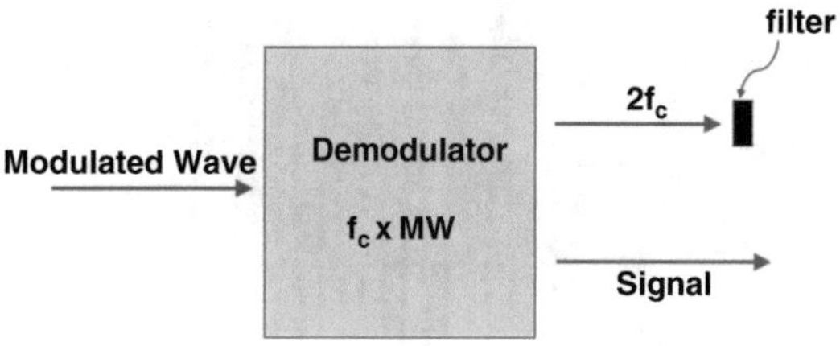

Fig. 8.11 Demodulation

8.3 AM Transmission

AM is an acronym for **amplitude modulation**—the transmission of signals via radio waves by modulating the amplitude of the carrier wave with the signal that is to be transmitted.

Consider the 10-Hz sine wave in Fig. 8.12.

Next consider the wave shown in Fig. 8.13, which is a graph of 1+ 0.1 $\boxed{\text{1-Hz}}$.

If we **multiply** together the previous two graphs, we will obtain the graph shown in Fig. 8.14. The 10-Hz sine wave is the **carrier wave** and is said to be **amplitude-modulated** by a 1-Hz sine wave.

$$\boxed{f(\text{t})} = \left(1 + 0.1\,\boxed{\text{1-Hz}}\right) \otimes \boxed{\text{10-Hz}} \qquad (8.8)$$

I am using low frequencies for both the carrier wave and the signal for display purposes only. Radio waves have frequencies that are much greater than the peak audio sound frequency, so that the two are not easy to represent together on the same graph. Generally, the modulating wave will be complex, as exemplified in Fig. 8.15.

The modulated radio wave transmitted by the radio station WEEI in Boston (with carrier wave frequency 850-kHz) carrying the signal $\boxed{f_s(\text{t})}$ is represented by

$$\boxed{\text{MW}} = \left(A + f_s(t)\right) \otimes \boxed{\text{850-kHz}} \qquad (8.9)$$

where A is the amplitude of the base carrier wave.

8.3.1 Sidebands

Suppose that the signal is a pure sine wave of frequency f. Then from Eq. (8.6) we have

$$\boxed{f} \otimes \boxed{\text{850-kHz}} = \frac{1}{2}\boxed{\text{850-kHz} + f} + \frac{1}{2}\boxed{\text{850-kHz - } f} \qquad (8.10)$$

Thus the modulated wave includes a **sum** of two sine waves, one of frequency (850-kHz + f), the other of frequency (850-kHz - f). We display this spectrum in Fig. 8.16, corresponding to a carrier frequency of 1,000,000-Hz and a signal frequency of 1-Hz. [I have assumed a carrier wave with a very narrow line-width.] The **sideband frequencies** lie at 999,999-Hz and 1,000,001.

Now consider a typical signal to be transmitted that might have huge number of frequency components, ranging from 20-Hz to 20-kHz. Each component has a pair of sideband contributions. The full frequency spectrum for WEEI will range from 830-kHz to 870-kHz. The full bandwidth is 40-kHz. In order for there not to be interference with the waves transmitted by WEEI, no other radio station should be allowed to transmit a range of frequencies that lies within the range 830-kHz to 870-kHz. Generally, AM radio stations should have carrier wave frequencies that are at least 40-kHz apart.[4]

8.3.2 AM Demodulation

The pattern of the sound wave that we want to transmit is the analog of the complex wave above. The radio station transmits a modulated radio wave that is akin to the above modulated wave. The radio receiver must now extract the complex

[4]We see that the ideal carrier wave frequency spacing of 40-kHz is **not** always present with the radio stations that were less than 45 miles away from Boston on December 12, 2018: **in kHz**: 590, 630, 650, 680, 740, 800, 830, **850**, 890, 920, 950, 980, 1030, 1060, 1090, etc.

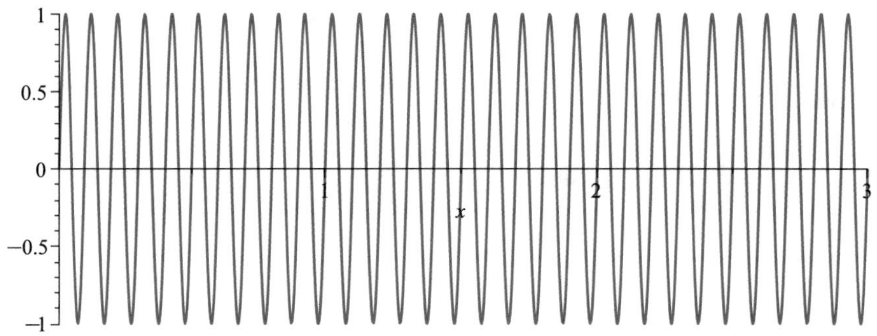

Fig. 8.12 10-Hz sine wave

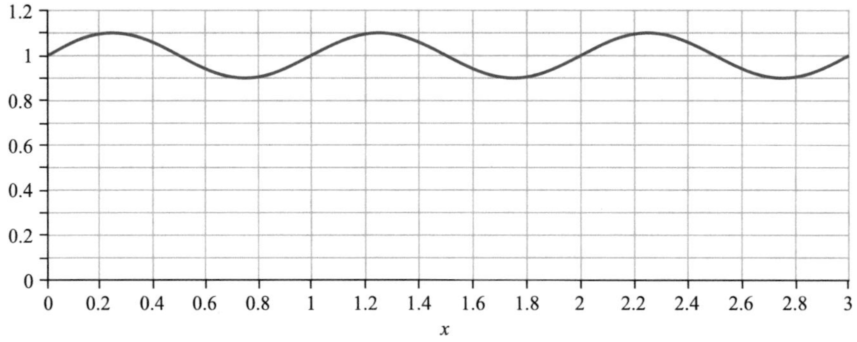

Fig. 8.13 Sum of a constant and a 1-Hz sine wave

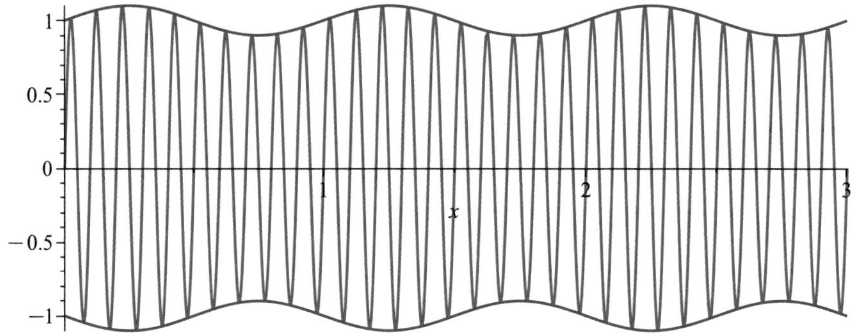

Fig. 8.14 AM modulation by a sine wave

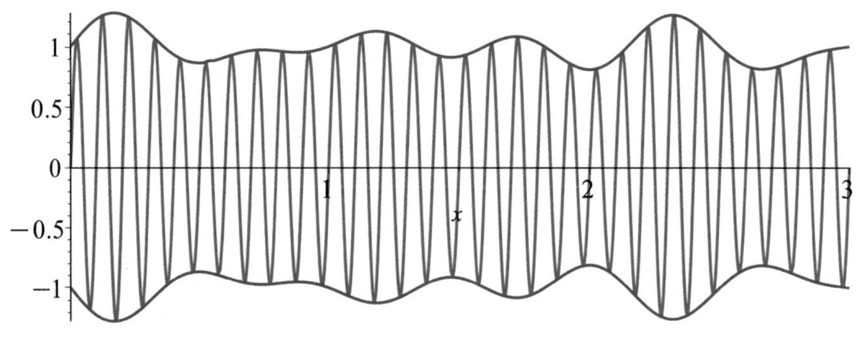

Fig. 8.15 AM modulation by a complex wave

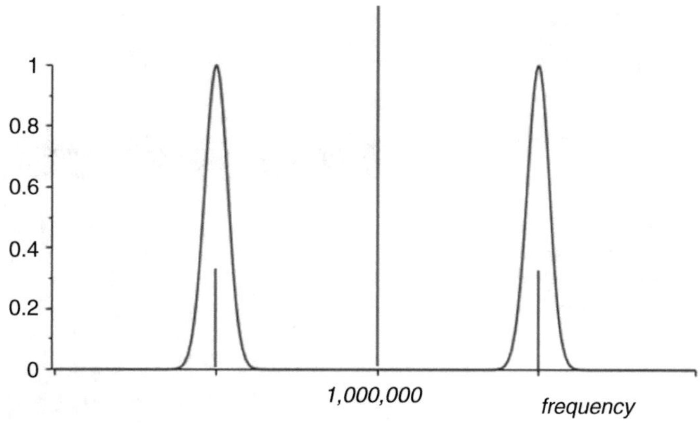

Fig. 8.16 Sidebands of a single modulating frequency

wave—which is a pattern we wish for the sound output of our loudspeaker—from the modulated wave. The process is known as **demodulation**.

The first step is to **multiply** the MW by a sine wave having the carrier frequency, resulting in the expression:

$$\left(\mathrm{A} + \mathrm{f_s(t)}\right) \otimes \boxed{\text{850-kHz}} \otimes \boxed{\text{850-kHz}} \quad (8.11)$$

We make use of Eq. (8.6)[5]:

$$\boxed{f_c} \otimes \boxed{f_c} = \frac{1}{2}\boxed{f_c + f_c} + \frac{1}{2}\boxed{f_c - f_c}$$
$$= \frac{1}{2} + \frac{1}{2} + \boxed{2f_c} \quad (8.12)$$

Thus

$$\boxed{\text{850-kHz}} \times \boxed{\text{850-kHz}} = \frac{1}{2} + \frac{1}{2}\boxed{\text{1700-kHz}} \quad (8.13)$$

To extract out the signal $f_s(t)$, the electronics of the radio wave system filters out all but this low frequency component, by discarding the constant and using a **low-pass filter** to discard the high frequency component 1700-kHz. The extracted signal in the radio receiver is then amplified and can be sent to a loudspeaker to produce the desired sound.

[5]From trigonometry: $\sin^2(\theta) = (1/2)[1 - \cos(2\theta)]$.

8.4 FM Transmission

FM refers to **frequency modulation**. The amplitude of the carrier wave remains constant while the frequency is modulated by the signal. The following symbol represents a carrier of frequency f_c (e.g. 89.7-MHz for WGBH-FM in Boston) modulated by a sine wave of frequency f_s, with an amplitude Δf. Note that there are a total of three frequencies in this wave: f_c, Δf, and f_s.

$$\boxed{f_c + \Delta f \boxed{f_s}}$$

Here is a nice way to understand an FM wave of this simple sort. Consider the **vibrato** of a violinist. The carrier frequency is produced by the central position of the violinist's finger on the black fingerboard. The length of string that is free to vibrate determines the fundamental frequency—$f = v/2\ell$. To produce a vibrato, the violinist "rolls" her finger around the fingerboard periodically so as to vary the length of string that is free to vibrate. There are two frequencies involved in the rolling motion: The amplitude of the variation of position and therefore of the variation of the change in frequency, Δf, and the frequency at which she rolls her finger, f_s. In the case of an actual violin vibrato, the variation of the position along the fingerboard is not at all sinusoidal. See Fig. 8.17.

In FM transmission the ratio of the two frequencies, $\beta = \Delta f / f_s$, is called the **modulation index**.

Fig. 8.17 Violin vibrato

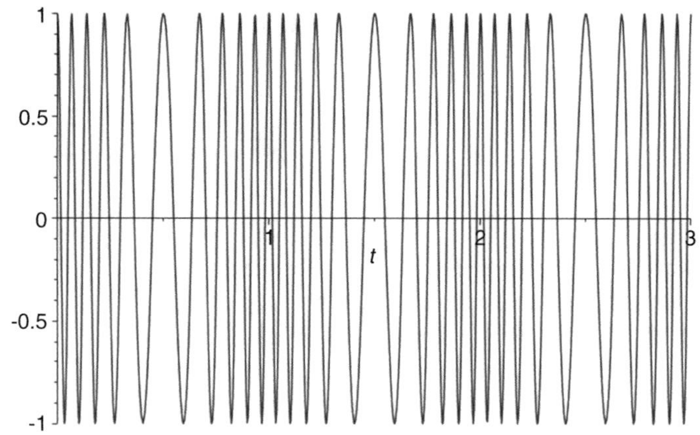

Fig. 8.18 FM carrier wave of frequency 10-Hz modulated by signal of 1-Hz, with $\beta = 5$

$$\boxed{\beta = \frac{\Delta f}{f_s}} \tag{8.14}$$

In Fig. 8.18 we see an FM modulated wave. The carrier wave has a frequency of 10-Hz, the signal frequency is 1-Hz, and the modulation index is $\beta = 5$.

An FM wave has an interesting frequency spectrum. First there is the carrier wave frequency f_c. Then there are an infinite number of **sideband frequencies**, $f_c \pm f_s$, $f_c \pm 2f_s$, $f_c \pm 3f_s$, ... The amplitudes of the sidebands decrease overall with distance from the central carrier wave. The larger the modulation index, the more slowly do the amplitudes decrease.

At the top of Fig. 8.19, we see an FM wave with carrier wave with a frequency of 1000-Hz. The signal frequency is f_s=50-Hz and the modulation index is β=5.[6] Thus, the maximum deviation of the frequency from the carrier wave frequency f_s is $\Delta f = \beta f_s$ = 5(50)=250-Hz. At the bottom of the figure we see the frequency spectrum of the wave. The FM wave in Fig. 8.18, with carrier wave frequency of 10-Hz modulated by a frequency of 1-Hz, has a frequency spectrum with sideband frequencies 10 ± 1, $10 \pm 2 \ldots$, or 9, 11, 8, 12, ...

There is an approximation for the bandwidth BW given by "**Carson's Rule**":

$$\boxed{\textbf{Carson's Rule}: \ BW \approx 2(\Delta f + f_s) = 2(\beta + 1) f_s} \tag{8.15}$$

For the FM wave in Fig. 8.19, the rule gives BW=2(50+250)=600-Hz, which compares well with the width of about 600-Hz of the spectrum.

8.4.1 FM Demodulation

Electronics engineers devised an insightful way to obtain the signal from an FM wave. The first step is to insert a **high-pass filter**. Only components with frequencies **above** the carrier wave frequency f_s are passed through.[7] For display purposes we will discuss the results for the wave in Fig. 8.18, which has a carrier wave frequency

[6] The figure was produced from the Mathematica website: http://demonstrations.wolfram.com/PowerContentOfFrequencyModulationAndPhaseModulation/.

[7] A **high pass filter** allows only frequency components above a certain value to pass through, while a **band pass filter** allows only frequency components lying between two frequencies to pass through.

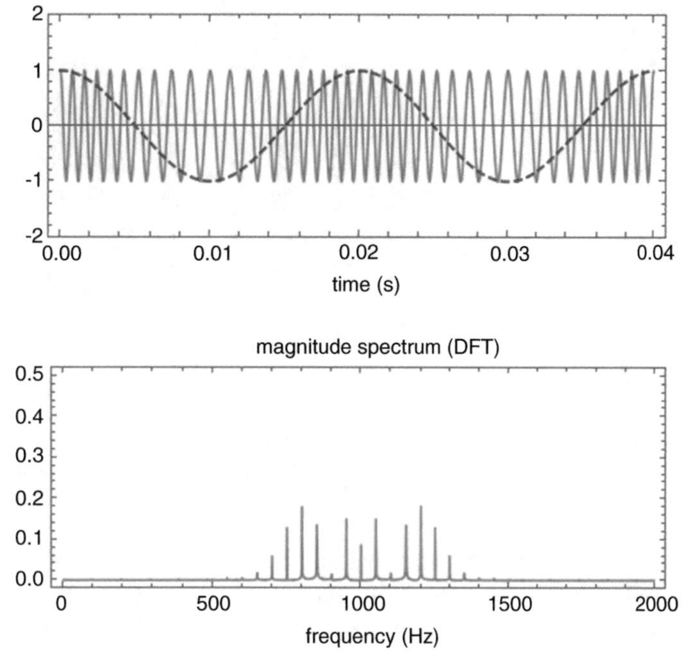

Fig. 8.19 FM carrier wave of frequency 1000-Hz modulated by signal of 50-Hz, β=5. Note the peaks at 1000 ± 50, 1000 ± 100,... in the frequency spectrum below

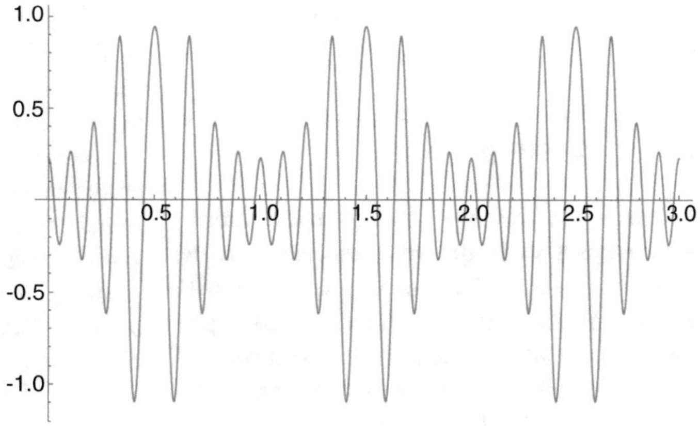

Fig. 8.20 Stage one in demodulating an FM wave with carrier frequency 10-Hz, modulated by a signal of 1-Hz

of 10-Hz and a signal of frequency 1-Hz. Then only the components with frequencies 11, 12, 13, ...-Hz pass through the filter. The resulting wave is shown in Fig. 8.20. If we look carefully we will note that this wave is essentially the carrier wave of frequency 10-Hz **amplitude-modulated** by the signal wave of frequency 1-Hz. The signal can then be extracted by removing the frequencies of carrier wave.

In the case of the FM wave in Fig. 8.19, the first stage will lead to a wave of frequency 1000-Hz that is **amplitude-modulated** by a wave of frequency 50-Hz. We would then insert a low-pass filter that passes only frequencies below 10-Hz to obtain the signal.

8.4.2 Bandwidth Limitation Imposed on FM Radio Frequency Separations

We noted that for AM radio, in order to avoid interference, no local radio station can broadcast at a frequency closer than about 40-kHz, the total bandwidth. In the case of FM broadcasting, the total bandwidth will be about BW=$2(\beta + 1)$20-kHz. If $\beta = 5$, we obtain BW=240-kHz. In this case, FM radio stations ideally need a carrier frequency separation of 240-kHz=0.24-MHz. If we examine the frequencies of FM radio stations, we find that their frequencies are typically separated by 0.2-MHz=200-kHz. Some FM stations broad-

Fig. 8.21 Lark sparrow waveform—The full song
The horizontal axis is the time in seconds; the vertical axis is the relative amplitude

cast at the same frequency but can handle this equality by broadcasting with very low power and not being too close to each other. Examples are WMFO (Tufts University) and WMLN (Curry College), both with a frequency of 91.5-MHz.

8.5 Spectrogram

In Sect. 4.7 we discussed the **spectral intensity with respect to frequency** , especially in the context of sound. The spectral intensity with respect to frequency was introduced as representing a sound wave that has a distribution of frequencies that is *constant over all time*. How could the distribution *not* be constant in time? Next, let us consider that **pitch** is associated with **frequency**, and frequency, strictly speaking, is associated with a sine wave having an infinite duration in time. Yet we refer to a pitch during a certain interval of time and a *changing pitch*. Clearly we need to modify the meaning of our terminology. In this section we will introduce a **spectral intensity that is defined over a finite time interval** and our terminology will become better defined.

How well can I describe the song of a bird in this book, such as the song of a **lark sparrow**? I couldn't do well with just words. Most bird songs might be described as having a changing pitch. Often, bird songs don't even have content that reflects a sense of pitch. In Fig. 8.21 we see the waveform of the bird song of a lark sparrow, which I downloaded from the software **Raven Lite**.[8] The waveform is quite complex such that I don't expect to gain much insight into the sound of the bird song by examining the waveform. Such is also the case if I were to examine the waveform of an audio recording of a musical performance. Yet, I do know that a waveform of sound contains all the information related to the sound that reaches my ears.[9]

In Fig. 8.22 we see an image of the waveform of the bird song of a lark sparrow over the course of a short time interval of about 14-ms. A close study of the wave reveals that over this time interval, the wave is close to sinusoidal, having a frequency of about 9000-Hz. In fact, given that there is also a modulation with a period of about 13-ms, we deduce that the wave is **amplitude-modulated** at a frequency of about $1/(13 \times 10^{-3}\text{-s}) \sim 80$-Hz.

However, an examination of the waveform at a later time reveals that the waveform is sinusoidal but with a different frequency. We are led to

[8]The software **Raven Lite** is available free of charge from the website http://ravensoundsoftware.com/raven\discretionary-downloads/.

[9]Recall that our musical perceptive and esthetic experience depend quite a bit upon the processing that occurs in our brains.

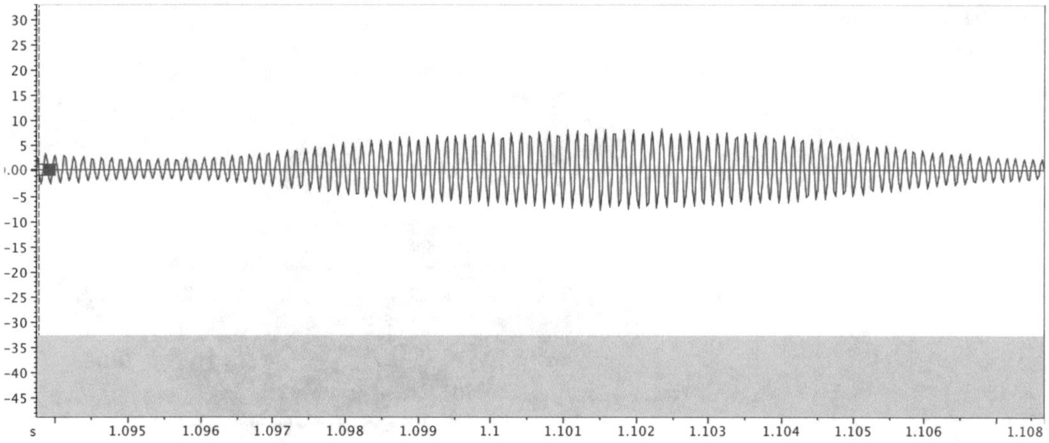

Fig. 8.22 Lark sparrow waveform—A short segment

Fig. 8.23 Line of musical score with frequency-time axes

consider a function that is an expansion of the concept of a spectral intensity with respect to frequency, that is, to a function that we will refer to as a **time-varying spectral intensity**. We will represent it by the symbol $I(f, t)$. Note that a time-varying spectral intensity would have to be plotted as a surface in a three-dimensional space, referred to as a **3-D plot**. If we color the surface, we will be able to display this surface as a two-dimensional plot in the **frequency-time plane**, as we will see in Fig. 8.25. In the following figures, we have focused upon the simple case when there is only frequency, and the intensity increases from zero to a maximum and then decreases back to zero.

Figure 8.23 exhibits a line of music, with a time axis and a frequency axis. There are often a number of notes played simultaneously, with varied durations in time. Since notes represent pitch and not frequency, and we will see in Chap. 12 that pitch is proportional to the logarithm of the frequency, the frequency axis is not linear but rather logarithmic. The **durations** of each note (hence pitch and frequency) are displayed in a discrete sense, omitting shaping in time. Moreover, the intensity is left for the musician to choose, as well as the timbre, which reflects the **admixture of overtones**. All these additional attributes are reflected in the function $I(f, t)$.

8.5.1 Understanding the Content of a Spectrogram

Let us study Fig. 8.24. For each instant of time we can make a slice through the "mound" from the right to left and a bit upwards—from a direction of 5-o'clock to 11-o'clock. Each slice produces a bell-shaped curve of the time varying spectral intensity vs. frequency at fixed time; these slices are peaked at the center, at a frequency of 440-Hz. Correspondingly, we can cut the mound

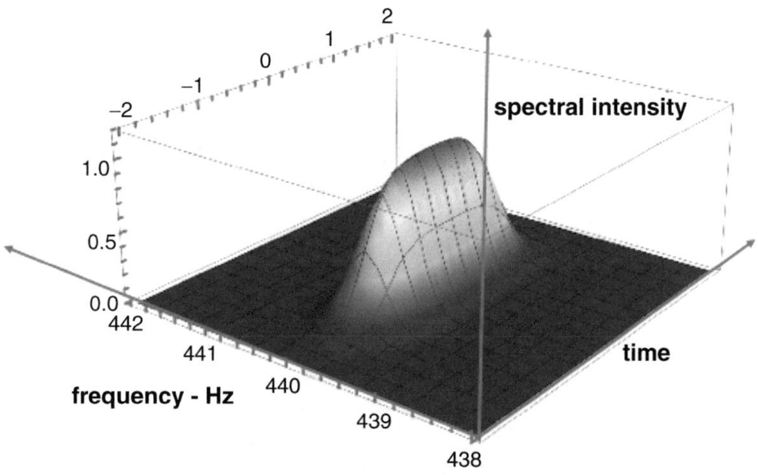

Fig. 8.24 An evolving spectral intensity

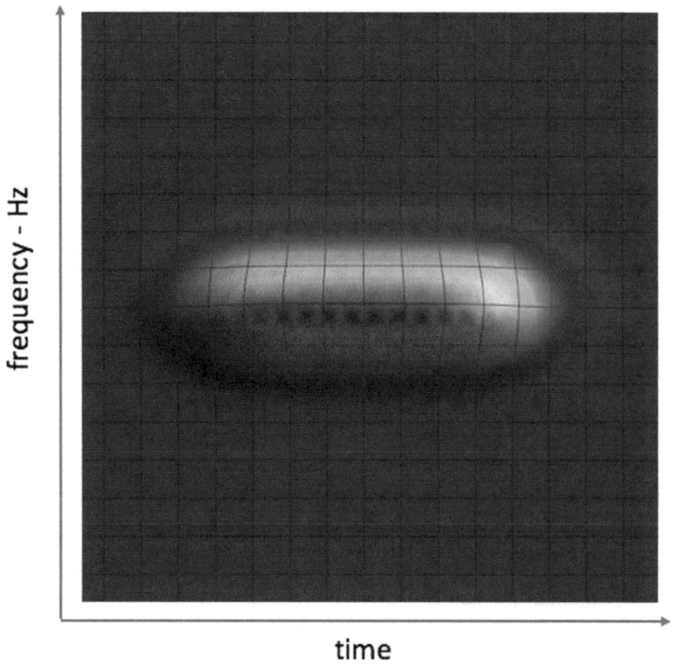

Fig. 8.25 The evolution of the spectral intensity displayed as a spectrogram

with slices in a direction of seven-o'clock to one-o'clock. Each slice shows us the variation of the spectral intensity over time for a given fixed frequency. The range of times is from -2 s to $+2$ s.

The color coding for the magnitude of the spectral intensity is from violet to red—a *reverse* of the familiar "ROYGBV" of a rainbow. *Violet* represents a vanishing spectral intensity. Thus, for early negative times, the spectral intensity vanishes; it peaks at $t = 0$, and vanishes for later positive times.

We can now rotate the mound so as to look down on the mound directly from above, as shown in Fig. 8.25.

1. The horizontal axis is the time axis.
2. The vertical axis is the frequency axis.
3. And, every point in the frequency-time plane is color-coded so as to represent the magnitude of the spectral intensity.

Fig. 8.26 A chirp

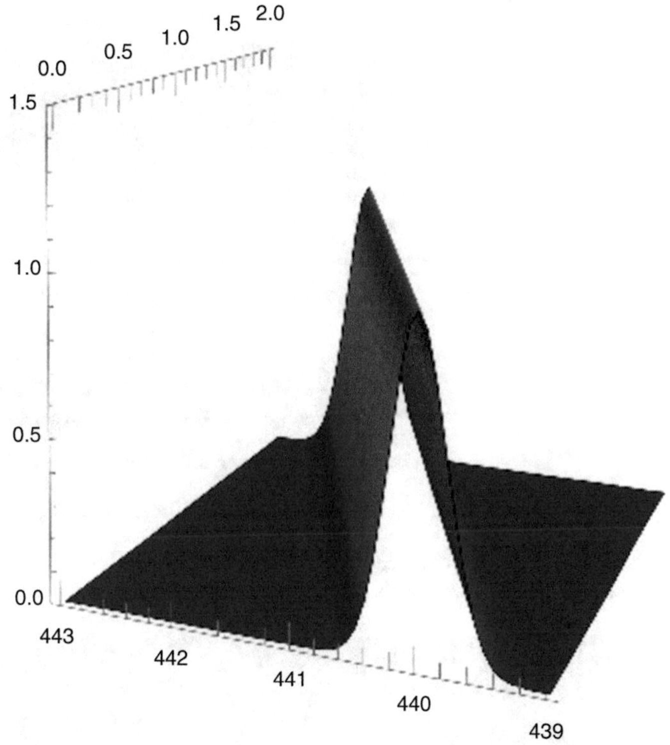

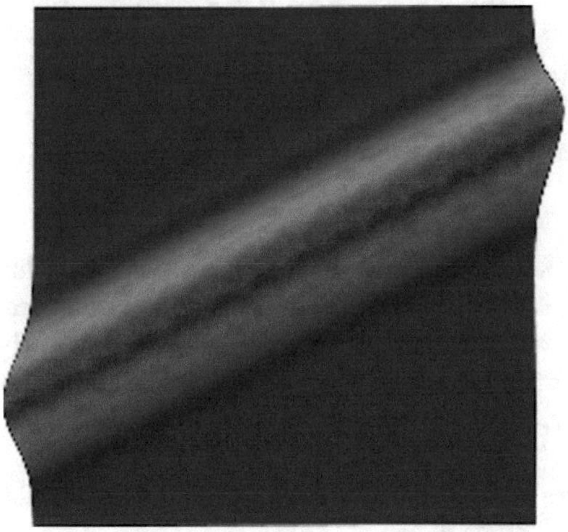

Fig. 8.27 The chirp displayed as a spectrogram

We see how the spectrogram allows us to replace the 3-dimensional plot by 2-dimensional plot by using color coding.

In Figs. 8.26 and 8.27 we see a 3D plot and a 2-D spectrogram of a **chirp**, which is often used in signal processing. The chirp consists of a wave that can be described as having a frequency that increases linearly starting with a specific initial frequency. The chirp begins at a time $t = 0.0$-s at a frequency of 439.5-Hz and increases linearly to a frequency of 440.5-Hz as at a time $t = 2.0$-s.

In Fig. 8.28 we see a spectrogram of a frequency modulated wave. We see a 2000-Hz carrier wave modulated with a signal having a period of 1-s and a frequency amplitude of 1000-Hz.

The image in Fig. 8.29 displays the **spectrogram** of the bird song of a lark sparrow, corresponding to the waveform in Fig. 8.22. It was obtained from the software **Raven Lite**. In the spectrogram for the song of the lark sparrow, shown in Fig. 8.29, aqua represents zero spectral intensity. As the spectral intensity increases, the color ranges from yellow to orange to red to blood-red. The spectrogram provides us with a **visual representation** of the bird song.

8.5.2 The Short Time Fourier Spectrum

In the previous section I stated that the spectrogram does not precisely represent the evolution in time of a spectral intensity. If a spectrogram were

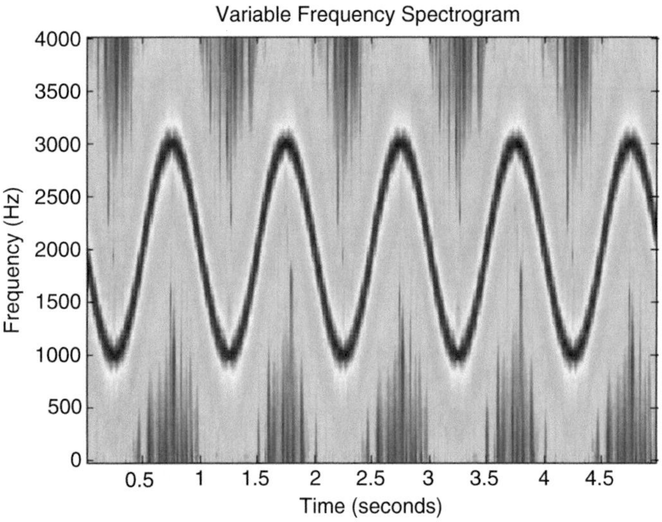

Fig. 8.28 Spectrogram of a frequency modulated wave.
Source: https://upload.wikimedia.org/wikipedia/commons/2/2b/VariableFrequency.jpg

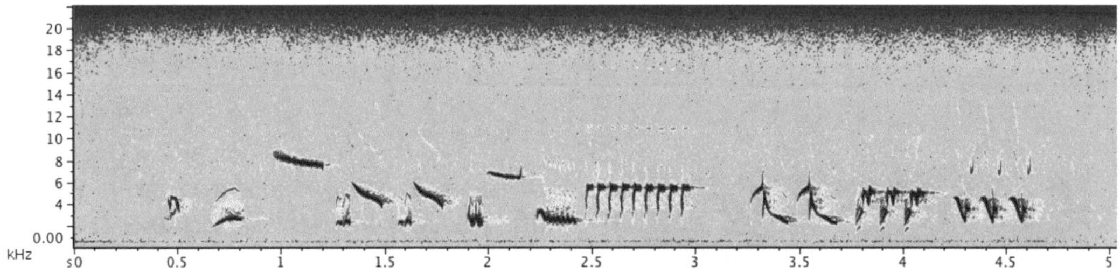

Fig. 8.29 **Lark sparrow spectrogram**. The horizontal axis is the time in seconds; the vertical axis is the frequency in kHz

precisely so, we would have to be able to define a spectral intensity at a specific instant of time. Similarly, we would have to be able to define a frequency at a given instant of time. In this section we will describe the **short time Fourier spectrum**.[10] The STFT produces the extended frequency-time intensity $I(f, t)$.

[10]In Sect. 4.7 we mentioned in a footnote the "Fourier transform." This transform is the spectrum obtained by the mathematical method referred to as "Fourier transformation." The technical term for the short time Fourier spectrum is that which is obtained by the Short Time Fourier Transform, of STFT for short. For more details see the Wikipedia article on the STFT: https://commons.wikimedia.org/wiki/File:STFT_colored_spectrogram_1000ms.png Note that this article is mathematically advanced.

Note While this section is focused upon the relation between the STFT and the spectrogram, I would like to point out the following extremely important role it plays in the modification of an audio recording. The two are:

- changing the pitch without changing the tempo.
- changing the tempo without changing the pitch.

Both of these modifications can be produced on an **.mp3 audio file** or an **.aiff audio file** using the software **QuickTimePlayer 7**, with controls in the A/V window. Pitch shifts don't produce much of a change in timbre or distortion. Speed changes do produce significant distortion if the

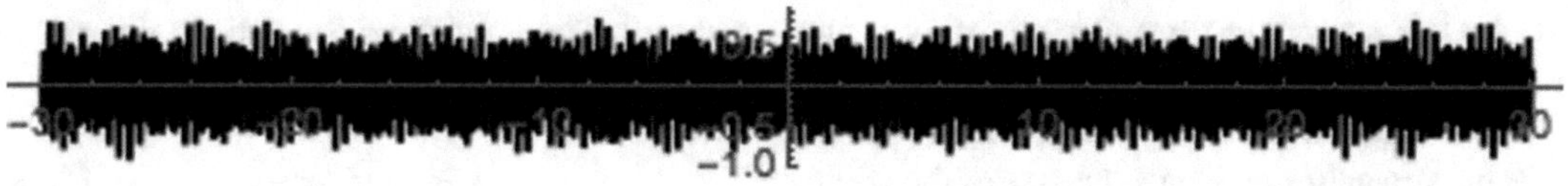

Fig. 8.30 A complex wave with such high frequency components that the individual oscillations cannot be seen a the scale of the image

Fig. 8.31 Clip of the complex wave—four seconds long

speed is decreased by more than a factor of about 3/4 or increased by more than a factor of about 4/3.

Sample Problem 8.3 A **reel-to-reel tape recorder** was very commonly used years ago but is still in use because of its level of fidelity. The wave pattern of sound is stored through the orientation of the North/South poles of microscopic **magnetic particles** that are laid down on the tape. One way to read the wave is to use a coil that is held above the tape as it moves by. The oscillations of the magnetic field produce an **induced EMF**, as discussed in Sect. 5.11. This EMF can be amplified so as to drive a loudspeaker.

One common speed of rotation of the tape is 7-1/2 inches per sec. Over time, a problem can arise: A motor produces the rotation of the tape, and ultimately, a rubber gear that drives the axle of the reel can wear down a bit, thereby slowing down the rate of rotation.

The problem for us is the following: Suppose that the rate of rotation decreases from 7.5 inches per sec to 7.0 inches per sec. What would be the resulting frequency of a sound that was originally recorded at 440-Hz, the usual frequency for the note "A" to which musicians tune their instruments?

Solution The frequency of sound that is sent to the amplifier is proportional to the rate that magnetic particles move past the coil and therefore proportional to the rotation rate of the reel. Therefore, the new frequency will be

$$f' = \frac{7.0}{7.5} \times 440 = 436.9\text{-Hz} \tag{8.16}$$

The change in frequency is very close to a **half-step** in music and therefore corresponds close to the note "$A_\flat$".[11]

Defining the Short Time Fourier Transform—The STFT

The basis of the STFT is as follows.

Suppose that we have a complex wave shown in Fig. 8.30 that extends over a period of 60-s, from −30-s to +30-s. Let us clip out a segment of the wave that is 4-s long (Fig. 8.31). We next determine the Fourier spectrum of this segment of a wave.

To understand the nature of the Fourier spectrum of such a windowed wave, it is very helpful to examine a windowed sine wave, which is the simplest of waves. Consider a wave that is sinusoidal, with a frequency of 1-kHz. We can determine the spectral intensity of the windowed wave analytically using the mathematical software called **MAPLE**.[12] MAPLE carries out a **Fourier analysis**; the *spectral intensity is the absolute square of the* ***amplitude*** determined by MAPLE. The results are shown in Fig. 8.32. We see the **Spectral Intensity** associated with the windowed sinusoidal wave of frequency 1-kHz.

[11]The drop in frequency is close to 7% and corresponds to a drop in the radius of the gear of the same percentage.

[12]See https://www.maplesoft.com/products/Maple/features/index.aspx.

To the left the window has a width of 4-ms; to the right the window has a width 32-ms.

Generally, the width Δt of the window in time is approximately equal to the inverse of the width Δf of the spectral intensity:

$$\boxed{\Delta f \sim \frac{1}{\Delta t}} \tag{8.17}$$

1. To the left, in Fig. 8.32, we see the spectral intensity of the wave with a window of width 4-msec. Correspondingly, there are four oscillations of the sinusoidal wave of frequency 1-kHz. The width of the peak of the spectral intensity is about 0.2-kHz, which is close to the estimate of Eq. (8.17): 1/(4-ms).
2. To the right, we see the spectral intensity of the wave with a window of 32-ms, corresponding to 32-oscillations. The peak is correspondingly much narrower, with a width of about 0.04-kHz, which is so small as to be difficult to estimate. According to Eq. (8.17) this value should be compared with (1/32-ms)~0.03-kHz.

There is an alternative way to express the relation between the width Δf of a peak in the spectral intensities in Fig. 8.32, associated with a sinusoidal wave of frequency f, and the width Δt of the window in time over which the wave is analyzed. The number of oscillations during the interval of time of the window is given by

$$N = f\Delta t \tag{8.18}$$

Using Eq. (8.17) we obtain

$$N = \frac{f}{\Delta f} \tag{8.19}$$

Alternatively,

$$\boxed{\frac{\Delta f}{f} = \frac{1}{N}} \tag{8.20}$$

Now imagine if a computer program could determine the spectral intensity for a window of a wave, once for each instant of time. We will have an infinite number of spectral intensities. This collection of spectral intensities constitutes the **short time Fourier Transform**, represented by the function $I(f, t)$.

Above, we discussed a rectangular window of the wave—which simply clips out a specific interval of time. We have seen that the rectangular window produces oscillations around the central peak. The oscillations result from the discontinuity at the boundaries of the window. These sharp drops don't reflect the smooth sinusoidal oscillations of the sine wave. To avoid such oscillations, other windows are used to analyze waves and

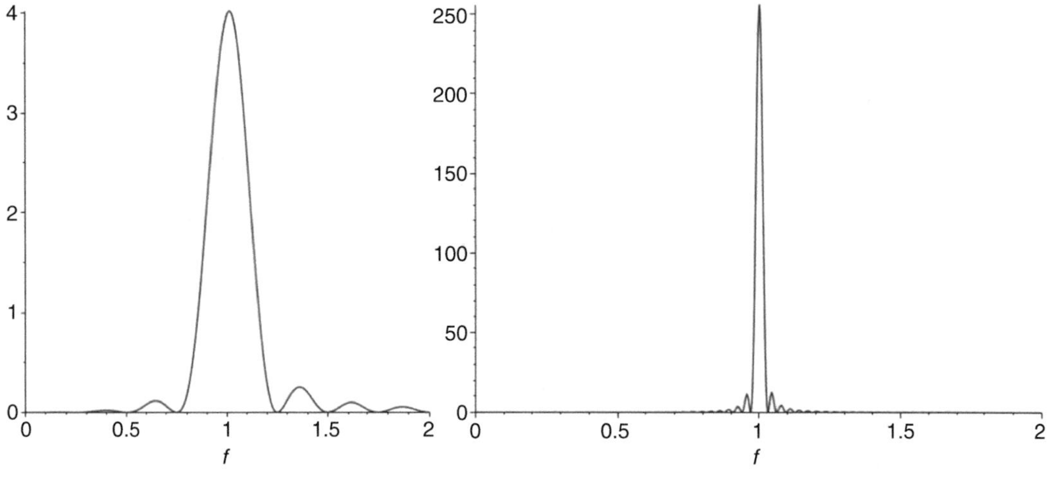

Fig. 8.32 Spectral intensities associated with windows of a sinusoidal wave of frequency 1-kHz. To the left, a window width=4-ms; to the right, a window width of 32-ms

produce STFTs. Generally, a window is peaked, with a maximum at the center of unit and vanishes outside the window. Each window has its advantages for reasons that are quite beyond the level of this book. Here is the **Hann window**, shown in Fig. 8.33. If one uses a window that goes to zero gently at the boundaries, such as the Hann window, the oscillations go away. The effect of applying a Hann window on a segment of the complex wave in Fig. 8.30 is shown in Fig. 8.34.

The spectral intensity of a sine wave of frequency 1-kHz with a Hann window of width 32-ms is shown in Fig. 8.35. We see how the oscillations at the edges of the peak have disappeared.

So far we have studied a wave that is sinusoidal, with one frequency. What if the wave is sinusoidal for one interval of time and is sinusoidal for a subsequent interval of time but with a different frequency? If we have a single window width (remember that this is a time interval), that width must be much larger than the inverse of the smallest frequency.

$$\text{window width} \gg \frac{1}{\text{smallest frequency}} \tag{8.21}$$

In addition, as the window moves along the pattern with respect to time, the window will necessarily overlap the instant of transition and the spectrogram will not display the discontinuity of this transition with much resolution. In general, rapid variations in frequency require narrower windows, which in turn reduce resolution.

$$\text{resolution in time} \sim \frac{1}{\text{resolution in frequency}} \tag{8.22}$$

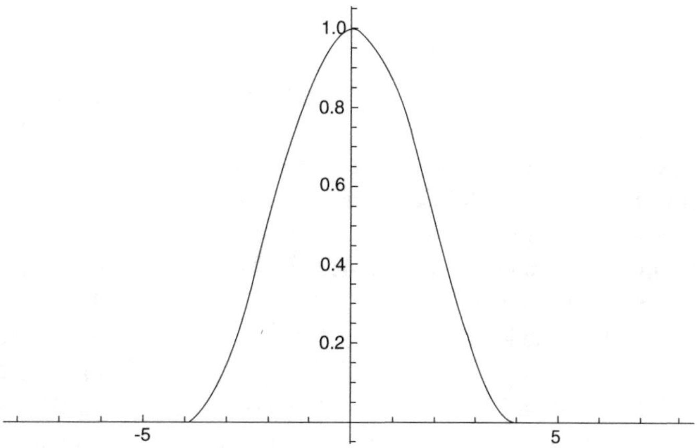

Fig. 8.33 Hann window The horizontal axis is the time axis

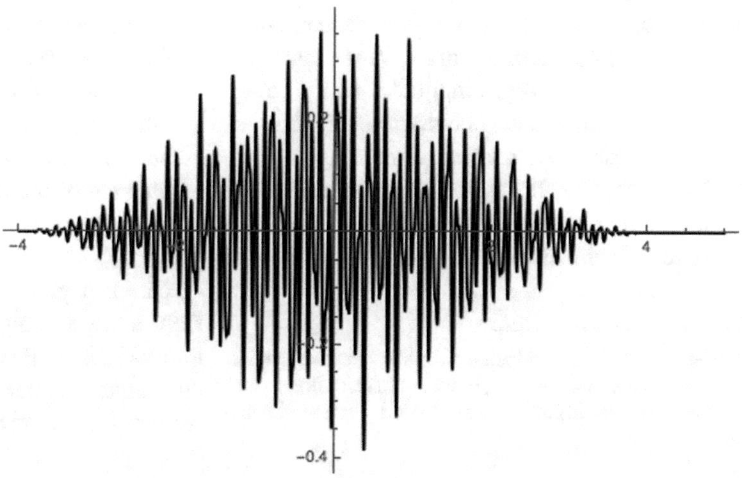

Fig. 8.34 Waveform with Hann window

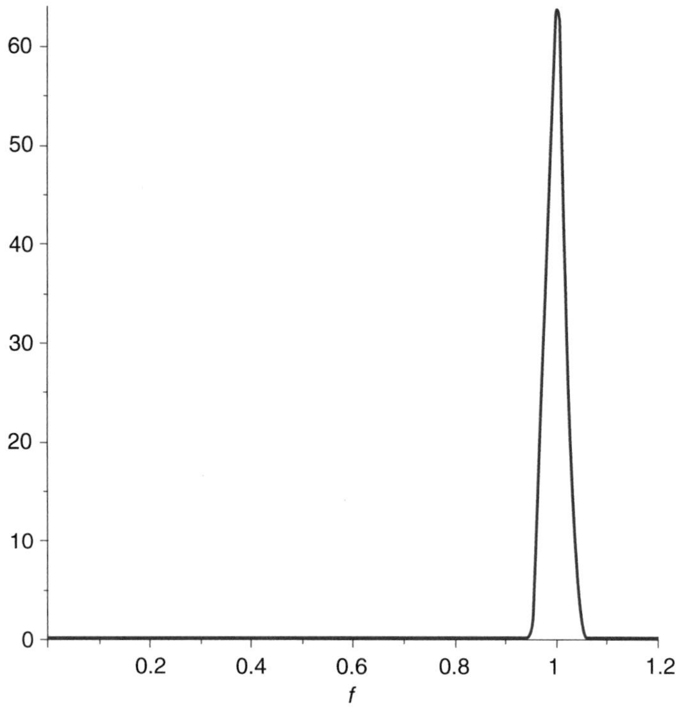

Fig. 8.35 The spectral intensity of a sine wave of frequency 1-kHz with a Hann window of full width 32-ms

We can see this conflict in the following two figures.[13]

In each of the following two figures, (Fig. 8.36) and (Fig. 8.37), we see a spectrogram of a sequence of four segments of sine waves, having frequencies 10-Hz, 25-Hz, 50-Hz, and 100-Hz, respectively.

These are the highlights:

- Each segment has an interval of five seconds. Note that for this spectrogram, the frequency increases in the downward direction.
- The color bar at the right indicates the relation between color and amplitude. We see that dark blue represents zero amplitude, while going across the rainbow of colors towards the red end represents an increase of intensity.
- Note, too, that the segments are not infinitesimally thin horizontal lines. That is, the lines have a width in frequency as well as a variation in color, corresponding to a peak of the spectral intensity at the respective frequencies.
- According to the website, the window for the spectrogram in Fig. 8.36 has a width in time of 375-ms. The numbers of oscillations within a window are, respectively, 3-3/4, 9-3/8, 18-3/4, 37-1/2.
- In comparison to the spectrogram in Fig. (8.36), the segments in Fig. (8.37) are much broader with respect to frequency. This increased breadth does not represent the actual distinctness of the sinusoidal behavior of the wave but rather a broadening due to the decreased window width. The lowest frequency, 10-Hz, suffers most in frequency blurring, as expected. Furthermore, we see how the decreased window width leads to a blurring of the *transition* from one frequency to another (Fig. 8.37).

One can produce an STFT and spectrogram from a wave using the mathematical procedure outlined above that makes use of a sequence of an infinite number of windows. As an example, we can thereby produce the spectrogram of the

[13]The source is a Wikipedia site https://en.wikipedia.org/wiki/Short-time_Fourier_transform. According to the website, the spectrogram was produced by Alessio Damato.

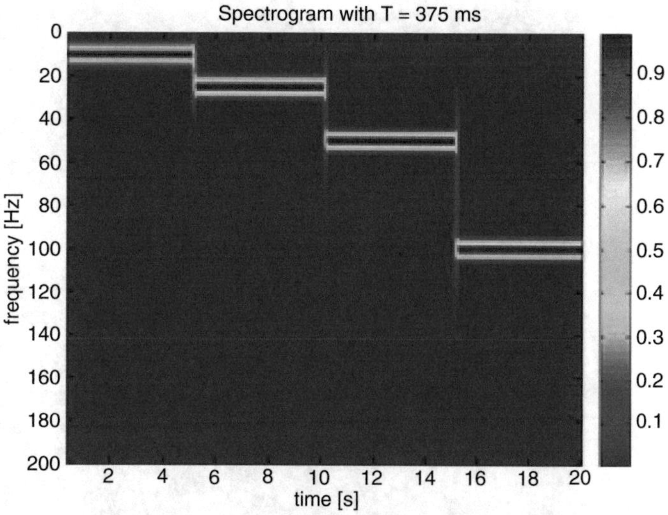

Fig. 8.36 Spectrogram for a sequence of sinusoidal wave segments with frequencies 10-Hz, 25-Hz 50-Hz, and 100-Hz, produced by a window with width 375-ms

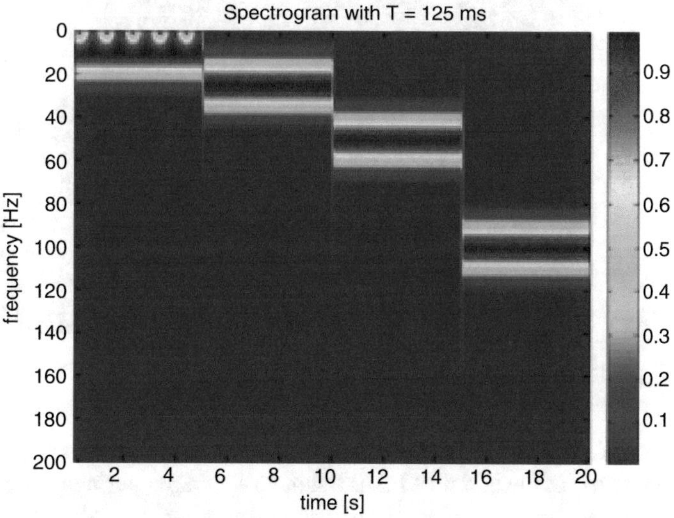

Fig. 8.37 Spectrogram for a sequence of sinusoidal wave segments with frequencies 10-Hz, 25-Hz 50-Hz, and 100-Hz, produced by a window with width 125-ms

song of a Lark Sparrow shown in Fig. 8.29 from the waveform of its song in Fig. 8.21.

In Fig. 8.38 we see a spectrogram of the author's recitation of the vowels A, E, I, O, U, obtained using the iOS app "Spectrum Analyzer RTA by Onyx".[14] The frequencies of the vowel "I" are approximately 100-Hz, 200-Hz, 300Hz, 400-Hz, 500-Hz, . . . , and are therefore members of a **harmonic series**, with **fundamental frequency**100-Hz. Note that the vertical frequency scale is logarithmic, not linear. As a result, the bands are not equally spaced. The other vowels have marked changes in the pitch over time. Finally we see a spectrogram (Fig. 8.39) of two notes voiced by the author, the first with a fundamental of 125-Hz, the second that follows is a note with a fundamental that is greater than the first by a factor of 3/2 and hence has a frequency of 187.5-Hz. The spectrogram exhibits the first seven harmonics of both notes, with the seventh harmonic of the second note having quite

[14]Here are a few apps that you can use to obtain a spectrogram of a sound input using a mobile device: For an Android mobile device you can download **SpectralPro Analyzer** from the website **https://play.google.com/store/apps/details?id=radonsoft.net.spectralviewpro&hl=en_US**. For an iOS mobile device you can download **SpectrogramProSpectrum Analyzer RTA by Onyx** or **SpectrumView** from the App Store.

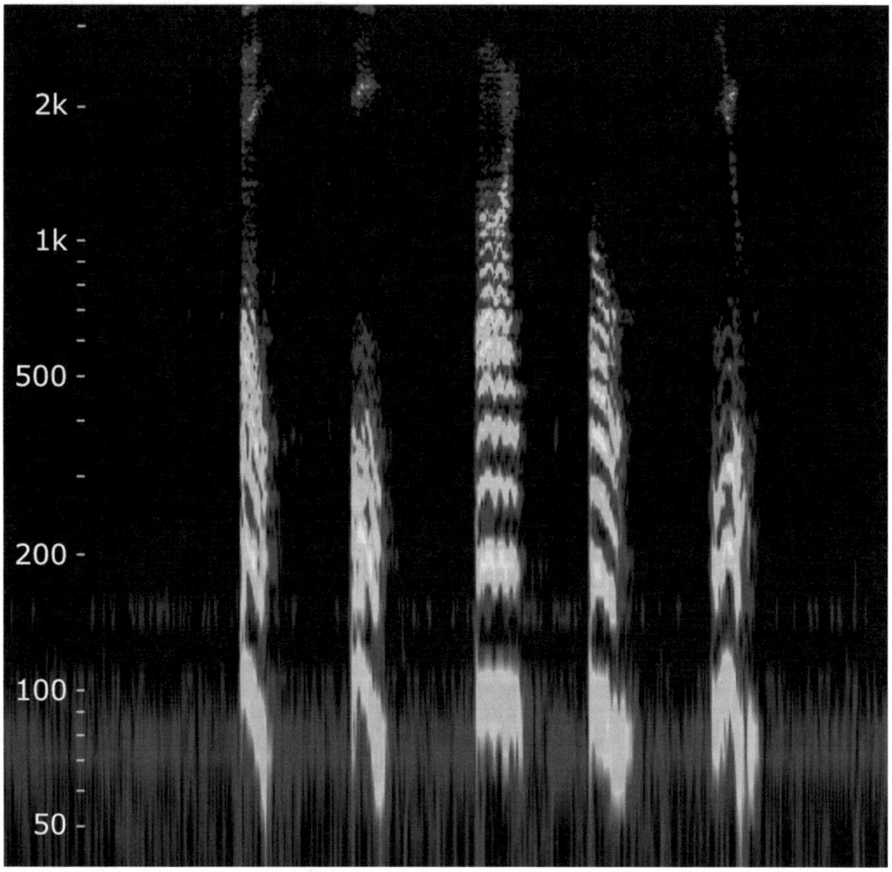

Fig. 8.38 Spectrogram of voiced vowels—A, E, I, O, U

a weak intensity. We can clearly see that the third harmonic of the first note has the same frequency as the second harmonic of the second note, as expected. See Chap. 2.

Spectrograms are used to study the sound that is produced by mechanical and electronic systems in a wide range of ways. For me personally, it played an important role. Around 2005, my ENT specialist sent me to a voice specialist to examine my throat. An endoscopic video[15] revealed that I had a cyst on one of my **vocal folds**.[16] Vocal folds are akin to the vibrating strings discussed in Chap. 2, but clearly are more complex. Like strings, they have **modes of vibration**.[17]

My voice therapist, Barbara Arboleda, asked me to sing a single note, holding it for as long as I could; about ten seconds was adequate. I used the video she gave me to produce the image of my vocal folds in Fig. 8.40. The vocal folds are the two beautiful white bands in the image! You can see them in motion on the video that I have posted on a YouTube.[18]

[15] An endoscope is a long flexible tube that is introduced down the pharynx, larynx and sometimes as far down as the digestive tract, along with a light and camera at the far end.

[16] Known by the layperson as **vocal cords**.

[17] See http://www.ncvs.org/ncvs/tutorials/voiceprod/tutorial/modes.html.

[18] See https://www.youtube.com/watch?v=lDyvMEyGp30. The YouTube contains two video segments of my vocal folds: one from the year 2005, the other from the year 2013. I am deeply grateful to Barbara Arboleda for sharing the former and to Dr Matthew Naunheim, for sharing the latter.

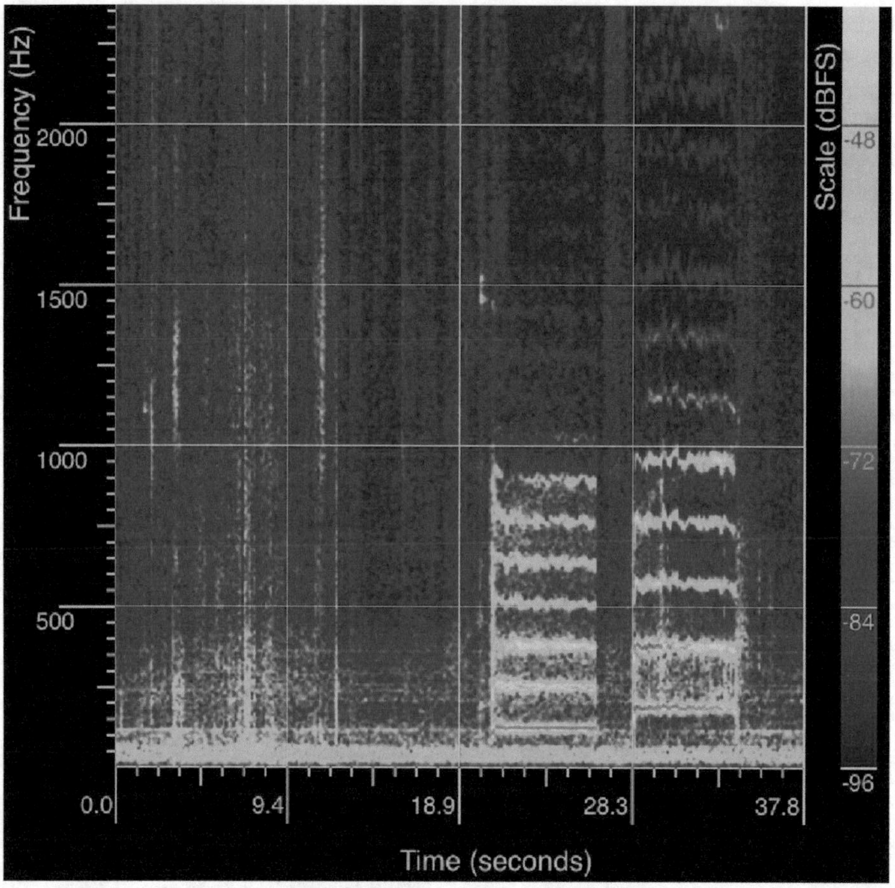

Fig. 8.39 Spectrogram of the author singing two notes—125-Hz and 187.5-Hz

Fig. 8.40 Vocal folds of the author in 2005

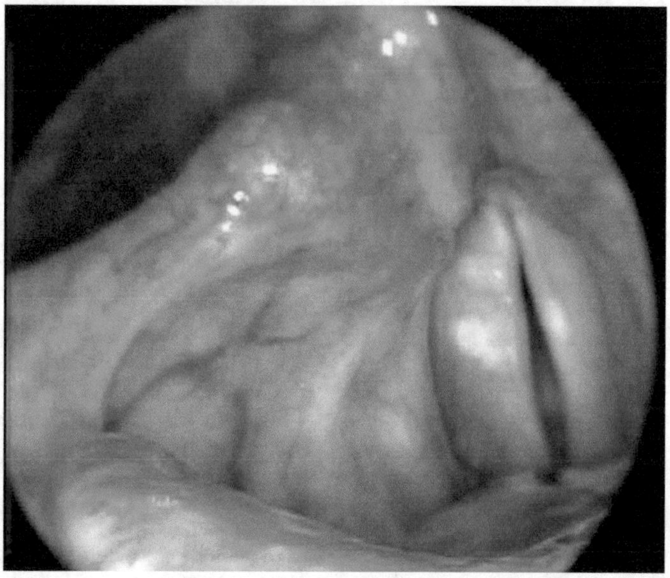

From the sound I produced she was able to produce a spectrogram. The spectrogram meant nothing to me. However, the voice therapist was able to read quite a bit about the behavior and physical state of my vocal folds that were not clearly visible in the video.[19]

8.5.3 The Relationship of Spectrograms with Measurement and Perception

The Measurement of Frequency

It takes time to measure a parameter, let alone a frequency. Moreover, measurements are always accompanied with a level of precision. Devices that measure frequency, in particular, with a specification of a numerical value are called **frequency meters**. There are devices that are dedicated to measuring the frequency of a sound wave or an electromagnetic wave (such as radio wave, of a light wave). They have intrinsic limitations in precision expressed by the "readout." The level of precision is usually expressed as follows. For a radio wave we might see

$$f = 580\text{-kHz} \pm 1\text{-kHz} \tag{8.23}$$

The level of precision in this case would be $\Delta f = 2$-kHz. It can be shown that the level of precision is inversely proportional to the interval of time τ over which the frequency is measured. This time interval is also referred to as the **observation time.**[20]

$$\Delta f \sim \frac{1}{\tau} = \frac{1}{\text{observation time}} \tag{8.24}$$

[Another common pair of terms used are **resolution in time** and **frequency resolution**.]

In the above case, the observation time would have to be at least

$$\tau \sim \frac{1}{\Delta f} = \frac{1}{2\text{-kHz}} = 0.5\text{-ms} \tag{8.25}$$

The relation between frequency resolution and resolution in time is analogous to the relation that is reflected in sonograms. The **widths of the peaks** of the spectral intensities represent the **uncertainty of the frequency**. In measurements of the frequency of a wave, the level of uncertainty is directly related to the number of oscillations over the course of the measurement.

To have a relatively small error in measuring the frequency, we need $\Delta f/f \ll 1$. Therefore, the number of oscillations needed to have a small error must be large. Repeating Eq. (8.20),

$$\boxed{\frac{\Delta f}{f} = \frac{1}{N}} \tag{8.26}$$

we see that need $N \gg 1$.

As an example: Suppose that you are measuring a frequency of 400-Hz and you sample the sound for ten seconds. You can then rely on the result to within 1/10 Hz. For some devices the reading will simply be 400.1-Hz—and therefore a result with four significant figures.

The Perception of Pitch

While an electronic device measures the frequency spectrum of an electronic signal, our psycho-acoustical/physiological system (including our brain) measures and interprets a sound wave incident upon our ears. In particular, we have a perception of pitch. We have pointed out in Chap. 2 that pitch is associated with a sine wave of infinite extent in time. And, if a wave has a frequency spectrum that forms a harmonic series, we will have a sense of pitch. Since a musical note has a finite duration, the note does not have a well-defined frequency spectrum. Section 11.6 discusses the relationship between the ability to perceive a sense of pitch in a sine wave and the duration of time that the sound is heard. If we let N_m be the minimum number of

---

[19] See website of Rob Hagiwara—**https://home.cc.umanitoba.ca/~robh/howto.html**, which describes how the information contained in a voice spectrogram can be extracted.

[20] Interestingly, this relation is analogous to what is called the **Heisenberg Uncertainty Principle** in Quantum Theory. In this case, we cannot pin down the details of a wave at a specific time without losing information about a frequency. At the level of quantum systems, such as electrons, we cannot pin down an electron's position without losing information about its momentum or velocity. Specifically, the uncertainty of the position Δx is related to the uncertainty of the velocity Δv via the relation $\Delta x \sim \hbar/mv$, where m is the mass.

oscillations necessary for the determination of pitch, minimum duration of time τ_m to determine a pitch will be

$$\boxed{\tau_m = \frac{N_m}{f}} \tag{8.27}$$

The results are totally in line with our discussion of spectrograms and a measurement of frequency. Consider that we can sense a sequence of pitches in a piece of music. We have pointed out in Chap. 2, pitch is associated with a sine wave of infinite extent in time. Since a musical note has a finite duration, the note does not have a well-defined frequency.

Note The perception of pitch requires a process that is akin to the **short time frequency analysis** used for a spectrogram. There is a major difference: The time associated with the window that is used for a spectrogram is located at the center of the window. On the other hand, for pitch perception, the window can only *monitor the past*—it must end no later than the time that the pitch is perceived.

8.5.4 Spectrogram of the Gravitational Wave (GW) from a Collapsing Binary Neutron Star

Recently, **gravitational waves** were detected by an incredibly designed apparatus that was designed to detect the waves amidst a background that is much greater in overall magnitude. The waves were the result of the collapse of a pair of neutron stars—referred to as a binary star—billions of years ago. That longtime span is connected to the fact that gravitational waves travel at the speed of light—c—in vacuum. You can see a simulation of the spiraling of the binary neutron stars about each in Fig. 8.41.[21] This motion is referred to as **inspiraling**. As the stars inspiral, the time for each revolution—the **period**—accelerates.

Maxwell's equations, which were discussed in Chap. 5, account for the emission of electromagnetic waves by an accelerating electric charge. An electric charge that revolves emits EM waves with a frequency equal to the frequency of revolution.
Einstein's Theory of Gravitation predicts the emission of gravitational waves by an **accelerating mass**. For this reason, few physicists were utterly surprised that gravitational waves were detected except for the fact that the signal is incredibly small and there is extremely strong noise that masks the signal. As with a revolving electric charge, a revolving mass emits gravitational waves with a frequency equal to the frequency of revolution of the mass.

The inspiraling ends when the two neutron stars merge into each other, releasing a myriad of matter, emitting extremely intense light.

The gravitational waves that are detected on earth can be analyzed so as to produce a spectrogram. We see the GW signal from the binary star **GW170817** in Fig. 8.42.[22] The frequency starts from about 25-Hz and reaches a limit as high as 1-kHz at the point of the final merger. Notice how sinusoidal the signal is. This is so because the period doesn't change much over the course of a few revolutions except for a time close to the merger, when the frequency changes rapidly.

Note that since the range of frequencies is in the audible range of sound, the signal can be used to produce a sound wave that we can hear. The spectrogram of the signal is shown in Fig. 8.43.[23]

Note Let us examine the range of frequencies and corresponding range of periods of the inspiraling. In writing this book, I have had a deep

[21] Source NASA: See https://svs.gsfc.nasa.gov/vis/a010000/a010500/a010543/ . You can see a simulation of the collapse in this link.

[22] Source: https://www.ligo.org/science/GW-Inspiral.php.

[23] Source: https://www.ligo.caltech.edu/system/avm_image_sqls/binaries/98/original/gw170817clean.jpg?1508108648. You can hear an excellent audio replication on this website: https://www.youtube.com/watch?v=X6dJEAs0-Gk.

Fig. 8.41 Collapsing binary neutron stars

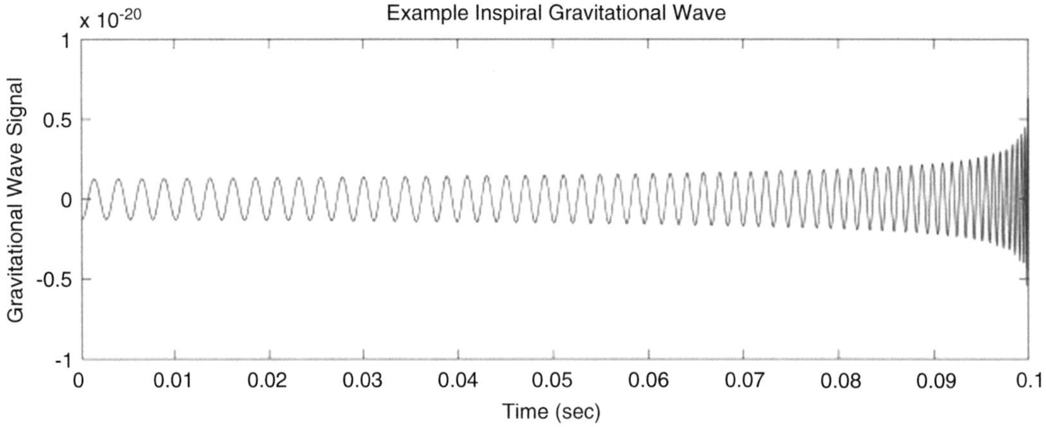

Fig. 8.42 Gravitational wave signal

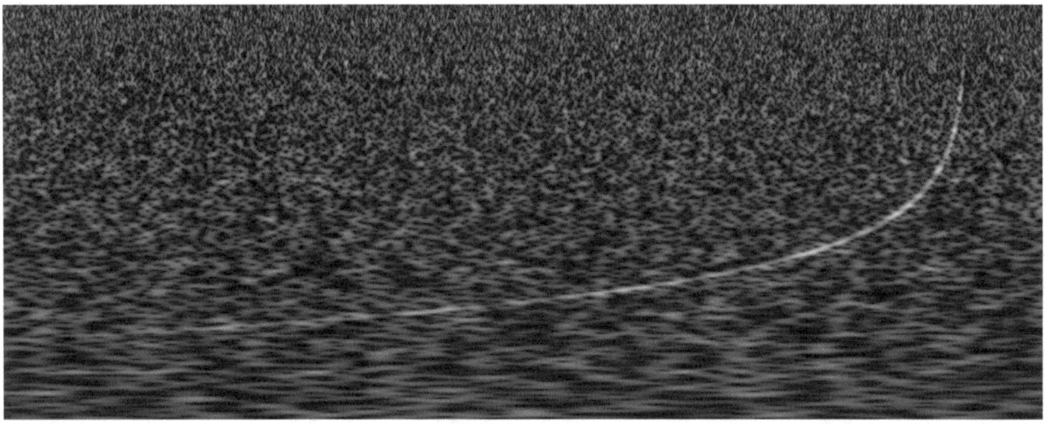

Fig. 8.43 Spectrogram of gravitational wave signal from collapsing binary neutron stars

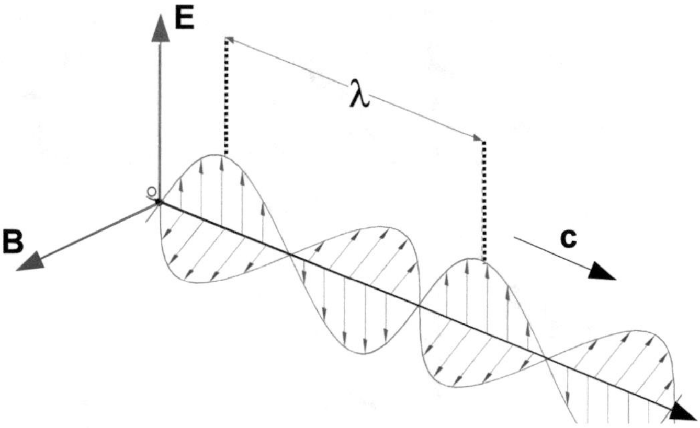

Fig. 8.44 Light as an electromagnetic wave (source: http://en.wikipedia.org/wiki/Wave#Electromagnetic_waves)

goal to show how the physics of music and the physics of color share common physical principles and concepts. And, to show how they share common **physical laws**. In a different realm, it is not difficult to appreciate that the motion of the binary neutron stars is related to the motion that is much more common to us: the motion of the earth and other planets about the Sun.

Given, this relationship, we can turn to Kepler's Third Law, which is discussed in Sect. A.4.1 of Appendix A. The Law is not a law in our current terminology but rather a mathematical reflection of the relationship between the period P of a planet's orbit about the Sun and its distance from the Sun, expressed as the radius R of the orbit.[24]

Kepler's Third Law can be expressed as a proportionality:

$$P \propto R^{3/2} \tag{8.28}$$

Alternatively, we can write, $P = bR^{3/2}$, where b is a constant. If we have two planetary orbits, with periods $P_2 = P_1$ and corresponding radii $R_2 = R_1$, the parameters are related by the equation

$$P_2 = P_1\left(\frac{R_2}{R_1}\right)^{3/2} \tag{8.29}$$

We will use the earth for #1 and a neutron star for #2. We have P_1 = 1-yr = 3.2×10^7-s and R_1 = 150×10^6-km = 1.5×10^8-km. For a neutron star's orbit we will use its estimated initial theoretical radius, of about 100-km. The result is

$$P_2 = P_1\left(\frac{R_2}{R_1}\right)^{3/2} = 3.2 \times 10^7\text{-s} \times \left(\frac{100\text{-km}}{1.5 \times 10^8\text{-km}}\right)^{3/2} \sim 0.016\text{-s} \tag{8.30}$$

The corresponding frequency is 1/0.016=60-Hz. This result is very much on the order of magnitude of the observed initial frequency of about 25-Hz of the gravitational waves.[25]

8.6 Polarized Light

In this section we will be discussing the nature of **polarized light**. It is of importance not only for fundamental physics reasons and practical applications but also because some artists make use of polarized light in their art creations. We have mentioned that a simple EM wave is *transverse*, with a direction of displacement that is determined by the direction of oscillation of the electric field. In Fig. 8.44 we see both the electric

[24]More precisely, the **semi-major axis** of its **elliptical orbit**. The orbits are very close to being circular and I will refer to the radius of a nearly circular orbit for simplicity and therefore without significant error.

[25]In the case of the earth and Sun, the Sun is very nearly stationary because its mass is so much greater than the earth's mass. On the other hand, in the case of the binary neutron stars, the masses have equal orders of magnitude. The two stars spiral about each other. If they have equal masses, Kepler's expression is missing a factor of two and our value of 60-Hz should be replaced by 30-Hz.

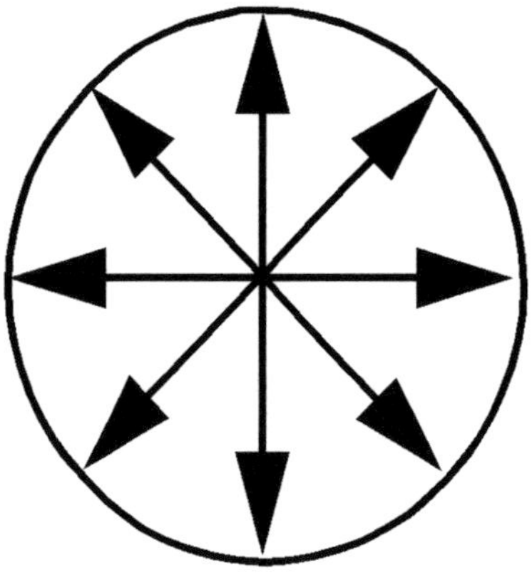

Fig. 8.45 Symbol for unpolarized light

field and magnetic field as they vary in space at some moment in time. The wave is propagating in the direction of the arrow labeled 'c', the velocity of light in a vacuum.

Light that is emitted from typical light sources consists of a mixture of waves having a random distribution of electric field orientations. We say that the light is **unpolarized** and will indicate this state by the symbol shown in Fig. 8.45. The symbol represents a distribution of axes of polarization. *It is important to keep in mind that we view these axes with respect to the beam of light coming towards us.*

8.6.1 How Can We Obtain a Beam of Polarized Light?

There are numerous ways to obtain a beam of **polarized light**. Laser light is polarized. So is the light from excited atoms whose quantum states have been properly selected. However there is a way to obtain a polarized light beam by passing the beam through a polarizer.

Ideal Polarizer

An **ideal polarizer** has a **polarization axis**. It has the property that when a beam of light passes through the polarizer, whatever the polarization state of the incoming beam—it could be unpolarized or polarized along an axis—the outgoing beam is polarized with an axis parallel to the polarization axis of the polarizer.

We will use the following notation, as shown in Fig. 8.46.[26] The **direction of the axis of polarization** of both the polarizers and the beams will be specified by its angle with respect to the horizontal, *as observed with the beam traveling* ***towards*** *the viewer*. Thus, the horizontal axis has an angle of 0°—at "3-o'clock." The angle θ is measured **counter-clockwise** as shown. Since the electric field oscillates from a direction at 3-o'clock to 9-o'clock, a polarization angle of 180° is equivalent to an angle of 0°. The angle of the polarization axis of a *beam* will be shown within a circle; the angle of the polarization axis of a *polarizer* will be shown within a square.

The action of a polarizer is shown in Fig. 8.47. In the figure, the axis of the incoming beam is labeled with an "X," representing the fact that the outgoing beam is polarized with an axis along the axis of the polarizer whatever the polarization state of the incoming beam.

In Fig. 8.48 we see the action of an ideal polarizer on an incident polarized beam.

The following general relation between the incoming intensity I_1 and the outgoing intensity I_2 is

$$\boxed{I_2 = I_1 \cos^2(\theta_2 - \theta_1)} \qquad (8.31)$$

This equation is referred to as **Malus' Law**.

In Fig. 8.49 we see the action of an ideal polarizer on an unpolarized beam.

8.6.2 Calcite

An important method of obtaining beams of polarized light is to use a crystal like **calcite**. If a beam of light is passed through a single crystal of calcite which is properly cut according to the crystal axes of the crystal, the output consists of two, physically separated, polarized beams, polarized perpendicular to each other, as shown in (see Fig. 8.50).

- The ray that progresses straight through the crystal without deflection is referred to as the

[26] You might find a different choice for the angle of a polarization axis—for example, the angle with respect to the vertical.

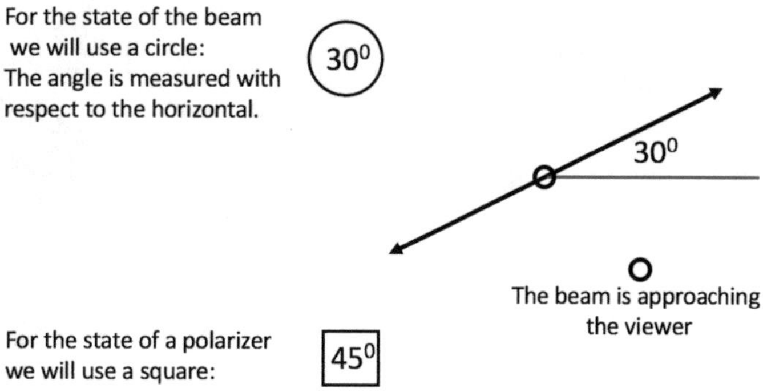
For the state of the beam
we will use a circle:
The angle is measured with
respect to the horizontal.
30⁰
30⁰
The beam is approaching
the viewer
For the state of a polarizer
we will use a square:
45⁰

Fig. 8.46 Symbols for the polarization axes of a polarized beam and a polarizer

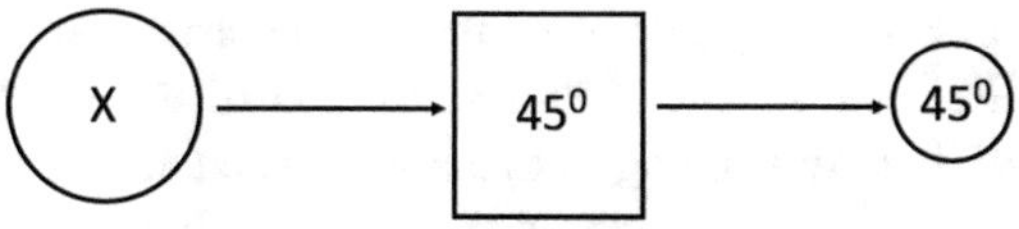
X
45⁰
45⁰

Fig. 8.47 Action of an ideal polarizer

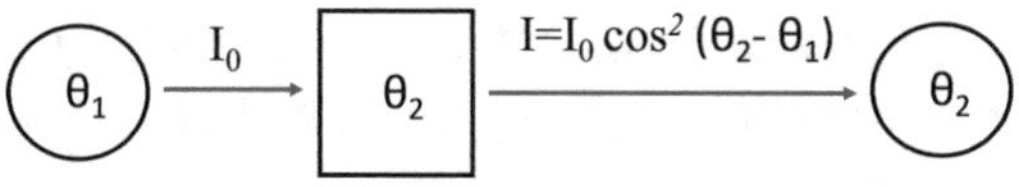
θ1
I0
θ2
I=I0 cos2 (θ2- θ1)
θ2

Fig. 8.48 Action of an ideal polarizer on a polarized beam: Malus' Law

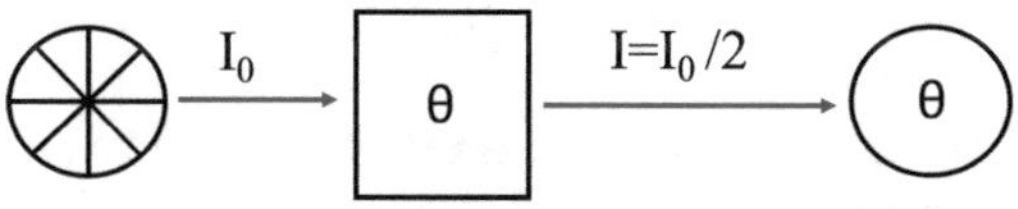
I0
θ
I=I0 /2
θ

Fig. 8.49 Action of an ideal polarizer on an unpolarized beam

ordinary ray. Note that the outgoing ordinary ray is *horizontal* as shown in the figure.

- The ray that progresses through the crystal with gradual deflection is referred to as the **extraordinary ray**. The degree of net deflection is proportional to the thickness the crystal.

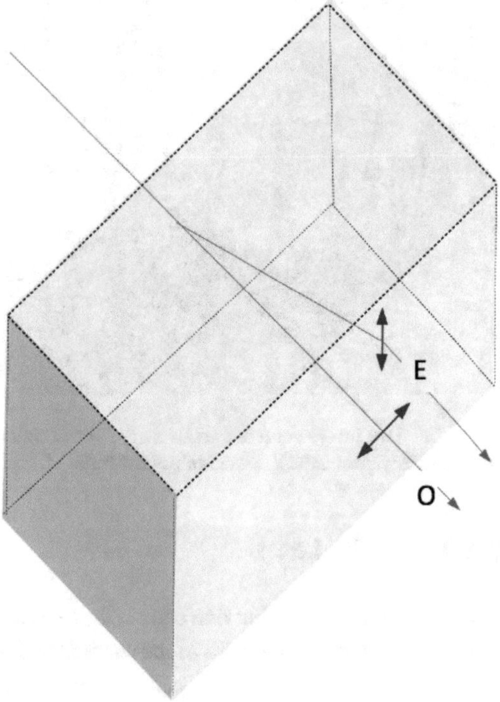
E
O

Fig. 8.50 Action of calcite
O=ordinary ray, E=extraordinary ray

In Fig. 8.51 we see a schematic drawing of the action of a calcite crystal.

The relative deflection is due to the crystal's being **birefringent**: The term means that the crystal has a different **index of refraction**[27] for the two different axes of polarization with respect to the crystalline structure. If the crystal is rotated about an axis along the ray of light, the ordinary ray will remain unchanged in its path while the extraordinary ray will rotate with the crystal.

In Fig. 8.52 we see a photo of an actual crystal of calcite, lying over a single black dot. We see two images of the dot. If we rotate the crystal, one image will remain fixed, while the other image will revolve about the first image. The light beams from the two dots are polarized according to which beam is the ordinary ray and which beam is the extraordinary ray. Their polarization axes can be determined using a polarizer.

[27] **Refraction** is discussed in detail in Sect. 9.3.

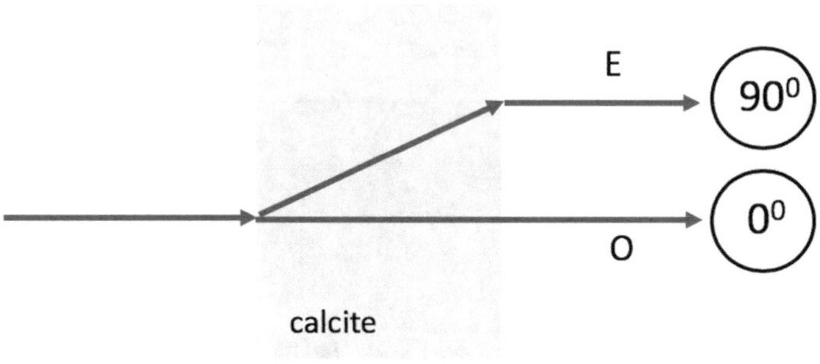

Fig. 8.51 Schematic drawing of the action of a calcite crystal

Fig. 8.52 Two images of a dot under a crystal of calcite, with mutually perpendicular axes of polarization

8.6.3 Calcite Loop

Above, we mentioned that one can rotate a calcite crystal so as to rotate the path of the extraordinary ray. In Fig. 8.53 we see two calcite crystals of equal thickness in tandem. A beam, X, enters the first crystal and is split into the two rays—the ordinary and the extraordinary. The crystals can, in principle, be cut so that the ordinary ray and extraordinary ray that exited the first crystal are reconstituted so as to produce the original incoming beam, X, whatever X is.

8.6.4 Polaroid

A second material that can produce a beam of polarized light is **Polaroid**, which is a plastic material that was discovered by Edwin Land. A sheet of Polaroid has what is referred to as an **axis of polarization**. The absorption of a ray of polarized light through the sheet depends upon the angle between the axis polarization of the light and the polarization axis of the sheet of Polaroid, being a maximum when they are parallel. This property of **selective absorption** is referred to as **dichroism**.[28] Polaroid is a **non-ideal polarizer**, as we will see.

The **transmittance** T of a beam is defined as the ratio of the outgoing intensity to the incoming intensity. Thus

$$\boxed{\mathrm{T} = \frac{\mathrm{I}_{\mathrm{out}}}{\mathrm{I}_{\mathrm{in}}}} \tag{8.32}$$

The transmittance of a sheet of material ranges for zero to unity.

Without loss of generality, we will assume that the axis of a Polaroid sheet is at an angle of 0^0 and refer to this direction as the *horizontal direction*. Let T_h be the transmittance of an incoming polarized beam with an axis that is horizontal, and therefore *parallel* to the axis of the Polaroid sheet. Let T_v be the transmittance of a polarized beam with an axis that is vertical, and therefore *perpendicular* to the axis of the Polaroid sheet. For Polaroid, the two transmittances are unequal, with $\mathrm{T}_h > \mathrm{T}_v$. [For an ideal polarizer, $T_h = 1$ and $T_v = 0$.]

Suppose now that the axis of the Polaroid sheet is at an angle 0^0, we will refer to this direction as the horizontal direction. We will further assume that an incident beam is polarized with an axis at angle θ.

$$\mathrm{T}(\theta) = \mathrm{T}_h \cos^2\theta + \mathrm{T}_v \sin^2\theta \tag{8.33}$$

[28]See Hecht, E., Optics, 5th ed.

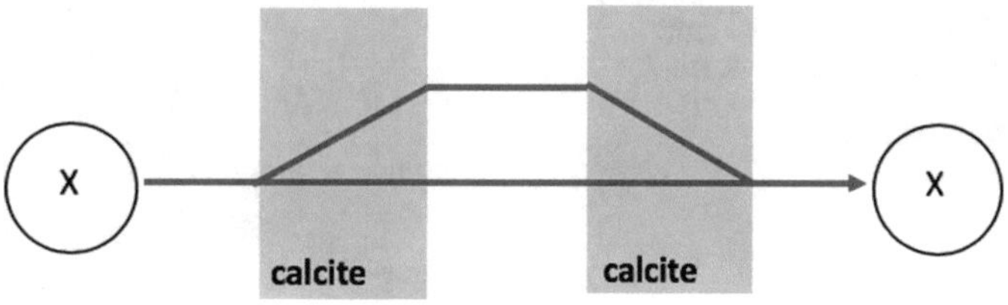

Fig. 8.53 Calcite loop

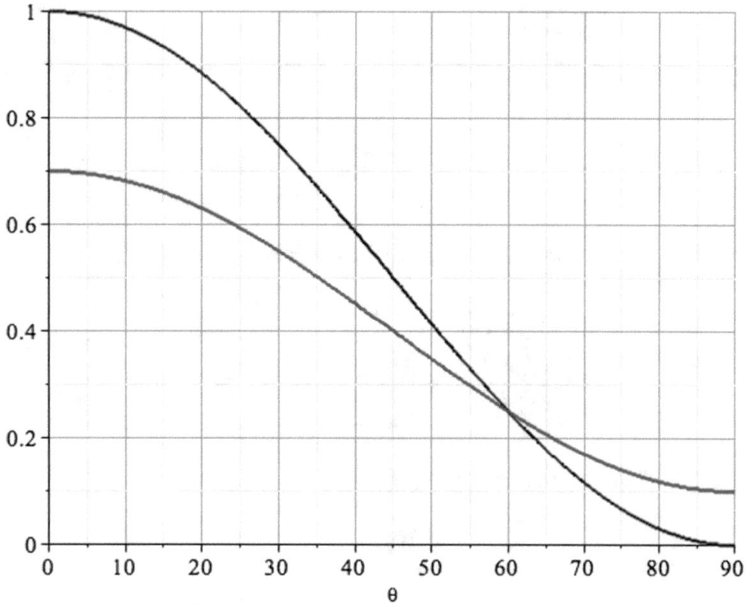

Fig. 8.54 Transmittance of incident polarized light Black: ideal polarizer; Red: non-ideal polarizer like a polaroid sheet; $T_h = 0.7$; $T_v = 0.1$

A comparison between the transmission of incident polarized light at an angle θ through Polaroid vs. an ideal analyzer is shown in Fig. 8.54.

Note that if the transmittances T_h and T_v were equal, $T = T_h$ for all angles θ, and thus be independent of the angle; there is no polarization. If T_v=0, the outgoing beam is linearly polarized. In what follows, we will assume that $T_h = 0.7$ and T_v=0.1 .

Transmission of **polarized light** at an angle θ through a sheet of Polaroid whose axis is horizontal ($\theta = 0^0$) produces a polarized beam with an axis at an angle $\theta' \neq 0$, as shown in Fig. 8.55.[29]

Transmission of **unpolarized light** through a sheet of Polaroid produces a **partially polarized** beam, as shown in Fig. 8.56. The total transmittance is equal to $[T_h + T_v]/2$. The bow tie shaped boundary and two-headed arrows represent the intensity as a function of the transmission angle θ'.

8.6.5 Series of Ideal Polarizers

A light beam can pass through a series of polarizers. Below, we will consider a number of examples that indicate how polarizers affect light beams.

We will let I_0 be the incident intensity and $I_1, I_2, \ldots$ be the intensities at various subsequent stages.

In Fig. 8.57, we depict what happens to an unpolarized beam of intensity I_0 after passing through two sequential polarizers set at angles 0° and 45°, respectively. We note that an angle of 45° between the beam and the polarizer axes leads to a reduction in beam intensity of one-half. What about other angles?

[29]The angle θ' is related to the input angle θ through the equation $\tan(\theta')=\sqrt{T_v/T_h}\,\tan(\theta)$.

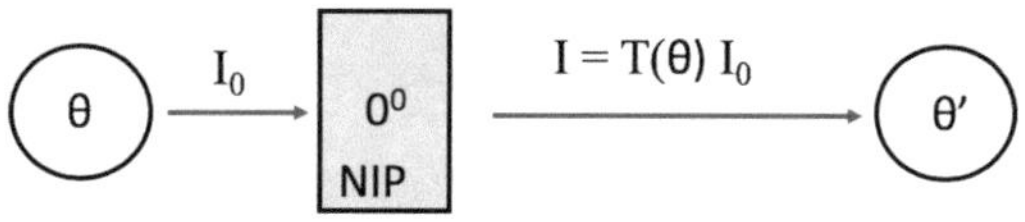

Fig. 8.55 Polarized light through a polaroid sheet

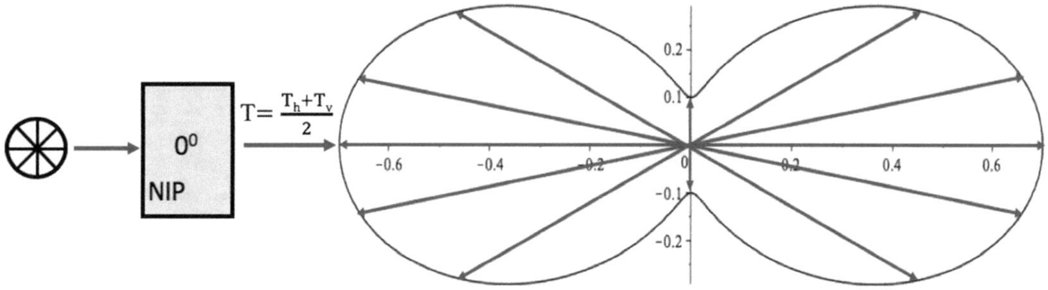

Fig. 8.56 Unpolarized light through a non-ideal polarizer; T_h=0.7, T_v=0.1

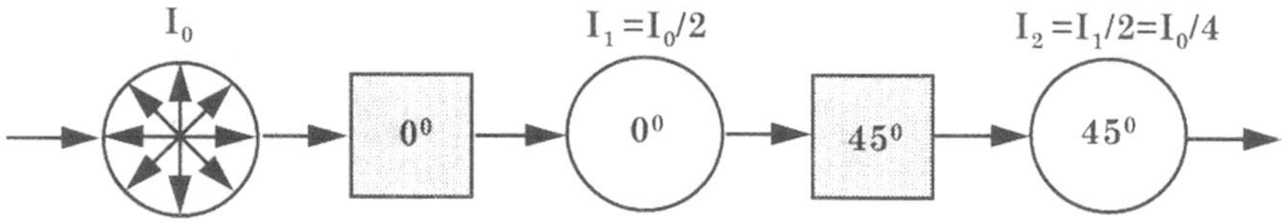

Fig. 8.57 Two polarizers in series

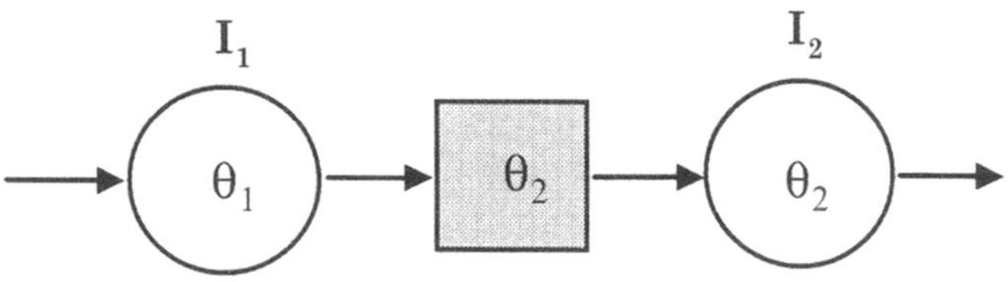

Fig. 8.58 General series of two polarizers

Here is the central rule that you need to know about the action of **ideal polarizers**:

1. **Whatever the state of the incoming beam, the outgoing beam has the State of Polarization of the Polarizer** (Fig. 8.58).

In the applications that follow we will use the following *exact* values for the cosine so that, for example, we don't end up with a number like 0.49 when the exact answer, 1/2, is significant.

$$\cos 45^\circ = \frac{1}{\sqrt{2}}, \qquad \text{so that} \quad \cos^2 45^\circ = \frac{1}{2}$$

$$\cos 30^\circ = \frac{\sqrt{3}}{2}, \qquad \text{so that} \quad \cos^2 30^\circ = \frac{3}{4}$$

$$\cos 60^\circ = \frac{1}{2}, \qquad \text{so that} \quad \cos^2 60^\circ = \frac{1}{4}$$

8.6.6 Sample Problems

We now turn to some examples in order to illustrate how we apply the two basic rules.

Example 8.1 In the situations shown in Fig. 8.59, the intensities of the beams are indicated in the figures themselves since they are relatively simple to determine from our discussion above.

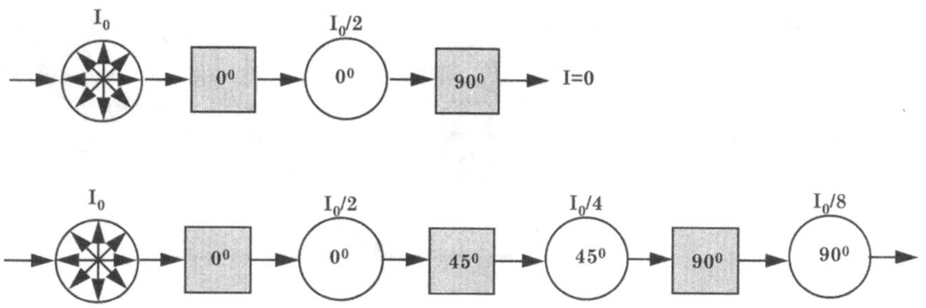

Fig. 8.59 Polarizer Example 8.1

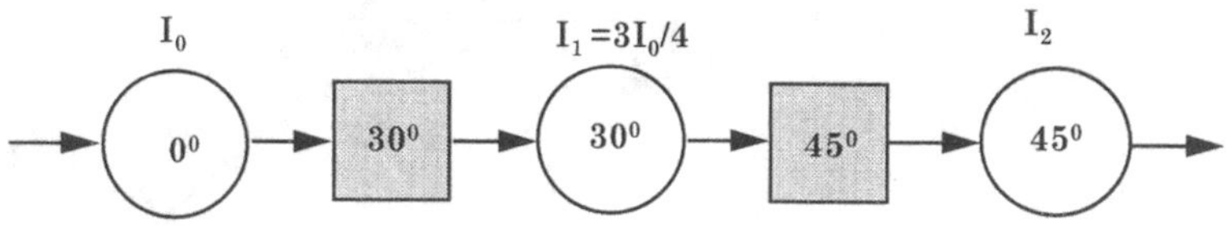

Fig. 8.60 Series of polarizers for Example 8.2

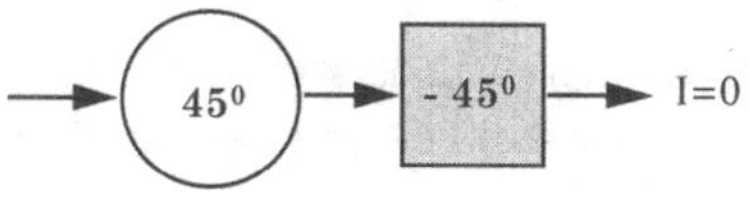

Fig. 8.61 Setup for Example 8.3

Example 8.2 In Fig. 8.60

$$I_1 = I_0 \cos^2 30° = \frac{3}{4} I_0$$

and

$$I_2 = I_1 \cos^2(45° - 30°) = \frac{3}{4} I_0 \times 0.93 = 0.70 I_0$$

Example 8.3 In Fig. 8.61 , $\theta_1 - \theta_2 = 90°$, so that the outgoing intensity is zero.

8.6.7 Partial Polarization of Reflected Light

Light that is reflected off a surface is **partially polarized** along an axis that is *perpendicular to the plane determined by the incident and reflected rays*, as shown in Fig. 8.62.

This phenomenon can be used to determine the axis of polarization of an isolated polarizer, as follows: Look at an unpolarized beam of light that is reflected off a flat surface (*not* a mirror!) through the polarizer, as in the above figure. Rotate the polarizer until the image on the surface is brightest. Then the axis of polarization of the polarizer is currently horizontal. The phenomenon also demonstrates that the polarizer material is not isotropic as one would at first assume.

We have here an explanation for why Polaroid sunglasses are so useful for cutting down the glare of sunlight reflecting off lake and ocean surfaces. By looking at a shiny floor through a piece of Polaroid one can see the varying degree of shininess as one rotates the Polaroid.

8.6.8 The Polarization of Scattering Light

We pointed out in Sect. 9.1 that the atmosphere scatters light preferentially towards higher frequencies; as a result the sky is blue. Another interesting property of scattered light is that *it is partially polarized along an axis which is perpendicular to the plane determined by the incoming and scattered rays*. The polarization is a result of the variation of the intensity of the scattered waves with incident polarization, being a maximum for a polarization that is perpendicular to the page. The geometry is depicted in Fig. 8.63. (Compare the geometry here with that of partially polarized reflected light.)

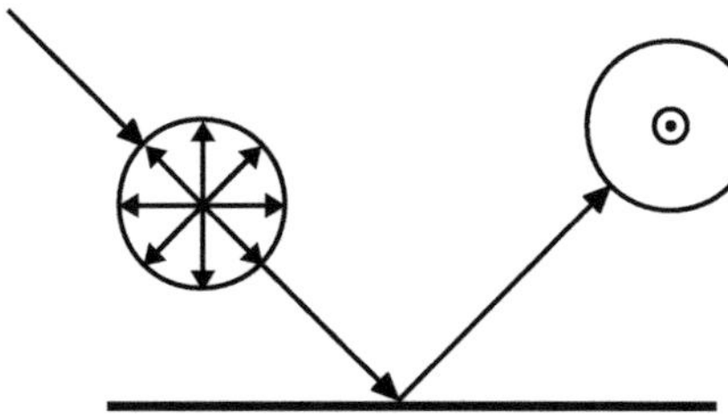

Fig. 8.62 Polarization of reflected light

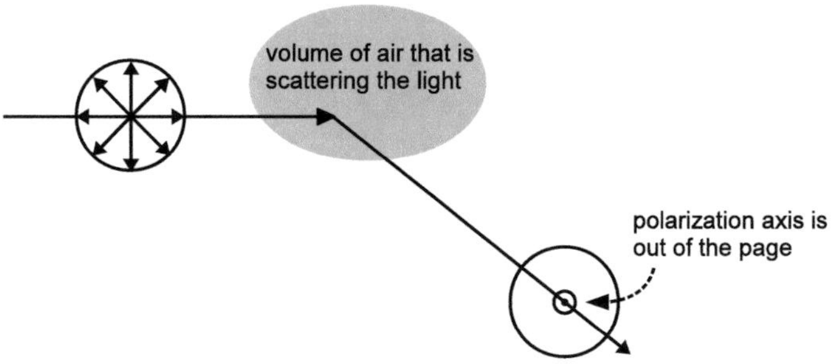

Fig. 8.63 Polarization of scattered light

Can you see why Polaroid sunglasses are not useful for cutting down the glare of a bright sky in all situations?

8.6.9 The Polarizer Eyes of Bees

The eyes of bees each have a circular array of eight polarizers whose axes are oriented at angles $360°/8 = 45°$ apart, as shown in the figure below. As a result, there are varied degrees of transmission of the polarized sunlight from the sky. From the intensity pattern of the polarizers, the bee is able to determine the orientation of its body with respect to the sun and hence its bee-hive! A bee has a built-in analog for a magnetic compass. This deduction is based upon experiments pioneered by Karl von Frisch wherein he changed the polarization of the light seen by bees and observed the bees changing their direction of flight in response. Figure 8.64 is a schematic of what a bee's eye might see in response to a change in its orientation.[30]

[30]The figure is based upon the discussion of Karl von Frisch in his book, *The Dance Language and Orientation of Bees*, Belknap Press of Harvard University Press, Cambridge, MA, 1967.

8.6.10 Using Polarization of EM Radiation in the Study of the Big Bang

According to cosmologists, our Universe evolved with a **Big Bang** from an extremely dense concentration of energy about ten billion years ago. It has been expanding ever since. A mere 400, 000 years after the start of the Big Bang, the radiation that filled the entire Universe became decoupled from the matter in the Universe as Black Body radiation. It thus contains a record of the situation at the time of decoupling. Currently, the radiation is at a temperature of 3 K. This "3 K" radiation has experienced much scattering from one region to another and is therefore polarized. The variation of temperature and direction of polarization of the universe is exhibited in Fig. 8.65, taken from the website http://map.gsfc.nasa.gov/m_mm.html. Colors indicate "warmer" (red) and "cooler" (blue) spots. The white bars show the "polarization" direction of the oldest light.

Will the currently expanding Universe expand forever or will it eventually reach a maximum expansion and then collapse? This is probably the most important question not yet resolved (in the year 2008). Detailed information such as is provided by the above figure will help answer this question.

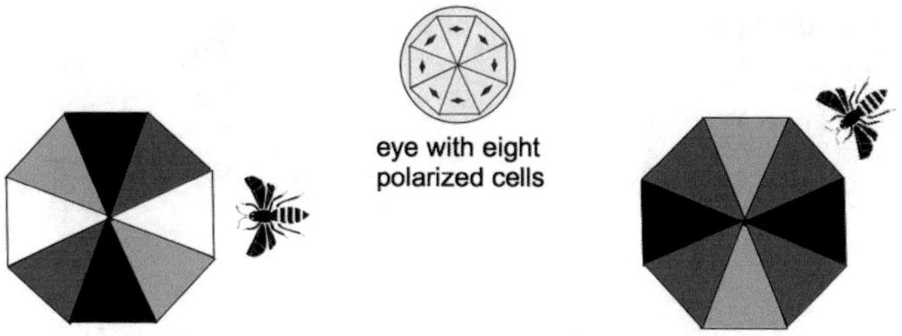

Fig. 8.64 The Frisch experiment of the polarizing eyes of bees

Fig. 8.65 Variation of temperature and polarization in the universe as viewed in various directions from the earth. (source: http://en.wikipedia.org/wiki/Big_Bang)

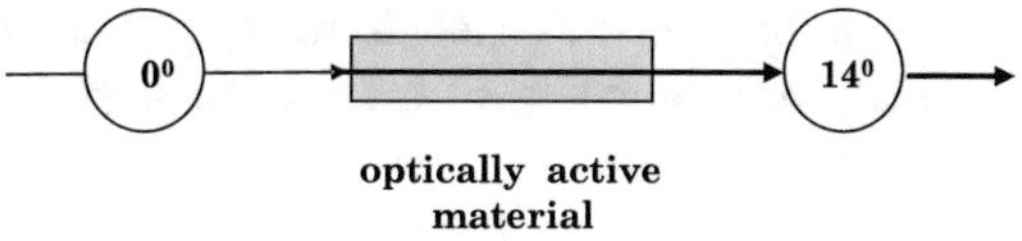

Fig. 8.66 Rotation of the axis of polarization by an optically active material

8.6.11 Optical Activity

Certain materials have the remarkable property that when a polarized light beam passes through the material, the axis of polarization of the beam is *rotated*. Such materials are said to be **optically active**. Thus, in Fig. 8.66 we have an example of a rotation by 14°.

Generally, the angle of rotation, $\Delta\theta$, is proportional to the distance ℓ of material through the beam passes: $\Delta\theta \propto \ell$.

If $\Delta\theta$ is positive, corresponding to a **clockwise rotation**, the material is said to be **right-handed**. Examples of such materials are solutions of dextrose sugar, quartz, and camphor.

If $\Delta\theta$ is negative, corresponding to a **counter-clockwise rotation**, the material is said to be **left-handed**. Examples of such materials are solutions of levulose sugar (also called "fruit sugar" or **fructose**), nicotine, menthol, and turpentine.

Note Corn syrup, sold as KARO syrup, is a mixture of dextrose and levulose; it is a right-handed material.

It can be shown that in order for a material to be optically active, its **molecular structure must be such as to differ from its mirror image**. Generally, a system that differs from its mirror image is said to be **chiral**. Systems whose mirror images differ from each other are said to possess **chirality**. The molecule and its mirror image are called **enantiomorphs**.

In the figure below, we exhibit two molecules that have the same set of four atoms, A, B, C, and D. Atoms A and C are in the foreground, while atom D is in the rear. They are mirror images of each other. Note that they are different in that one

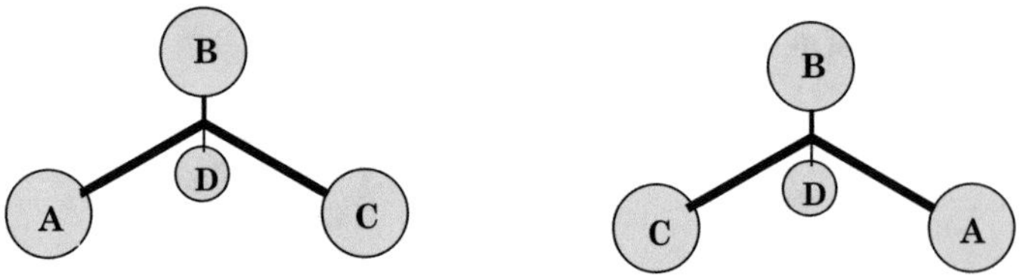

Fig. 8.67 Enantiomorphs

Table 8.1 Table of specific rotatory powers—optical activity and the asymmetry of biological systems

Material	Wavelength of light	Specific rotary power σ or σ'
Quartz	410 nm	$\sigma = 475°$ *per* cm
Quartz	589 nm	$\sigma = 217°$ *per* cm
Sucrose in water solution	410 nm	$\sigma' = 15°$ per cm per gm/cm^3 concentration
Sucrose in water solution	589 nm	$\sigma' = 6.5°$ per cm per gm/cm^3 concentration
Levulose in water solution	546 nm	$\sigma' = -10.5°$ per cm per gm/cm^3 concentration

cannot be rotated into the other, as is the case with our right and left hands (Fig. 8.67).

Notes:

1. The mirror image of a material is left-handed and vice versa. As an example, consider glucose: Dextrose is simply d-Glucose. The "d," for "dextrorotatory," means that dextrose rotates the polarization clockwise as you look at the polarized beam; dextrose is also said to be "right-handed." ℓ-Glucose is the corresponding mirror image, "ℓ" referring to its being "levorotatory," that is "left-handed" or counter-clockwise rotating.
2. The constant of proportionality in the relation $\Delta\theta \propto \ell$ is called the **specific rotatory power**, for which we will use the symbol σ. Thus
$$\Delta\theta = \sigma\ell \tag{8.34}$$
3. **Rotatory power** depends upon the wavelength of the light. Also, for solutions, it depends upon the concentration of the solution, being proportional to the concentration. Consequently, the concentration of a solution can be determined from the angle of rotation that the solution produces on a polarized beam that is passed through it.

In Table 8.1, we present the specific rotary power of quartz and of a water solution of sucrose (commonly known as cane sugar). In the first case, the specific rotary power is expressed in degrees rotation/cm. In the case of a solution, optical rotation depends upon how much sugar is in the solution so that specific rotatory power is expressed in degrees rotation/cm *per* unit concentration of one $gram/cm^3$. Thus,

$$\sigma = \sigma' \times c \tag{8.35}$$

where $sigma'$ is the specific rotation per unit concentration and c is the concentration.

Notice the *negative* specific rotary power of levulose, corresponding to a *counter-clockwise* optical rotation.

Sample Problem 8.4 Suppose that 589 nm polarized light is passed through 5 *mm* of quartz. Through what angle will the axis of polarization be rotated?

Solution

$$\Delta\theta = \sigma\ell = 217 \times 0.5 = 109°$$

Sample Problem 8.5 Suppose that 410 nm light is passed through 30 cm of a solution of sucrose in water having a concentration of 50 gm/liter. Through what angle will the axis of polarization be rotated?

Solution One liter = 1000 cm^3, so that 1 gm/liter =1 gm/10^{3}cm^3 = 10^{-3} gm/cm^3. The specific rotatory power is given by

$$\sigma = 150 \cdot 50 \times 10^{-3} = 0.750 \text{ per cm} \tag{8.36}$$

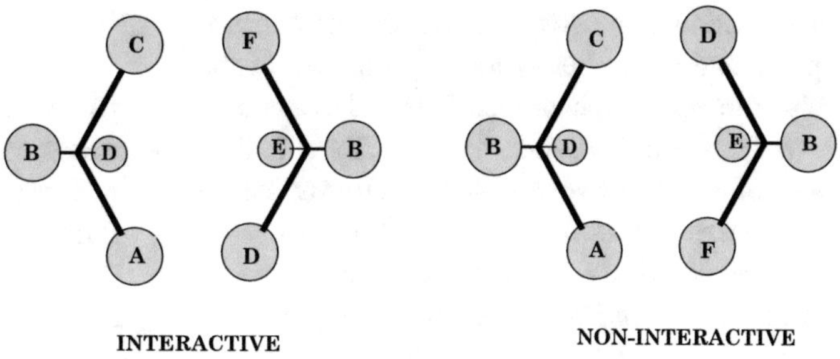

Fig. 8.68 An interactive pair of appropriate enantiomorphs vs. a non-interactive pair

Then the angle of optical rotation is given by

$$\Delta\theta = \sigma\ell = 0.75 \times 30 = 23^\circ \tag{8.37}$$

8.6.12 Our Chiral Biosphere

Our entire **biosphere** relies on the chemical interactions within a huge system of optically active molecules. Only one member of each enantiomorphic pair is represented in this system. Some are dextrorotatory while others are levorotatory. A mirror image of representative members of our biosphere will not react at all or not react correctly with the system of molecules of our biosphere.

In Fig. 8.68, we see two schematics of a pair of molecules approaching each other, an ABCD molecule and a BDEF molecule. The configuration of the ABCD molecule is the same in both schematics. However, the BDEF molecule is represented by the mirror image enantiomers. Consider the left schematic. We see the following pairs of atoms lined up next to each other such that an interaction can take place: AD, CF, and DE. On the other hand, in the right schematic this pairing cannot take place simultaneously: we see the pairing AF, CD, and DE. As a result there is no chemical interaction.

For example, ℓ-Glucose is not digestible. Levulose, which is ℓ-Fructose (or commonly referred to as "fructose" or "fruit sugar") is digestible, while its mirror image, d-Fructose, is not. It is interesting to note that the asymmetry of the molecules of our biosphere is matched and may be connected with the asymmetry of our bodies [e.g., the heart is on the left side].

In principle, one could have an entire biosphere of animals and plants looking outwardly absolutely identical to ours, except that the chiral molecules are mirror images of ours.[31] Yet, any

[31]The strikingly different chemical properties of enantiomorphs have played an extremely important role in pharmaceuticals. I will present two different scenarios, one which led to wonderful pharmaceutical benefits and the other to more expensive and probably unnecessary medications. (1) The first scenario is represented by **thalidomide**. Among its early uses was its treatment of morning sickness in pregnant women. Unfortunately, the drug was prone to producing severe birth defects and was withdrawn from the market. Subsequently, it was found that one of its enantiomers is responsible for the birth defects while the other provides the desired pharmaceutical effects. Separating out the desired enantiomer made the drug available for numerous diseases and medical issues. (2) The second scenario is the role of patents in the pharmaceutical industry. Consider, **Prilosec**, which is a drug used to treat heartburn. When the patent owned by the pharmaceutical company AstraZeneca ran out, the company produced a form that had the pure drug-effective enantiomer and called this new drug **Nexium**. The company was able to obtain a new patent on the drug and sell it at a much higher price than Prilosec. I have researched the web for many studies that compared the two drugs and have yet (7-1-2019) to find one that reported a significant difference in their effectiveness—that is, more than a few percent in whatever way improvement can be measured. You can read information on this subject on the following websites (1-29-2011): http://en.wikipedia.org/wiki/AstraZeneca#Nexium; http://en.wikipedia.org/wiki/Esomeprazole#cite_note-12; http://www.medscape.com/viewarticle/481198_8. In spite of the negative responses towards AstraZeneca's actions, you should always be on the lookout for contrary opinions about the Nexium-Prilosec controversy. Beware about the significance of a claim that drug A is more effective than drug B. The comparative effectiveness might mean, in simple terms, that A is 5% more effective than B. If so, A might be 90% effective while B might be 85% effective. The ratio is a mere 1.05. On the other hand, the respective effectivenesses might be 10% and 5%, in which case A is twice as effective as B!

attempted mating between a member of our biosphere and a member of the opposite sex of the mirror image biosphere would be unsuccessful.

A fascinating question arises: **Why is it that only one of the two systems of mirror image biospheres is found on earth?** Why is life on earth **homochiral**? If life arose in many places on earth *independently*, one would expect equal probability for the two types of biospheres to have developed. Are there external factors—e.g., polarization of electromagnetic radiation from outer space [there is such a thing as "circularly" polarized electromagnetic waves which might have been effective] or other cosmic radiation with a handedness [distinguishing left from right]—which might have favored our type of biosphere over its mirror image?

Recently, it has been shown[32] that a combination of unpolarized light and a magnetic field can produce an excess of one enantiomorph over another starting with a non-chiral medium. This phenomenon is believed to be a possible explanation for the **homochirality** of life on earth. Much further research remains to be done in order to give this possibility strong support.

What if there are no external factors that are responsible for our biosphere's being homochiral? In this case there are at least two possibilities: [1] A single homochiral biosphere (ours) appeared with an extremely low probability. And, the mirror image biosphere simply didn't appear. [2] Life began with two biospheres having opposite chiralities. The two biospheres competed with each other for survival. It can be shown that even a state of equal populations of the two biospheres is unstable. A minuscule chance inequality of populations will lead to the ultimate evolution of a single biosphere along with the extinction of the other. We might nevertheless expect the two biospheres to have left distinguishable sets of fossils. You can see how this idea might raise more questions that it answers.

[32]See the two articles in the June 22, 2000 issue of the science journal NATURE.

8.7 Terms

- Amplitude modulation
- AM demodulation
- AM wave transmission
- Bandwidth
- Beat frequency
- Beats
- Big Bang
- Biosphere
- Calcite
- Chirp
- Demodulation
- Enantiomorphs
- Frequency modulation
- FM demodulation
- FM wave transmission
- Homochiral
- ℓ-Glucose
- Modulated wave
- Modulation index
- Modulator
- Hann window
- Optical activity
- Partial polarization
- Polarization axis
- Polarized light
- Polarizer
- Polaroid
- Rectangular window
- Resolution time
- Short time frequency spectrum
- Sidebands
- Specific rotatory power
- Spectrogram
- Unpolarized light
- Vocal folds

8.8 Important Equations

Beat frequency:

$$f_B = |f_2 - f_1| \tag{8.38}$$

The product of two sine waves of frequencies f_1 and f_2, respectively:

$$\boxed{f_2} \otimes \boxed{f_1} = \boxed{f_2 + f_1} + \boxed{f_2 - f_1} \tag{8.39}$$

Modulation index, where Δf is the amplitude of the frequency modulation and f_s is the signal frequency of the frequency modulation:

$$\beta = \Delta f / f_s \tag{8.40}$$

Carson's Rule

$$\text{band width} \sim 2 \times (\text{modulation index } \beta + 1) \times \text{signal frequency } f_s \tag{8.41}$$

With Δf the width of the peak in the frequency spectrum corresponding to a window of N-oscillations of a sine wave of frequency f:

$$\frac{\Delta f}{f} \sim \frac{1}{N} \tag{8.42}$$

Malus' Law, where I_1 is the intensity of polarized light at an axis θ_1 incident upon a polarizer at an axis θ_2 and I_2 is the outgoing intensity:

$$I_2 = I_1 \cos^2(\theta_2 - \theta_1) \tag{8.43}$$

With θ the angle of rotation at an axis of polarization, σ the specific rotatory power, and ℓ the distance of chiral material through which a polarized beam passes:

$$\Delta\theta = \sigma\ell \tag{8.44}$$

8.9 Questions and Problems for Chap. 8

1. A piano tuner finds that two strings produce a beat frequency of 3-Hz, when one of the strings has a known frequency of 440-Hz. What can the tuner conclude about the frequency of the second string?
2. (a) Suppose that you want to determine your heart's pulse rate. You are seated next to a digital clock that is blinking at a rate of one blink per second. You can't read the change in time. You find that when the clock has blinked 11 times, your heart has beat 13 times. Determine your pulse rate.[33] We will now see how we can increase our accuracy even when we never observe a beat coincide with a blink of the clock. In Fig. 8.69 we see marks representing the blinks of the clock, starting from the first blink at the initial time at $t = 0$. Below these marks are small marks representing the beats of my heart; we have assumed that the first beat is at $t = 0$.
   (b) Suppose that we estimate that the events (blinks and beats) coincide at four seconds. What would be the corresponding estimate of my pulse?
   (c) Suppose that we estimate that the events coincide at seven seconds. What would be the corresponding estimate of my pulse?
   (d) Suppose that we estimate that the events coincide at eleven seconds. What would be the corresponding estimate of my pulse?
   (e) In fact, the marks were made using one-half inch spacings for the clock blinks and one centimeter spacings for the beats of my heart. What would be my pulse on this basis?
   Notice how the results of parts (b) through (d) approach this last actual value.
   (f) In the above example, a pair of events eventually coincides after 11 s. It is

[33] The technique used in this problem is a simple application of the technique that Galileo is conjectured to have used to study the motion of a ball down an inclined plane. See the applet on the website (2-11-2011): http://www.joakimlinde.se/java/galileo/.

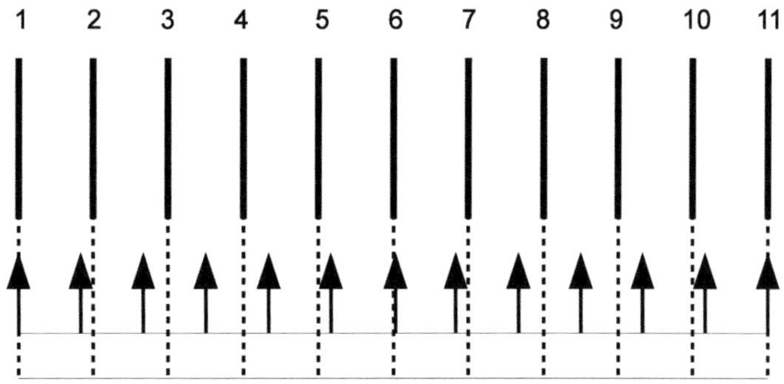

Fig. 8.69 Heart beats vs. Clock blinks

possible that a pair of events will never coincide.

Can you determine the condition on the frequencies of the two sequences of events that is necessary for events to eventually coincide?

This exercise and study has the purpose of giving us a bit of insight into the nature of the phenomenon of **beats**. Whether or not the events of two periodic sequences do coincide as displayed above, the formula for the beat frequency holds

$$f_B = |f_1 - f_2| \tag{8.45}$$

(g) What is the beat frequency for the two sequences above?

3. Express the product

$$\boxed{1} \otimes \boxed{440}$$

as a sum of two sine waves.

4. Express the following sum of two sine waves as a product of two sine waves

$$\boxed{663} + \boxed{657}$$

5. What does the **spectrogram** in Fig. 8.70 tell us about the corresponding wave?
6. What does the **spectrogram** in Fig. 8.71 tell us about the corresponding wave?
7. Suppose that a 45 rpm (rpm=rotations per minute) record is played on a **turntable** set at 33-1/3 rpm. What will be the frequency heard on the loudspeaker of a tone set in the recording at a frequency of 440-Hz?

If you wish, you can look ahead in Chap. 12: Find the musical interval corresponding to the change if rpm. Then given that 440-Hz corresponds typically to an "A,", find the note heard due to the change in rpm.

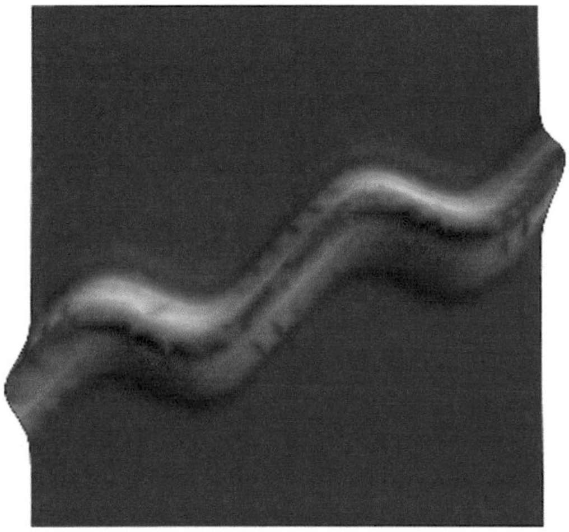

Fig. 8.70 Spectrogram for Problem 5

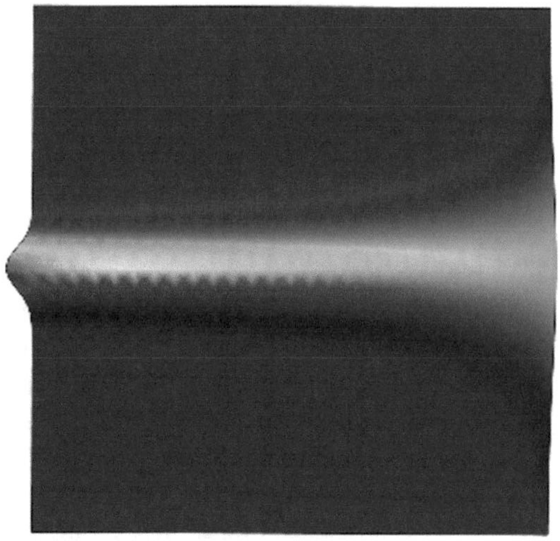

Fig. 8.71 Spectrogram for Problem 6

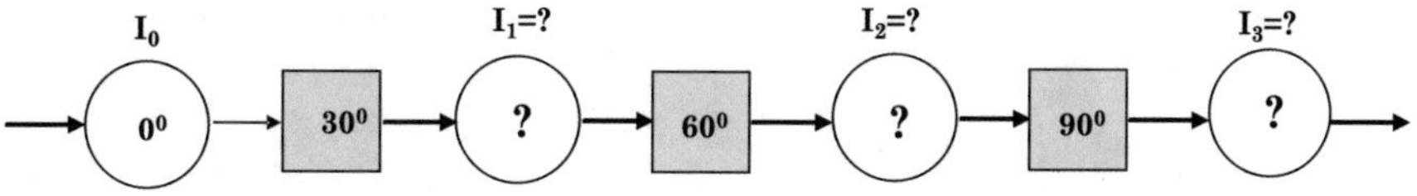

Fig. 8.72 Sequence of three polarizers

Fig. 8.73 Sequence of nine polarizers

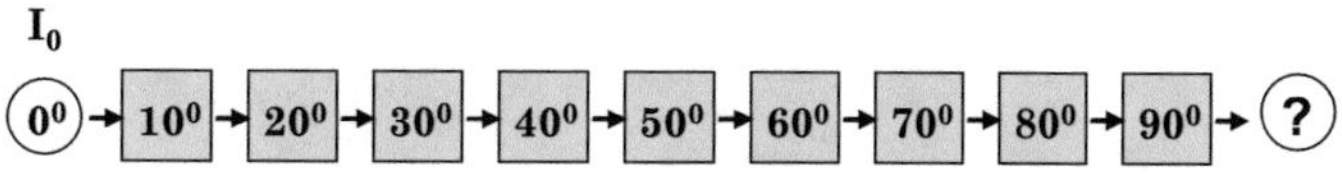

Fig. 8.74 Polarizing of light reflected off a puddle of water

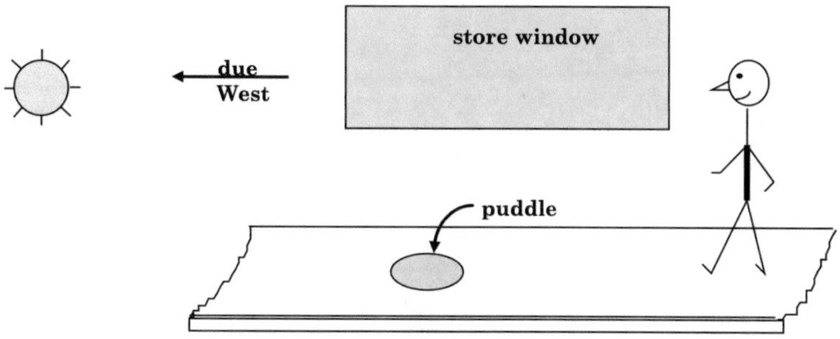

8. (a) Fill in the circles and determine the intensities for the sequence of polarizers in Fig. 8.72.
   (b) Repeat the following sequence of nine polarizers, shown in Fig. 8.73, determining only the final state of the beam.
   (c) Can you see how one could use ideal polarizers to rotate the axis of polarization by 90° without any loss in intensity? Explain.
9. Suppose you are walking due West down a street at sunset, as shown in Fig. 8.74.

   Determine the direction of the axis of partial polarization of sunlight which you see:
   (a) reflected off the window of a store;
   (b) reflected off the water puddle on the sidewalk;
   (c) scattered by the atmosphere directly above you.
10. Polarized light of wavelength 410-nm is passed through a distance of 7-mm of quartz. Find the angle through which the axis of polarization is rotated.
11. This problem shows us how optical activity can be used to quickly determine the concentration of a solution of chiral molecules.

    Polarized light with a wavelength of 589-nm is passed through a distance 10cm of a solution of sucrose in water. The axis of polarization is found to be rotated by an angle of 19°. Determine the concentration of sucrose.
12. The following problem is not a trivial one at all. You are presented with this problem mainly to get you thinking about it so that you can appreciate how difficult the problem is:

    *Suppose that you were establishing communication with an **extraterrestrial** being. You describe yourself in broad terms and succeed to establish that you have an organ (your heart) that pumps fluid (blood) through your body. You now want to indicate that the heart is essentially on the **left** side of your body. How might you establish the difference between **left** and **right** using only radio communication and thus without identification of and reference to various celestial bodies? Might your ability to refer to celestial bodies help?*

9 Propagation Phenomena

Sound and light propagate through space from their sources and are ultimately detected by a receiver such as a person or a device. The subject of the principle of superposition in Chap. 7 dealt with the waves produced by more than one source. In this chapter we will deal with effects on waves when their propagation is complex. The phenomena to be studied are:

- **Diffraction**, which refers to the way waves bend around obstacles. In particular, we will deal with how it is that we can hear sound from a loudspeaker without being able to see the loudspeaker. We cannot see something that is blocked by an opaque object. There is a great difference between sound and light. Why is this so?
- **Reflection** of waves off interfaces between two media (such as sound off a wall or light off a mirror or rough surface). What are the conditions for a surface to be shiny? Why does polishing a surface make it shiny?
- **Refraction**, which refers to the way waves behave when they are transmitted (pass on) from one medium to another (such as light from air to glass or sound from air to water). The operation of lenses, which are used in eyeglasses, microscopes, and telescopes, relies on the phenomenon of refraction.
- **Scattering** of waves by a tenuous distribution of obstacles, such as light off air molecules. In particular, we will learn why the sky is blue in color.
- **The Doppler Effect**, which characterizes the effect on the frequency of a sine wave that is observed by a receiver that is moving with respect to the source of the wave. We will understand why the frequency of sound emitted by the horn of an approaching train starts off at a large value and gradually decreases as the train moves away from us. This phenomenon allows an astrophysicist to determine the speed of distant star towards or away from us.

It is important to keep in mind that the above phenomena are exhibited by **all types of waves**, including sound waves, EM waves, and waves propagating along the surface of a liquid, such as ocean waves.

L. Gunther, *The Physics of Music and Color*,
https://doi.org/10.1007/978-3-030-19219-8_9

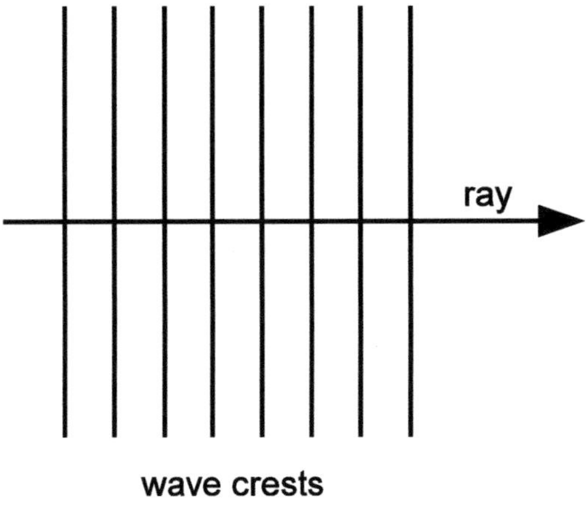

Fig. 9.1 Wave crests

9.1 Diffraction

In order to study the phenomenon of diffraction, we will focus our attention on the simplest possible wave in three-dimensional space, the **plane wave**. A plane wave moves in a straight line. As a result, in filling three-dimensional space, a single crest occupies an entire plane, as seen in Fig. 9.1.

Suppose that a plane wave is incident on an opaque sheet of material. In the sheet we cut a round hole of diameter d (Fig. 9.2). What will be the nature of the wave that progresses on the opposite side of the hole?

If the wave consisted of a beam of particles streaming along, we would observe a beam with a circular cross-section having a sharp boundary (Fig. 9.3):

Our normal experience with light agrees with this prediction. However, careful observation reveals this prediction to be false, or at least, in these normal circumstances, only approximately true. What we actually observe is a "fanning out" of the beam. This phenomenon is referred to as **diffraction** and is exhibited in Fig. 9.4.

The angle θ in the figure is called the **diffraction angle**. It depends upon the wavelength λ and the diameter d of the hole. Suppose that we express the angle θ in radians. Then for small λ such that $\lambda \ll d$ it can be shown that

$$\boxed{\theta \sim \frac{\lambda}{d}} \tag{9.1}$$

Recall that 2π radians equals 360°, so that one radian is about 57°. Thus an angle θ that is about one degree or less is much less than one radian.[1]

Notice that the larger the wavelength is, the greater is the amount of diffraction.

Also, the smaller the hole is, the greater the amount of diffraction.

If the beam is cast on a screen, the image will consist of a set of concentric circles, as shown in Fig. 9.5.

While the boundary fades away asymptotically to zero at infinity, the *essential* image does have a size, with a diameter given by

$$\boxed{d_{im} \approx d + 2\lambda \frac{L}{d}} \tag{9.4}$$

where L is the distance from the hole to the screen.

d_{im} is the total diameter of the image. A sharp image, free of diffraction, would have $d_{im} = d$. The increase of the diameter by an amount $2\pi L/d$ increases with increasing wavelength and decreasing hole diameter.

The diffraction profile described above also holds for the very beam of light emitted by a laser. Such a beam is typically regarded as being straight and of essentially constant cross-section as it propagates. In fact, it fans out as in Fig. 9.4.

---

[1]It is interesting to note that the above relation is similar to the expression for the angle for first order constructive interference for two sources of waves. [See Eq. (7.6).] As long as the angle θ is much less than one radian,

$$\sin\theta \sim \theta \tag{9.2}$$

where θ on the right-hand side is expressed in radians. Then the equation for the angle for first order interference becomes

$$\frac{\lambda}{d} = \sin\theta_1 \sim \theta_1 \tag{9.3}$$

In this last equation, the parameter d refers to the distance between two sources, whereas in this chapter d refers to the diameter of a hole.

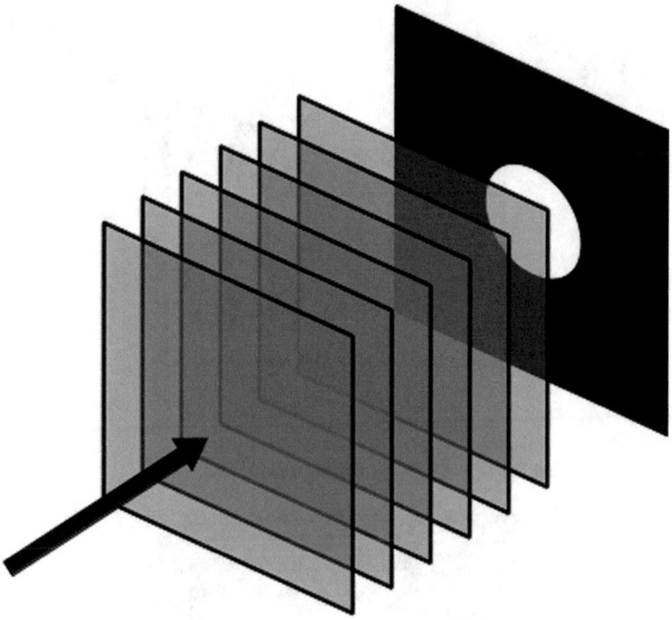

Fig. 9.2 Plane wave incident upon a hole

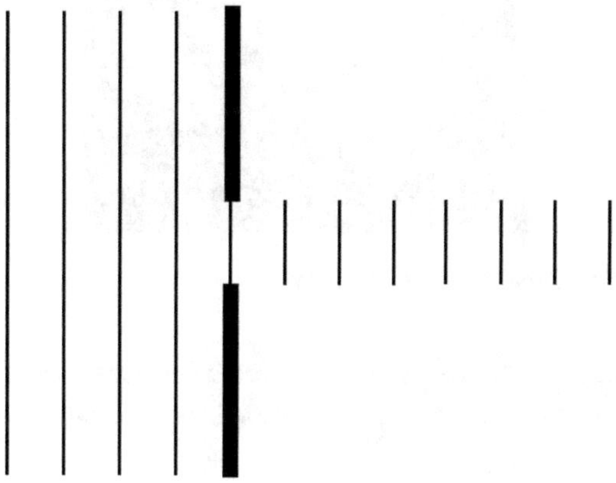

Fig. 9.3 Wave going through a hole with negligible diffraction—side view

Suppose that the laser beam has a wavelength of 600 nm (= 600×10^{-9} m; "n" = "**nano**" $\equiv \times 10^{-9}$) and a diameter = 1-mm when it leaves the laser. In the table below, we list the calculated beam diameter for various distances from the laser along the beam.

L (cm)	$2L/d$ (mm)	$d_{im} = d + 2L/d$ (mm)
1	0.012	1.012
100 (=1m)	1.2	2.2

With an initial beam diameter of 1 mma, there is a significant broadening of the beam, with the beam more than doubling in diameter at a distance of one meter from the laser.

Consider now a more common sized beam, with an initial diameter of 10 cm and the same wavelength.

When $d = 10$ cm

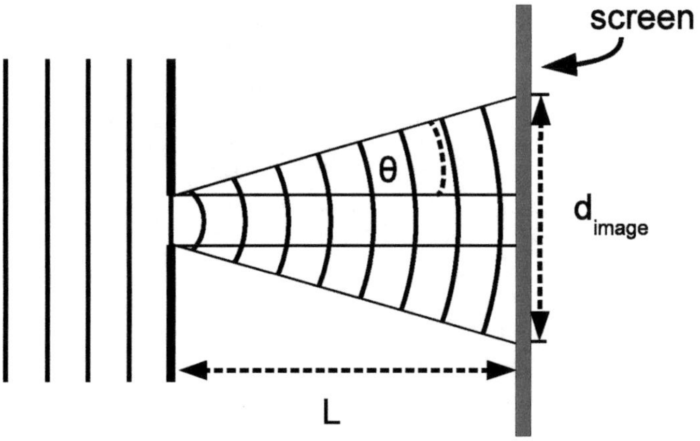

Fig. 9.4 Schematic of a wave going through a hole exhibiting diffraction

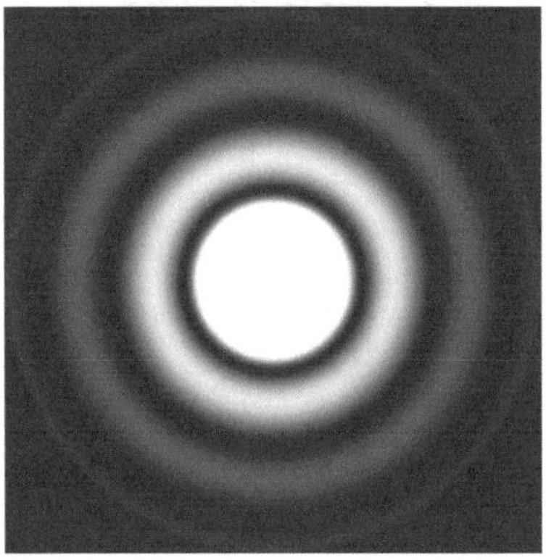

Fig. 9.5 Image produced as a result of diffraction by a hole

L (cm)	$2L/d$ (mm)	$d_{im} = d + 2L/d$ (mm)
1	0.00012	1.00012
100 (=1m)	0.012	1.012

Since **diffraction broadening** is only a few percent or less of the total image size, it is no wonder that we are not aware of diffraction effects of light.

We now turn to an interesting question: Suppose we have a source of light with the above frequency and we wish to cast an image with the smallest possible diameter on a screen that is a given distance of 2 m away, by varying the diameter d of the source. Without detailed thought, one might think that there is no limit to how small an image we can make. We merely have to shrink d down to as small a value as we want. However, diffraction broadening increases with decreasing source diameter and at some point this broadening dominates the image diameter, as we can see from the table below:

d (mm)	$2L/d$ (mm)	$d_{im} = d + 2L/d$ (mm)
3	0.8	3.8
2	1.2	3.2
1	2.4	3.4
0.5	4.8	5.3

We see that as the source diameter is decreased, the image diameter first decreases and then increases. In Fig. 9.6 we present a graph of d_{im} versus d.

Notice that d_{im} has a minimum value for a certain value $d_{\min}$ of the *source*. This is value of the source that leads to a minimum image diameter, $mind_{im}.$. It can be shown that $d_{\min}$ corresponds

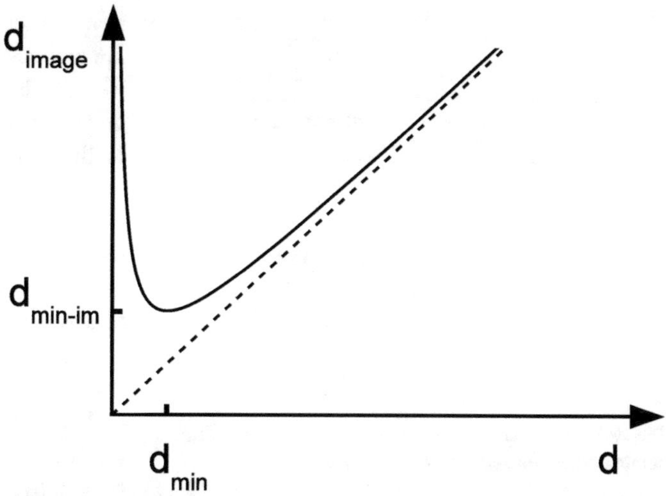

Fig. 9.6 The diameter of the image, d_{im}, versus the diameter of the source, d

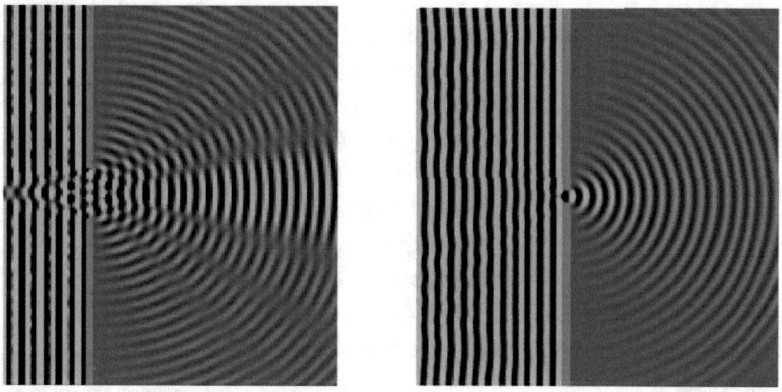

Fig. 9.7 Diffraction by a small slit

to the source diameter and diffraction broadening being equal; that is, $d = 2\lambda L/d$. Solving this equation for d, we obtain $d_{\min}^2 = 2\lambda L$. Thus the value of d that produces the minimum size of an image is

$$\boxed{d_{\min} = \sqrt{2\lambda L}} \tag{9.5}$$

According to Eq. (9.4), the corresponding **minimum image diameter** is given by twice $d_{\min}$. Thus,

$$\boxed{\text{minimum image diameter} = min\ d_{im} = 2\sqrt{2\lambda L}} \tag{9.6}$$

Sample Problem 9.1 Find the minimum possible image size for the values $\lambda = 600$ nm $= 6 \times 10^{-7}$ m and $L = 2$ m.

Solution We obtain

$$\min d_{im} = 2\sqrt{2 \cdot 6 \times 10^{-7} \cdot 2} = 3.10 \text{ mm}$$

which corresponds to a source diameter of 1.55 mma.

When the source diameter is less than the wavelength, the wave beyond the hole is "fanned out," producing the spherical waves of a point source. In Fig. 9.7 we see two images: to the left a small slit, to the right a minute slit with slit width much less than the wavelength.[2]

[2]The figures were produced with applet on the website (2-11-2011): http://www.falstad.com/ripple/.

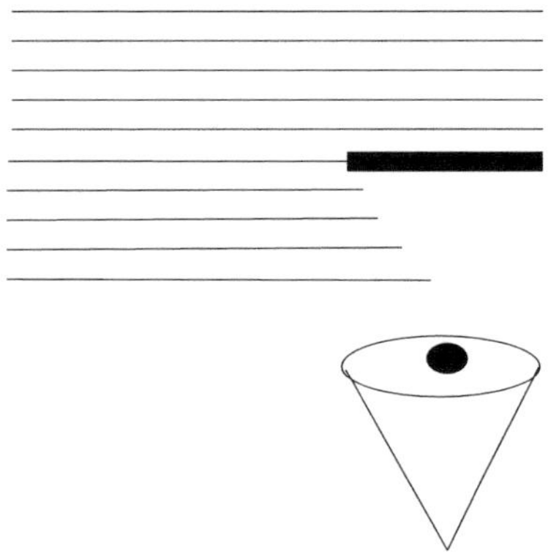

Fig. 9.8 Schematic drawing of a wave that diffracts around a wall that occupies a half space

Dark regions represent low intensity while bright regions represent high intensity. Look at the figure to the left, for which the diameter of the hole is comparable to the wavelength of the wave. If we examine the intensity of the wave along an arc of a circle at a specific distance from the hole, we see oscillations of the intensity along the arc. On the other hand, these oscillations are absent in the figure to the right, when the diameter of the hole is very small compared to the wavelength of the wave.

Sample Problem 9.2 Suppose that a sound wave in air with a frequency of 3400-Hz is incident upon a wall having a door that is open to a width of 1-cm. Describe the sound wave that emanates to the other side of the door. Restrict your solution to a description that is as detailed as we have used in this section.

Solution The wavelength is approximately λ=v/f=340-(m/s)/3400-Hz=0.1-m. Since the wavelength is ten times the width of the door opening, the sound wave will fan out as in the image at the right in Fig. 9.7. A person who is standing outside the room with the source of sound will be able to hear the sound without being exactly in front of the door opening.

Diffraction effects are also relevant when one half of space is blocked off by a wall or a mountain, as is indicated in Fig. 9.8. In the schematic drawing, we see a light wave incident downwards towards the viewer below. A screen (the black rectangle in the schematic drawing) is set up to block the light behind the screen. In the absence of diffraction, there would be a sharp shadow. The schematic figure shows how the wave proceeds into the shadow region behind the screen.

The image in Fig. 9.9 is an actual photograph of a wave that is diffracted by a wall occupying a half space, taken at a distance of about one wavelength from the screen. A sharp boundary is depicted by the dotted line, with a value $y = 2.0$. Instead, light extends a bit into the shadow region. In addition, along the negative x-axis, the intensity oscillates, with a maximum intensity above $y = 2$ at the maxima of the oscillations. As indicated in the schematic drawing, the further away we are from the screen, the more will the light extend into the shadow region.

In Fig. 9.10 we see an actual photograph of the shadow of a razor blade. The blurriness due to diffraction is quite evident.

Questions to ponder:

1. Which voices will be heard better through a crack in a door, high pitched or low pitched ones?
2. Which radio waves will be more easily picked up at large distances over hilltops, AM or FM waves?

9.1.1 Scattering of Waves and Diffraction

Consider a plane wave incident upon an object as shown in Fig. 9.11. The object acts as an **obstacle** to the wave. These are the extreme cases. Let d be the diameter of the object. Then,

1. $\lambda \ll d$: The obstacle produces a sharp shadow, with mild diffraction effect that increases with increasing distance. We can see that the wave is "collapsing" around the sphere.
2. $\lambda \gg d$: The incident wave is barely affected by the obstacle. The obstacle produces a weak

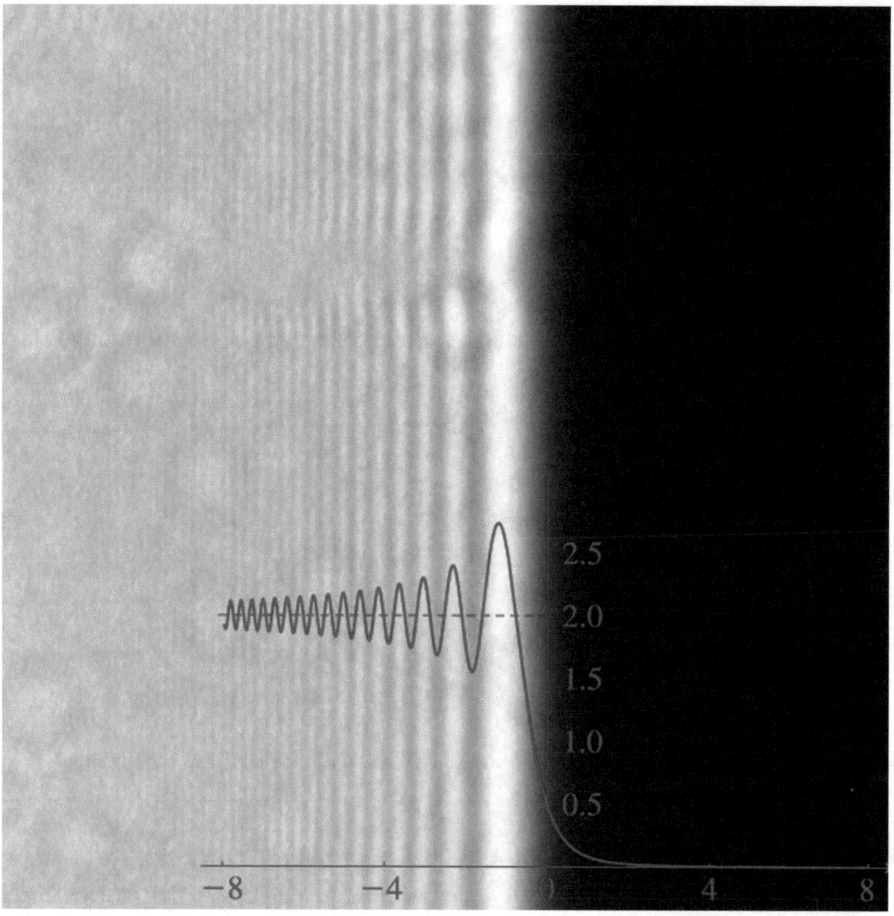

Fig. 9.9 Photograph of a wave that diffracts around a wall that occupies a half space—(Source: http://dlmf.nist.gov/7.SB1)

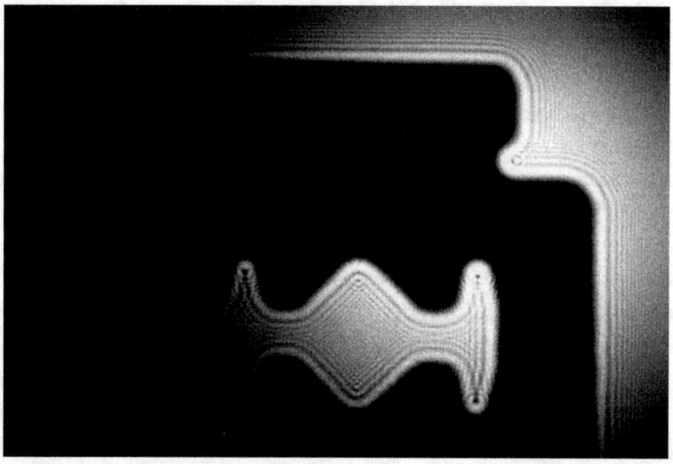

Fig. 9.10 Actual diffraction of light by a razor blade (source: Photo courtesy of Harvard Natural Sciences Lecture Demonstrations. Copyright 2011, President and Fellows Harvard College, All Rights Reserved)

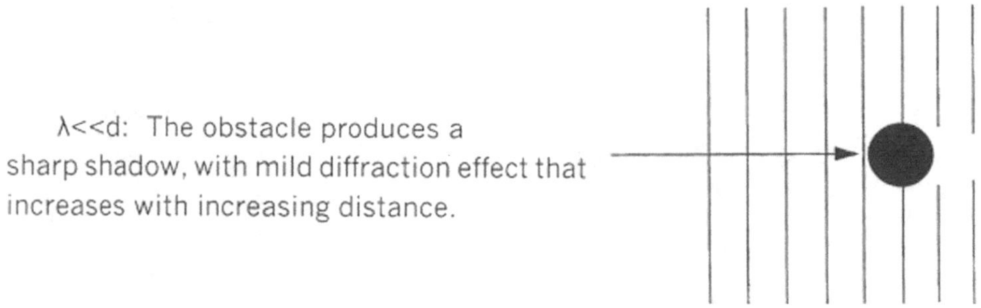

λ<<d: The obstacle produces a
sharp shadow, with mild diffraction effect that
increases with increasing distance.

Fig. 9.11 Shadow created by a spherical object—wavelength comparable to the diameter of the object

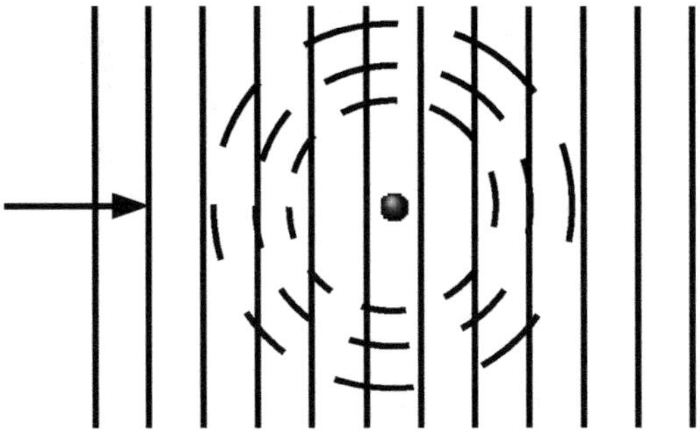

Fig. 9.12 Schematic of a wave scattered by an object with diameter much smaller than the wavelength

"**scattered wave**." This case is exhibited in Fig. 9.12.

The two cases above that are represented schematically in the above figures are shown below in Fig. 9.13 produced by a simulation of a surface wave in water.[3]

Note that in (a) there is strong blockage yet we can easily see the effects of diffraction; in (b) there is strong diffraction, so that the wave is barely affected by the presence of the object. Still, if you look carefully you should see the weak spherical scattered wave.

Home exercise: Observe the effect your body has on large ocean waves.

The above discussion allows us to understand why it is that ordinary laboratory microscopes that use light for illumination cannot allow us to clearly see objects that are on the order of the wavelength of light ($\sim 5 \times 10^{-7}$ m or less). Diffraction produces blurry boundaries. We can also appreciate why AM radio waves can "cross" mountains more easily than FM radio waves, which have a higher frequency and hence a smaller wavelength.

9.1.2 Why Is the Sky Blue?

Our sky is blue because of the scattering of sunlight by the molecules of air of the atmosphere. How can we understand this phenomenon? After all, the diameter of air molecules is on the order of a few Ångstroms, while the wavelength of light is much greater $\lambda \sim 5,000 Å$. That amounts to a ratio of about one thousand to one. As a result, diffraction effects are very strong so that *one would expect that very little scattering of light by a single molecule would take place*. (See

[3]The figure was made from the applet on the website (2-5-2011): http://www.falstad.com/ripple/.

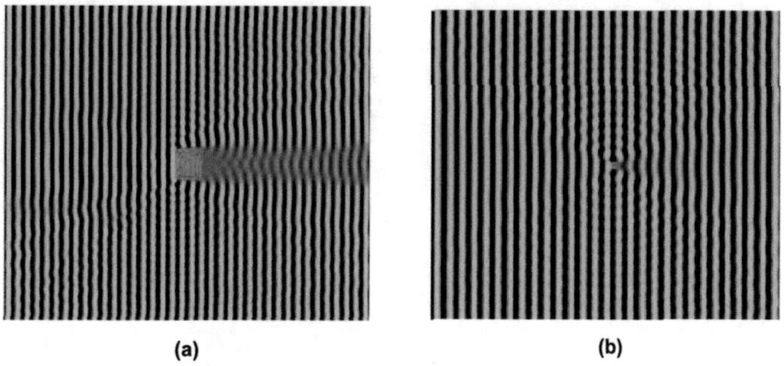

Fig. 9.13 The scattering of a wave: (**a**) object diameter a few times the wavelength; (**b**) object diameter much smaller than the wavelength

Fig. 9.12.) The sky is bright on account of the sum of scattering by a vast number of molecules.[4]

We can now understand why the sky is blue. A detailed analysis was provided about 100 years ago, when Lord Rayleigh (alias John William Strutt) showed that the intensity of scattered light behaves like:

$$\boxed{I_{\text{scatt}} \propto \frac{1}{\lambda^4}} \tag{9.7}$$

This equation shows us the relationship between the intensity of scattered light and the wavelength of the light.

We should expect the degree of scattering to increase with decreasing wavelength since there will be decreasing diffraction. As a consequence, the violet end of the visible spectrum, which has the shortest wavelength, is scattered most. For a numerical comparison, we will calculate the ratio of the scattered light for a wavelength $\lambda = 4 \times 10^{-7}$-m (violet) to that for $\lambda = 7 \times 10^{-7}$-m (red), or 400-nm to 700-nm:

Now sunlight is whitish, consisting of a mixture of wavelengths which extends from the infrared to the ultraviolet. Since the violet end of the visible spectrum is scattered most, scattered sunlight looks bluish. On the other hand, when we look directly at the sun through the atmosphere or observe the horizon in the west at sunset, we are seeing light which has started out white (from the sun) and has had the violet end of the spectrum removed most by scattering. Such light looks reddish. (See Chap. 15, for details on color perception.)

[4]Interestingly, one can show that if the density of air molecules were to be perfectly uniform, this sum would result in no net scattering; the sky would be perfectly transparent! It is the modest degree of non-uniformity of the density that is responsible for the scattering. In fiber optics communication, the glass is so pure, that is, free from impurities and inhomogeneities, that it is the small degree of non-uniformity in the molecular density associated with the random thermal motion of the molecules, that is responsible for the small attenuation in the fibers.

9.2 Reflection

The characteristics of a wave that is reflected off an object can be, surprisingly, quite complicated, whether it be a light wave, sound wave, ocean wave, or any other wave. We will discuss three relatively simple cases; they can be described in terms of the appearance of a surface under the reflection of light: **shiny surface** (like a mirror), **dull surface**, and **sparkling surface**.

Let us first consider the reflection of *light* off a painted wall. Some walls are dull; others, with a concentrated enamel paint are shiny. What is the physical difference between the two surfaces? We know that rubbing a surface often polishes the surface, meaning that the surface is made shiny. We recognize that polishing involves making a surface smoother. But how smooth must a surface be to be shiny? How do we characterize smoothness?

The central factor is what we will refer to as the **length scale of roughness**, with a symbol ℓ_r. The smaller this length is, the smoother a surface is. The degree of shininess is obtained by comparing this length to the wavelength λ of the light. [Recall that the range of wavelengths of light λ is about 4000–7000 $\AA$.]

We can describe the surface of the wall with varying degrees of detail. The surface might be

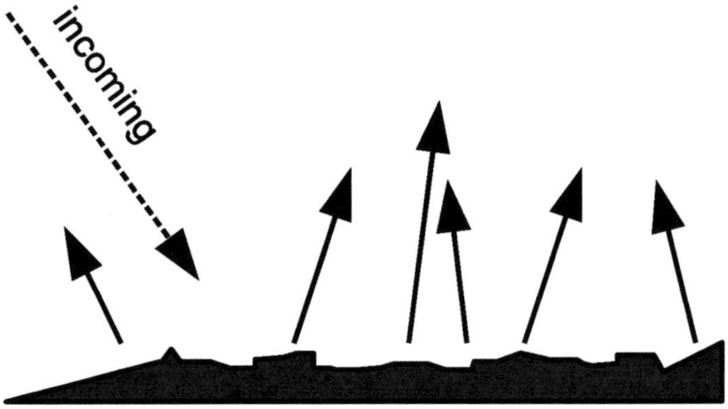

Fig. 9.14 Diffuse reflection off a rough surface

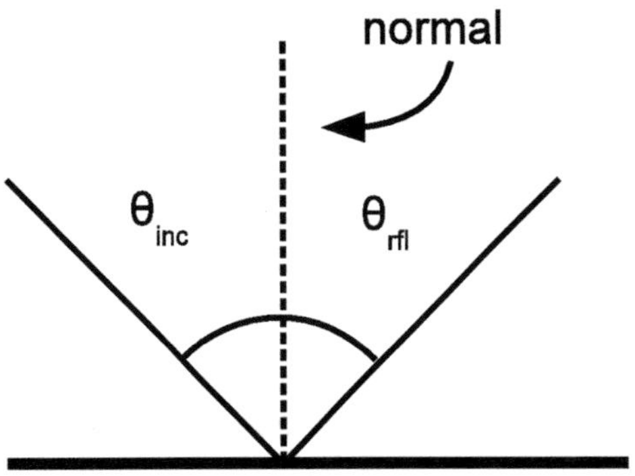

Fig. 9.15 Reflection by a smooth surface

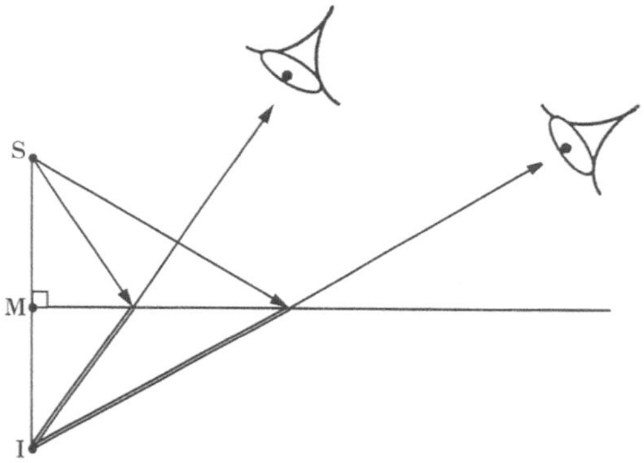

Fig. 9.16 Two observers, each receiving its own respective set of reflected waves from the source S

smooth to touch, so that without close scrutiny, we would simply describe the surface as being smooth. However, if we examine the surface with a microscope, we might be able to see bumps on the order of 0.1 millimeter in size. We would say that the length scale of roughness ℓ_r is about 0.1 millimeter. In this case, $\lambda \ll \ell_r$ and the surface will appear *dull*. Examination will reveal that a beam of light is reflected from the wall in many random directions. We have what is referred to as **diffuse reflection**, as exhibited in Fig. 9.14.

Note If a laser beam casts a spot on a *dull wall*, everyone in the room can see the spot, since there exists a ray that reaches the observer's eye.

In the case of a wall painted with *enamel* paint, the length scale of roughness is on the order of a wavelength, so that the surface is shiny. When the length scale of roughness is much smaller than a wavelength, the surface acts like a mirror and we have what is referred to as **specular reflection** or **mirror reflection**. A given ray of light produces a single reflected ray as exhibited in Fig. 9.15.

We see that an incident ray of light produces a reflected ray, with the **angle of incidence** θ_i equal to the **angle of reflection** θ_{rfl}. Notice that these angles are measured relative to a line perpendicular to the interface. This line is called the **normal**, since the word "normal" means perpendicular.

Suppose that two viewers, labeled V_1 and V_2, respectively, look at a point source through a mirror. The point source emits rays in all directions, but each viewer receives through their own eye only a small *set* of rays concentrated around the respective rays in Fig. 9.16.

In the figure, S is the point source while I is the apparent position of S – as seen by V_1 and V_2. I is said to be the **mirror image** of S. Note that $\overline{SM} = \overline{IM}$ and that $\overline{SMI}$ is perpendicular to the interface. The arrowed line segments represent

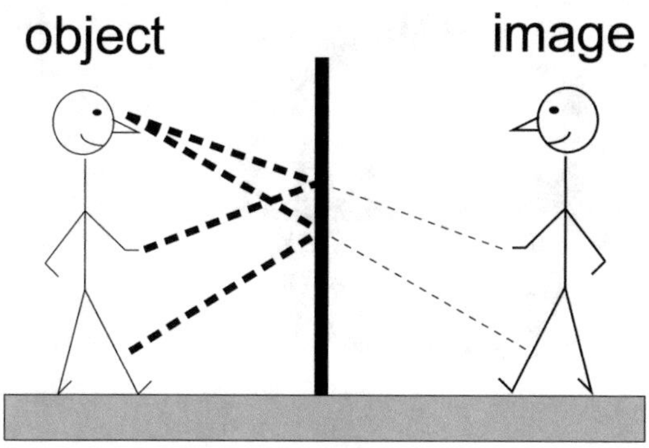

Fig. 9.17 The rays of light from a hand and from a leg that reflect from a mirror into the eye

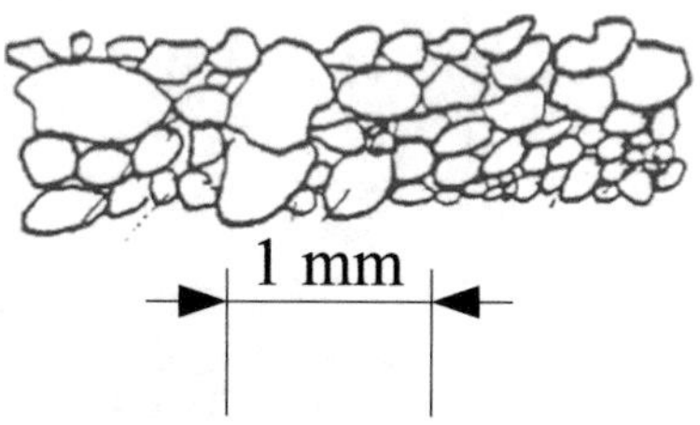

Fig. 9.18 Magnified surface of sand

the actual rays traveling towards the viewers. The key result is that the viewers' brains will interpret the incoming rays as coming from S and perceive the source as actually being at I. Since the source is not really at I, the image is said to be a **virtual image**.

In Fig. 9.17 we see a person standing in front of a mirror. The particular rays that reflect off a hand and off a leg and which strike the mirror and reach the eye (as dashed lines) are indicated. The visual system, whose main components are the eye and the brain—the **eye-brain system**—assumes that the ray left a point on the leg of the image (as a fine-dashed line).

Note that if a laser beam is reflected off the wall under specular reflection, only an eye—regarded approximately as having an iris the diameter of one ray of light—that is in line with the single reflected ray will be able to see the light.

Consider now a light beam incident on a dull painted wall. Its dullness reflects the fact that you can see an image of the light beam on the wall no matter where you stand. This situation indicates that the reflection is diffuse, as shown in Fig. 9.14. Since the light has a wavelength lying between 4×10^{-7} m and 7×10^{-7} m, in order to have specular reflection the length scale of roughness must be on the order of 10^{-7} m or more.

In sum, if $\lambda \ll \ell_r$, we have **diffuse reflection**; while if $\lambda \gg \ell_r$, we have **specular reflection**.

These results hold for the reflection of any type of wave off a surface. Examples are sound waves off walls, a concert audience, or wall tapestry, as well as radio waves off a forest of trees or a suburban houses. Since the smallest wavelength of audible sound (corresponding to the largest audible frequency) is $\lambda = v/f = (340\text{m/s})/(20,000\text{Hz}) = 0.17$ m $= 170$ mma, we see that *audible sound waves reflect specularly off walls*. [See also the problems at the end of the chapter.]

A complex surface: a sand particle

Let us consider how we would regard a *flat* surface of **sand**, flat on a scale of centimeters, as depicted below. We know that the sand consists of a multitude of sand particles, of varied shapes and sizes. The range of particle diameters certainly does not exceed a value on the order of 1 mma. Depending upon the sample of sand, it may not fall below 0.1 mma.

For our purposes, a grain of sand has at least three important **length scales**:

First, there is the *size* (average diameter) ℓ_g of the *grain of sand*. Next, if one looks closely at the grain of sand, one would find the surface rough and bumpy, as seen in Fig. 9.18. The second length scale is the average size of the bumps on the grain's surface, ℓ_r in the above figure. This is the length scale of roughness. Finally, there is the size ℓ_a of the individual *atoms*, at the Ångstrom level. Certainly, $\ell_g > \ell_r \gg \ell_a$. We will assume, for simplicity, that $\ell_g \gg \ell_r$; that is, the bumps are much smaller than the grain size.

Let us now discuss the reflection of light off the surface of the sand depicted in Fig. 9.19. Certainly, λ is much smaller than the size of a

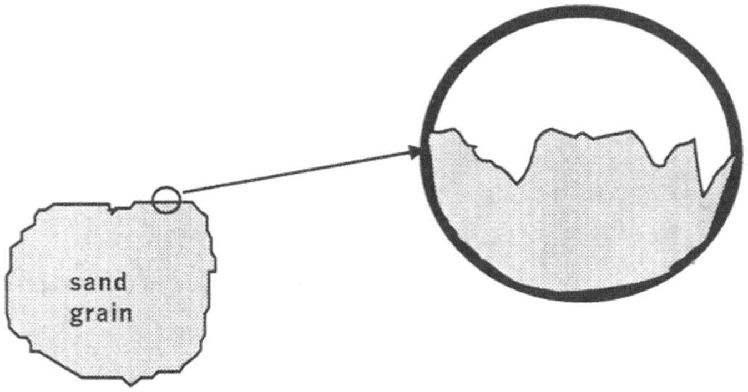

Fig. 9.19 Surface of a grain of sand—magnified

grain. As a consequence, the surface of the sand appears rough. Furthermore, the wavelength λ of the light is much greater than the average atomic diameter ℓ_a. [$\ell_a \ll 5000\text{Å}$.] The only remaining question is how λ compares with the size of the bumps, ℓ_r. If $\lambda \gg \ell_r$, the particle surface will appear shiny and the sand as a whole will sparkle. Otherwise, the sand will appear dull at all levels.

Questions:

How would sound waves reflect off an audience of people?

How would AM or FM radio waves reflect off a forest of trees or a suburban neighborhood of homes?

9.3 Reflection and Refractance

Suppose that a fisherman is standing at the edge of a lake and is concerned that if he talks the fish will hear him. To answer this question, we need to determine what fraction of sound energy that is incident on the surface of the water is transmitted into the water. Or, suppose that a light beam is propagating in a transparent medium and is incident upon a second transparent medium, How does the intensity of the transmitted beam compare to the intensity of the incident beam?

In Chap. 4, we discussed absorption and attenuation of waves as they propagate in a medium. Often one talks about the fraction of an incident wave that is **reflected**, **transmitted**, and **absorbed**. When a sound wave traveling in air is incident upon a dense material such as acoustic tile, we focus on absorption as a process rather than transmission. For our purposes, we will disregard the details of what happens to a transmitted wave once it enters the second medium.

The ratio of the intensity of the reflected wave to the intensity of the incident wave is referred to as the **reflectance** with a symbol R; the ratio of intensities of the transmitted wave and the incident wave is referred to as the **transmittance**, with a symbol T. The incident wave is replaced by the reflected and the transmitted waves. Thus,

$$\begin{aligned} R &= \frac{I_{\text{rfl}}}{I_{\text{inc}}} \\ T &= \frac{I_{\text{trans}}}{I_{\text{inc}}} \end{aligned} \tag{9.8}$$

Since the total intensity must equal the total resulting intensity. That is,

$$I_{\text{rfl}} + I_{\text{trans}} = I_{\text{inc}} \tag{9.9}$$

Alternatively,

$$R + T = I \tag{9.10}$$

Note In this book, we will restrict ourselves, for simplicity, to a plane wave that is incident **perpendicularly** on a plane surface, as shown in Fig. 9.20.

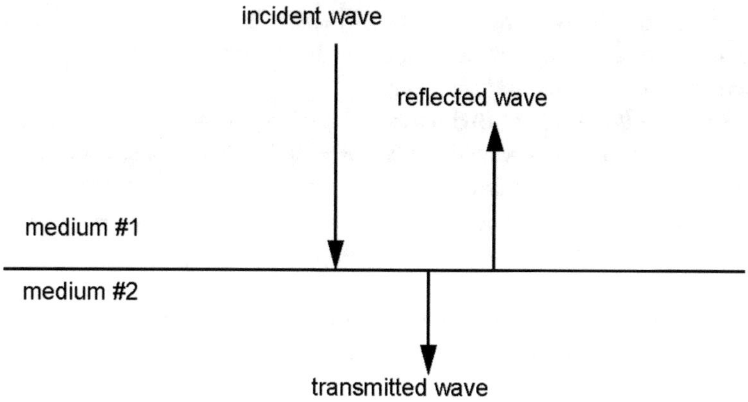

Fig. 9.20 Reflectance and transmittance of an incident wave

9.3.1 The Reflectance for a Light Wave

The reflectance of light at an interface between two transparent materials depends upon what is called the **index of refraction**, which is given the symbol n. In a vacuum, the speed of light is $c = 3.00 \times 10^8$ m/s. In a transparent medium it is given by

$$v = \frac{c}{n} \tag{9.11}$$

Clearly, $n = 1$ in a vacuum. In a medium, the speed of light is less than c, so that $n > 1$.

Below, we list the values of the index of refraction for some materials:

Material	Index of refraction n
Air (STP)	1.0003 (the denser the air, the greater the index of refraction)
Water	1.33 ~ 4/3
Diamond	2.42
Crown glass	1.52
Flint glass	1.66

Unless otherwise stated, we will set $n = 1$ for air as an excellent approximation.

The reflectance R of a light wave is determined solely by the respective indices of refraction, n_1 and n_2 of the two media. It is given by:

$$R = \left[\frac{n_1 - n_2}{n_1 + n_2}\right]^2 \tag{9.12}$$

Note If $n_1 = n_2$, then $R = 0$. There is no reflected wave if the indices are equal—even if the two materials are otherwise quite different! Only the indices of refraction determine the reflectance.

Note The reflectance is the same whether the incident ray is in medium #1 or in medium #2.

The basis for this result can be seen as follows: The result of interchanging media can be determined by interchanging the symbols n_1 and n_2 in Eq. (9.12). The value for the reflectance does not change since

$$\frac{(n_1 - n_2)^2}{(n_1 + n_2)^2} = \frac{(n_2 - n_1)^2}{(n_2 + n_1)^2}$$

This result will be shown to hold for the reflection of sound waves, too. [See below.]

Sample Problem 9.3 Find the reflectance for light at an air–water interface, given that $n_1 = 1$ and $n_2 = 4/3$.

Solution

$$R = \frac{\left(1 - \frac{4}{3}\right)^2}{\left(1 + \frac{4}{3}\right)^2}$$

$$= \left[\frac{-\frac{1}{3}}{\frac{7}{3}}\right]^2 = \left(-\frac{1}{7}\right)^2 = \frac{1}{49} \approx 0.02$$

Thus only 2% of the energy is reflected. Most of the energy is transmitted.

Note Without being quantitative, we will say something about the case when the incident wave is *not* perpendicular to the surface. **The reflectance increases as the angle of incidence increases.** Check this by examining reflection off a surface. Note that the shininess is dramatic if you view a surface at a grazing angle even if the surface is dull!

Problem for the reader: Using the relation $v = c/n$, show that the reflectance for an EM wave is given by

$$R = \left[\frac{v_2 - v_1}{v_2 + v_1}\right]^2 \tag{9.13}$$

where v_1 and v_2 are the wave velocities of the respective media.

9.3.2 The Reflectance for a Sound Wave

Here the different mass densities ρ_1 and ρ_2 must be taken into account:

$$R = \left[\frac{\rho_2 v_2 - \rho_1 v_1}{\rho_2 v_2 + \rho_1 v_1}\right]^2 \tag{9.14}$$

Note that the interchange subscripts in the expression leave the reflectance unchanged as in the case of light waves. Note too that Eq. (9.13) for light waves can be obtained from Eq. (9.14) by assuming that the media through which light propagates all have the same mass density.

In Eq. (9.14), the product "ρv" is called the **impedance** of the medium. The typical symbol for the impedance is Z, so that

$$Z \equiv \rho v \tag{9.15}$$

Then we can write:

$$R = \left[\frac{Z_2 - Z_1}{Z_2 + Z_1}\right]^2 \tag{9.16}$$

Sample Problem 9.4 Find the reflectance for an air–water interface at STP, given that

Air: $Z = \rho v = 1.3\ \text{kg/m}^3 \times 345\ \text{m/s} = 450\ \text{kg/m}^2\text{s}$

Water: $Z = \rho v = 1000 \times 1500 = 1.5 \times 10^6\ \text{kg/m}^2\text{s}$

Solution Substitution into Eq. (9.14) leads to $R = 0.999$. Thus, only 0.1% of the sound energy is transmitted! On the basis of this result, a fish should have difficulty hearing a person who is on the shore talking.

Sample Problem 9.5 How many dB corresponds to 0.1%?

Solution $\Delta SL = 10\log(0.001) = -30$ dB.

9.4 Refraction

Suppose that a plane wave is traveling in one medium and is incident upon an interface of this medium with a second medium through which the wave can propagate. Examples include light in air that is incident upon an air/glass interface, or a sound wave in air, incident upon an air/water interface.[5] We have both a reflected wave and a **transmitted wave**. We focus here on the transmitted wave.

In Fig. 9.21 we exhibit a wave incident on an interface. Notice that the transmitted wave is

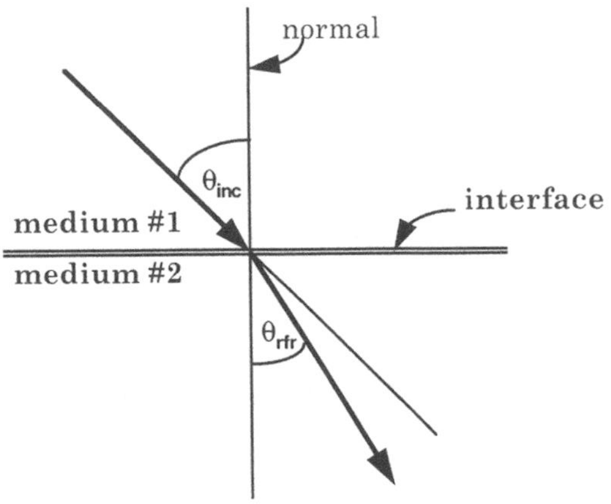

Fig. 9.21 Refraction at the interface of two media

[5] We are used to referring to the interface between air and water as the **surface** of water. However, how should we refer to the boundary between water and oil? The word **interface** is a neutral term, clearly superior to the term **surface**.

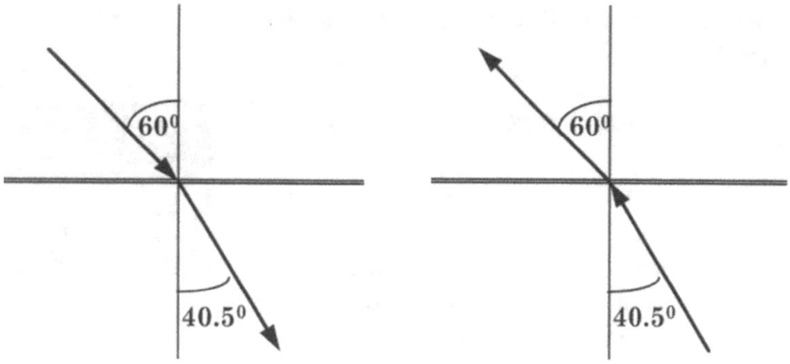

Fig. 9.22 Refraction for two opposite directions of a ray of light

not in the same direction as the incident wave. This change in direction is called **refraction**. Of greatest interest is the relationship between the **angle of incidence** θ_i and the **angle of refraction** θ_r.[6]

In 1621, Willebrord Snell discovered a relationship, known as **Snell's Law**, between these two angles for light beams. They are related by the indices of refraction of the two media.

$$\boxed{n_i \sin\theta_i = n_r \sin\theta_r} \tag{9.17}$$

Applications of this law follow below:

Sample Problem 9.6 Given $n_i = 1$, $\theta_i = 60°$, and $n_r = 4/3$, what is θ_r?

Solution We have $1 \times \sin 60° = (4/3)\sin\theta_r$, so that $\sin\theta_r = (3/4)\sin 60° = 0.650$. We then obtain

$$\theta_r = \arcsin(0.650) = 40.5°$$

Sample Problem 9.7 Given $n_i = 4/3$, $\theta_i = 40.5°$, and $n_r = 1$, what is θ_r?

Solution It should be clear from Snell's Law that $\theta_r = 60°$. This example describes the path of the ray of the previous problem when reversed in direction! See Fig. 9.22.

Note It follows generally from Snell's Law that if an incident ray were to have the direction of the original refracted ray, the resulting refracted ray would be in the direction of the original incident ray. Physicists refer to this behavior as **reversibility**. Another way to apply this behavior is to imagine a single photon [See Sect. 6.7] incident on an interface between two transparent media. Its path in time would follow that described by refraction. Suppose we were able to take a video of the photon's path. If were to reverse the video in time, we would see the photon retrace the same path in the opposite direction. As we will see later in this chapter, reversibility occurs in the behavior of lenses.

Comments

1. Suppose $\theta_i = 0°$, then $\theta_r = 0°$: Thus, an incident ray which is perpendicular to the interface is not refracted.
2. If $n_i > n_r$, the beam is refracted *away from* the normal. If $n_i < n_r$, the beam is refracted *towards* the normal.

Because of this reversibility, it is simple to remember Snell's Law in the form that makes no reference to which media the subscripts refer:

$$\boxed{n_1 \sin\theta_1 = n_2 \sin\theta_2} \tag{9.18}$$

9.5 Total Internal Reflection

Consider the following problem.

Sample Problem 9.8 Given that $n_i = 4/3$, $n_r = 1$, and $\theta_i = 60°$, find θ_r.

[6]Not to be confused with the angle of reflection θ_{rfl}.

Solution From Snell's Law, $n_r \sin\theta_r = n_i \sin\theta_i$ so that

$$1 \times \sin\theta_r = (4/3)\sin 60^\circ = 1.15$$

This last equation leads to $\sin\theta_r$ having to be equal to 1.15, which is *impossible*; *this equation has no solution* since the sine of an angle cannot be greater than unity. Then, what does it mean to have an equation (Snell's Law) that has no solution?

What happens is that we have *no transmitted refracted beam.* We have what is called **total internal reflection**. Generally, refraction is always accompanied by reflection; the fraction of the intensity that is reflected versus the fraction that is refracted depends upon the two indices of refraction and the angle of incidence. In this case, there is no transmitted, refracted ray.

Total internal reflection can happen only if the index of refraction of the medium of the incident ray is greater than the index of refraction of the medium of the transmitted ray. Thus, a ray incident from air onto an air–water interface cannot be totally reflected. This conclusion should be evident from Snell's Law:

As θ_i is increased, so is θ_r. But $\theta_r > \theta_i$. Therefore, θ_r will reach 90° before θ_i does. And, if θ_i were to be further increased, there is no solution to Snell's equation. Then we would obtain

$$\sin\theta_r = (n_i/n_r)\sin\theta_i > 1$$
$$\sin\theta_i > n_r/n_i$$

The angle for which this last inequality is replaced by an equality is called the **critical angle**, which we give symbol θ_c. It satisfies the equation

$$\boxed{\sin\theta_c = \frac{n_r}{n_i}} \qquad (9.19)$$

In sum, we have *total internal reflection when the angle of incidence exceeds the critical angle.* For this situation to be possible, we must have $n_i > n_r$.

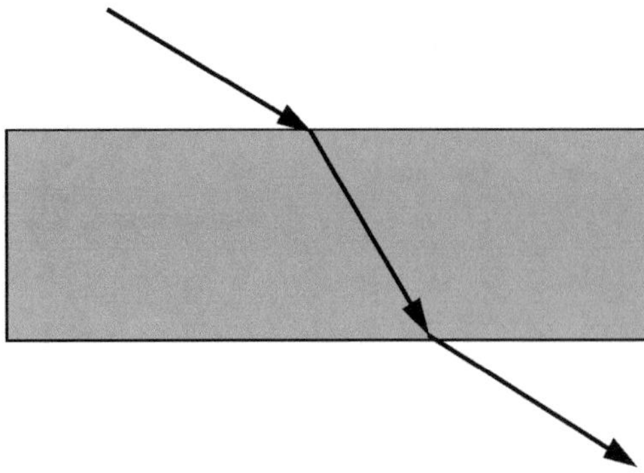

Fig. 9.23 Refraction through a slab of glass

For a water–air interface (with the incident ray in the water) we obtain a critical angle of 48.6°:

$$\sin\theta_c = 1/(4/3) = 3/4$$
$$\theta_c = \arcsin(3/4) = 48.6^\circ.$$

NOTE: Even when there is a refracted ray, there is a reflected ray. In the case of total internal reflection, there is no refracted ray.

Sample Problem 9.9 Prove that when a light ray is refracted twice by a block of glass with parallel faces, as shown in Fig. 9.23, that the outgoing ray is **parallel** to the incoming ray.

Solution In Fig. 9.24 we see the angles associated with the two refractions. The dotted lines are parallel to each other and both perpendicular to the horizontal surfaces of the slab. The first angle of refraction is equal to the first angle of incidence—that is θ_2. Snell's Law, expressed as in Eq. (9.18) proves that the horizontal surfaces are parallel.

9.6 The Wave Theory of Refraction

There were two theories of light presented to account for refraction: Newton's was based upon a particle theory, while **Christiaan Huygens'** theory of refraction was based upon his wave theory of propagation. See Fig. 9.25.

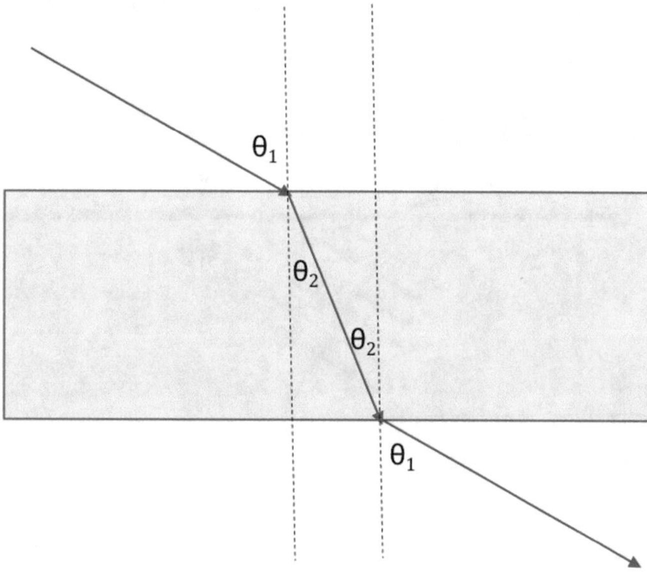

Fig. 9.24 Refraction through a slab of glass

Fig. 9.25 Christiaan Huygens (source: http://http://upload.wikimedia.org/wikipedia/commons/6/6a/Christiaan_Huygens.gif)

Here is Huygens' theory: We assume that a wave is traveling in medium #1 and is then transmitted into medium #2, in which the wave velocity is smaller. The most important thing to note is that the *frequency of a wave is unchanged upon transmission.*[7] As a result the wavelength is smaller in medium #2, as seen in Fig. 9.26. The result also follows from the equation $\lambda = v/f$, with reduced wave velocity and constant frequency.

Next, in Fig. 9.27, we see that because the two triangles, ABC and ABD, in the offset are *similar*.

Finally, recall that $\lambda_1 = v_1/f$ and $\lambda_2 = v_2/f$, where v_1 and v_2 are the wave velocities in the two respective media. Then,

$$\frac{\sin\theta_1}{\sin\theta_2} = \frac{v_1/f}{v_2/f} \tag{9.20}$$

or

$$\frac{\sin\theta_1}{\sin\theta_2} = \frac{v_1}{v_2} \tag{9.21}$$

[7]To see this, suppose that we have observers at points P and Q, in the two respective media. The wave proceeds in a continuous manner. Thus, the rate f_1 at which crests pass point P must equal the rate f_2 at which crests pass point Q. Thus, we will replace the two symbols f_1 and f_2 by the common symbol f.

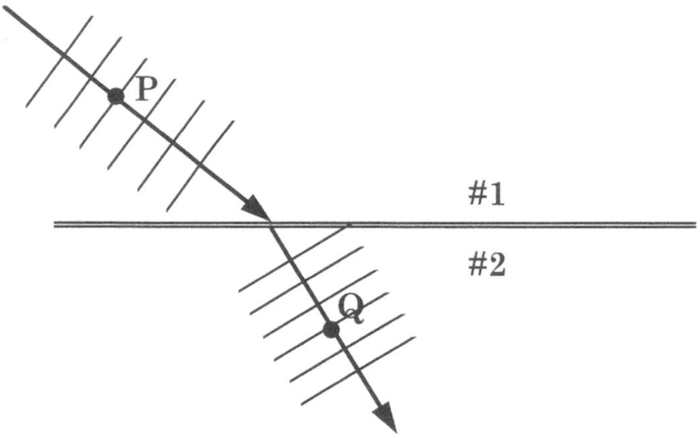

Fig. 9.26 Refraction of waves

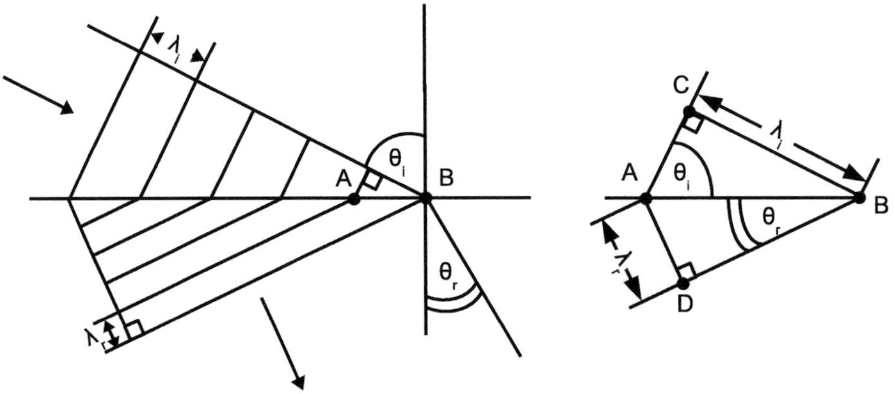

Fig. 9.27 Detailed schematic of the Huygens wave theory of refraction

This last equation is a general expression for the refraction of waves. Note that it makes no reference to the index of refraction. In fact, **it can be applied to sound waves** as well as light waves. For example, consider the following problem:

Sample Problem 9.10 A sound wave in air is incident at an angle of 10° on a water–air interface. Find the angle of refraction.

Solution We have $v_i \cong 340$ m/s and $v_r = 1,400$ m/s. Thus,

$$\sin\theta_r = \frac{v_r}{v_i}\sin\theta_i = (1,400/340) \times \sin 10° = 0.72$$

so that

$$\theta_r = 46°$$

We now return to light waves and **derive Snell's Law** from Eq. (9.21) above: We have

$$v_i = \frac{c}{n_i} \qquad \text{and} \qquad v_r = \frac{c}{n_r} \tag{9.22}$$

so that

$$\frac{\sin\theta_1}{\sin\theta_2} = \frac{\frac{c}{n_1}}{\frac{c}{n_2}} = \frac{n_2}{n_1} \tag{9.23}$$

This is Snell's Law in an algebraically rearranged form.

Note For "refraction" of a beam of **particles**, it can be shown that (cf. Eq. (9.21))

$$\frac{\sin\theta_1}{\sin\theta_2} = \frac{v_2}{v_1} \tag{9.24}$$

Then, if $\theta_2 < \theta_1$, we also have $v_2 > v_1$. That is, a beam should be refracted **away** from the normal on passing from air into glass. The experimental finding about 150 years ago, that for *light waves* the opposite is true, (ie. $v_2 < v_1$), gave further confirmation (after the interference experiments described in Chap. 7), that light propagates as a wave.

9.7 Application to Mirages

On a very hot day, the ground will be heated up to temperatures greatly exceeding the temperature of the air above. As a consequence, the air close to the ground will be hotter than the air above. The hotter air has a lower density than the cooler air above and hence has a lower index of refraction. Now imagine a ray of light that originates from a region where the index of refraction is larger than that at ground level and that propagates downwards towards the ground. We can have an occurrence of **total internal reflection**. The light ray can strike someone's eye, thus producing an effect that is referred to as a **mirage**. This phenomenon is exhibited in Fig. 9.28.

According to Snell's Law applied to refraction at the boundary between two homogeneous media, the product $n \cdot \sin\theta$ is the same for the two media in which a ray travels. In the case of the hot air, the index of refraction varies continuously. It can be shown that Snell's Law applies too in the same form: **The product $n \cdot \sin\theta$ is constant all along the path of the ray**, where θ is measured from the vertical. The angle θ corresponding to a point along the dotted ray in Fig. 9.28 is measured relative to the vertical.

The ray will continuously become ever more horizontal as it propagates and the index of refraction decreases. If it becomes absolutely horizontal (that is, $\theta = 90°$), thereafter it will propagate upwards. The ray experiences total internal reflection.[8]

A photograph of a mirage in a desert[9] is exhibited in Fig. 9.29.

9.8 The Prism

In Fig. 9.30 we depict a transparent **prism**, with a ray of light incident on one face and a refracted ray leaving a second face.

Note in the figure how the directions of the two refracted beams, one inside the prism and the other outside the prism, are each determined with respect to their respective normals, being towards the normal #1 for the first one, since the ray is going from air into the prism and away from the normal #2 for the second one is going from the prism into the air.

9.9 Dispersion

In the case of waves along a taut string, without stiffness, any wave pattern will propagate along the string without any change in shape, to the extent that attenuation can be neglected. Furthermore, all patterns propagate at the same velocity (for a given tension). The same holds true for sound waves and light waves in vacuum.

However, in the case of a light wave in a medium such as glass or plastic, or even air to a very small extent, **only sine waves propagate without a change in shape.** This fact is directly connected to another characteristic of such waves: The **wave velocity of a sine wave depends upon the wavelength and hence the frequency.**

For example, in the case of light propagating through *light flint glass*[10]

$$v = 1.88 \times 10^8 \text{ m/s @ } \lambda = 434 \text{ nm}$$

[8]It is often thought that the ray described above originates from the sun. A bit of analysis shows that this is impossible: In outer space $n = 1$, so that $n \cdot \sin\theta = \sin\theta \leq 1$. At the turning point, $\theta = 90°$, so that $n \cdot \sin\theta > 1$. The only way we can have total internal reflection is for the ray to originate from light scattered by the atmosphere.

[9]https://commons.wikimedia.org/w/index.php?curid=10842357.

[10]Reference: Handbook of Chemistry and Physics, 65th edition, (Chemical Rubber Comp., Boca Raton, FL, 1984).

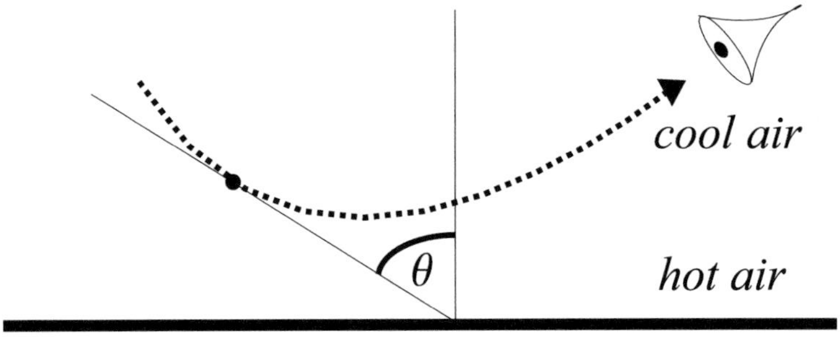

Fig. 9.28 A mirage

Fig. 9.29 A desert mirage

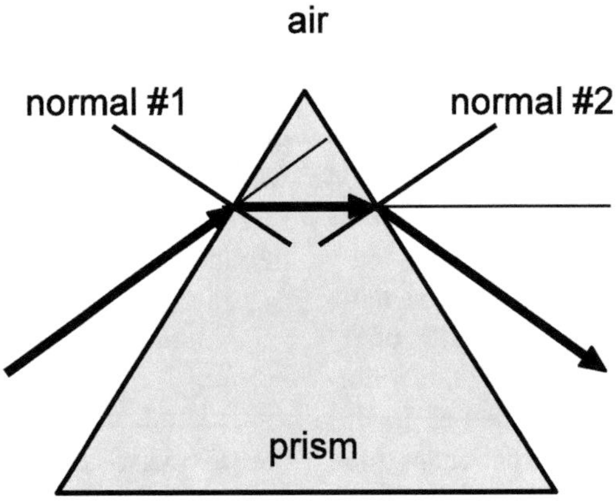

Fig. 9.30 Schematic of a prism

while

$$v = 1.91 \times 10^8 \text{ m/s} @ \lambda = 768 \text{ nm}$$

The variation in wave velocity with wavelength is referred to as **dispersion**. A wave having this property is said to be **dispersive**. Generally the degree of variation is relatively small. Nevertheless, it is significant enough as to be responsible for the colors of the rainbow, which is produced when sunlight passes through microscopically small drops of water in the sky.

Light waves traveling in a medium and waves propagating along the surface of a liquid, such as ocean waves, are dispersive. Waves traveling along a string without stiffness, sound waves, and light waves in vacuum are **non-dispersive** . Waves along a string that has stiffness are dispersive. The very significant importance of dispersion in fiber optics is discussed in Sect. 9.9.2

9.9.1 Effect of Dispersion on a Prism

The speed of light in a medium depends upon the wavelength. Therefore, the index of refraction and the degree of refraction depend upon frequency.

We see that for the prism in Fig. 9.31, violet light is refracted more than red light. Alternatively stated, refraction decreases with increasing wavelength.

Question: How does the speed of light depend upon wavelength for this prism?

Compare this behavior with *diffraction*, which *increases* with increasing wavelength. However, we must note that the behavior of the prism depicted here is dependent upon the particular material that the prism is made of: I know of no physical principle that doesn't allow for an *increase* of refraction with increasing wavelength.

Here is a useful application of dispersion. As was pointed out in Chap. 6, the prism can be used to carry out a spectral analysis of a beam of light. [See Fig. 6.1.] If the incident beam is not monochromatic, each monochromatic component in the mixture will leave the prism at a different angle, as shown schematically in Fig. 9.31:

In Fig. 9.32 we see a rainbow at the left. In fact, if you look closely, you might notice a faint second rainbow above the first. The detailed theory behind the rainbow was published by René Descartes, the great philosopher, mathematician—the creator of analytical geometry and therefore Cartesian coödinates—and scientist. It is difficult to imagine how physics and mathematics would have developed without analytical geometry.

9.9.2 Effect of Dispersion on Fiber Optics Communication

Recall that the attenuation of a sine wave depends upon its frequency. Therefore, as a wave propagates in the presence of attenuation, its Fourier components attenuate at different rates. The tone quality of a sound wave depends upon the ratio of its Fourier amplitudes. As a result, not only does the intensity and overall loudness decrease as the sound wave propagates, its tone quality changes too.

The shape of a wave is determined only by the ratio of the amplitudes and the phase relations among the Fourier components. [If attenuation rates were the same for all frequencies, the *ratio* of the amplitudes would remain constant and the shape would not change.] From the above, we conclude that attenuation will change the shape of the waves.

What about the effect of dispersion? The amplitudes of the Fourier components do not change as a result of dispersion. However, since each Fourier component travels at a different speed, the peaks of each component will shift one relative to another. That is, the relative phases of the Fourier components will change as the wave propagates. As a result, ***dispersion** causes the shape of a wave to change.*

We are now in a position to understand why dispersion can cause a serious problem in fiber optics communication: *Analog* communication of sound converts the sound wave into an electric voltage whose variation in time is the same as the pattern of the sound wave. Fiber optics communication, on the other hand, transmits information by sending a sequence of light pulses down a glass fiber. The pattern of the sound wave is represented by a corresponding sequence of time intervals between the pulses. This system is an example of *digital* communication.

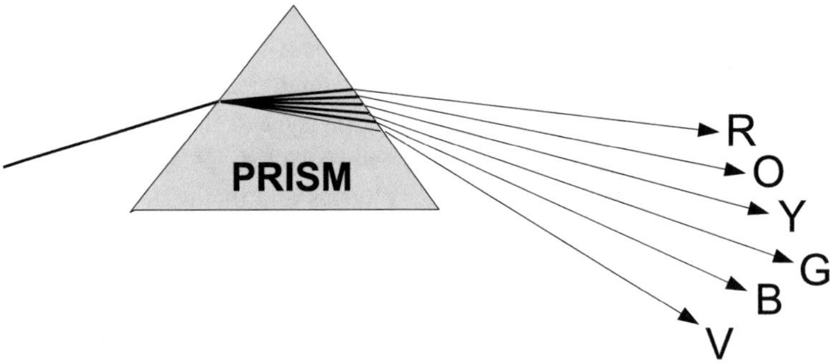

Fig. 9.31 Prism splitting colors as a result of dispersion

Fig. 9.32 A double rainbow Source: https://en.wikipedia.org/wiki/Rainbow#/media/File:Rainbow1.svg

As a sequence of pulses travels down the glass fiber, the amplitudes of the pulses are attenuated due to the intrinsic properties of the glass. This is a problem that has to be dealt with in fiber optics communication. Fortunately, in Chap. 4 we noted that the current attenuation is at a very low level of about 0.1-dB per kilometer. There is a more serious problem. A glass fiber has a significant degree of dispersion. As a consequence, the shape of any non-sinusoidal wave will change its shape as it propagates along the fiber. In particular, a pulse will either become broader or become narrower. It happens that the pulses produced in fiber optics communication become broader. In time, pulses proceeding along a long fiber will become so broad as to overlap with their neighbors to such an extent that they cannot be distinguished as individual pulses. The information contained therein will become non-discernible. See Fig. 9.33.

In order to deal with this problem, special devices are inserted in sequence along the fiber. They read the broadened pulses before they are indistinguishable and replace each pulse with a re-emitted narrower pulse. The original is "reconstituted." Thus the information contained therein is preserved.

9.10 Lenses

There are two major types of **lenses**, the **converging lens** and the **diverging lens**. They are commonly used to correct a person's vision or serve as the major component in cameras, telescopes, microscopes, and film and slide projectors. The human eye itself has a lens. (See Chap. 13 for its unique characteristics.)

9.10.1 The Converging Lens

Consider what would happen if a plane wave were incident on a pair of prisms as arranged in Fig. 9.34:

We have here represented an incident plane wave by a series of parallel rays because different parts of the wave strike different parts of the sys-

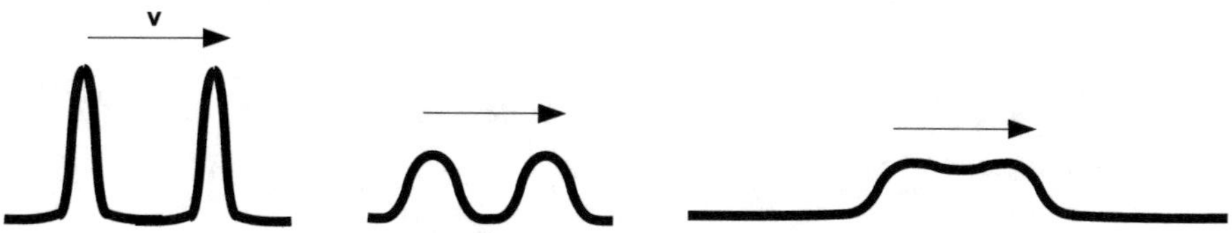

Fig. 9.33 Spreading of a pair of pulses due to dispersion

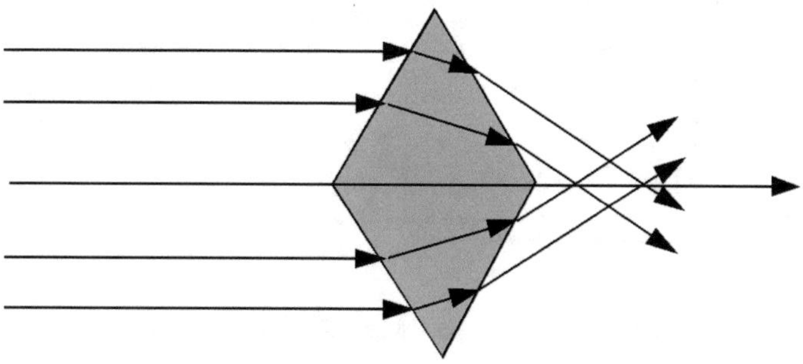

Fig. 9.34 A pseudo-converging lens from two prisms

tem. *Diffraction effects are completely neglected in this section.* Note how the corresponding rays exit parallel to each other.

A converging lens is a refined version of the above system, consisting of two faces which are sections of spheres of relatively large radius. This large radius corresponds to the lens being rather thin.[11]

A converging lens has the remarkable property that all incoming rays parallel to the **axis of the lens** *essentially* converge at a single point on the axis, call the **focal point**, labeled F. We see this property in Fig. 9.35.

The distance between the center of the lens and the focal point is called the **focal length**, with the symbol f. The greater the index of refraction, the more the convergence and the *smaller* the focal length.

[11]Another term for a converging lens is a **convex lens**, since both sides of the lens are convex. The diverging lens, discussed later, is also called a **concave lens**. There also exist lenses that are concave on one side and convex on the other. If these are possibilities for consideration, one must remove any unambiguity by referring to a **biconvex lens**, or a **biconcave lens**, or a **convex-concave lens**.

9.10.2 Lens Aberrations

- In the previous paragraph, I stated that all rays "essentially converge to a single point." The reason is that convergence to a single point is never perfect for an actual lens. The lack of perfect convergence is called **spherical aberration**, and decreases with decreasing lens thickness. This unavoidable property of a lens is shown in Fig. 9.36, where we see what actually happens.

  The degree of aberration is much reduced for rays that are incident close to the axis, as seen in Fig. 9.37.
- Since the index of refraction depends upon the wavelength of the light—the phenomenon referred to as **dispersion**—a beam of white light cannot be focused at a point even if **spherical aberration** were absent: The various monochromatic components of the beam will be focused at different points, as we observed in the context of the prism. We can simply say that the focal length of a monochromatic beam is dependent upon the wavelength. This defect in the lens is referred to as **chromatic aberration**. We see chromatic aberration exhibited in Fig. 9.38. The incoming beam is a mixture of two monochromatic components, one red and the other cyan.

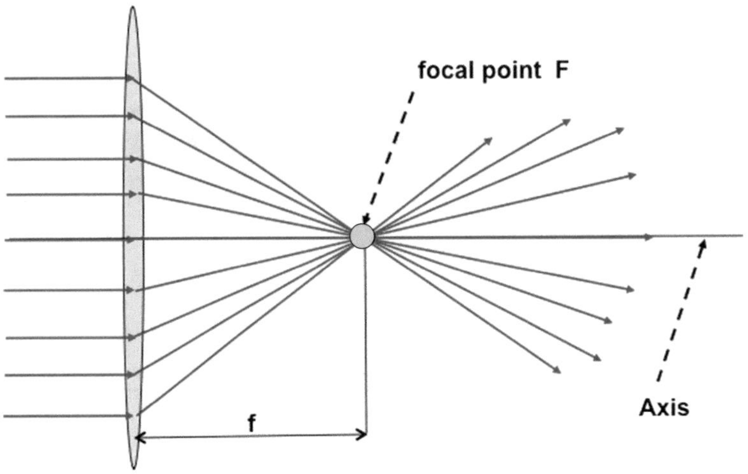

Fig. 9.35 Rays parallel to the axis meet at the focal point in an ideal lens

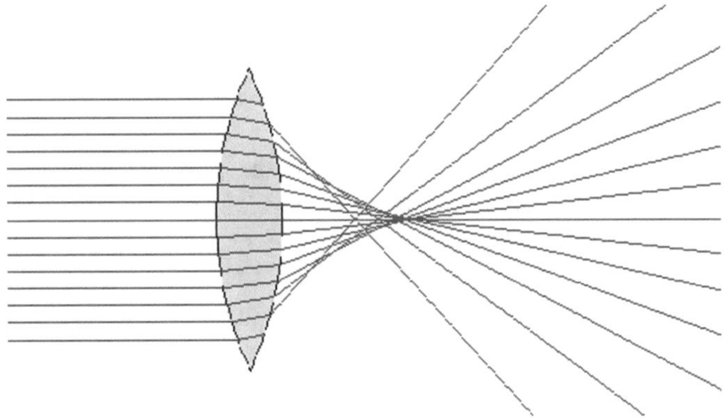

Fig. 9.36 Spherical aberration of a converging lens

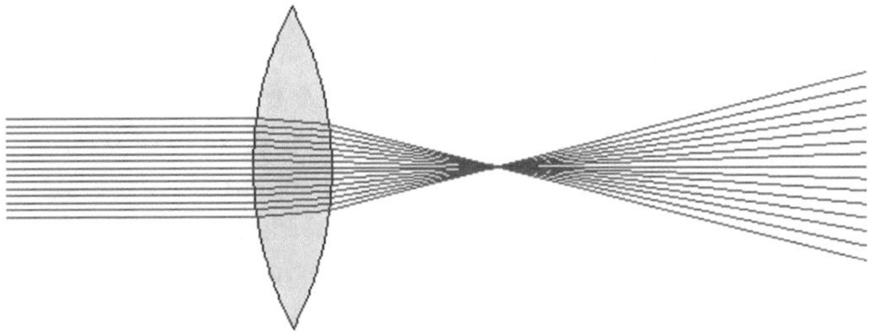

Fig. 9.37 Spherical aberration—weak for rays near the axis

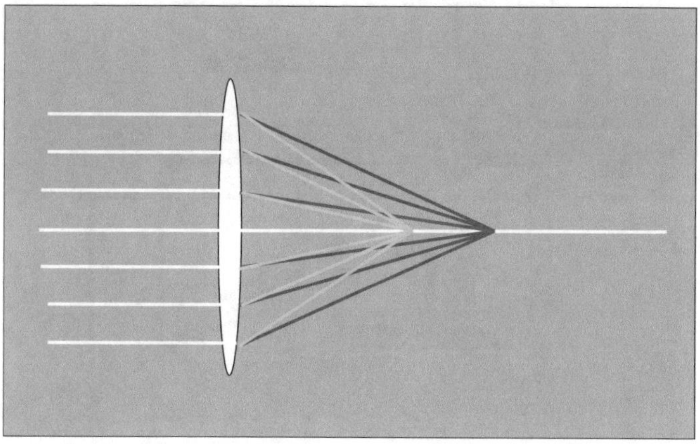

Fig. 9.38 Chromatic aberration

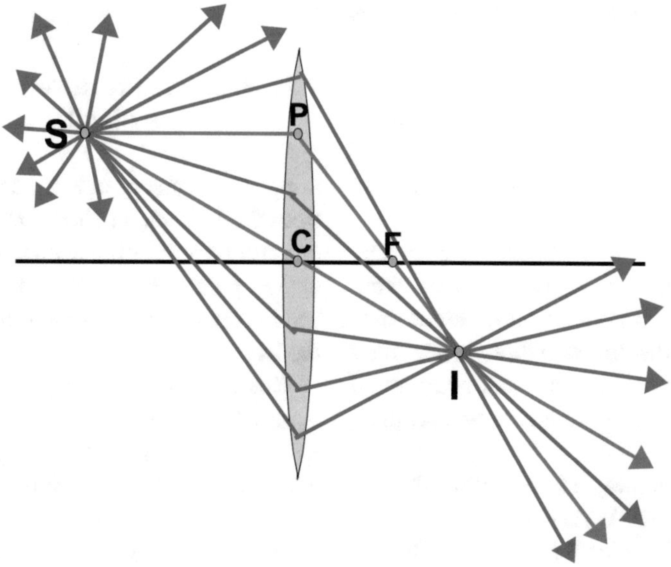

Fig. 9.39 Image of an object using a converging lens

They happen to be such as to produce a white beam when mixed. They are said to be **complements**. [See Chap. 15, where color vision is discussed in detail.][12]

[12]Polycarbonate is a material often used for eye lenses because of its strong shatter resistance and light weight. It has a drawback in having stronger chromatic aberration than glass. The so-called *Abbé* number is used as a material's level of dispersion. The larger the number, the lower the level is dispersion. Thus, while crown glass has an *Abbé* number of about 55 while polycarbonate has a value of about 32. See Wikipedia (1-6-2011): http://en.wikipedia.org/wiki/Abbe_number.

Note We will neglect aberration effects in the following discussion. The neglect of spherical aberration is referred to as the **thin lens approximation**.

Now suppose we have a point source S of light located to the left of a lens. The wave crests from a point source are spheres so that the wave is referred to as a **spherical wave**. The wave will therefore be represented by many rays directed radially outward, as seen below in Fig. 9.39. Most rays never strike the lens and go their merry way. However, a fraction pass through the lens.

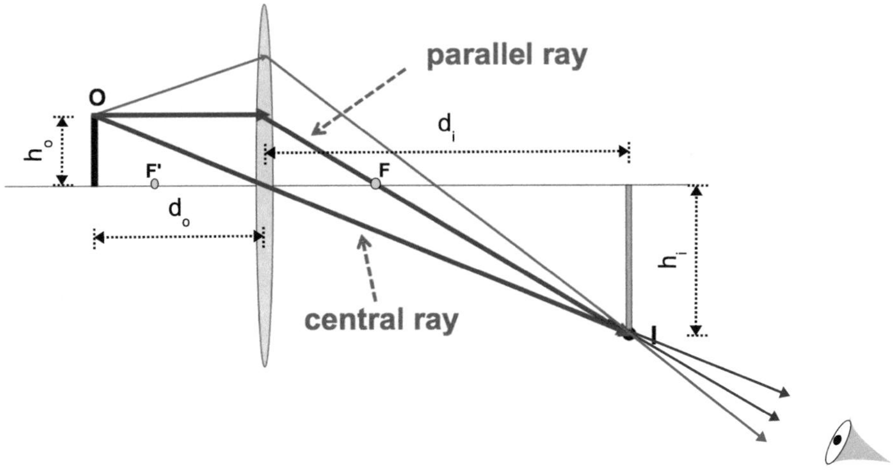

Fig. 9.40 Geometric analysis: real image of a converging lens

The remarkable property of an ideal (aberrations neglected) lens is that all rays from S, *which pass through the lens*, cross at the same point I—called the **image point**. I is said to be the **image** of S.

Spherical aberration will blur the image. Chromatic aberration will produce a different image point for each component of monochromatic light from the source point. White light from the source point that has a continuum of all visible wavelengths will produce a line segment with a rainbow of colors.

Here is a simple way to determine the position of the image: A very important ray in lens analysis is the **parallel ray**, SP. It will be refracted so that it passes through the focal point at F. Another important ray in lens analysis is the **central ray** SC, which goes through the center of the lens, unrefracted. These two rays—shown in blue—cross at the image point I. Thus, knowing the paths of these two rays determines the position of the image point.

Note Every source point has its own *unique* image point. Source points and image points are said to be in one-to-one correspondence. This is the fundamental property of a **thin lens**.

9.10.3 Image Produced by a Converging Lens

Suppose that an object is located at some distance from the lens, d_O, as shown in Fig. 9.40. We call this distance the **object distance**. You will note a blue point labeled F' to the left of the lens. This point is a focal length distance away from the lens.

It is significant that I placed the object more than a focal length distance from the lens; you will soon understand the significance of this placement We assume that the **object** is very thin and is lined up perpendicular to the axis.

We can locate the image of the "head" of the object at O using two special rays. They are the **central ray** and the **parallel ray**. The remaining image points of the extended object produce a vertical image as shown, at the **image distance** d_i—all pairs of points being in a one-to-one correspondence. In this figure we also exhibit the height of the object h_o and the height of the image h_i.

As pointed out in the previous section, any ray from the object that passes through the lens will pass through the same point at the position of the image. We show this for an additional ray passing through the top of the lens, as shown in Fig. 9.40.

Note

1. We assume that the lens is so thin that its thickness is negligible compared to the object distance or image distance. This unfortunately is not reflected in the figures of the text!
2. The eye to the right of the image observes rays emanating from the single point at the tip of the object. As a consequence, the brain perceives the image as an object located at the position of the image. This fact, or its detailed explanation thereof is seldom pointed out in textbooks.

Using elementary trigonometry we can derive a relation among the object distance d_o, the image distance d_i, and the focal length f. The relation is called the **thin lens equation**:

$$\frac{1}{d_o} + \frac{1}{d_i} = \frac{1}{\mathrm{f}} \qquad (9.25)$$

Note

1. If a screen is placed at the position of the "image," a clear image will appear on the screen. The image is said to be a **real image**. Any other placement of the screen will produce a blurred image.
2. Note that if someone were to be looking at the object through the lens, it would appear to be as the image shown in Fig. 9.40.
3. As the object approaches the focal point of the lens, that is, as d_o approaches f, it can easily be shown from the thin lens equation that d_i approaches infinity.

Sample Problem 9.11 Given f = 2-cm and d_0 = 4-cm, find d_i. Note that the object distance is greater than the focal length.

Solution We have $1/4 + 1/d_0 = 1/2$ so that

$$\frac{1}{d_i} = \frac{1}{2} - \frac{1}{4} = \frac{1}{4}$$

$$d_i = 4\ cm$$

Sample Problem 9.12 Given f = 2-cm and d_0 = 2.1-cm, find d_i. Notice that in contrast to the previous sample problem, the object distance is just a little greater than the focal distance.

Solution We have $1/d_i = 1/2 - 1/2.1 = 0.024$ or d_i = 1/0.024 = 42-cm. There is a dramatic increase in the image distance.

Sample Problem 9.13 We next consider an object that is located less than a focal length distance from the lens: $d_o < \mathrm{f}$. Given f = 2-cm and d_0 = **1-cm**, find d_i.

Solution We have $1/d_i = 1/2 - 1/1 = -1/2$, so that $d_i = -2$ cm.

The **negative** value for d_i means that the image is on the same side of the lens as the object.

What is the meaning of this result? In this situation, all rays from an object point which pass through the lens, *appear to be coming* from a single point behind the lens. See Fig. 9.41 below. The dashing of two of the line segments indicates that in fact there is no light ray along these segments.The perception by the visual system (consisting of the eye and brain) is based upon the existence of these virtual rays.

The image is said to be a **virtual image**: No image will appear on a screen placed at the image position. In this arrangement, the image is larger than the object and the converging lens is serving as a **magnifying glass**.

In the first two problems, the object distance was to the left of the focal point, which results in the image being an **inverted and real image**. In the third problem, the object was between the focal point and the lens, which results in the image being an **upright and virtual image**.

Note Note that all rays from the top of the object, which pass through the lens and thereby reach the eye, **appear** to be emitted by the top of the image. As a consequence, the brain perceives the image as an object at the position of the image. While the image is referred to as being "virtual," for the brain it appears to be as real as any other image!

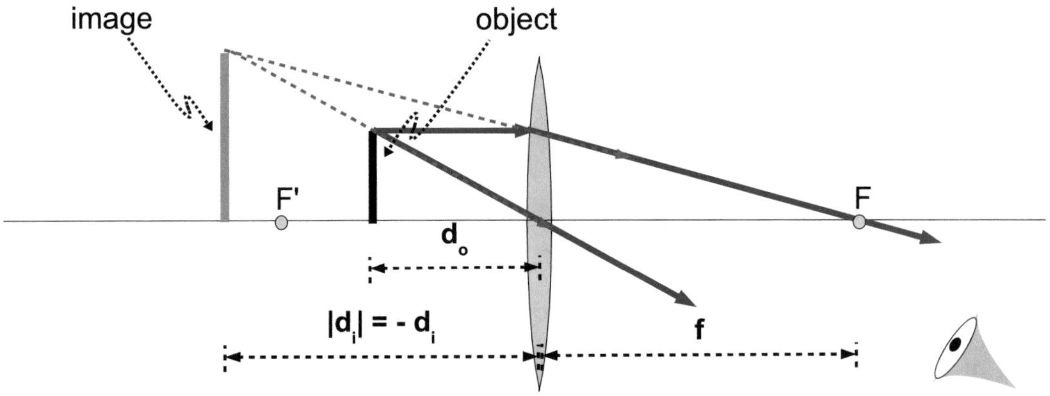

Fig. 9.41 Geometric analysis of the converging lens producing a virtual image

9.10.4 Magnification

How does the size of the image compare to the size of the object?[13] Let us redraw Figs. 9.40 and 9.41 as simple figures to highlight the plane geometry.

For both the real image [seen in Fig. 9.42] and the virtual image [seen in Fig. 9.43] we find the following:

Let h_o be the height of the object and h_i be the height of the image. The ratio of the two is called the **magnification**, with a symbol M. Since the two triangles, △ [CAO] and △ [CBI] are *similar*, we have

$$M \equiv \frac{h_i}{h_o} = \left|\frac{d_i}{d_o}\right| \tag{9.26}$$

The absolute value takes into account the case when the image distance d_i is negative.

Numerical examples:

For problem 1, $M = |4/4| = 1$

For problem 2, $M = |42/2.1| = 20$

For problem 3, $M = |-2/1| = 2$

[13]In Appendix H we discuss **magnifying power**, which is a property of a lens or instruments such as a telescope or a microscope that consist of a series of lenses - referred to as a **compound lens**. Magnifying power represents the ability of an optical instrument to increase the image size on the retina that is produced by an object. In order to appreciate this material, it is necessary to understand how the eye works, as discussed in Chap. 13.

The Real Image of a Converging Lens as a Secondary Object

To an observer who is to the right of the real image of a converging lens, *all rays* from an object, which have gone through the lens, *appear to be coming from the real image*. For this reason, if you look through a converging lens at an object that is at a distance from the lens greater than the focal length so as to produce a real image, you will see an inverted image that appears closer to you than the lens. (There is a requirement that the rays of the image reach your eye.) This property of lenses is manifest for any series of lenses in sequence, whatever their number. Together, the set of lenses form what is referred to as a single **compound lens**. The eye itself has a number of interfaces between two media, as we will see in Chap. 12. At each interface, we have refraction. The image produced by each interface becomes the object of the next interface. Other examples of compound lenses are **microscopes** and **telescopes**.[14] The microscope is discussed in some detail in the Appendix on the **Magnifying Power of an Optical System**.

[14]To determine the ultimate position of the image produced by a compound lens, one must apply the thin lens equation sequentially. For the effect of each lens, one must make sure to use the distance from that lens of the image produced by the previous lens as the object distance of that current lens. In the case of eyeglasses, the distance of the eyeglasses from the eyes is so small that one usually can assume that the eyeglasses are coincident with the center of the compound lenses of the eyes.

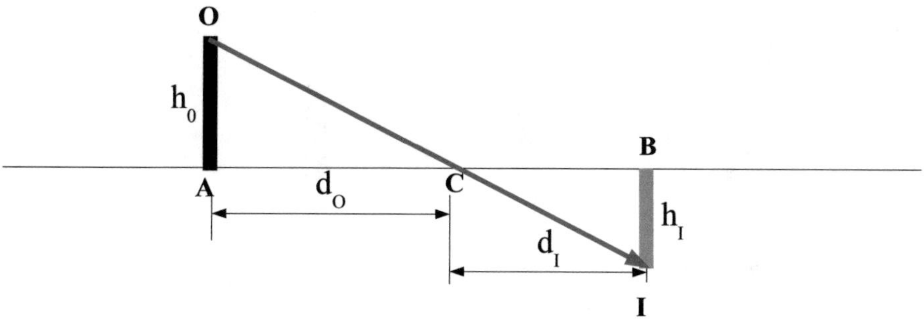

Fig. 9.42 Simplified geometric analysis: real image of a converging lens

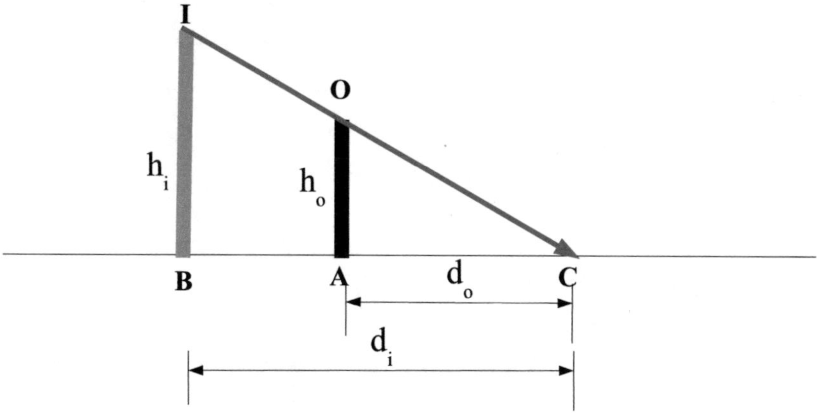

Fig. 9.43 Simplified geometric analysis: virtual image of a converging lens

9.10.5 The Compound Lens

In Fig. 9.44 we see a compound lens consisting of two convex lenses. We have indicated the object and the respective focal points of each lens.

We now discuss how we can determine the ultimate image of the compound lens. The first step is to determine the position of the image that would be produced by the first lens alone, as shown in Fig. 9.45. The figure exhibits the case when this first image is to the left of the second lens. [We will discuss the more complex case when the first figure is to the right of the second lens.] While we can display the position of this first image, a **precise determination** is determined by using the thin lens Eq. (9.25).

To determine the position of the final image, we treat the first image as the object of the second lens, using the usual two rays, the parallel ray and the central ray. See Fig. 9.46. To determine the precise position of the final image we again apply the thin lens Eq. (9.25).

Suppose that the distance between the centers of the two lenses is d. During the course of above calculation, we will have an object distance d_{o1} for the first lens, the image distance d_{i1} for the first lens, the object distance $d_{o2} = d - d_{i1}$ for the second lens, and the position of the final image that is a distance d_{i2} to the right of the second lens.

We now discuss the situation when the image of the first lens is to the right of the second lens, as shown in Fig. 9.47.

The use of a compound system of lenses in a **telescope** reveals a bit of ingenuity: We saw above that a magnifying glass, consisting of a single lens, requires that the object distance of the object be very close to the focal length. For a distant object, the magnifying glass therefore fails to magnify. However, we can place a second

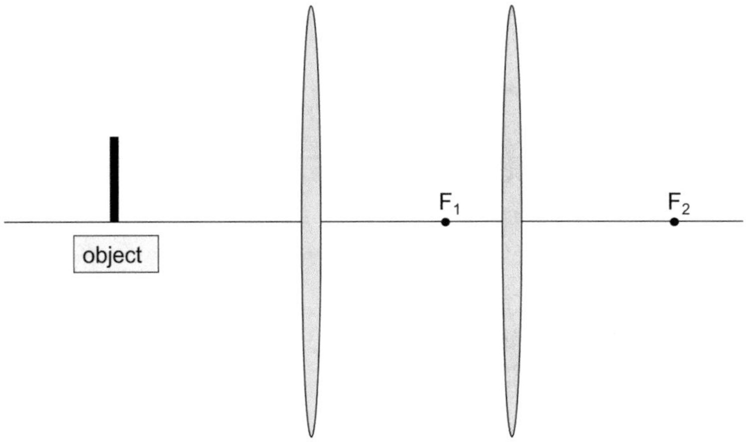

Fig. 9.44 A compound lens with two lenses in sequence

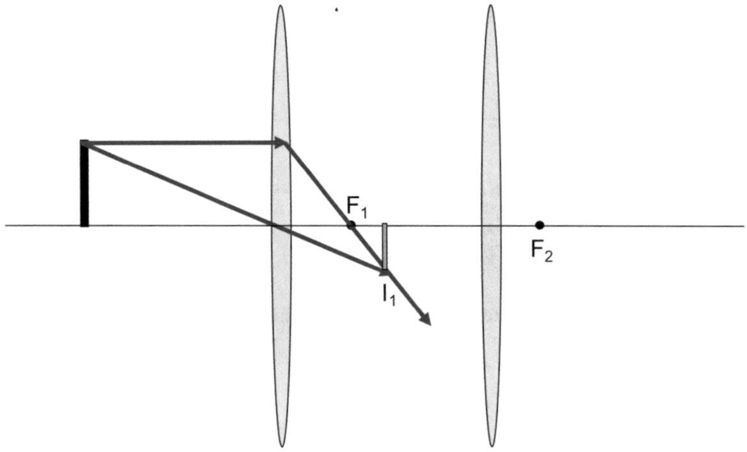

Fig. 9.45 A compound lens with two lenses in sequence

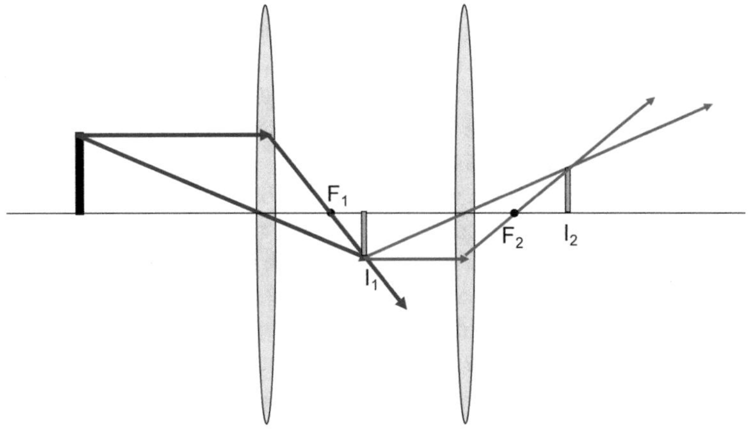

Fig. 9.46 Final image of the compound lens

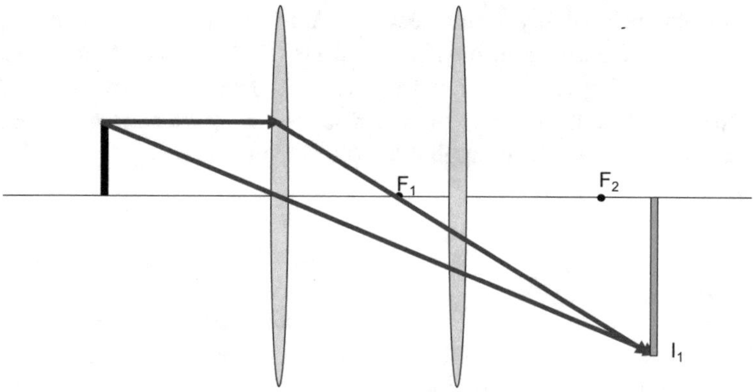

Fig. 9.47 A compound lens with two lenses in sequence

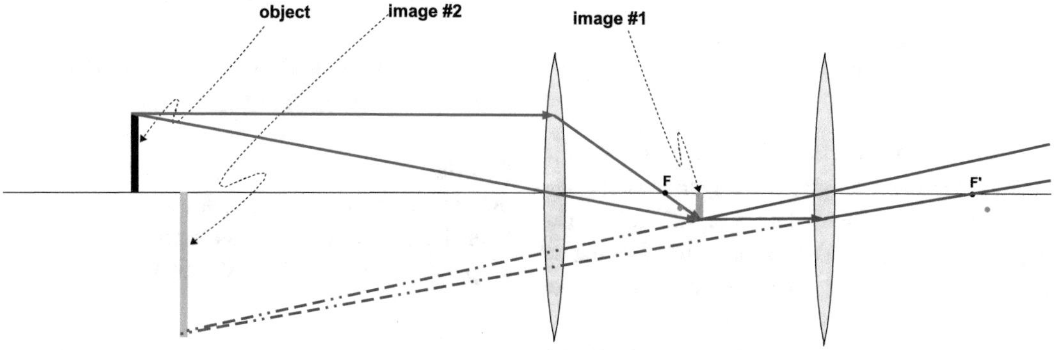

Fig. 9.48 Schematic representation of a telescope (not to scale)

lens in front of the magnifying glass in a position such that the image of this second lens is at an appropriate close position to the magnifying glass. Then the magnifying glass can serve its original function and magnify the distant object. The two lenses together constitute a telescope!

See Fig. 9.48.

We see the original black object, upright at the far left. The first lens produces a small inverted gray image between the two lenses. The position of this image is determined by the pair of red rays. The blue rays show us how to determine the final inverted image that is produced by the second lens to the far right. The dashed segments are the continuations of these rays into a region where there are actually no light rays at all. An eye-brain situated to the right of the second lens will believe that the object is located at the final inverted image that is located a bit to the right of the object. For an actual telescope, the object is very far away from the telescope—perhaps light years away—while the images are in the vicinity of the telescope.

9.10.6 Reversibility of Rays—Interchange of Object and Image

If we examine the thin lens Eq. (9.25), we note a **symmetry** in the two distances, the object distance and the image distance: For a given focal length, if the object has the position of the image, the image must have the original position of the object. Alternatively, if a ray leaves the object and passes through the lens, it must pass through the image position. Conversely, if this ray is reversed in direction at the image position, it will pass through the lens and then pass through the object position. This behavior is referred to as

reversibility. Note how reversibility is manifest for the single refraction of a ray at an interface between two materials owing to the symmetry in Snell's Law. Note too that this reversibility is exhibited by all of the rays that pass through the lens in Fig. 9.39.[15]

9.10.7 The Diopter

We note that the greater the focal length the weaker is its ability to bring an object into focus. This characteristic is represented mathematically in the lens equation, wherein the **inverse** of the focal length appears.

$$\frac{1}{\mathrm{f}} = \frac{1}{d_o} + \frac{1}{d_i} \tag{9.27}$$

According to this equation, for fixed d_o, the smaller the focal length, the larger must be the left-hand side of the equation and therefore the larger $1/d_i$ must be. The result is that d_i must be smaller; that is, the lens focuses the image more strongly.

A **diopter D** is simply the inverse of the focal length **with the focal length expressed in meters** and therefore reflects the above characteristic of the focal length. Note that it appears naturally in the thin lens equation. We have

$$\mathbf{D} = \frac{1}{\mathrm{f}} \tag{9.28}$$

[15]Reversibility is manifest in the orbit of a planet about the Sun: If a planet were stopped dead in its tracks and its path reversed so that at that point the original direction is reversed while the speed is the same, the planet would retrace its path into the past, where it came from. What we would observe could be seen by taking a movie of the planet's motion and then running the movie backwards. Reversibility is manifest in the basic laws of physics. The consequence is that every sequence of events has a possibility of occurring. Yet, there are movie scenes that are hilarious if they are run backwards. Why? Because the reversed sequence is regarded as impossible. [Imagine someone shown jumping off a ladder onto the ground.... Now reverse the sequence.] Such sequences are referred to as being **irreversible**. One of the challenges of physics is to understand how such extremely unlikely, irreversible sequences are never seen and yet have a possibility of occurring, in principle.

A focal length of 1-m has a diopter value of D=1/1=1-diopter. A focal length of 0.1-m has a diopter value of D=(1/0.1)=10-diopters. Clearly the smaller the focal length, the greater the diopter value.

Sample Problem 9.14 Suppose that the focal length is 50-cm. Find the diopter value.

Solution First let us express the focal length in meters. We have f=0.5-m.

We obtain

$$\mathbf{D} = \frac{1}{(-0.5)} = 2 \text{ diopters} \tag{9.29}$$

Sample Problem 9.15 Suppose that the focal length is f=−25-cm. Find the diopter value.

Solution First let us express the focal length in meters. We have −0.25-m. Next we note that the focal length is negative. As a consequence, the diopter value is negative. We obtain

$$\mathbf{D} = \frac{1}{(-0.25)} = -4 \text{ diopters} \tag{9.30}$$

An interesting case obtains for a compound lens with lenses that are so close to each other that we can regard them as sharing a common center. This is the situation we have with **eyeglasses**: The effective focal length of an eye can vary. [See Chap. 13.] Correspondingly, it has a particular diopter value. The focal length and diopter value must be positive in order to produce an image on the retina. Furthermore, the effective focal length of an eye might need correction by an eyeglass because of the eye's limited ability to adjust the focal length, that is, to **accommodate** using its flexible eye lens.

We will consider the case of two simple lenses, for which the image of the first lens lies to the left of the common center. Then the object distance for the second lens is the image distance of the first lens. The position of the final image lies at the object distance of the second lens. We have

$$\frac{1}{d_{i1}} = \frac{1}{\mathrm{f}_1} - \frac{1}{d_{o1}} \tag{9.31}$$

We also have

$$\frac{1}{d_{i2}} = \frac{1}{\mathrm{f}_2} - \frac{1}{d_{o2}} \tag{9.32}$$

Now we have to take into account that the value of the image distance d_{i1} is negative, whereas the object distance d_{o2} is a positive value. Thus, we must set $d_{o2} = -d_{i1}$

We then find

$$\frac{1}{d_{i2}} = \frac{1}{\mathrm{f}_2} + \frac{1}{d_{i1}} = \frac{1}{\mathrm{f}_2} + \frac{1}{\mathrm{f}_1} - \frac{1}{d_{o1}} \tag{9.33}$$

The final result is that

$$\frac{1}{d_{i2}} + \frac{1}{d_{o1}} = \frac{1}{\mathrm{f}_2} + \frac{1}{\mathrm{f}_1} \tag{9.34}$$

The effective focal length of the compound lens is then simply

$$\frac{1}{\mathrm{f}} = \frac{1}{\mathrm{f}_2} + \frac{1}{\mathrm{f}_1} \tag{9.35}$$

Symbolically, the diopter value of the compound lens is the **sum** of the diopter values of the two component lenses:

$$\mathbf{D} = \mathbf{D}_1 + \mathbf{D}_2 \tag{9.36}$$

In words, we can say that placing two lenses next to each simply adds their strengths. In the case of an eyeglass, $\mathbf{D}_1$ would be the diopter value of the eye and $\mathbf{D}_2$ the diopter value of the eyeglass, being positive or negative.

While we derived this result assuming that the image of the first lens is to the left of the compound lens, it is worthwhile for the reader to think about why the same result obtains when the image of the first lens is to the right of the lens.

9.10.8 The Diverging Lens

The diverging lens is a refinement of the following arrangement of two prisms shown in Fig. 9.49:

Look at Fig. 9.50. Note that incident *parallel* rays emerge so as to appear to be coming from a common point, the focal point F of the lens. Note that now the focal point is on the same side of the lens as the source of light—to the left of the lens.

Next, in Fig. 9.51, we display the image of an object as produced by a diverging lens. Images are **always erect and virtual** and appear on the same side as the object, both between the lens and the focal point.

The thin lens equation can still be used, with the focal length f in the equation set equal to a **negative** number! The image distance parameter d_i obtained is always negative, indicating that the image is on the same side as the object, to the left of the lens. Furthermore, $|d_i| < d_o$ always. This indicates that the magnification M is always less than unity. **Thus, the image is always smaller than the object**.

Note Note that the diopter value of a diverging lens is negative.

Sample Problem 9.16 Given a diverging lens with a focal length of −2 cm, and an object distance, $d_0 = 3$ cm, find the image distance and the magnification.

Solution We have

$$\frac{1}{d_i} = \frac{1}{\mathrm{f}} - \frac{1}{d_o}$$
$$= \frac{1}{-2} - \frac{1}{3} = -\frac{1}{2} - \frac{1}{3} = -\frac{5}{6}$$

so that $d_i = -6/5$ cm.

The magnification is given by $M = |d_i/d_o| = (6/5)/3 = 2/5 = 0.4$.

9.10.9 Determining the Focal Length of a Diverging Lens

We can determine the focal length of a **converging lens** by measuring the distance from the lens to the image of a far off source or by comparing the image distance with the object distance. Unfortunately, a **diverging lens** does not produce a **real image**, so that it is not immediately clear how its focal length can be determined. We will

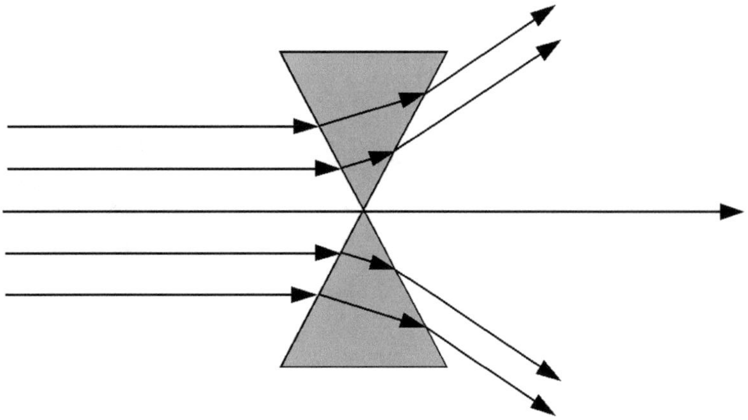

Fig. 9.49 A crude diverging lens using prisms

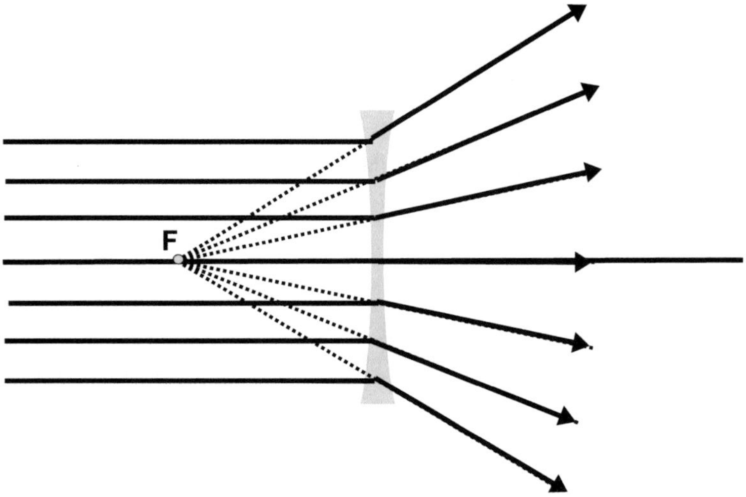

Fig. 9.50 The focal point of a diverging lens

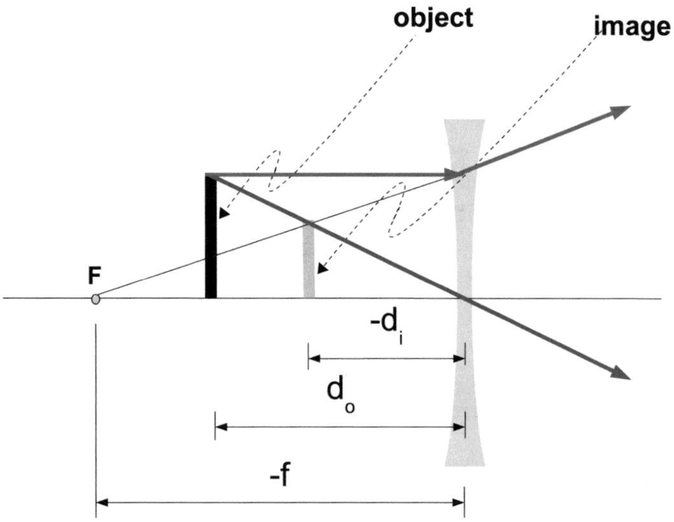

Fig. 9.51 Geometric analysis: the virtual image of a diverging lens

outline a method of doing so: It involves placing a second, converging lens in tandem with the diverging lens, as shown in the complex Fig. 9.52 below. The two lenses form a compound lens, which we discussed in Sect. 9.10.5. The diverging lens produces a virtual image. Then the converging lens produces a real image whose position can be determined. See Fig. 9.52 below.

The position of virtual image (#1) is obtained as follows: We draw the red ray from the top of the object parallel to the axis. The second red ray is determined by connecting its origin at the top of the diverging lens back towards the focal point at F. The first blue ray passes through the center of the diverging lens; its intersection with the second red ray continued to the focal point F determines the position of the top of image (#1).

This first, virtual, image serves as an effective object for the converging lens. We find the position of the ultimate image (#2) as follows: We draw the black parallel ray from the top of image (#1) to the top of the converging lens. This ray is continued through the focal point F' of the converging lens. We draw a second central ray from the top of image (#2); its intersection with the previous ray determines the position of the bottom of the final image (#2). The figure shows a third red ray from the converging lens to image (#2). The three red rays together form one actual path that starts from the object and ends up at the ultimate image. Another actual path is represented by the two blue rays. *Can you determine a third actual path?*

Here are the complex mathematical details: We use $(-\mathrm{f}_d)$ and $(-d_i)$ for the distances in the figure and Eq. (9.37) below because both are negative.

We can measure the parameters d_0, ℓ, f_c, and d_i' that are shown in the figure. The object distance of the converging lens is given by $(-d_I+\ell)$, so that

$$\frac{1}{-d_i+\ell}=\frac{1}{\mathrm{f}_c}-\frac{1}{d_i'} \tag{9.37}$$

We must solve this equation for the parameter d_i. Next, since we now know d_o and d_i, f_d can be determined from the lens equation applied to the diverging lens:

$$\frac{1}{\mathrm{f}_d}=\frac{1}{d_o}+\frac{1}{d_i} \tag{9.38}$$

9.11 The Doppler Effect

Have you ever paid attention to the sound of a car racing past you while you stand at the side of a

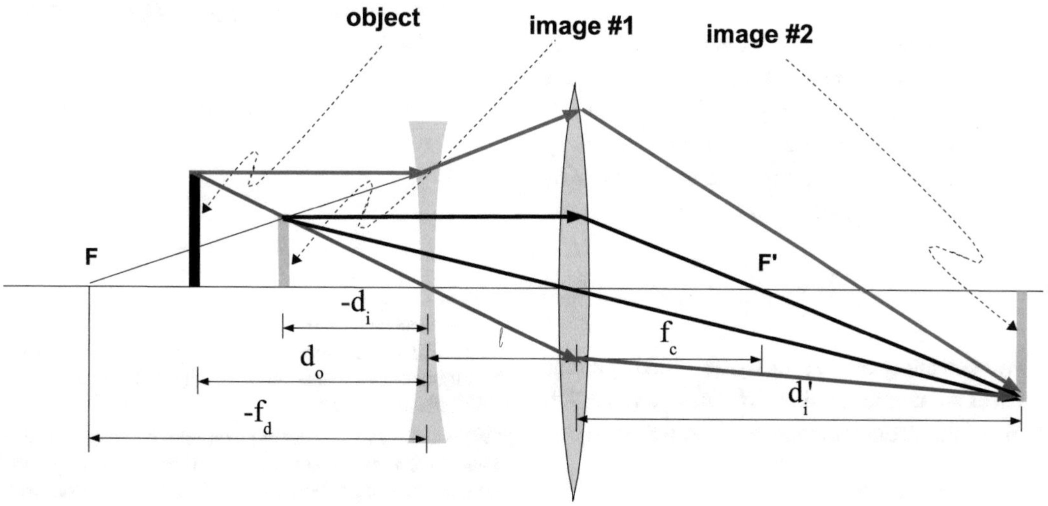

Fig. 9.52 Setup for determining the focal length of a diverging lens

highway? The whirring sound is noisy; nevertheless, one can discern a distinct pitch. This pitch steadily decreases as the car approaches and then recedes away. This effect is due to the **Doppler effect**. The Doppler effect occurs whenever there is relative motion between a source of waves and an observer of the waves, whatever the type of wave.

In essence, if the source emits a sine wave of frequency f, the observed frequency f' differs from the source frequency f. The ratio f'/f depends upon the motion of both the source and the observer.

We will deal only with the case that the source and observer are moving along a line joining the them. Thus, they are moving towards each other or away from each other. We will let their relative speed be u.

Suppose that the relative *speed* u of the source and observer is small compared to the wave velocity v. If they are **moving towards** each other, the approximate change in the frequency, $\Delta f \equiv f' - f$, is given approximately by

$$\Delta f \approx f\frac{u}{\mathrm{v}} \tag{9.39}$$

We then have for the relative change in frequency, $\Delta f/f$,

$$\frac{\Delta f}{f} \approx \frac{u}{\mathrm{v}} \tag{9.40}$$

Thus we see that the frequency increases. In particular, if the relative speed[16] is 1% of the wave velocity, the frequency will increase by 1%.

If the source and observer are **moving away from** each other, we merely have to replace u by $-u$ in the above equation, and obtain

$$\frac{\Delta f}{f} \approx -\frac{u}{\mathrm{v}} \tag{9.41}$$

We can apply our results to the whirring sound we hear as an automobile passes us by: When it is approaching from far away, we hear the relatively high frequency of fu/v. As it passes us by, its approach *towards* us is steadily reduced. And after it passes us by, it is moving more *away* from us, until its motion is essentially *directly away* from us. Thus, the observed frequency will begin at fu/v and decrease down to a final frequency of -fu/v.

[16]See the Appendix A for the definition of the **relative change of a parameter**.

9.11.1 Doppler Effect for Waves in a Medium

Consider the Doppler effect of sound waves in air. The source and/or the observer could be moving with respect to the air. We will see below that the observed frequency depends upon the motion of the source and the observer with respect to the medium.[17] We will discuss two simple cases.

Case (i): The Source Is at Rest with Respect to the Medium, While the Observer Is Moving with Respect to the Medium

A point source of sine waves will emit a wave with wave crests that are concentric spheres, as shown below. The distance between neighboring spheres is equal to the wavelength. The wave crests are traveling at a speed v with respect to the air, which is the medium in this situation.

We begin with the case that the observer is **moving towards** the source (Fig. 9.53).

The crests therefore **approach** the observer with a speed[18]

$$\mathrm{v}' = \mathrm{v} + u \tag{9.42}$$

[17]Since there is no medium for the propagation of light, the formulas below are not relevant for light. This case will be discussed later.

[18]We add the two speeds: For example, if you are running towards me at a speed of 10ft/s with respect to the ground and I am moving towards you at a speed of 1ft/sec with respect to the ground, you would be moving towards me at a speed of (10+1)=11-ft/s.

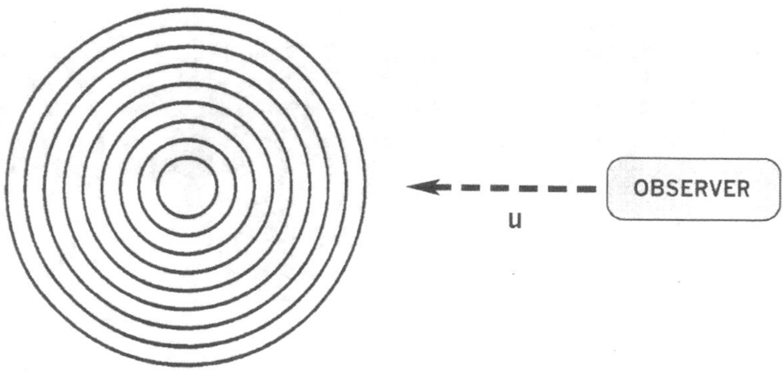

Fig. 9.53 An observer moving towards a stationary source of sound waves

The observed frequency f' is then given by

$$f' = \frac{v'}{\lambda} = \frac{v + u}{\lambda} = \frac{v}{\lambda}\left(1 + \frac{u}{v}\right) \tag{9.43}$$

Since $v/\lambda = f$, we have for a moving observer and stationary source

$$\boxed{f' = f\left(1 + \frac{u}{v}\right)} \tag{9.44}$$

We see that the observed frequency decreases when the observer is moving away from the source.

If the observer is moving away from the source, we merely have to replace u by $-u$ in the above equations, obtaining

$$v' = v - u \tag{9.45}$$

and

$$\boxed{f' = f\left(1 - \frac{u}{v}\right)} \tag{9.46}$$

Sample Problem 9.17 Suppose that a source is emitting a sound with a frequency of 1,000-Hz and is moving at a speed of 34.0-m/s with respect to the air, in which the sound velocity is 340-m/s. Find the observed frequency if the observer is moving directly towards the source and also if the observer is moving directly away from the source.

Solution

$f = 1{,}000\text{-Hz},\quad u = 34.0\text{ m/s},\quad v = 340\text{ m/s}$

We have $u/v = 0.1$.
Then if the motion is "towards":

$$f' = 1,000(1 + 0.1) = 1{,}100\text{-Hz}$$

If the motion is "away":

$$f' = 1,000(1 - 0.1) = 900\text{-Hz}$$

Case (ii): The Source Is Moving with Respect to the Medium, While the Observer Is at Rest with Respect to the Medium

Crests travel at the wave velocity with respect to the medium. However, because of the motion of the source, the wave crests are not concentric spheres. In Fig. 9.54, we see two sources, one moving at a speed $u < v$, the other at a speed $u > v$. In front of the source, the wavelength is decreased, while behind the source, the wavelength is increased.

The crests move at a speed v, the wave velocity, with respect to the medium and the observer. As a result, we can show that

$$\lambda' = \lambda\left(1 - \frac{u}{v}\right) \tag{9.47}$$

$$\lambda' = \lambda\left(1 - \frac{u}{v}\right) \quad \text{and } f' = \frac{v}{\lambda'} \tag{9.48}$$

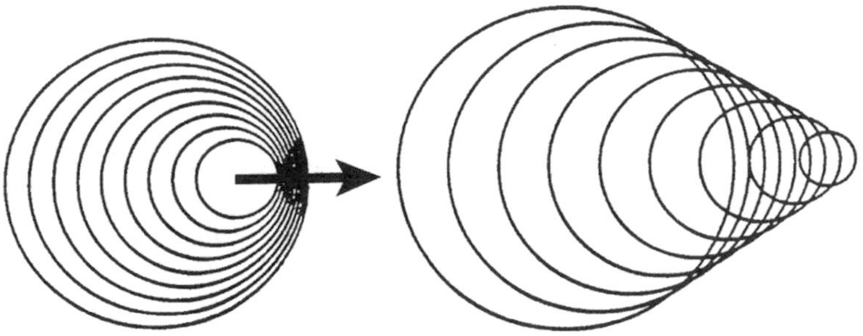

Fig. 9.54 The waves from a source moving at two different speeds, u. At the left, $u < v$; at the right, $u > v$

so that for a source a stationary observer

$$\boxed{f' = \frac{f}{1 - \frac{u}{v}}} \tag{9.49}$$

For a **source moving away** from a stationary observer we let $u \Rightarrow -u$ above and obtain

$$\boxed{f' = \frac{f}{1 + \frac{u}{v}}} \tag{9.50}$$

Sample Problem 9.18 Repeat the previous sample problem with the change that now the source is moving and observer is stationary. It is interesting to compare the results for the two cases. You should expect the difference to decrease as the ratio u/v decreases.

Solution Given f = 1,000-Hz, u = 34.0-m/s, and v = 340.0-m/s, find f' for the source moving away and coming towards you.

We have $u/v = 0.100$, so that

- If towards: $f' = 1000/(1 - 0.1) = 111$ Hz
- If away: $f' = 1000/(1 + 0.1) = 909$ Hz

Notes:

1. Note that the difference between these results and the previous results is on the order of 10%, corresponding to the ratio u/v=0.1.
2. It is interesting to compare the above numerical results. We have $u = 0.1$ v, so that speed $u \sim$ v. We can see that the frequencies f' do depend upon which of the two, the source or the observer, are moving with respect to the medium.
3. There is a **mathematical catastrophe** in the formula (9.49) for the case when the source is moving towards the observer at a speed which exceeds the wave velocity. In this case, we see that the frequency is negative and so is meaningless. [Moreover, when the source is moving at the wave velocity itself, the expression for the frequency diverges ($f \Rightarrow \infty$)!] What happens, in fact, is that there is a **shock wave** and no wave is perceived! Clearly, there cannot be a wave in front of a source if the source is moving faster than the wave would move. This situation is shown in Fig. 9.54b above.

There are numerous interesting phenomena connected with shock waves. We will discuss a few below.

1. When a jet plane accelerates through the speed of sound (known as **Mach one**), thus breaking the sound barrier, the production of the shock wave is accompanied by a loud sound. This sound was regarded as a great nuisance when commercial jet planes were first introduced.
2. The triangular trailing pattern of a boat moving along the surface of a lake is a beautiful example of a shock wave. One can sometimes observe swarms of minuscule, miniature insects swimming in a jagged manner along the surface of a pond leaving a feathery, impressionistic pattern of waves. Without these miniature shock waves, the insects would barely be noticeable!
3. The production of a **shock wave** is the principle behind the **whip**. A whip is made with a long piece of leather whose diameter is

tapered. In cracking a whip, a pulse is sent down the length of the whip from the thicker end. Now recall that the wave velocity along a taut string increases with decreasing linear mass density. [The process is complicated by a change in the tension too.] As the pulse proceeds, the wave velocity increases. By the time the pulse reached the tip of the whip, the wave velocity exceeds the speed of sound in the air. The resulting shock wave is responsible for the "crack" of the whip![19]

4. When a charged particle, such as an electron, is moving through a medium at a speed faster than the speed of light in that medium, light is given off within the conical trail of a shock wave of electromagnetic radiation. This light is called **Čerenkov radiation.**[20] See Fig. 9.55, wherein we see the blue Čerenkov radiation from charged fundamental particles moving in a nuclear reactor.

9.11.2 Doppler Effect for Electromagnetic Waves in Vacuum

Cases (i) and (ii) above cannot have meaning here because there is no medium. Only the **relative velocity** of the source and observer can matter. The correct formulas are obtained using **Einstein's Theory of Special Relativity**. They are (remember that v = c):

$$\boxed{f' = \sqrt{\frac{1 \pm \frac{u}{c}}{1 \mp \frac{u}{c}}}} \tag{9.51}$$

We use the upper signs if the source and observer are moving towards each other; we use the lower signs if the two are moving away from each other. The examples will make this point clear.

Sample Problem 9.19 Given:

$f = 1,000$ Hz , $u = 3.00 \times 10^7$ m/s, so that $u/c = 0.100$.

If "away"

$$f' = 1000\sqrt{\frac{1 + 0.100}{1 - 0.100}} = 1106 \text{ Hz}$$

If "towards"

$$f' = 1000\sqrt{\frac{1 - 0.100}{1 + 0.100}} = 905 \text{ Hz}$$

Note Suppose that we compare the results for EM waves with those for sound waves that neglect the effects of special relativity. We see that Eq. (9.51) is the square root of the product (hence the *geometric mean*) of the corresponding two expressions that hold for a medium, one when the observer alone is moving (9.46) and one when the source alone is moving (9.49). A similar relation holds when the source and observer are moving away from each other. The geometric mean of two numbers always lies between the two numbers.

Note For small relative velocities ($u/\text{v} \ll 1$) the value of f' is approximately the same whether we have a moving source or moving observer with sound waves or EM waves in a vacuum; that is, $\Delta f/f \approx u/\text{v}$.

9.11.3 Applications of the Doppler Effect

In the problems at the end of the chapter the reader will be shown how the Doppler effect can be used:

1. Have you ever wondered how police can determine the speed of vehicles by using radar? See problem 9.20 to learn how the Doppler effect and beats are involved.

[19]See http://www.hypography.com/article.cfm?id=32479 for a summary of recent research on the cracking of a whip. Also, see the following website for a discussion of how shock waves are the clue behind the trick for cracking a piece of wood with one's bare hand: http://www.worldkungfu.com/whip.html#WHIPS.

[20]The letter Č is pronounced like "ch" in "cheer."

Fig. 9.55 Blue Čerenkov light from a high flux isotope reactor (source: http://en.wikipedia.org/wiki/ Cherenkov_radiation)

2. Recall that we mentioned that astronomers and cosmologists have discovered that the Universe is expanding. How can they know this? They make use of the well-known spectra of atoms, whose wavelengths are known upwards of eight significant figures. The determination of the velocity of a star or galaxy with respect to the earth can made by measuring the shift in its atomic spectra. If a star is moving away from the earth, the frequency is lowered, so that colors change towards the red end of the visible spectrum. We have what is referred to as a **redshift**. An example is shown in the figures below from Palomar Observatory. In Fig. 9.56 we see the quasi-stellar radio source (quasar) 3C273.[21] It appears as a large bright star. In Fig. 9.57 the light spectrum from the quasar is shown above the spectrum of a stationary source of hydrogen and helium. The spectral lines labeled H_α, H_β , and H_γ are seen to be shifted to the right in the figure, corresponding to larger wavelengths. Calculations indicate that the quasar is moving away from the earth at about one-seventh the speed of light.

[21] This quasar is estimated as having a mass equal to about one-billion solar masses.

Note that in addition to a redshift due to a Doppler effect, there is also a **gravitational redshift**, which is akin to the slowing down of an object when we throw it up into the air.

Sample Problem 9.20 A train passes you by while sounding a whistle whose frequency varies from 550 Hz to 500 Hz . That is

$$f'_{\text{towards}} = 550\ \text{Hz}$$
$$f'_{\text{away}} = 500\ \text{Hz}$$

Assume that the speed of sound in air is 340 m/s. From this information you will be able to determine the speed of the train. We will show (see below) that it is given by

$$\boxed{u = \mathrm{v}\frac{\left(\frac{f'_{\text{towards}}}{f'_{\text{away}}} - 1\right)}{\left(\frac{f'_{\text{towards}}}{f'_{\text{away}}} + 1\right)}} \tag{9.52}$$

Thus the ratio of the two frequencies determines the speed of the train.

Note In Chap. 12 we will learn that the ratio of the two frequencies is directly connected to the corresponding **musical interval**. Thus, if you

Fig. 9.56 Quasar 3C273 (source: http://en.wikipedia.org/wiki/3C_273)

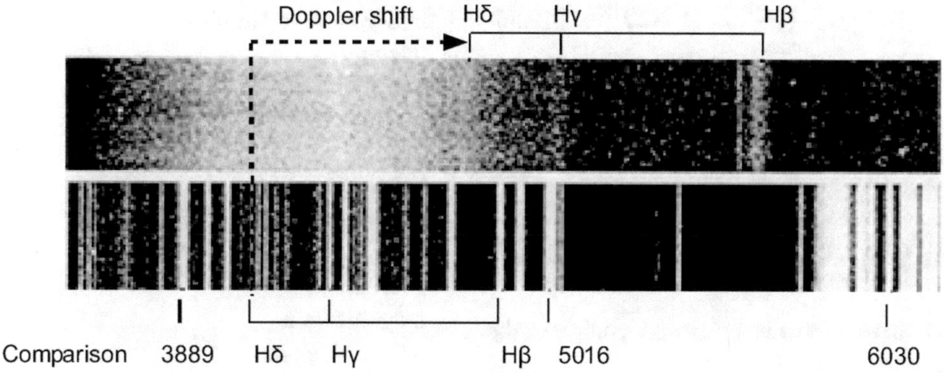

Fig. 9.57 Spectrum of the quasar (source: adapted from http://chandra-ed.harvard.edu/3c273/quasars.html)

recognize the musical interval between the maximum and minimum frequencies that you hear in the train whistle, you can determine the speed of the train! For example, a musical interval of a "**fifth**" has a frequency ratio of very close to 3/2. As a consequence, the train's speed is v/5.

$$\text{musical interval} \Longrightarrow \frac{f'_{\text{towards}}}{f'_{\text{away}}} \Longrightarrow \text{speed of train}$$

We want to

1. derive Eq. (9.52).
2. determine the speed u of the train.
3. determine the frequency f of the train's whistle.

Solution

1. **This is a complex algebraic problem** that requires many steps.

We will start by showing that

$$\frac{f'_{\text{towards}}}{f'_{\text{away}}} = \frac{(1+u/v)}{(1-u/v)} \quad (9.53)$$

We have for a moving source (the train)

$$f'_{\text{towards}} = \frac{f}{(1-u/v)} \quad (9.54)$$

and

$$f'_{\text{away}} = \frac{f}{(1+u/v)} \quad (9.55)$$

If we take the ratio of the left-hand sides of Eqs. (9.54) and 9.55 by each other, we obtain Eq. (9.53).

Next we multiply both sides of Eq. (9.53) by (1 − u/v) and obtain

$$(1-u/\text{v})\frac{f'_{\text{towards}}}{f'_{\text{away}}} = (1+u/\text{v}) \quad (9.56)$$

Let us now gather all terms that multiply u/v:

$$(u/\text{v})\left(\frac{f'_{\text{towards}}}{f'_{\text{away}}}+1\right) = \left(\frac{f'_{\text{towards}}}{f'_{\text{away}}}-1\right) \quad (9.57)$$

I leave it to the reader to complete the algebra to obtain Eq. (9.52).

Now we can obtain the speed u of the train. The ratio of frequencies is 550/500=1.10. Thus

$$\frac{u}{\text{v}} = \frac{\frac{f'_{\text{towards}}}{f'_{\text{away}}}-1}{\frac{f'_{\text{towards}}}{f'_{\text{away}}}+1} = 0.10/2.10 = 0.05 \quad (9.58)$$

2. Finally we obtain the speed of the train, u=(0.05)(340)=17-m/s.
3. From Eq. (9.55)

$$f = (1+u/\text{v})f'_{\text{away}} = 1.05 \times 500 = 525\text{-Hz} \quad (9.59)$$

9.12 Terms

- angle of incidence
- angle of reflection
- angle of refraction
- birefringence
- central ray
- chromatic aberration
- concave lens
- converging lens
- convex lens
- diffraction grating
- diffraction
- diffraction angle
- diffuse reflection
- dispersion
- diverging lens
- Doppler effect
- fiber optics communication
- focal length
- handedness of our biosphere
- image
- image distance
- image point
- impedance
- index of refraction
- length scale of roughness - l_r
- lens
- lens axis
- magnification
- magnifying glass
- minimum image diameter
- mirage
- mirror image
- object distance object point
- one-to-one correspondence
- parallel ray plane wave
- prism
- real image
- reflectance
- reflection of a wave
- refraction
- scattered wave
- scattering of a wave
- Snell's Law
- specular reflection

- spherical aberration
- thin lens approximation
- thin lens equation
- transmittance
- virtual image
- wave crests

9.13 Important Equations

Condition for specular reflection:

$$\lambda \gg \ell_r \tag{9.60}$$

Condition for diffuse reflection:

$$\lambda \ll \ell_r \tag{9.61}$$

Speed of light in a medium of index of refraction n:

$$v = \frac{c}{n} \tag{9.62}$$

Snell's Law:

$$n_1 \sin\theta_1 = n_2 \sin\theta_2 \tag{9.63}$$

Generalized law for refraction:

$$\frac{\sin\theta_1}{\sin\theta_2} = \frac{v_1}{v_2} \tag{9.64}$$

Equation for the critical angle for total internal reflection:

$$\sin\theta_c = \frac{n_1}{n_2} \tag{9.65}$$

Thin lens equation:

$$\frac{1}{d_O} + \frac{1}{d_I} = \frac{1}{f} \tag{9.66}$$

Magnification vs. object/image distances:

$$M \equiv \frac{h_I}{h_O} = \left|\frac{d_I}{d_O}\right| \tag{9.67}$$

Reflectance vs. indices of refraction:

$$R = \left(\frac{n_2 - n_1}{n_2 + n_1}\right)^2 \tag{9.68}$$

Reflectance in terms of the wave velocities:

$$R = \left(\frac{v_2 - v_1}{v_2 + v_1}\right)^2 \tag{9.69}$$

Reflectance of sound between two media

$$R = \left[\frac{Z_2 - Z_1}{Z_2 + Z_1}\right]^2 \tag{9.70}$$

Approximate equation for the Doppler effect:

$$\frac{\Delta f}{f} \sim \frac{u}{v} \tag{9.71}$$

9.14 Questions and Problems for Chap. 9

1. When the dimensions of scattering particles are smaller than the wavelength of blue light, [CHOOSE ONE]
   (a) red light is scattered more effectively than blue light.
   (b) both red and blue light are scattered about equally and better than green light.
   (c) both red and blue light are scattered about equally but not as well as green light.
   (d) blue light is scattered more effectively than red light.
2. (a) Red light diffracts :more/less: than blue light.
   (b) A soprano's voice diffracts more/less: than a bass's voice.
3. Which voices will be heard better through a crack in a door, high pitched or low pitched ones? Explain.
4. Which radio waves will be more easily picked up at large distances over hilltops, AM or FM waves?
5. The animal called the **bat** is not blind[22] However, it hunts at night and makes use of **echo-location**, wherein an ultrasonic series

[22] See Wikipedia (1-11-2011): http://en.wikipedia.org/wiki/Bat.

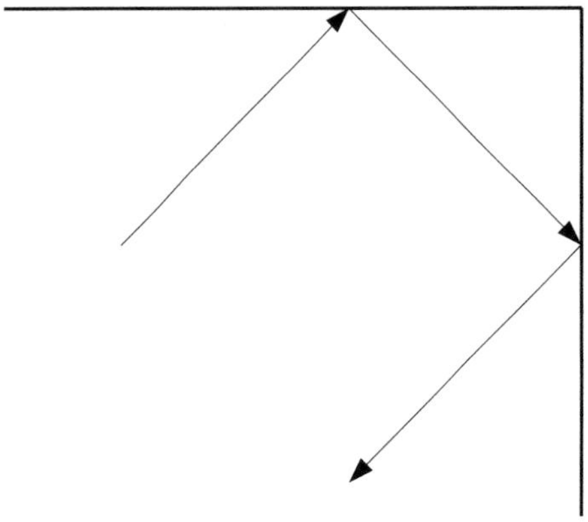

Fig. 9.58 Light ray reflecting off two perpendicular mirrors

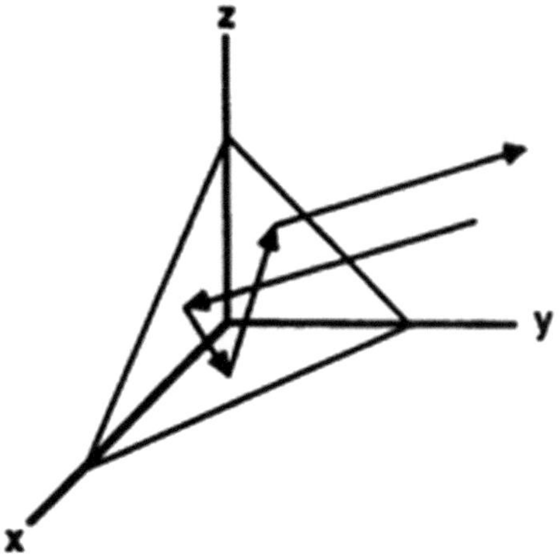

Fig. 9.59 Light ray reflecting off three perpendicular mirrors

of pulses are emitted and their reflection used to locate a prey.

(a) Why must the wavelength of the sound be much smaller than the object size for the bat to have a clear "image" of the object?

(b) Estimate the minimum frequency of sound necessary for the bat to discriminate objects with a resolution on the order of 1 mma.

6. What two physical phenomena account for the ability of a prism to analyze light?
7. Will an FM radio wave with frequency 89.3MHz be reflected **specularly** or **diffusely** off a field of corn?
8. A laser beam starts out on earth with a diameter of 2 mm. Find the diameter of the image of the laser beam on the moon, a distance 400, 000 km away. The laser's wavelength is 6300 $\AA$.
9. A laser beam is to produce the smallest possible image on a screen one km away. Its wavelength is 5500 $\AA$. What should the size of the aperture be? What will be size of the image?
10. A light ray is reflected off two mirrors that are at right angles with respect to each other. Prove that the reflected ray is parallel to the incident ray. See Fig. 9.58.
11. A light ray is reflected off three mirrors that are mutually perpendicular to each other, as shown in Fig. 9.59. Prove that the outgoing ray is parallel to the incoming ray.

Retroreflectors are extremely important in a number of areas. An example is shown in Fig. 9.60, where we see two bicycle reflectors. Another example is the retroflection of a radar beam off an object in space, even as far as the Moon. It can be used to determine the distance to an object to an extremely high accuracy. This application is called **laser ranging**.[23] The earth–moon distance has been measured to 1/10 of a km out of a total of 385,000-km.

12. A person is standing at a distance of five feet from a mirror. What is the apparent distance of the person's image from the person?
13. Suppose we have a laser beam that has a power of 5-mW and a beam area of 10-mm^2.

(a) Calculate the intensity of the beam.

Now suppose that the beam is incident upon a wall that reflects the light **diffusely**. In this case all people looking at the wall will see the image of the beam cast upon the wall. Estimate the intensity of the beam that you would see if you were to look at the image from a distance of 2-meters from the wall.

[23] See https://en.wikipedia.org/wiki/Lunar_Laser_Ranging_experiment for a fascinating description of the application.

Fig. 9.60 Bicycle retroreflectors.Source: https://en.wikipedia.org/wiki/Retroreflector#/media/File:BicycleRetroreflectors.JPG

(b) Proceed by assuming the reflected light can be approximated by the light emitted by a point source at the wall, uniformly in all **forward** directions.

(c) What is the ratio of the intensity of light at the point of observation to the incident intensity?

Note There are guidelines for avoiding damage to our eyes from an incident laser beam. A 5-mW laser should not be viewed head-on—so-called **intra-beam viewing of a laser beam**—for more than the duration of the blink of an eye. That interval of time is about 0.25-s. An increased power or exposure time can cause permanent damage. Nevertheless, a 5-mW laser with a beam diameter of a few millimeters is readily available in toy stores and can be purchased for a few dollars! Unfortunately, such toys are given to children without knowledge as to how dangerous they can be.

14. (a) Suppose that you look at the number "2" directly on a piece of paper. Now look at the number "2" as seen in a **mirror** by placing the paper in front of the mirror.

Examine Fig. 9.61. Which of the images—(a), (b), (c), or (d)—will be the image of the number "2"?

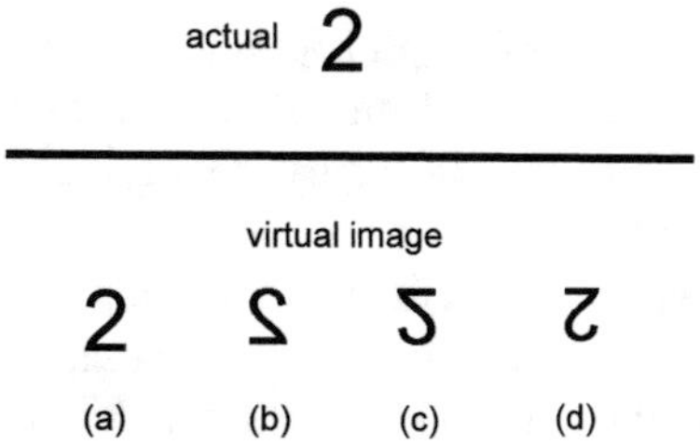

Fig. 9.61 Images of the number 2

(b) Now look at the number "2" as seen by placing a **converging lens** between the paper and you so as to produce a **virtual image**. Which of the four images in Fig. 9.61 will be the image of the number "2"?

15. A popular device for producing an image of an infinite series of replications of actual objects uses a mirror plus a sheet of glass that is coated so as to have a very low transmission. We see one of these devices in Fig. 9.62. The only lights that are actually present are those on a circle near the perimeter. All the other concentric circles of lights are virtual.

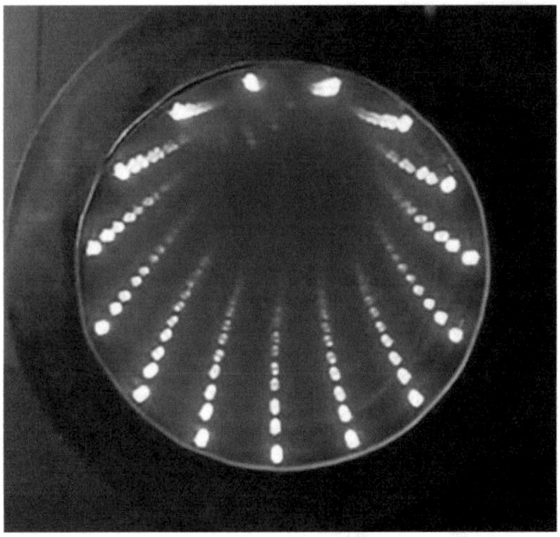

Fig. 9.62 A device that produces an infinite series of images (source: http://www.youtube.com/watch?v=VTONKZkaVX4&NR=1&feature=fvwp)

Use physical principles to account for the infinite series of images as well as the fact that the images appear to be increasingly distant.

16. Suppose that a light beam in air is incident on an air–glass interface. Suppose, also, that the glass has an index of refraction of 1.5.

    Show that the reflectance R at the interface is 4%.

17. Consider an aluminum rod that is struck at one end so as to produce a sound wave that travels down the rod. When the sound meets the other end, there is reflection and transmission. Here, the interface between the rod and the air might be a square 1-cm on a side. For this problem, as an approximation, you are to use (9.16), which is valid for an infinite interface.

    Calculate the reflectance and transmittance at the steel–air interface. The sound impedance Z of air is given by $Z = \rho v$=(1.3kg/m^3) (340m/s)=440 kg/m^2-s. The sound impedance of steel is[24]

$$Z = \rho v = (2700\text{kg/m}^3)(5100\text{m/s})$$
$$= 1.4 \times 10^7\text{kg/m}^2 - \text{s}.$$

18. Let I_{trans} = transmitted intensity and I_{inc} = incident intensity. Then the transmittance T and reflectance R are given by:

$$T = \frac{I_{\text{trans}}}{I_{\text{inc}}} = 1 - R \tag{9.72}$$

    (a) **Express** the transmittance of a light wave in terms of the **indices of refraction**. See Sect. 9.3.1.
    (b) **Express** the transmittance of a sound wave in terms of the **impedances**.
    (c) Suppose that my voice produces a sound level of 40 dB at the surface of a lake.
        i. What is the reflectance? Assume a normally incident sound wave, so that you can use the results of (a) and (b) above.
        ii. Find the sound level of the transmitted sound.

19. (a) Prove that the reflectance R of light can be expressed as:

$$R = \left[\frac{v_2 - v_1}{v_2 + v_1}\right]^2 \tag{9.73}$$

    Note that this expression can be rewritten in terms of the *ratio* v_2/v *alone*:

$$R = \left[\frac{v_2/v_1 - 1}{v_2/v_1 + 1}\right]^2 \tag{9.74}$$

    (b) **Prove** that the reflectance for **sound** can be written as:

$$R = \left[\frac{\frac{Z_2}{Z_1} - 1}{\frac{Z_2}{Z_1} + 1}\right]^2 \tag{9.75}$$

[24] As discussed in Chap. 3, the speed of sound in a bulk sample of a medium, that is having no boundaries, is given by $\sqrt{B/\rho}$, where B is the bulk modulus. When sound travels down a solid pipe of material, the walls of the pipe are free to expand outwards without any material. As a result, one has to use a difficult modulus, called the Young's modulus. For aluminum, this modulus differs less than 10% from the bulk modulus.

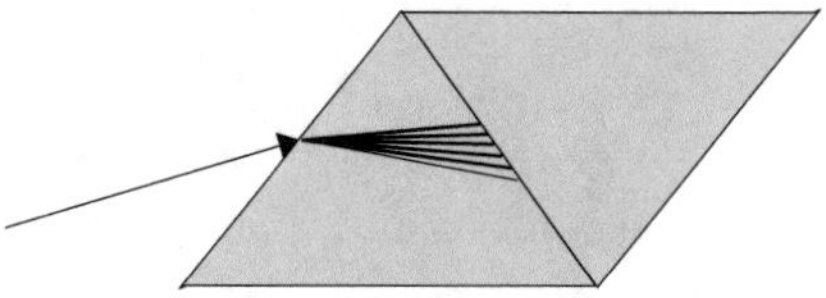

Fig. 9.63 Dispersion with two prisms in tandem

Only the ratio of the impedances, $Z \equiv \rho v$, appears in the relation.

20. For flint glass, the critical angle is 37°. Thus,
    (a) light incident on the glass from the air with an incident angle larger than 37° will be totally **refracted**.
    (b) light incident on the glass from the air with an incident angle larger than 37° will be totally **reflected**.
    (c) light incident on the air from the glass with an incident angle larger than 37° will be totally **reflected**.
    (d) light incident on the air from the glass with an incident angle smaller than 37° will be totally **reflected**.
    (e) light incident on the glass from the air with an incident angle smaller than 37° will be totally **reflected**.
21. We know when a ray of white light is incident upon a prism, dispersion will lead to an outgoing beam with a rainbow of colors, each component wavelength traveling in a different direction. Suppose that a ray of white light is incident upon a pair of prisms in tandem, as shown in Fig. 9.63. Describe the outgoing beam of light.
22. A light ray, traveling in water, is incident on a water–glass interface at an angle of incidence of 30°. Find the angle of refraction if the index of refraction of the glass is 1.5.
23. Find the critical angle for total reflection of a light wave in a glass having an incident of refraction of 1.7 at an interface with air.
24. Find the **critical angle** for total reflection of a sound wave in air incident on a water surface.
25. Consider a laser beam in air, region #1 with wavelength λ=600-nm incident normally on an interface with **water**, having an index of refraction 1.33. Under the water, region #2, there is a diffraction grating with 6000 grooves per cm. See Fig. 9.64.

    Determine the angle θ_1 of the first order of interference.

    Beware that you will need the wavelength of the laser light in the water, not the air. Keep in mind:

    Note The frequency of a wave doesn't change upon passing from one medium to another. A justification for this fact is given in footnote 7 of Sect. 9.6.

26. Figure 9.65 is a diagram of an object (the vertical black rectangle) that is to the left of a lens.

    We see two rays that leave the top of the object and strike the lens at two positions along the lens. The parallel ray is shown passing through the focal point. A second ray in red is shown striking the lens.
    (a) Is the lens converging or diverging? Explain.
    (b) Complete the diagram by showing where the image is formed.
    (c) Show where the red ray progresses after it exits the lens and where it appears to be coming from the top of the image.
    (d) Is the image real or virtual?
    (e) Suppose that the object distance is $0.5|f|$, where f is the focal length. Determine the image distance in terms of f.

        Why did I insert an absolute value sign around f?
27. Figure 9.66 shows an "air lens" under water. It consists of a balloon in the shape of a convex lens and filled with air.

    Where would the image of the object be located? To answer this question, think of the way a light beam will be refracted in passing from water into air and from air into water. Here are your choices:
    (a) Between the object and the focal point, F to the left of the lens.

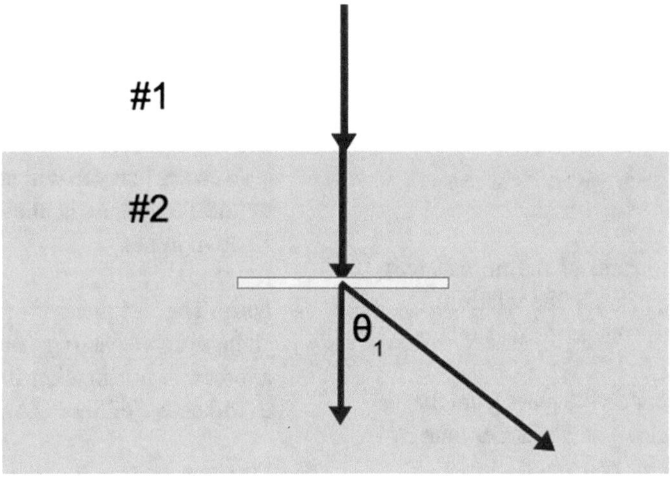

Fig. 9.64 UnderwaterGrating

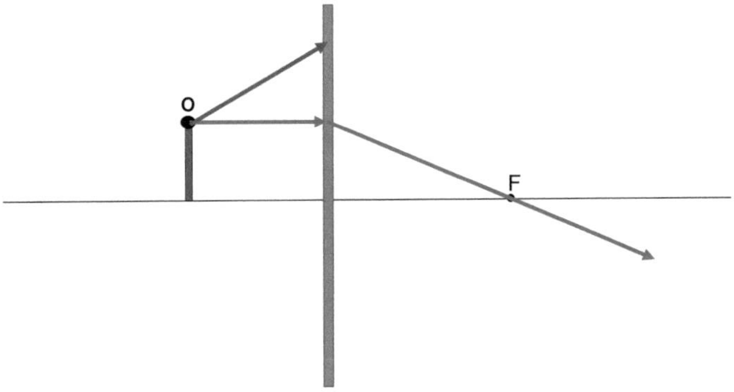

Fig. 9.65 Ray diagram of a lens

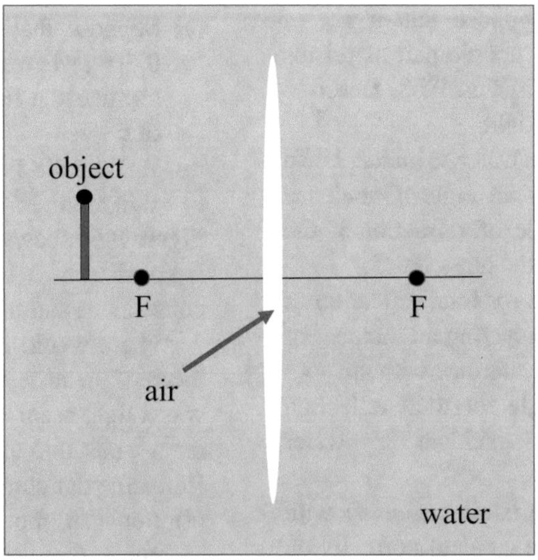

Fig. 9.66 Air lens in water

(b) Between the focal point F to the left of the lens and the lens
(c) On the right side of the lens.
(d) To the left of the object.
(e) There would be no image for this type of lens

28. Suppose that a piece of opaque material is placed just in front of the lens of a slide projector, thus blocking the light emanating from the *top half* of the lens. Describe the effect on the image on the screen. Choose from the following:
    (a) The top half of the image be removed.
    (b) The bottom half of the image be removed.
    (c) Neither of the above be true. [Describe very qualitatively what one should observe].

    HINT: Use the fact that there is a one-to-one correspondence between an object point and an image point. In fact, the image point is produced by the set of all rays that leave an object point and pass through the lens.
29. Suppose that you look at an object through a **converging lens** having a focal length of +4 cm. The object is placed at a distance of 7 cm from the lens.
    (a) Find the image distance.
    (b) Is the image erect or inverted? Real or virtual?
30. Explain how reversibility applies to refraction, the thin lens equation, and to reflectance.
31. Suppose that you look at an object through a **converging lens** having a focal length of +4-cm. The object is placed at a distance of 3-cm from the lens.
    (a) Find the image distance.
    (b) Is the image erect or inverted? Real or virtual?
32. A convex lens of focal length 20-cm receives light rays from an object with object distance 30-cm. Find the image distance. Is the image real or virtual? Erect or inverted? Can a person see the image viewed from the side opposite to the object?
33. Suppose that you look at an object through a **diverging lens** having a focal length of −4 cm. The object is placed at a distance of 7 cm from the lens.
    (a) Find the image distance.
    (b) Is the image erect or inverted? Real or virtual?
34. Suppose that a person five feet in height is at a distance of 25 feet from a convex lens having a focal length of 0.4 feet. Find the image distance and the image height. Is the image real or virtual? Erect or inverted?
35. Snell proposed his empirical law without mention of the sine function. Keen observation and study led him to the following result: In Fig. 9.67 we see an incident ray refracted when passing from medium #1 into medium #2. We see the usual normal through the point of entry of the ray at point P, along with a second normal to the right.

    According to Snell, the ratio of the lengths PA to PB is the same for all pairs of rays and a given pair of materials.

    Prove that this statement is equivalent to the expression for Snell's Law in terms of sines.
36. Discuss how **reversibility** applies to **refraction**, the **thin lens equation**, and **transmission** of sound or light from one medium to another.
37. Suppose that you are submerged in a pool of water with an absolutely calm surface. Above the surface the room is full of light. The walls and floor are perfectly black, so that they absorb all light completely, and do not reflect any light back into the water. What will you see when you look up at the surface? [You will *not* see a fully lit surface.]

    HINT: Use the **reversibility** property of rays that are refracted. Consider the paths of all rays of light that can emanate from the eye and emerge from the water into the air above. This set of rays is the same as the set that can enter from the air outside the water and reach the eye (Fig. 9.68).

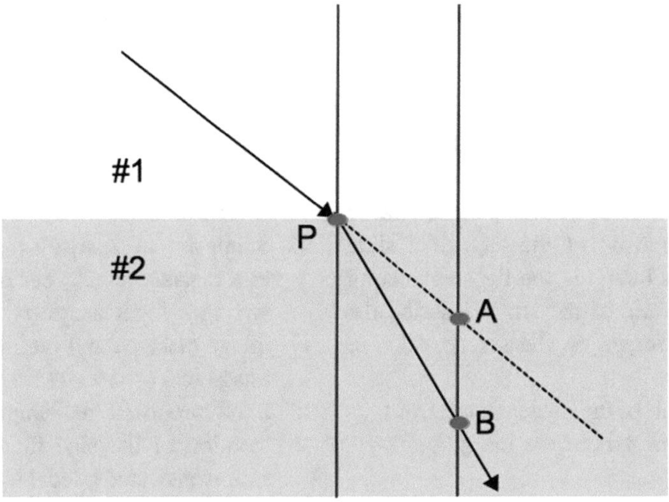

Fig. 9.67 Snell's analysis of refracted rays

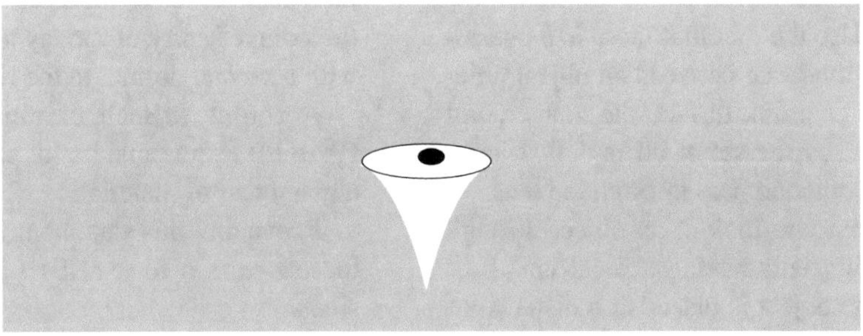

Fig. 9.68 Eyeball underwater looking up towards the surface of a swimming pool

38. On a road, John drives toward Marsha tooting his horn. Marsha immediately notices that
    (a) John looks slightly bluer than normal.
    (b) John looks slightly redder than normal.
    (c) The horn is pitched a little higher than normal.
    (d) The horn is pitched a little lower than normal.
39. The atomic spectrum of hydrogen is observed in the light coming from a star. If the star were not moving (relative to the earth), the *red* line would be observed to have a wavelength of 6560Å. Instead, the *observed wavelength* is 6562Å.
    (a) Is the star moving towards or away from the earth?
    (b) Find the speed u of the star relative to the earth, assuming it is moving along a line joining it to the earth. You may assume that $u \ll c$ and therefore use Eq. (9.41).
40. A train is moving *towards* you at a speed of 100-km/h while sounding a whistle having an intrinsic frequency of 600-Hz..
    (a) Using a sound velocity of 340-m/s and the small speed approximation (Eq. (9.41)), calculate the frequency of sound that you will perceive.
    (b) Repeat the above if the train is moving away from you.
41. Repeat the previous problem with the change that the train is stationary and the observer is a person driving in a car **towards** the train or, in the second calculation, **away from** the train at 100-km/h.

42. Here is how a radar device is used to determine the speed of a car: The device sends a radar signal of frequency 10-GHz. (= 10^{10}-Hz.) towards the car. Given that the car is moving directly towards the source, the car receives a Doppler shifted frequency f' with respect to the car. In turn, the car becomes a source in sending back radar waves to the device. Since the car is moving towards the device (which has a radio transmitter and a radar receiver), the frequency observed (measured) by the receiver will be a Doppler shifted frequency f'' of f'. We have **two** Doppler shifts. Given that $u << c$, we can use Eq. (9.41) except that we must double the shift. Thus, we obtain

$$\Delta f = f'' - f = (f'' - f') + (f' - f) = \frac{2u}{c} f \tag{9.76}$$

Now suppose that f'' and f produce a beat frequency of 2000-Hz.

Find the speed of the car.

43. The Doppler effect of ultrasound is used to determine the flow velocity of blood in a blood vessel; this is especially useful in determining whether there is a blood clot impeding flow in an artery of a leg.
    (a) Assuming that the sound velocity in the body is about 1500 m/s and that one needs a resolution of an image to be about 1mm, what should the minimum wavelength of the ultrasound be and the corresponding frequency?
    (b) Suppose that the frequency used is that obtained in the previous problem. Assume also that the beat frequency is 1Hz between the frequency of the incident sound and the frequency of the reflected sound. Determine the velocity of the blood in the artery if the blood is flowing directly towards the source of sound.

The Ear 10

SOUND

Insects one hears
and one hears the talk of men—
with different ears

Haiku by **Masaoka Shiki** (1867–1902)

We have studied the nature of sound and how sound waves are produced and propagate through media. However, the focus of this text is on sound as experienced by people. Sounds of insects, of men, and of a multitude of sources reach our ears, perhaps providing us with our principal means of communication with the outside world. How can we hear these sounds "differently"? What happens to the sound that enters our ears? What is the essence of hearing? What is the source and explanation for the pleasure we have in hearing beautiful music or for the annoyance at hearing loud or dissonant noise? Many a reader might hope that science can arrive at answers to such questions. Unfortunately, science is severely limited in this domain.

Hearing begins with the ears and ends with the **brain**. The ear is the organ that is used to gather a sound wave and convert the waveform of the sound wave, as faithfully as possible, into **nerve signals** that travel to the brain to be analyzed and interpreted. Our mode of hearing is determined in most instances by the manner in which the brain analyzes the auditory nerve signals it receives.

In addition, the brain has been shown to have the remarkable capacity to alter the physical state of the ear and hence its manner of converting sound into nerve signals. In any case, the physical distinction between the brain and the ears exists: The poetical "ears" in Shiki's Haiku include the brain, while the physicist's "ears" do not.

This chapter is concerned with the physics of the **human ear** – that is, the physical processes whereby the ear converts a sound wave into nerve signals. We do not discuss how the brain analyzes these nerve signals. This subject is beyond the scope of this text. Briefly, the ear can be compared to a highly sensitive microphone, capable of responding to a range of frequencies from 20 to $20,000$ Hz and to a range of intensities spanning twelve orders of magnitude, with a high efficiency and extremely low level of distortion. A knowledge of how the ear functions enables one to partially understand how what we hear is related to the sound incident upon our ears. This is the subject of Chap. 11. In particular, we will be able to qualitatively account for the response characteristics of the ear and our ability to discriminate pitch. Most significantly, in Chap. 11, we will be able to provide a qualitative basis for the existence of consonant and dissonant musical intervals and the related perception of **combination tones** which are certain sounds that a person can hear even though they are not present in the sound wave incident upon the ear.

L. Gunther, *The Physics of Music and Color*,
https://doi.org/10.1007/978-3-030-19219-8_10

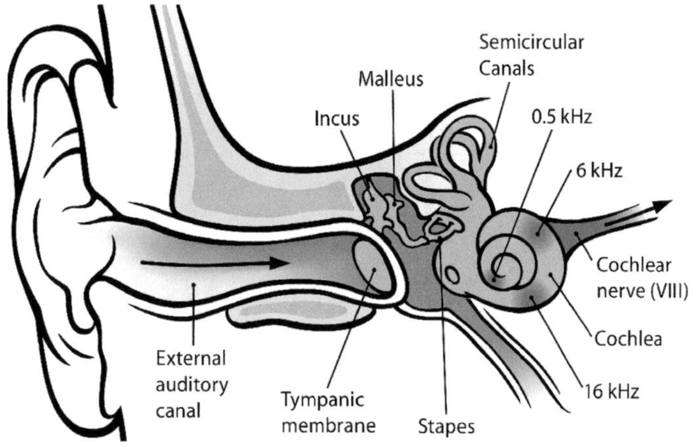

Fig. 10.1 The human ear (source: Chittka L, Brockmann A (2005) Perception Space. The Final Frontier. PLoS Biol 3(4): e137. doi:10.1371/journal.pbio.0030137; Creative Commons Attribution 2.5)

10.1 Broad Outline of the Conversion Process

Figure 10.1 is a drawing of the human ear.[1] The **pinna**, or outer ear, serves to funnel a sound wave into the auditory canal. The wave travels down the canal, at the end of which it sets the **tympanic membrane** (also referred to as the **eardrum**) into motion. The eardrum in turn sets a system of three bones—the **ossicles**—into motion. At the other end of the ossicles is the **footplate**, which is the base of the **stapes** (also referred to as the **stirrup**). [See Fig. 10.5 for more details.] In fact, the footplate covers the hole shown in the snail shaped **cochlea**. This hole, the **oval window**, leads into the inner chamber of the cochlea that is filled with a fluid—the **cochlear fluid**—that is set into motion by the vibrations of the footplate. The cochlea contains the neural sensors that transmit the information about the sound to the brain.

In Fig. 10.2 we see a drawing of the ossicles, while in Fig. 10.3 we see a closeup photograph of a plastic replica of the actual ossicles alongside a centimeter ruler.

To view the replica directly with your eyes can be quite breathtaking: These bones, which are an essential instrument for transmitting the wonderful sounds we hear, over an incredible range of intensities and with great fidelity, are puny and delicate. Equally amazing is the strong evidence that the three bones evolved from a combination of parts of the gills of a fish and the jawbones of reptiles.[2]

In Fig. 10.4 we see a schematic of the entire ear, along with a cut away to see inside the cochlea. This figure was produced for the purpose of showing how a cochlear implant works. See later in this chapter for more information.

One may well wonder why so many steps are involved in the conversion process from sound to nerve signals. Why hasn't the ear evolved so that

[1]The figure has been reproduced in black and white.

[2]Here is an excerpt from the website http://museum.utep.edu/archive/biology/DDossicles.htm:

> Hearing is a wonderful thing, able to translate vibrations of air into sound. Numerous desert creatures rely on sound more than on sight, for many are nocturnal, only active during the dark hours. Part of the great sensitivity in mammals is due to the three small bones in our middle ears, the auditory ossicles. These transmit and amplify the vibrations of the ear drum, conveying them to the inner ear.
> What is fascinating from an evolutionary viewpoint is their origins. Studies of embryos and fossils trace their origin far back in time. The ossicle next to the inner ear, the stapes, can be traced back to part of a gill arch in a very distant fish ancestor. The other two ossicles, the malleus and incus, are derived from bones that, in our reptilian ancestry, formed the joint between skull and lower jaw, as they do today in modern reptiles. After incipient mammals evolved a new jaw joint, those bones were in perfect position to be incorporated into a hearing device in evolution's favorite avenue–jury-rigging structures for new roles.

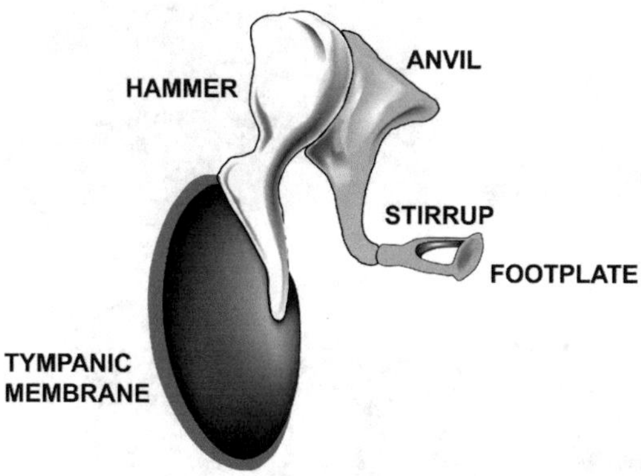

Fig. 10.2 Ossicles—details (source: courtesy of Russ Dewey)

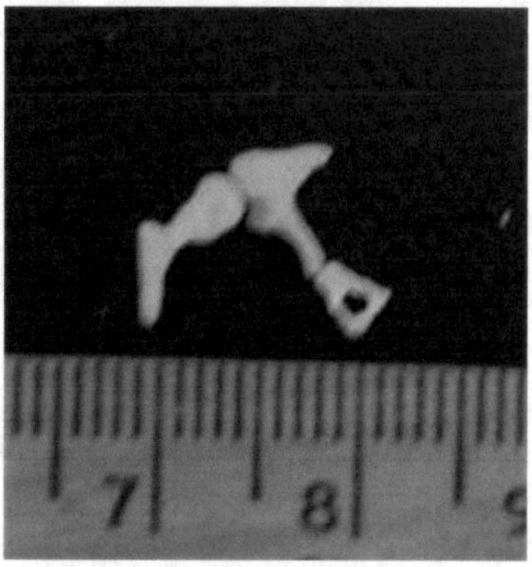

Fig. 10.3 Photo of replica of the ossicles (photo: Leon Gunther)

an incident sound wave will be incident directly upon the oval window? The primary reason is that of a very poor matching of the **impedances** of the air and the cochlear fluid.

The **impedance** Z is the product of the mass density ρ and the speed of sound v. Thus,

$$Z = \rho v \tag{10.1}$$

It determines the resilience of a medium to being disturbed by a change in the external pressure. In Sect. 9.3.2 we discussed its relevance in the reflection of sound at an interface between two media: When a sound wave is traveling in one medium and is then incident upon an interface with a second medium, a certain fraction of sound energy will be reflected and a certain fraction will be transmitted. In Eqs. (9.14) and (9.16) of Chap. 9 we have an expression for the reflectance R in terms of the impedances. With $Z = \rho v$, we have

$$R = \left(\frac{Z_2 - Z_1}{Z_2 + Z_1}\right)^2 \tag{10.2}$$

When the two impedances are equal, the reflectance vanishes and there is total transmission. Correspondingly, given that the transmittance T is given by

$$T = 1 - R \tag{10.3}$$

a bit of algebra leads to

$$T = \frac{4Z_2Z_1}{(Z_2 + Z_1)^2} \tag{10.4}$$

This equation can be rewritten in a different algebraic form that is illuminating. If we divide both numerator and denominator by Z_1^2 we obtain

$$T = \frac{4Z_2/Z_1}{(Z_2/Z_1)^2 + 1} \tag{10.5}$$

Thus we see that the **ratio of the two impedances determines the transmittance**. We already observed that the reflectance vanishes when the two impedances are equal, do that their ratio is one. For all other ratios, larger or smaller than one, the transmittance is less than one. The closer the ratio of the respective impedances of the media is to one, the closer will the transmittance be to unity. For example, a ratio of $\frac{1}{2}$ (or 2) leads to a fraction $8/9$ transmitted. The ratio for cochlear fluid to air is about $3300 : 1$, which results in only a fraction of about one part in $1,000$ of the sound energy transmitted, corresponding to a decrease in sound level of 30 dB.

We will see how the eardrum and the middle ear serve to increase the fraction of sound transmitted to the cochlear fluid.

10.2 The Auditory Canal

The **auditory canal** is a crooked tunnel (not as straight as Fig. 10.1 would suggest) one of whose functions is to protect the delicate eardrum from injury. It also serves as a resonator which aids in reducing the fraction of sound reflected back into

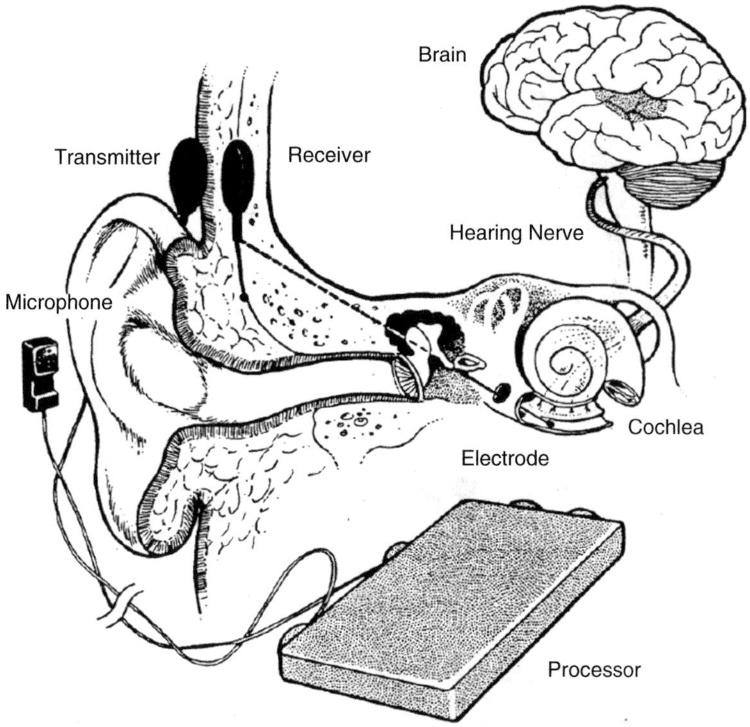

Fig. 10.4 Cochlear implant (source: Lahey Medical Center Journal, Burlington, MA)

the air. For an approximate calculation of its resonance frequencies it is adequately represented by an open-closed tube, with a length of about 2.7 cm and a diameter of about 7 mm. Suppose we assume that the temperature in the canal is 30°C (which lies between room temperature and a body temperature of 38°C). In problem 9.1 it is shown that the speed of sound at this temperature is 346 m/s and that the fundamental frequency f_1 of the tube is 3, 200 Hz. The overtone frequencies are $3f_1 = 9,600$ Hz, $5f_1 = 16,000$ Hz, etc.

It has been shown that the fundamental resonance provides an amplification of the intensity by a factor of 3–10 for frequencies between 2 kHz and 5 kHz. It is therefore no mere coincidence that the ear as a whole is most sensitive to sound waves with a frequency of about 3 kHz (see Sect. 11.1).

10.3 The Eardrum

Before we go into details, it is important to remember that it is the *difference* between the pressure on the outside, exposed side of the eardrum and the pressure within the ear, that leads to the net force on the eardrum. This difference is the **sound pressure**. The pressure within the ear is ideally maintained at the **ambient pressure**—that is, the pressure in the absence of the sound wave and therefore normally about one atmosphere. This is achieved by air inside of the ear being contiguous with the outside air via the **Eustachian tube**. [See Fig. 10.1.]

The eardrum is a very delicate membrane that serves to gather up sound energy and transmit it further on into the ear. It is oval in shape, having dimensions of approximately 9 mm by 12 mm. It also provides the major means of overcoming the **mismatch of impedances** by virtue of the fact that its area is about fifteen times that of the oval window. As a result of the eardrum alone, the **sound pressure** acting on the oval window would be about fifteen times that acting on the eardrum by the sound wave.

This effect can be understood qualitatively by considering the relative ease with which we can push a nail into the ground as opposed to pushing a block of wood into the ground. The force available is the same in both cases. However, we

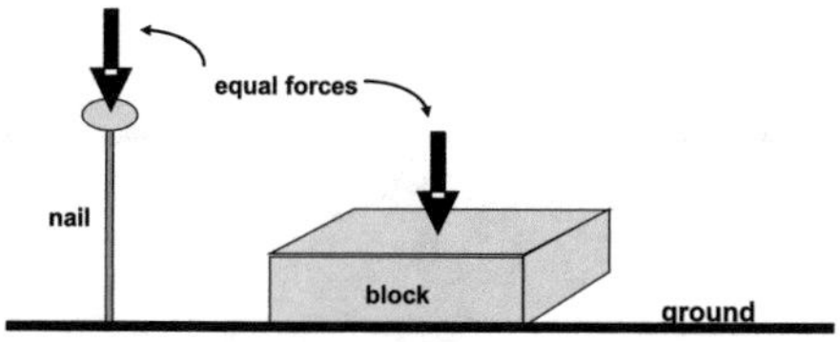

(a)

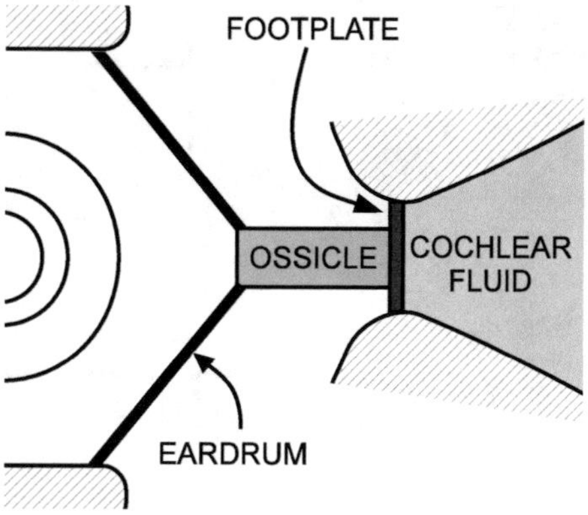

(b)

Fig. 10.5 Abstract diagram of the eardrum. (**a**) Though you have the same force, it is easier to push the nail into the ground, since it can assert more pressure. (**b**) The force of the sound wave is concentrated to a small area before entering the ossicle

get a greater pressure by the nail on the ground because the area of contact between the nail and the ground is much smaller than the area of contact between a block of wood and the ground. (See Fig. 10.5.)

10.4 The Ossicles

The ossicles, which are depicted in Fig. 10.2, consist of a set of three bones, known as the hammer, anvil, and stirrup (or, respectively, as the malleus, incus, and stapes, in medical terminology).

They serve as a further means of increasing the fraction of sound transmitted into the cochlear fluid. This increase is accomplished through the principle of **lever action**. Furthermore, they somewhat protect the inner ear from damage due to loud sounds. Obviously, they are inadequate in protecting the hearing of those who enjoy listening to loud rock bands. My own testing of students over a period of more than thirty years reveals a dramatic decrease in the highest frequency that can be heard. Around 1975, most students heard frequencies exceeding 20,000 Hz. By the year 2000, I have found that few students can hear frequencies above 18,000 Hz.

Consider the seven-foot see-saw illustrated in Fig. 10.6. A child of weight 40 lbs is seated to the left at end position, while a woman of weight 100 lbs is seated to the right at position B. The **fulcrum** (or pivot) at F is located at a position two feet from end B (that is $\overline{BF} = 2$ ft), so that the distance $\overline{AF}$ is five feet. The ratio $\overline{FB}/\overline{FA}$ is known as the **mechanical advantage**, which we will denote with the letter **r**. In the example above, the mechanical advantage is thus 2.5. In the case of the human ear, it has been reported to be only about 1.3. Generally, we have

$$\frac{F_B}{F_A} = \frac{\overline{AF}}{\overline{BF}} = \mathbf{r} \qquad (10.6)$$

which is known as **Archimedes' principle of lever action**.[3]

Lever action also leads to another essential means of increasing the fraction of sound energy transmitted to the cochlear fluid. A displacement of point A leads to a displacement of point B in ratio of the mechanical advantage **r**. [See Fig. 10.7.] This fact follows from simple trigonometry.

[3] **Archimedes**, who receives credit for this discovery, is purported to have stated, "Give me a long enough stick, a place to stand, and a pivot, and I'll move the earth!" The language here is quite loose. In fact, we need no such fancy system to move the earth. We do so every time we jump up into the air or walk or run—albeit by a minuscule unobservable amount.

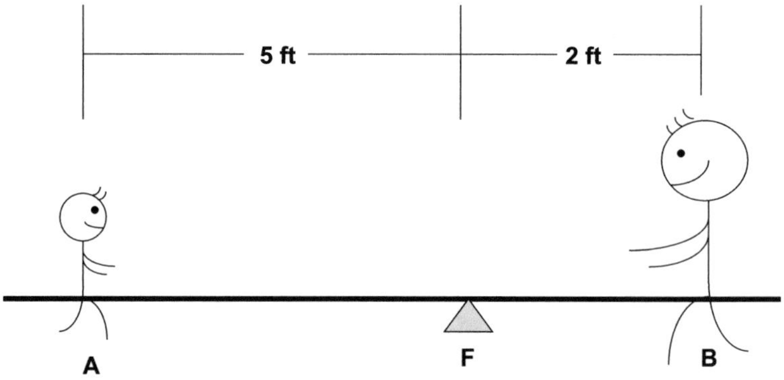

Fig. 10.6 A child and a woman on a see-saw

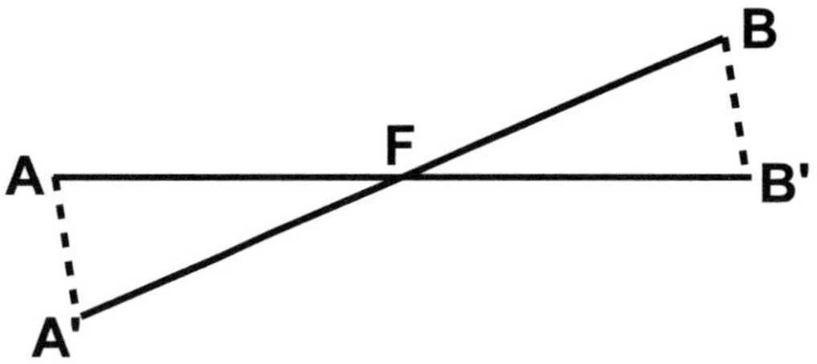

Fig. 10.7 Trigonometry of a lever

In the example of the see-saw, if end A is pushed down a distance of five inches, the woman at B will move up a distance of only two inches. In general, the ratio of the respective displacements, $\overline{AA'}$ and $\overline{BB'}$, at two ends of a lever is equal to the ratio **r** of the distances of the fulcrum from the two ends:

$$\frac{\overline{AA'}}{\overline{BB'}} = \frac{\overline{AF}}{\overline{BF}} = \mathbf{r} \tag{10.7}$$

Since the ratio of the **forces**, F_B/F_A, is equal to 1.3 for the ear, the ratio $\overline{AA'}/\overline{BB'}$ is equal to 1.3 also, by virtue of Eq. (10.6). Thus, if the air pushes the point at which the hammer of the ossicles is connected to the eardrum by a distance of one nanometer (about three or four diameters of a single atom!), the footplate will move a distance of only about $1/1.3 \sim 0.77$ of a nanometer.

Note that the product of the force and displacement on one end is equal to the same product for the other end:

$$F_A\,\overline{AF} = F_B\,\overline{BF} \tag{10.8}$$

In the example above, we have

$$40 \text{ lbs } \times 5\text{ft} = 100 \text{ lbs} \times 2\text{ft} \tag{10.9}$$

10.5 Improving on the Impedance Mismatch: Details

We will outline the basis for improving on the impedance mismatch. The expression for the transmittance in Eq. (10.5) in terms of air and fluid can be written as

$$T = \frac{4Z_a/Z_f}{(Z_a/Z_f)^2 + 1} \tag{10.10}$$

The transmittance is low because $Z_a \ll Z_f$. We have pointed out that two factors play a role in increasing the transmittance—the ratio of the areas of the eardrum and the oval window, A_d/A_w, and the **mechanical advantage**, **r**, of the lever arm of the ossicles. An analysis to be outlined below leads to an effective increase in the ratio of the two impedances:

$$\frac{Z_a}{Z_f} \rightarrow \mathbf{r}^2 \frac{A_d}{A_w}\frac{Z_a}{Z_f} \tag{10.11}$$

For the ear, the ratio of impedances is then increased by a factor of $(1.3)^2(15) \approx 25$, and therefore from $1/3300$ to $1/130$. The corresponding transmittance is 0.03 with a reduction of the sound level by 15 dB instead of the 30 dB reduction without the eardrum and ossicles.

In Fig. 10.8 we see a simplified but concrete description of the process. An incident wave of

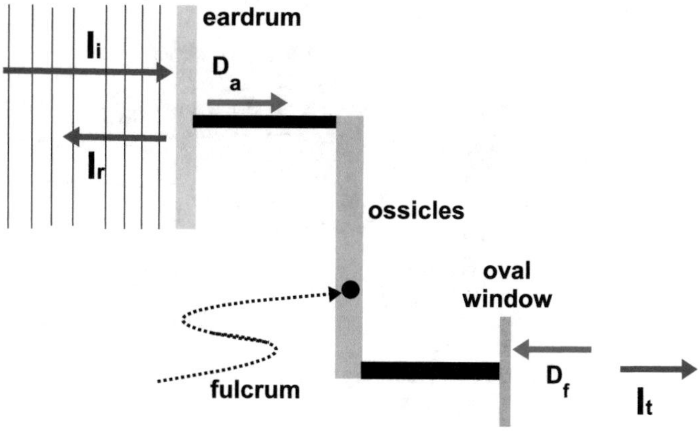

Fig. 10.8 Schematic of the operation of the ossicles

intensity I_i strikes the eardrum with area A_d. A reflected wave has an intensity I_r. The eardrum has a displacement equal to that of the air at the eardrum, D_a, represented by the red arrow to the right. We see a schematic of the ossicles that provide lever action, with the fulcrum represented by the black dot. The result is a displacement D_f of the fluid, which moves with the oval window. The oval window has an area A_w. I_t is the intensity of the sound wave that is transmitted into the fluid, which I have indicated with a direction to the right even though the oval window is moving to the left at the instant shown in the figure.

Let us begin by discussing the effect of having different areas. The transmittance ultimately tells us what fraction of energy of the incident wave is transmitted—or alternatively what fraction of **power** P is transmitted. Thus,

$$T = \frac{P_t}{P_i} \tag{10.12}$$

If the incident wave falls directly on the boundary with a second medium, the two media have a common area and

$$T = \frac{P_t}{P_i} = \frac{I_t A}{I_i A} = \frac{I_t}{I_i} \tag{10.13}$$

Since the areas are different we will have

$$T = \frac{P_t}{P_i} = \frac{I_t A_w}{I_i A_d} \tag{10.14}$$

Instead of conservation of energy being represented by $I_t + I_r = I_i$ we have

$$P_t + P_r = P_i \tag{10.15}$$

When a wave in a medium is incident upon a boundary with another medium, the amplitude of the displacement of the waves in the media must be equal at the boundary. In Fig. 10.8 the displacements D_a and D_f are the respective amplitudes of the displacement of air and fluid, respectively. With direct contact of the eardrum with the oval window of the cochlea we would need

$$D_a = D_f \tag{10.16}$$

However, as a result of the lever action,

$$D_a = \mathbf{r} D_f \tag{10.17}$$

Next, if there is a reflected wave, the wave in the first medium consists of two displacement amplitudes added together—one from the incident wave and one from the reflected wave. We have

$$D_a = D_i - D_r \tag{10.18}$$

Note the minus sign because the reflected wave is reversed in direction as when a wave reaches the end of a closed pipe. [See Sect. 3.6.]

The only remaining relation that is needed to derive the final result in Eq. (10.11) is the following relation between the intensity and the amplitude of displacement:

$$I = \frac{1}{2}(2\pi f)^2 Z D^2 \tag{10.19}$$

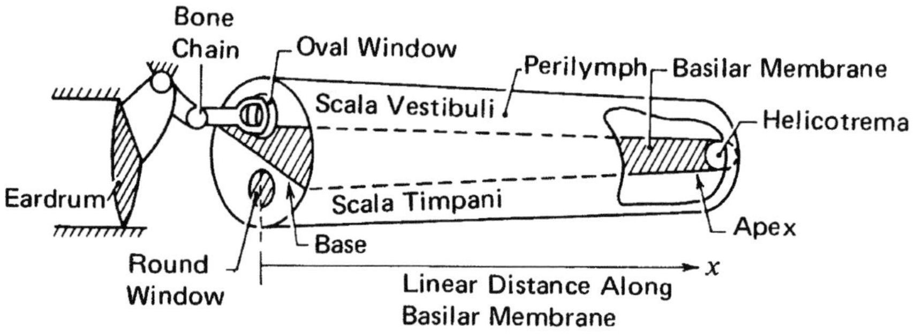

Fig. 10.9 Cochlear chamber, omitting the full complexity of the upper chamber (source: Roederer, op. cit.)

10.6 The Cochlea

The **cochlea** (or **inner ear**) is shaped like a snail. (See Fig. 10.1.) It is by far the most complicated part of the ear. No more than about one-half centimeter in width and almost entirely buried within a bony mass of the skull, it makes use of an intricate apparatus whereby the final conversion of the sound wave into nerve impulses is made. It also achieves a **partial frequency analysis**, which contributes to our ability to discriminate pitch.

The cochlear chamber is a spiral about 35 mm in length. Within is contained but a few drops of a liquid which has about the same density (1.03 gm/cm^3) as water and about twice the "thickness" (technically referred to as the **viscosity**) of water. In Fig. 10.9 we see a schematic of the interior of the full length of the cochlea if it were uncoiled. The center of the coil would be at the left. The chamber is divided into three sub-chambers—the **scala vestibuli**, the **scala tympani**, and the **scala media**. Only the first two are shown in this figure, for simplicity. Between the chambers is the **basilar membrane**. It has a width that varies from 0.08 mm near the oval window to 0.5 mm near the **helicotrema**. The basilar membrane serves as one of the two partitions between the three chambers.

In Fig. 10.10 we see a cross-section across the length of the cochlea. This figure reveals greater details, including **Reissner's membrane**, which is the partition between the scala tympani and scala vestibuli. The basilar membrane is set into motion by the motion of the cochlear fluid. It contains nerve endings within **hair cells**, which, through the motion of the basilar membrane, are stimulated into producing nerve signals that travel to the brain through the auditory nerve.

Experiments have revealed that when sound is exciting the ear, the oval window and round window move nearly in opposite directions (are nearly half a cycle out of phase). This indicates that the cochlear fluid moves essentially en-masse—that is, that there is negligible compression of the fluid. Thus, with negligible delay, while the oval window is moving to the right, the fluid in the scala vestibuli is moving to the right, fluid is flowing downward through the **helicotrema** and to the left in the scala tympani, and the round window is moving to the left. Directions are, of course, reversed when the oval window is moving to the left.[4]

Extensive experiments have been performed so as to study the detailed motion of the fluid and the basilar membrane when the footplate is set into oscillation. The pioneer in this field of research was **Georg von Békésy**, who received

[4]The physical basis for the fact that the fluid moves essentially en-masse is that the wavelength of a sound wave in the cochlear fluid ranges from 75 mm to 75 m for audio frequencies, and is therefore at least twice the length of the cochlear chamber.

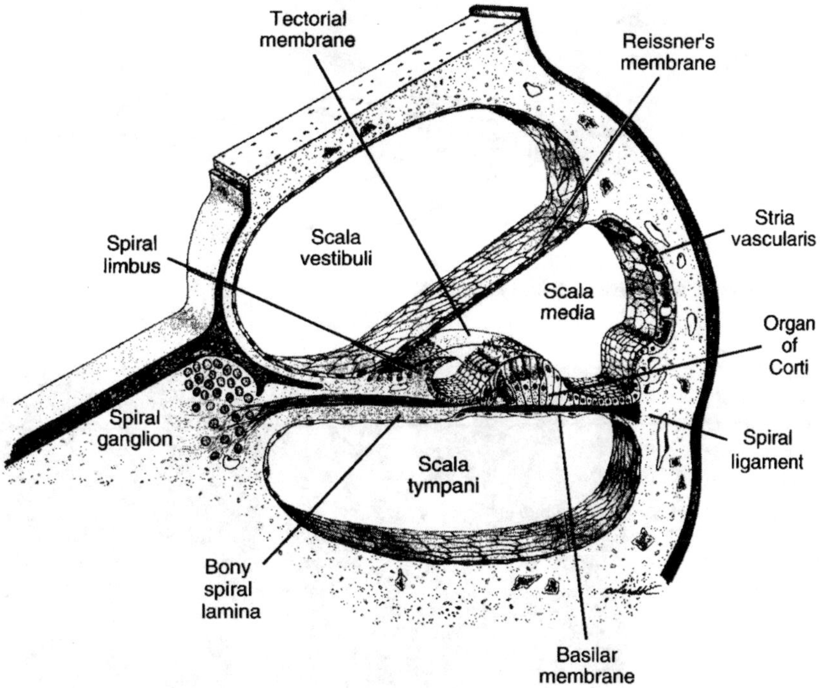

Fig. 10.10 Details of the cochlea (source: Roederer, op. cit.)

the Nobel Prize in Medicine in 1961 for his work.[5] The picture that has evolved follows.

As was pointed out above, the fluid in the scala tympani moves in a direction opposite to that of the fluid in the scala vestibuli. The moving fluid exerts a frictional force on the membrane, in opposite directions on the two sides of a given point on the membrane. As a result, the membrane is displaced in a direction normal to the membrane. In effect, the fluid sets up a transverse vibrational motion of the membrane.

In Fig. 10.11, the dashed curves represent the shape of the membrane at a number of different stages of a complete sinusoidal cycle of motion of the stirrups.

Curves 1–3 correspond to moments when the stirrup is moving to the left, while curves 4–6 correspond to moments when the stirrup is moving to the right. The solid curve is the **envelope**, which is the smallest smooth curve within which are contained all the curves representing the shape of the membrane at all stages of a complete cycle.

The basilar membrane is lined with two rows of **hair cells**, which are connected to **nerve fibers**. [See Fig. 10.12.] These fibers merge to form the **auditory nerve**, which leads to the brain. In the absence of sound, the nerve fibers emit impulses to the brain spontaneously and randomly.

If a sound is present, the relative motion of the basilar membrane is such as to produce a *bending* of the hair cells. This leads to an increase in the rate at which impulses are emitted by the nerve fibers and the perception of sound.

As the sound intensity increases, two changes occur which lead to an increase in the rate at which impulses are emitted. First, the **amplitude** of the motion of the basilar membrane increases, resulting in an increase in the rate of impulse emission by each nerve fiber. Second, the envelope of the motion of the basilar membrane

[5] See Georg von Békésy, *Experiments on Hearing* (McGraw-Hill Co., Inc., N.Y.,1960) which is an extensive treatise on the subject. His work has been of utmost importance in treating people with hearing difficulties.

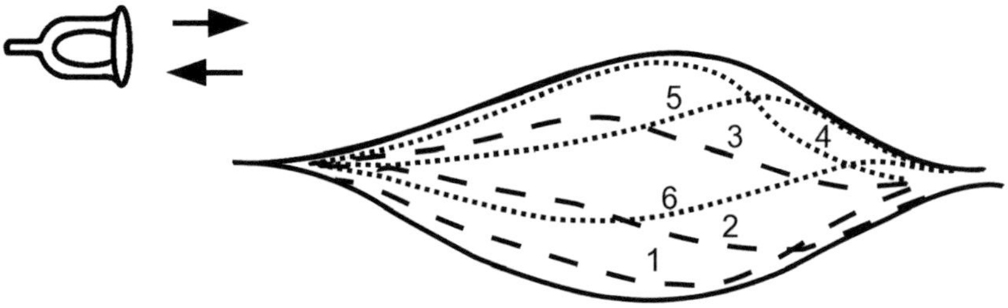

Fig. 10.11 Cochlea membrane wave with the envelope of the pulse

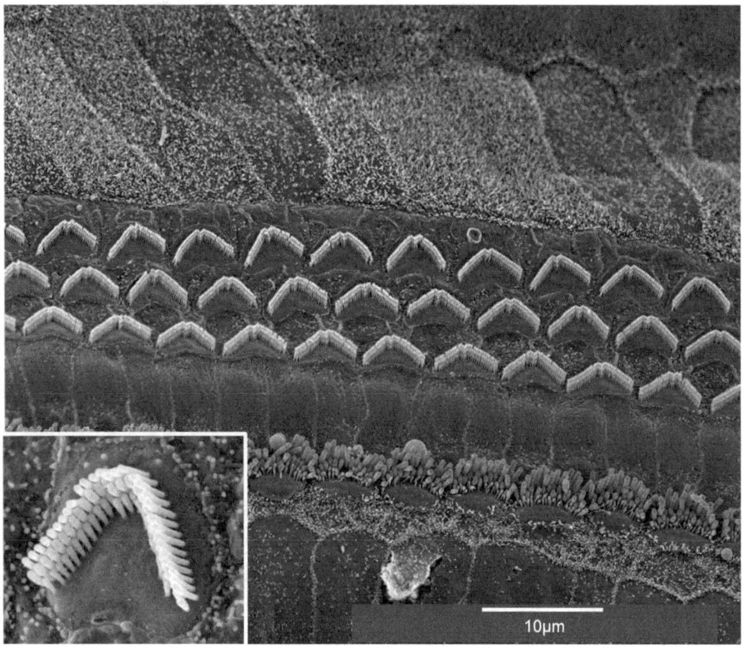

Fig. 10.12 **Scanning electron micrograph of a row of hair cells, each of diameter ca. 5 μm; inset is a closeup of one hair cell with its many cilia** (source: courtesy of Andrew Forge, University College of London Ear Institute)

widens, resulting in an increase in the *number of hair cells* which are significantly bent and whose nerve impulse emission rate is increased. The increased rate of impulse emission results in an increase of the loudness of the sound.

Summary We have seen how the eardrum and a set of three bones, which are far from beautiful in appearance nor simple in structure, serve to significantly increase the sound wave energy transmitted to the cochlea. Their beauty lies in their utter detail in the presence of puny size. As one grows older, these bones tend to become calcified (i.e., have excess calcium) and lose their flexibility at their joints. Consequently, the ossicles become an impediment to hearing and an operation is sometimes performed to allow the sound wave to by-pass the ossicles and flow directly to the oval window. A person then suffers a severe hearing loss because the sound wave has to be transmitted to the air behind the eardrum and then on to the oval window; therefore one loses the benefits of the ossicles as discussed above. An alternative is to have a microphone produce an amplified electronic signal that excites the nerves using a wire that is inserted into the cochlea. The device is called a **cochlear implant**. See Fig. 10.4. See later comment on their operation later on in this chapter.

10.7 Pitch Discrimination

A sense of pitch requires an ability to discriminate frequencies—technically referred to as **pitch discrimination**. Specifically, tones of different frequency must affect the ear and/or brain differently. *All nerve impulses are alike.* As a result, there are only two ways in which auditory signals can be differentiated: *first*, by the specific nerve fibers which are transmitting the impulses and *second*, by the way the pattern of nerve impulses transmitted to the brain varies in time. There is strong evidence that the ear makes use of both approaches. The first approach is referred to as the **Place Theory of Pitch Perception**. We will refer to the second approach as the **Rhythm Theory of Pitch Perception**. Since both are operative, it is misleading, though unfortunately common, to refer to them as competing theories.

We begin with a discussion of the Place Theory of Pitch Perception, an idea that was first developed by **Hermann von Helmholtz** (Fig. 10.13).[6] Helmholtz pictured the basilar membrane as being similar to a harp, consisting of a set of strings under tension, stretched transverse to the length of the membrane about a century ago.

The strings differed in length and in tension, and therefore differed in their fundamental frequencies.

Fig. 10.13 Hermann Ludwig Ferdinand von Helmholtz (source: http://en.wikipedia.org/wiki/Hermann_von_Helmholtz)

Then, a sound of definite frequency would excite only those strings having a fundamental frequency or an overtone frequency which is very close to the frequency of the sound wave. Very few strings would be excited, so that we would have a one-to-one correspondence between the sound frequency and the small set of strings excited. That one-to-one correspondence would be transmitted to the brain by the nerve fibers which were assumed connected to the individual strings, one-to-one, hence providing us with a sense of pitch.

Experiments by von Békésy and others showed that Helmholtz's idea was not far from the truth. In particular, von Békésy studied how the shape of the **envelope of the waves** (see Fig. 10.11) traveling along the basilar membrane varies with the frequency of pure tones. This variation in the shape of the envelope is depicted in Fig. 10.14.

We note that for frequencies above about 50 Hz, the envelope has a peak at some point along the membrane. The higher the frequency, the narrower the peak and the closer the peak is to the stirrup (stapes). In particular, it has been

[6]Hermann Ludwig Ferdinand von Helmholtz (1821–1894), German physicist, anatomist, and physiologist. He worked on acoustics, hydrodynamics, electrodynamics, thermodynamics, meteorology, optics, non-Euclidean geometry, and philosophy of natural sciences. He is known for his invention of the first ophthalmoscope, used by physicians to look into one's eye. In 1847 he formulated (independently of **Julius Robert Mayer** and **James Joule**) the law of conservation of energy. Very often more than one scientist independently makes essentially the same discovery about the same time. Egos can lead to competition, arguments, public battles, and disappointment. Even the great **Newton** tried to blot out the name of **Gottfried Wilhelm Leibniz** for the latter's co-discovery (along with Newton) of the **differential and integral calculus** in mathematics. See the sad but fascinating history of Mayer's work and his frustration from lack of recognition in the following website: http://www.uh.edu/engines/epi722.htm.

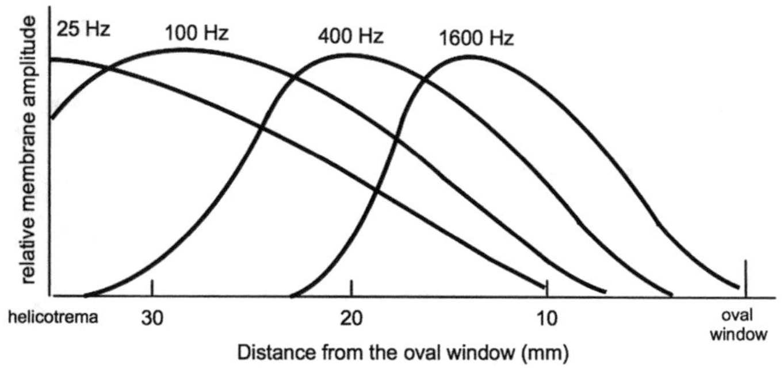

Fig. 10.14 Human basilar membrane response for various frequencies (figure based upon von Békésy, op. cit.)

shown that the **distance** x_p of the peak from the far end of the membrane at helicotrema is approximately **proportional to the logarithm of the frequency**.

In Chap. 12, it will be shown that **pitch is essentially proportional to the logarithm of the frequency**. Therefore, the distance x_p is proportional to the pitch. Since those hair cells and attached nerve fibers which lie in the peak are most strongly stimulated, a sense of pitch is transmitted to the brain.

There are deficiencies with the above theory. Our sense of pitch is very keen. We can discriminate frequencies which differ by only about one percent. On the other hand, the peak of the envelope is wide compared to the size of a hair cell, so that it is difficult to understand how the brain can discriminate between two envelopes which differ by only about one percent. The situation gets progressively worse as we get to lower frequencies. In fact, there is hardly any peak for frequencies below 50 Hz . Two solutions to these difficulties have been suggested[7]:

1. The first is an application of **Mach's Law of Simultaneous Contrast in Vision**, due to **Ernst Mach** [8] to hearing.
2. The second is the **Rhythm Theory of Pitch Perception**.

10.7.1 Some Mathematical Details on Pitch vs. the Peak of the Envelope

What of Helmholtz's strings? Recall (see Chap. 2) that the fundamental frequency of a string is given by $f_1 = v/2R = (\mathcal{T}/\mu)^{1/2}/2R$, where, in particular, $\mathcal{T}$ is the tension. Tension, if present in the membrane, is too small to be measurable to date. It appears that the only (or primary) restoring force is the **stiffness** (i.e., resistance to bending), so that $f_1 = v \times (\text{stiffness})^{1/2}$. In fact, Békésy showed that the logarithm of the stiffness was approximately proportional to the distance x_p; that is,

$$x_p \propto \log(\text{stiffness}) \qquad (10.20)$$

Now suppose we ignore the coupling along the membrane and treat the membrane as a set of rods of equal length and cross-section but of varied stiffness to bending. The fundamental frequency of vibration f_1 of a rod can be shown to be proportional to the square root of the stiffness:

$$f_1 \propto \text{stiffness}^{1/2} \qquad (10.21)$$

[7]There is a third factor that we mention here without details. Recently, Dennis Freeman has conducted research that reveals that the tectorial membrane (see Fig. 10.10) plays a considerable role in amplifying the effect of the frequency-dependent envelope in pitch discrimination. See MIT's *Technology Review*, Volume III, number 1, page M16.

[8]**Ernst Mach** (1838–1916), Austrian physicist and philosopher (Fig. 10.15). The basis of Mach's natural philosophy was that all knowledge is a matter of sensations, so that what people call "laws of nature" are only summaries of experience provided by their own fallible senses. He discovered that if a body moves through the air at a speed faster than the speed of sound, it must produce a shock wave. The so- called **Mach number** is the speed of a body relative to the speed of sound. Thus, a speed of "Mach 3" is equal to three times the speed of sound in air.

Fig. 10.15 Ernst Mach (source: Heliogravüre by H. F. Jütte Red, https://commons.wikimedia.org/w/index.php?curid=12030191)

so that

$$\text{stiffness} \propto {f_1}^2 \tag{10.22}$$

It follows that

$$x_p \propto \log({f_1}^2) = 2\log(f_1) \tag{10.23}$$

Thus, the distance x_p is proportional to the logarithm of the fundamental frequency of the string and therefore to the sense of pitch. According to von Bekesy's model, the peak in the response along the membrane that is displayed in Fig. 10.14 represents a **resonant response** of the rod at the location of the peak. Note that the quadruplings of the frequency from 100 to 400 and from 400 to 1600 lead to the same *change* in position along the membrane, corresponding to the logarithmic relation in Eq. 10.23.

The variation in the width of the membrane (from about 0.1 mm at the stirrup to about 0.5 mm at the helicotrema) will further increase the variation of the frequency f_1 with x_p. This effect does not make the model invalid. Calculations using models of real membranes with the above varied *stiffness and width* confirm the basic validity of the above model.

10.7.2 Mach's Law of Simultaneous Contrast in Vision

Mach's Law of Simultaneous Contrast in Vision is based upon the hypothesis (confirmed by experiment) that the nerve fibers emanating from different receptors on the retina of the eye (see Chap. 13) are not entirely independent. It is well known that a nerve impulse emitted by a receptor inhibits other receptors that are in its immediate vicinity from emitting nerve impulse. Consider then a simple situation wherein half of a region of mutually inhibiting receptors on the retina is uniformly stimulated by light, while the other half is stimulated at a lower intensity. In Fig. 10.16 we display the input and output of a string of six receptors. The inputs are 100 units for the first three and 50 units for the next three. The outputs are determined by the following inhibition: The output is equal to the input minus one-fifth the sum of the inputs of the two neighboring receptors. Thus, the output of the third receptor equals $100-(100+50)/5 = 70$. The output of the fourth receptor equals $50-(100+50)/5 = 20$.

We notice how the output exhibits a relative enhancement to the left of the boundary between the two regions (between #3and #4) and a reduction to the right of that boundary. Without inhibition, the ratio of the outputs of the two neighboring receptors at the boundary is $100/50 = 2$. With inhibition, the ratio of the outputs of the two neighboring receptors at the boundary is $70/20 = 3.5$. Thus, the contrast between the two regions is increased at their boundary. Figure 10.17 illustrates the effect of the Law of Simultaneous Contrast in Vision quite dramatically in the so-called **Mach bands**. The darkness seems to increase dramatically close to the left side of a given rectangle. In fact, each given band is uniform.

If we assume that the auditory nerves on the basilar membrane interact in the above inhibitory manner, it can be shown that the peak in the wave

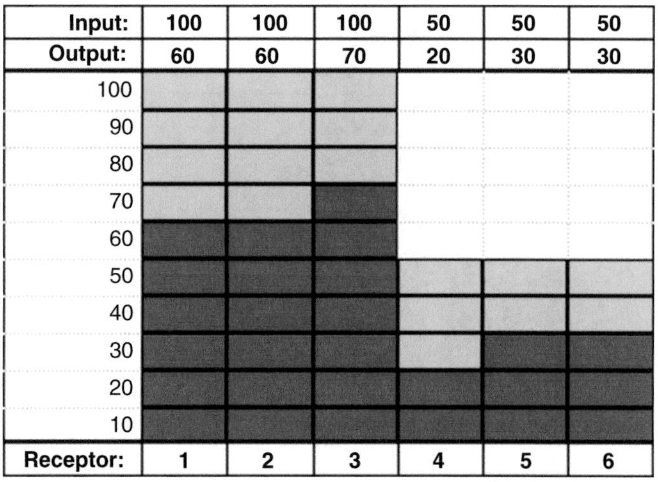
Input: 100 100 100 50 50 50
Output: 60 60 70 20 30 30
100
90
80
70
60
50
40
30
20
10
Receptor: 1 2 3 4 5 6

Fig. 10.16 Input and output of a string of six coupled receptors

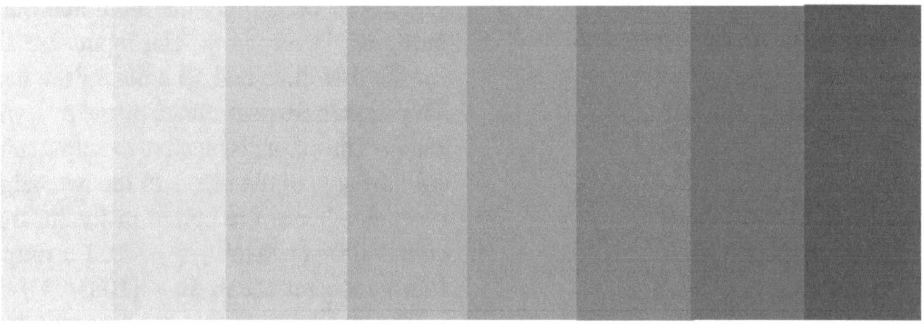

Fig. 10.17 Mach bands

envelope leads to a pattern along the membrane of the variation of the rate of nerve impulses which is more sharply peaked.

Mechanical explanations have also been proposed for a sharpening of the response curve. They are based upon the idea that it is the *relative* displacement of the basilar membrane that produces a bending of the hair cell and a consequent stimulation of the nerve fibers. That is, a hair cell is stimulated to produce nerve impulses not on the basis of the degree of bending itself but rather on the basis of the *variation* of the degree of bending along the membrane.[9]

[9] An analogy can be drawn between the **time** variation of the **displacement**, **velocity**, and **acceleration** of an automobile on the one hand, and the **spatial** variation along the length of the basilar membrane of its **displacement**, its **degree of bending**, and the above **variation** of the degree of bending.

10.7.3 Rhythm Theory of Pitch Perception

We now turn to the Rhythm Theory of Pitch Perception. This theory is based upon the experimentally proven fact that an increase in the nerve impulse rate occurs only during that part of a cycle of motion of the basilar membrane when the basilar membrane is moving towards the tectorial membrane. If a sinusoidal sound wave of frequency f is incident upon the ear, there will thus be an alternation at a frequency f between periods when the rate increases and periods when the rate does not increase. This situation is depicted in Fig. 10.18, wherein the vertical spikes of Fig. 10.18 represent the instants when a nerve impulse is emitted. We see from the figure that there is a resulting periodicity in the rate at which nerve impulses are emitted, which provides us with a sense of pitch.

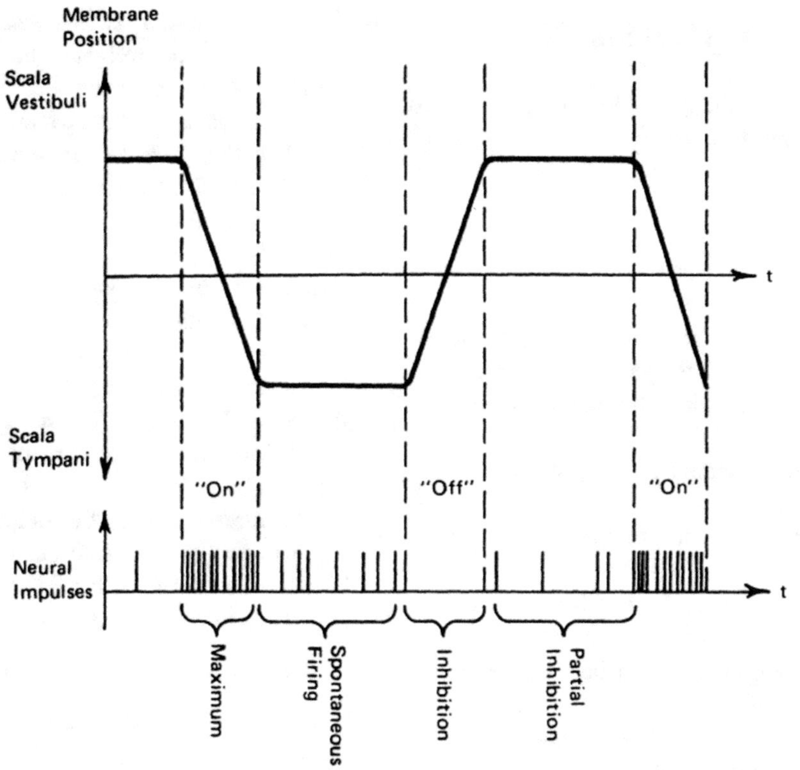

Fig. 10.18 Rhythm theory with nerve spikes (source: Roederer, op. cit.)

What is the resulting basic understanding of pitch discrimination by the ear? The above **Place Theory** accounts for pitch discrimination of frequencies above about 4000 Hz. Both the Place Theory and the Rhythm Theory are operative for frequencies between about 50 Hz and about 4000 Hz. For frequencies below about 50 Hz , only the Rhythm Theory is operative. Over the years, there have been alternative theories; nevertheless, they are in essence enhancements of the above theories. Finally, we should note that a **cochlear implant** by-passes the outer and middle ears in having sound impinge upon a microphone which sends an electronic signal down a wire into the cochlea; this signal excites the nerves at the site of the hair cells. The individual wires, with corresponding sites in the cochlea, can number about five to ten. Therefore there is quite limited pitch discrimination.

10.8 Terms

- Auditory canal
- Basilar membrane
- Cochlea
- Eardrum (tympanic membrane)
- Fulcrum
- Hair cells
- Law of Simultaneous Contrast
- Lever action
- Mach bands
- Mechanical advantage
- Ossicles
- Oval window
- Pitch discrimination
- Place theory of pitch perception
- Rhythm theory of pitch perception

10.9 Problems for Chap. 10

1. Mechanical vibrations are transformed to nerve impulses in the
   (a) middle ear.
   (b) Eustachian tube.
   (c) auditory canal.
   (d) semicircular canals.
   (e) cochlea.
2. The correct order in which mechanical vibrations pass through the parts of the ear is
   (a) eardrum, cochlea, hammer, stirrup, and oval window.
   (b) eardrum, hammer, anvil, stirrup, and cochlea.
   (c) eardrum, Eustachian tube, oval window, and round window.
   (d) eardrum, anvil, Eustachian tube, and organ of Corti.
   (e) eardrum, cochlea, basilar membrane, and middle ear.
3. What is the function of the ossicles of the ear?
4. What are the two theoretical mechanisms for pitch discrimination provided by the ear? Describe them in detail.
5. What are "Mach bands" and how do they relate to theories of pitch discrimination?
6. A mother of weight 120lbs wants to balance herself against the weight of her child whose weight is 40lbs. They are seated on opposite ends of a see-saw whose overall length is 12 feet. See Fig. 10.6. Where should she place the fulcrum? Start by determining the ratio of the distances from the fulcrum to each of the two of them.
7. Below is a table that is meant to illustrate the effect of Simultaneous Contrast. We have a set of inputs for a string of 14 receptors. The inputs are meant to represent the envelope of the wave that is traveling along the basilar membrane, with a broad peak at the ninth receptor.

   Calculate the respective outputs of the receptors using the recipe discussed in the text. That is, if we let I_n be the input to receptor n and O_n be the output from receptor n,

$$O_n = I_n - (I_{n+1} + I_{n-1})/5 \qquad (10.24)$$

   Produce two graphs—one for the input vs. the receptor number (1–14) and one for the output vs. the receptor number (1–14). Draw them, one below the other to emphasize the difference. Your result should indicate an enhancement of the peak of the response.

Receptor	1	2	3	4	5	6	7	8	9	10	11	12	13	14
Input	20	22	24.5	26	29	33	37	40	42	39	36	33	31	30
Output														

11 Psychoacoustics

In the last chapter, we learned about the last step in the trail from the source of sound to the brain via nerve signals from our ears. All these steps have been describable in what we refer to as *physical terms* and are *objective*. And yet we don't know in *physical terms* what it means when we say, **"I hear a sound."** This last step has eluded explanation and clarification. What is the *physical* nature of pain? Perhaps we are asking the wrong questions. Perhaps we are limited by language, which is after all a product of our conscious experience, and are therefore looking for the answer to a question that has no meaning and therefore is, shall we say, invalid as a question.[1]

The physicist knows how to characterize a sound *uniquely*. And yet, how people describe a given sound that is heard varies from one individual to another. Of course, a report must make use of the language of the person in relation to his/her own personal experience. Two people learn to associate the color red with a certain sensation they experience. What their individual experiences are, we don't know. Both may report that a given color is red, but what they actually experience may differ. We may both agree that two tones have the same or different pitch, but will never know the extent to which our perceptions of the sounds are similar.

In spite of the above difficult issues, we can ask people certain questions regarding their perceptive response to various sounds. For example: Which of the two sounds is louder? Or, which of the two sounds has a higher pitch? Or, sing a tone with a frequency of 440-Hz, recognizing that their ability to do so depends upon their previous exposure to this tone. Tests of individuals by psychologists have resulted, as you would expect, in a broad distribution of responses. Nevertheless, psychologists have summarized their results in terms of normal responses, and it is in these terms that we will discuss some of the characteristics of human **psychoacoustics**, which is the study of the relationship between the objectively characterizable sound incident upon a human ear and the corresponding perception of the sound.[2]

Note Before we get into the subject I need to point out that the subjects covered in this chapter

The original version of the chapter has been revised. Corrections to this chapter can be found at https://doi.org/10.1007/978-3-030-19219-8_16

[1]Such a situation exists in the regime of phenomena for which the specific nature of Quantum Theory is manifest. We have found, for example, that it is impossible to describe an atom in terms of images that we have amassed for describing the world at the macroscopic level. For further ideas into the issue of the nature of perception and what it means to think or feel, see the fascinating book by **Daniel Dennett Consciousness Explained**, [Penguin Press, UK , 1992].

[2]The following Wikipedia website is a useful resource of links to many psychoacoustic phenomena: http://en.wikipedia.org/wiki/Psychoacoustics.

L. Gunther, *The Physics of Music and Color*,
https://doi.org/10.1007/978-3-030-19219-8_11

are extremely limited in relation to the incredibly wide range of areas of study of psychoacoustics. The chapter focuses on a few subjects: (1) measures of loudness vs. intensity and frequency; (2) "combination tones": a phenomenon wherein we hear frequencies that are not actually present in the sound incident upon our ears due to the non-linear response of our ears; (3) Duration of a note needed for pitch discrimination; and (4) "fusion of harmonics": The sound of a musical instrument generally has a large ensemble of harmonics. Nevertheless, we don't hear the individual harmonics; rather, the sound appears to have one source.

In both this chapter on psychoacoustics and Chap. 15 on color vision, which is a branch of psycho-optics, we will focus on the basic elements of sound and color, respectively: the musical note, with its multitude of timbres and the color patch. We represent them below in Fig. 11.1 in order for us to focus our minds on these elements.[3]

It is interesting to compare these two images. The color patch displays a hue (red, green, etc.), saturation (degree of paleness), and brightness—the details about which you can read about in Chap. 15. On the other hand, the note symbol on the staff of a musical score has by itself only the content of the frequency chosen for the "A," which is currently typically 440-Hz. This component of information is the analog of the hue of the color patch. However, the actual sound produced by a musician has a huge range of timbres, vibratos, and dynamics that can each be varied throughout the duration of the note played. It is reasonable to compare these components of the note with the relatively small range of degrees of saturation. On the other hand, the level of research being carried out with respect to a color patch is many orders of magnitude greater than that for the performance of individual notes. In

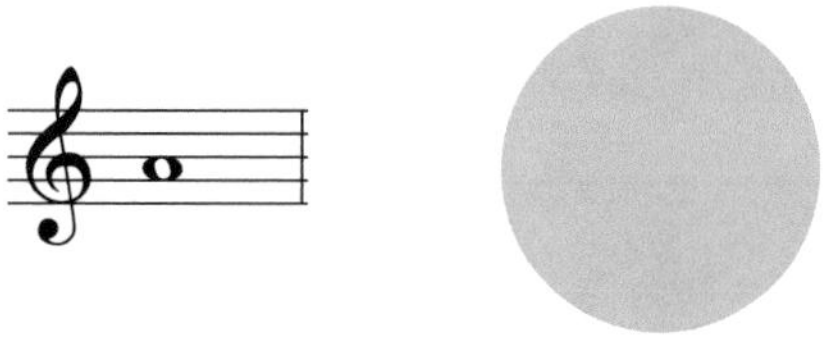

Fig. 11.1 A musical note—an A440—representing an element of a piece of music; A color patch—representing an element of color

parallel is the relative ease we have in remembering the appearance of a color patch in comparison with the characteristics of a performed note. Perhaps, we will learn how to pay more attention to these varied sound characteristics and find a clearer, more precise way to characterize their differences.

11.1 Equal Loudness Curves

For normal hearing, increasing the intensity always increases loudness. What is the quantitative relationship between intensity and loudness? To answer this question would require, for example, that we be able to clearly determine when one sound is twice as loud as another, a requirement that is impossible to meet. There are less demanding questions one could investigate. For example, if a person is exposed to two sounds of different frequency but equal intensity, they generally report that their sense of loudness of the two sounds differs. The person can then be asked to match the loudness of one tone by varying the intensity of the second tone. Back in the 1930s, extensive tests were carried out that resulted in the so-called **equal loudness curves**. A more recent set of curves is shown in Fig. 11.2. These curves reflect an **average of individuals' responses** to sound. Therefore, you should not assume that the curves apply to you. Furthermore, I haven't found information as to the spread of responses.

Let us study the significance and highlights of these curves. Note that the vertical axis at the left is labeled as the **Sound Level in dB**, while the intensities corresponding to the sound levels are marked off along the vertical axis at the right with various powers of ten. Neighboring

[3]The note is an A with a typical frequency of 440-Hz and a color patch with the color that the composer-pianist Alexandre Scriabin associated with the note A.

For the reader who is interested in a more comprehensive discussion of psychoacoustics, I highly recommend the book by **Juan Roederer**, *Introduction to the Physics and Psychophysics of Music—4th ed.* [Springer-Verlag, New York, 2008].

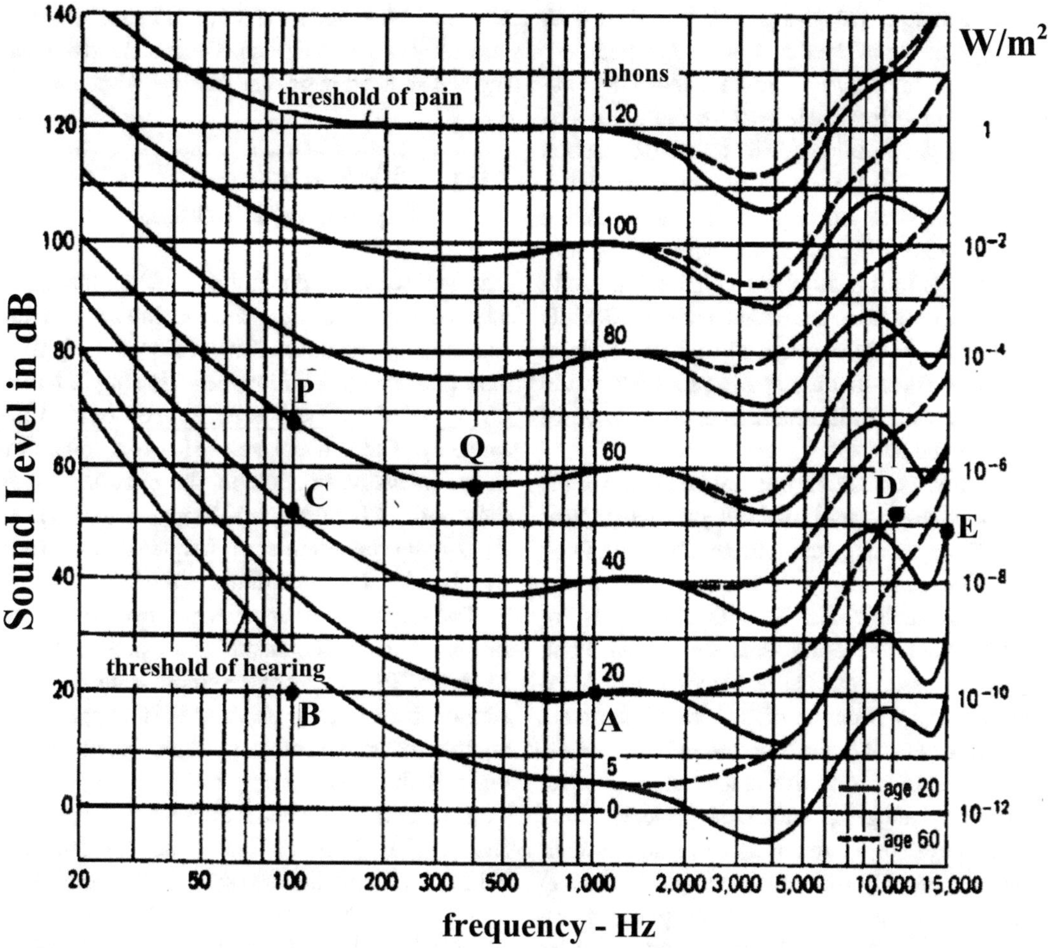

Fig. 11.2 Equal loudness curves
(source: H.E. White and D. H. White, *Physics and Music*, (Holt, Rinehart, and Winston, Philadelphia, 1980))

horizontal lines correspond to a 10-dB change in the sound level and a factor of ten change in intensity. The frequencies are marked off along the horizontal axis. Points *P* and *Q* refer to two tones at frequencies of 100-Hz and 400-Hz, respectively, and lie along the same equal loudness curve. According to the curve, 100-Hz at 68-dB will sound equal in loudness as 400-Hz at about 57-dB. The lowest curve is referred to as the curve of the threshold of hearing. While in Chap. 4 we stated that an intensity of 10^{-12} W/m^2, corresponding to a sound level of **zero-dB**, is the threshold of hearing, we see that in fact, this sound level is associated with a frequency of 2000-Hz. Moreover, one can hear a frequency of 4000-Hz at a sound level of −10-dB

The Phon Unit of Loudness

The **phon**, symbolized by $\boldsymbol{\phi}$, is a unit used to label the equal loudness curves. Suppose that a sound of given frequency has a certain intensity I (with a corresponding **sound level—SL**). The frequency and intensity determine a point in the figure. An equal loudness curve passes through that point. To determine the number of phons for that sound, we look along that equal loudness curve until we reach a frequency of 1000-Hz. The number of phons for the given sound is equal to the SL on the curve for a frequency of 1000-Hz.

For example, the two points **P** and **Q** both lie along the equal loudness curve labelled "60-phons" (on the 1000-Hz vertical axis). They correspond to the frequencies 100-Hz and 400-Hz, respectively. The number "60" refers to the sound level of a 1000-Hz sound at the given loudness. Zero phons corresponds to the threshold of hearing for all frequencies while 120-phons corresponds to the threshold of pain. The solid curves correspond to individuals at age 20 while the dashed curves correspond to individuals at age 60. We note that the hearing loss due to old age lies mostly in the high frequency range of 2000-Hz and above.[4]

The most outstanding qualitative feature of the equal loudness curves is that the ear is most sensitive to frequencies lying between about 1000-Hz and 4000-Hz. Typically, the maximum sensitivity lies at about 4000-Hz. This fact accounts for the following interesting aspects of music: First, while the bass drum has a power that can far exceed the power output of any other instrument in an orchestra, it doesn't drown out the orchestra because of its low frequency range. Second, professional singers, who must project a loud sound in spite of the relatively low range of their voices (especially that of basses and baritones) and a lack of amplification in the performance hall, as can be the case in opera, can do so by singing with what is called a **squillo**. Effectively, their voices contain a high amplitude of higher harmonics lying in the above range—from 2000-Hz to 4000-Hz if possible. In the **Estill method of voice development and theory**, the squillo is referred to as **twang**.

Finally, we should note that the region characterized by the equal loudness curves is bounded at the bottom and at the top. At the bottom we have the curve referred to as the **threshold of hearing**. As indicated above, for each frequency, we see the lowest possible intensity that is audible. The upper boundary is called the **threshold of pain**, above which sound is regarded as being "painful" and masks a sense of pitch. The threshold of hearing is discussed in greater detail in Appendix I.[5]

11.2 The "Sone Scale" of Expressing Loudness

It is well recognized that that doubling the intensity leads to *much less than a doubling* of the sensation of loudness. Therefore intensity is a very poor measure of loudness: In simple terms, our sense of loudness is not proportional to the intensity. The sound level—SL—is a logarithm of the intensity. In addition, the phon level $\boldsymbol{\phi}$ is merely the SL of a 1000-Hz sound that equals the sound level at the given frequency; therefore, the phon level is logarithmic in the intensity.

The logarithm **over**-compensates the inaccuracies of using the intensity as a measure of loudness. To obtain an even more accurate measure of loudness, the **phon** level is mapped onto another parameter called the **sone**, which we will represent by the letter **s**.

The relation between the **sone** and the **phon** is shown in Fig. 11.3 as well as being represented in Table 11.1. Note that the vertical scale in the plot is not linear; it is **logarithmic**. Each segment, running from 1 to 10 (a factor of ten increase) or from 10 to 100 (another factor of ten increase), and so on, is *equal in length* in the graph. The vertical axis is said to have a **logarithmic scale**. Over the course of each segment, the loudness increases in sones by a factor of ten (rather than adds to the loudness in sones by a fixed number). Because the horizontal axis is "normal," the graph is said to be a **semi-logarithmic graph**.

The solid curve on the graph represents the sone value for all values of phons. The mathematical relation between sones and phons for $\phi > 40$ phons is given by

$$s = 2^{\frac{\phi-40}{10}} \tag{11.1}$$

[4]There is a revised international standard (ISO 226 2003) to be found on the following website (1-22-2011): http://en.wikipedia.org/wiki/FletcherMunson_curves. I kept the older figure below because it includes the effect of aging. One reason for the difference are improved testing procedures.

[5]You can test your own hearing within the limitations of your level of training by using the applet on this website (1-22-2011): http://www.phys.unsw.edu.au/jw/hearing.html.

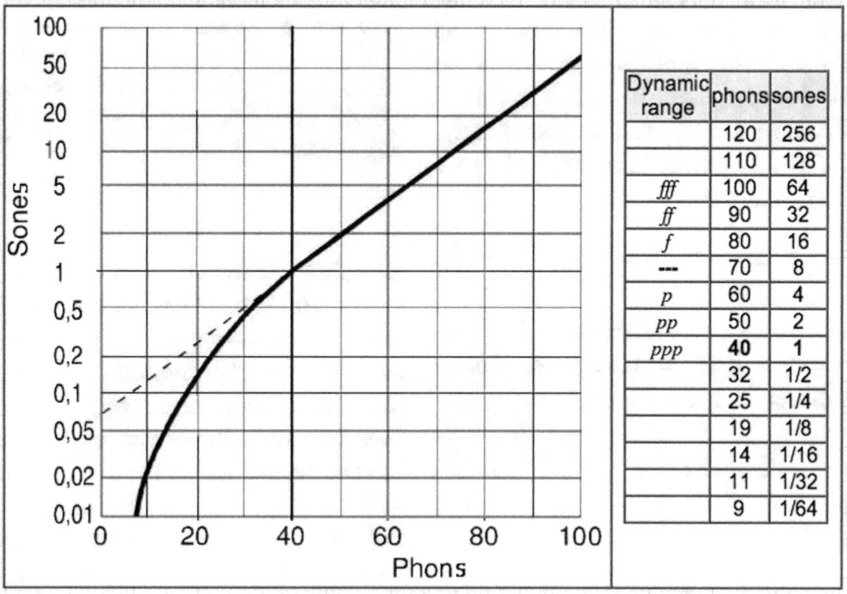

Fig. 11.3 Sones vs. Phons; Table Relating Sones to Dynamics (loudness) Symbols in Musical Scores (source: http://upload.wikimedia.org/wikipedia/commons/c/c7/Akustik_phon2sone3.jpg)

Table 11.1 Phons vs. sones

Phon level	40	50	60	70
Loudness in sones	1	2	4	8

Because the sone level is an exponential function of the phon level ($\propto$ a power of two), the sone level is a straight line in the **semi-logarithmic graph**.

For $\phi < 40$-phons, the relationship is no longer exponential and is, instead, linear:

$$s = (\phi/40)^{2.86} - 0.005 \qquad (11.2)$$

As a result, the graph is curved on the semi-logarithmic plot. The dashed curve down to the zero phon level is merely a continuation of the solid straight line that holds above 40 phons.

We will restrict our discussion to the simple regime when $\phi > 40$-phons. The behavior is based upon the observation of psychoacousticians that **each increase in the number of phons by ten doubles the loudness in sones**. Thus, the loudness in sones will double if the number of phons increases from 40-phons to 50-phons *or* from 50-phons to 60-phons. The loudness in sones will quadruple if the number of phons increases from 50-phons to 70-phons. See a sample of values in Table 11.1.

It can be shown that the mathematical relation between the sone level and intensity is given by[6]

$$s = \left(\frac{I'}{I'_{40}}\right)^{0.3} \qquad (11.3)$$

Here I' is the intensity of a tone at 1000 Hz that is equal in loudness to that of the given tone. I'_{40} is the intensity of a 1000 Hz tone that has a sound level of 40 dB, that is, 10^{-8} W/m^2. The sone level increases more slowly than linearly but more rapidly than logarithmically with respect to the intensity I'.

Note Both the number of phons and the sound level in sones that correspond to a given sound wave depend not only upon the intensity, but also upon the frequency!

[6]The number 0.3 in the equation is actually an approximation for log 2=0.3010... .

The relationships described above are quite complicated. Therefore, we will present as an example the following situation: Let us consider a sound of frequency 200 Hz and sound level 30 dB. According to Fig. 11.2, the sound corresponds to about 23 phons. From Eq. (11.1) we find that the loudness is

$$s = 2^{\frac{(23-40)}{10}} = 2^{-1.7} = 0.31 \text{ sones} \quad (11.4)$$

Suppose that we want a sound that is doubly loud, that is $s = 2 \times 0.31 = 0.62$ sones. Let us determine the required number of phons.

$$s \equiv 2^{\frac{\phi-40}{10}} = 2 \times 0.31 = 2 \times 2^{\frac{(23-40)}{10}} \quad (11.5)$$

Then we have

$$2 = 2^{\frac{\phi-23}{10}} \quad (11.6)$$

so that $(\phi - 23)/10 = 1$ and ϕ = 33-phons.

Note that we could obtain this result more directly by recalling that doubling the loudness in sones requires an addition of 10 phons to the phon level. Then we require ϕ = 23+10 = 33-phons.

To repeat,

Generally, in order to double the loudness, measured in sones, one must add 10 *phons to the phon level.*

Sample Problem 11.1 What is the sound level corresponding to the above 200-Hz sound having 33-phons?

Solution From Fig. 11.2, we find SL=37-dB.

Note that the change in sound level is from SL=30-dB to SL=37-dB, corresponding to a change of 7-dB and an increase in intensity by factor of $10^{\Delta SL/10} = 10^{0.7} = 5$. In sum, in order to double the loudness of this particular sound, one must increase the intensity by a factor of five.

The above results for a specific frequency of 200-Hz are summarized in Table 11.2. (The results generally depend upon the frequency.)

Table 11.2 Comparison: intensity, sound level (SL), phon level ϕ, and loudness in sones of a 200-Hz pure tone

Intensity I in W/m^2	Sound level SL	Phon level	Loudness in sones
10^{-9}	30	23	0.31
5×10^{-9}	37	33	0.62

Sample Problem 11.2 Consider a sound of frequency 1000-Hz. Let us compare two loudnesses in phons: ϕ =4-phons and ϕ =120-phons corresponding to the threshold of hearing and the threshold of pain, respectively.
Find the ratio of the intensities, I_2 and I_1, and the ratio of the loudnesses in sones s_2 and s_1.

Solution Since the frequency is 1000-Hz, the phon level is equal to the sound level. Therefore

$$\frac{I_2}{I_1} = \frac{10^{120/10}}{10^{4/10}} = 10^{116/10} = 3.3 \times 10^{11} \quad (11.7)$$

The ratio of the sound level in sones is

$$\frac{2^{(120-40)/10}}{2^{(4-40)/10}} = 2^{11.6} = 3,100 \quad (11.8)$$

The relative small ratio of the sones explains why the hearing system can tolerate such an incredible range of intensities.

11.3 Loudness from Many Sources[7]

Suppose that we have a number of sound sources. We address the question of how we would compute the resulting SL and the loudness in sones. To simplify the discussion, we will discuss only two sound sources; the generalization to more than two should be obvious.

There are a number of cases that distinguish the results:

[7] See the website (1-9-2011): http://home.tm.tue.nl/dhermes/lectures/SoundPerception/\05Loudness.html.

1. The two sources have **the same frequency and are coherent**—that is, have a definite relative phase at the wave level. In this case, we determine the resulting amplitude as discussed in Chap. 7. Squaring the total amplitude gives us the resulting intensity, from which we can calculate the resulting phon level and then the resulting number of sones.

   As a simple example, we will consider two such sources that have the same frequency of 1000-Hz, the same amplitude and are in phase. The total amplitude is doubled. Therefore, the intensity is quadrupled, resulting in an increase of 6-dB in the sound level, and hence six phons because the frequency is 1000-Hz. Therefore, the loudness in phons will change from ϕ to $\phi' = \phi + 6$. From Eq. (11.1) we see that the loudness in sones will increase by a factor of $2^{6/10} = 2^{0.6} \approx 1.52$
2. The two sources are independent and have different frequencies. Let Δf be the magnitude of the difference in the two frequencies. In this case, there are two sub-cases that depend upon the two frequencies. We need to consider the **critical bandwidth**, which refers roughly to the range of frequency differences such that the two pitches cannot be distinguished.[8]
   (a) Suppose that Δf < critical bandwidth. Then we find the intensity of each of the individual sources. We add the intensities to obtain the total intensity. We then calculate the corresponding number of phons for the total intensity based upon the average frequency. And finally, we calculate the number of sones based upon this phon level.
   Example: As in the previous example we assume that the frequencies are both close to 1000-Hz and that both have one sone, which corresponds to 40-phons. If we double intensity, we add 3 phons, which results in a total of 40+3 = 43 phons. Therefore the resulting loudness in sones is $s = 2^{(43-40)/10} = 2^{0.3} = 1.23$ sones.
   (b) Now suppose that Δf > critical bandwidth. This case is simple, we simply add the number of sones. In the above example we obtain 2 sones.

Sample Problem 11.3 Consider a swarm of 1000 mosquitos, each producing a sound with SL=10-dB. Let us estimate the resulting sound level SL and the loudness in sones.

We can easily obtain the sound level since ΔSL = 10 log(1,000) =30-dB. Therefore the resulting sound level is 10+30=40-dB.

To determine the loudness is sones is more complicated since we need to take into account the frequency spectrum of the sound. For simplicity, we will represent the spectrum by the most prominent frequency, which is about 300-Hz. From Fig. 11.2 we see that one mosquito's sound amounts to $\phi = 0$ phons. This corresponds to a loudness of $2^{-40/10} = 1/16$ sone. To compute the total loudness we note from Fig. 11.2 that 40-dB at 300-Hz corresponds to about 42-phons. Thus, we obtain

$$s = 2^{(42-40)/10} = 2^{0.2} \approx 1.1 \text{ sones} \qquad (11.9)$$

Note To close this section we should note the following:

The intensity and the sound level are both **objective** measures of loudness.

On the contrary, the phon level and the loudness in sones are **subjective** measures of loudness in the sense that they are based upon the results of testing the hearing of a number of individuals.

11.4 Combination Tones and the Non-Linear Response of the Cochlea

The phenomenon discussed in this section is quite unusual in that it is responsible for the perception of tones, the so-called **combination tones**. They are not at all present in the sound

[8]For details on critical bandwidth, see https://www.sfu.ca/sonic-studio-webdav/handbook/Critical_Band.html and https://community.plm.automation.siemens.com/t5/Testing-Knowledge-Base/Critical-Bands-in-Human-Hearing/ta-p/416798?lightbox-message-images-416798=39834iF612EC3E4156DD71.

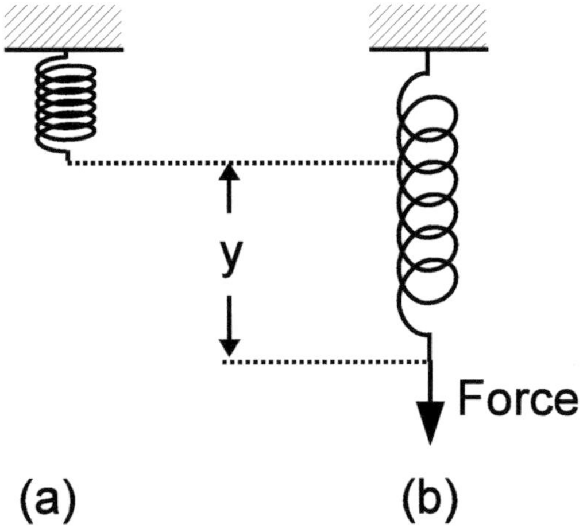

Fig. 11.4 Spring system

wave incident upon one's ears and may partially account for one's sense of musical consonance. In hi-fidelity terminology it is responsible for **harmonic distortion**. In the early days of radio, it was a favorable characteristic of poor audio speakers: The radio could not respond well to low frequencies (say, about <400-Hz), so that the fundamental and possibly some overtones of a low-pitched tone of a musical instrument might be missing or very weak in the electrical signal which excited the speaker. However, because of the **non-linear response** of the speaker, the higher harmonics would cause the speaker to produce a tone with the missing fundamental and overtones included as components. The result might be likened to enriched, bleached flour.

To understand **non-linear response**, it is helpful to first appreciate **Linear Response** , which we have thus far taken for granted. [Read Chap. 2 on the SHO, as a preparation for what follows.] Consider the spring illustrated in Fig. 11.4. A downward force is applied to the spring, which consequently increases in length by an amount y in Fig. 11.4.

If the spring were pushed upward by a force of the same magnitude, the spring would move upward by the same displacement, y. If the force is doubled, the displacement y will be doubled. Generally, the distance is proportional to the force.

$$\text{displacement } y \propto F \quad \text{or} \quad y = \frac{F}{k} \tag{11.10}$$

where k is the **spring constant**. If we graph Eq. (11.10), we obtain a straight line. **Positive** forces represent downward forces—hence a stretching of the spring and increasing y. **Negative** forces represent upward forces—hence a contraction of the spring and decreasing y. Because of the straight line graph, one says that the spring responds **linearly** to an external force. (The displacement is on the y-axis while the force is on the x-axis.). The technical term describing this behavior is that there is **linear response** . [Alternatively, one says that "the response of the spring to the force is linear."]

Suppose the force varies sinusoidally in time with a frequency f. Then the response, y, is just a sine wave multiplied by the constant $1/k$, and is therefore a sine wave of the same frequency f but with a different amplitude. Generally, whatever the pattern of variation of the force with time will be reproduced by the displacement.

Consider the process of a sound wave incident upon a microphone that is connected to an amplifier that is, in turn, connected to a loudspeaker. If there is linear response throughout, the pattern of the incident sound wave in time will be reproduced by the electrical signal from the microphone, by the electrical signal from the amplifier, and finally by the sound coming out of a loudspeaker. Thus

linear response is associated with fidelity!

Now consider the response of a **real** spring as depicted in Fig. 11.5. The curve is not a straight line. The response is said to be **nonlinear**.[9]

[9]It is important to note that for small enough forces, the response of a real spring is essentially linear. That is, we can assume that the displacement of the spring is proportional to the force to a good approximation. Thus, the so-called **ideal spring** is an abstraction whose behavior is approached by a real spring for small forces. It is a remarkable fact that the bulk of physical theories and concepts are based upon abstract models of the real world which assume linear response as an approximation, with deviations from linearity being second order effects which may or may not be essential to the phenomena of interest. Furthermore, the Principle of Superposition, which was discussed in Chap. 7, is dependent on a linear response of

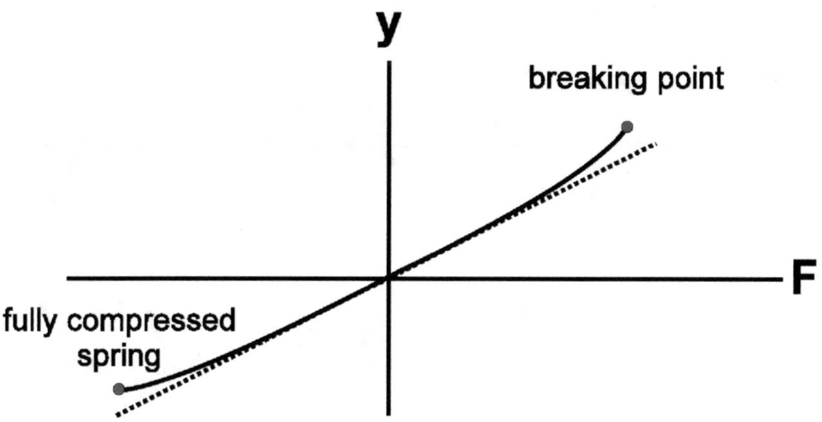

Fig. 11.5 Non-linear displacement of a spring in response to a force

Table 11.3 Various systems—generalized force vs. the response

System	Generalized force	Response
Spring	Force on spring	Displacement of spring
Stereo receiver	Amplitude of radio wave	Electrical voltage output of stereo receiver
Speaker	Electrical voltage input to speaker	Displacement of the diaphragm of the speaker
Wishbone	Pull on a wishbone	Change in the angle between the "legs" of the wishbone

In Table 11.3 we list a number of systems upon which a force or an analog of a force, a so-called **generalized force**) is exerted. The quantity which measures the response of the system to the generalized force is listed in the right-hand column.

We now consider a special case known as **quadratic non-linear response**. In this case the displacement is proportional to the $(force)^2$, or

$$y = bF^2 \tag{11.11}$$

where F is the force and b is a constant of proportionality.

the system. Thus, to the extent that the response is non-linear, this principle breaks down.

Suppose now that the force F is a pure sine wave of frequency f, as in Fig. 11.6. According to Eq. (11.11) the displacement will be the square of a sine wave and hence is always positive. The two functions, a sine wave and the square of a sine wave, are depicted in Fig. 11.6.

In Fig. 11.6 we see that the force F (the upper curve) is periodic and has twice the period of the square of the force F^2 (the lower curve). Alternatively, F^2 has doubled the frequency of F.[10] If the force has a frequency of 400-Hz, the output will be 800-Hz. If the response is a sum $y = aF + bF^2$, the output will be two frequencies, 400-Hz and 800-Hz.

We get a more interesting result if the force is a mixture of two Fourier components, say, with frequencies f_1 and f_2. Let us represent the component forces by the symbols F_1 and F_2, respectively. Then the total force F is given by the sum of F_1 and F_2:

$$F = F_1 + F_2,$$

while the square of F is given by

$$F^2 = (F_1 + F_2)^2 = F_1^2 + F_2^2 + 2F_1F_2$$

According to our preceding result, F_1^2 has a frequency component of $2f_1$, while F_2^2 has a

[10]This result is obtained from the trigonometric identity is $\sin^2\theta = 1/2 - (1/2)\cos 2\theta$.

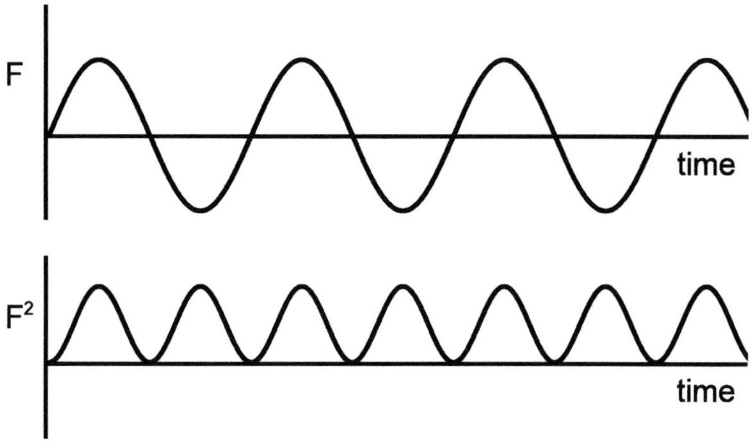

Fig. 11.6 A sine wave; the square of a sine wave

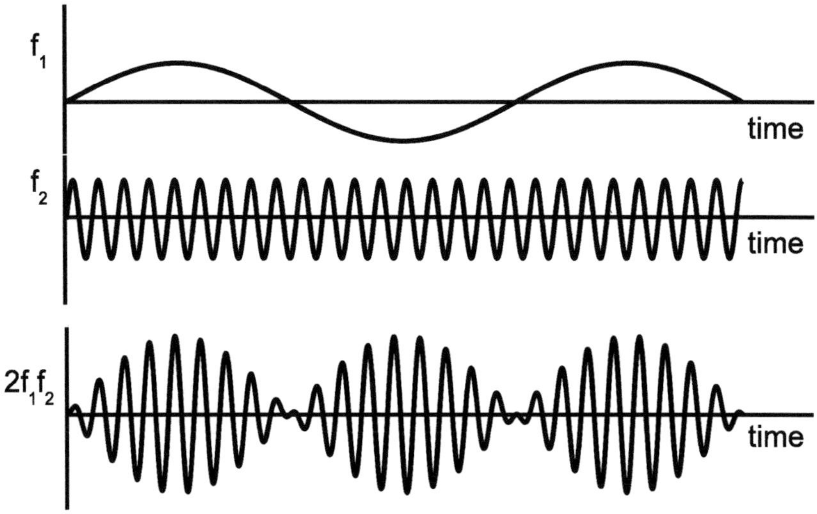

Fig. 11.7 Two sine waves, with $f_2 \gg f_1$, and the cross term of the square of their sum

frequency component $2f_2$. Lastly, we need to know the frequency composition associated with the product $F_1 F_2$.

It can be shown[11] that the product $2F_1F_2$ has two components, with frequencies $(f_2 - f_1)$ and $(f_2 + f_1)$, respectively.

We can make this result somewhat plausible by considering the case when $f_2 >> f_1$. Then, in the product $F_1 F_2$, the factor F_1 produces an envelope of the sine curve F_2. The behavior of F_1, F_2, and the product $2F_1F_2$ are depicted in Fig. 11.7.

[11]The result is based upon the trigonometric identity: $2\sin\theta_1 \times \sin\theta_2 = \cos(\theta_2 - \theta_1) - \cos(\theta_2 + \theta_1)$.

The pattern $2F_1F_2$ is that of the **beating** of two components, with a **beat frequency**

$$f_B = (f_2 + f_1) - (f_2 - f_1) = 2f_1.$$

In sum, if $y \propto bF^2$ and $F = F_1 + F_2$, y has Fourier components with frequencies

$$2f_1,\ 2f_2,\ (f_2 + f_1),\ \text{and}\ (f_2 - f_1).$$

Let us finally consider a second example of non-linear response, wherein the displacement is proportional to cube of the applied force: $y = cF^3$, with c as a constant. Then, if $F = F_1 + F_2$, it can be shown that y has components with frequencies $f_1, f_2, 3f_1, 3f_2, 2f_1 - f_2, 2f_2 - f_1,$

$2f_1 + f_2$, and $2f_2 + f_1$. (The reader can try to guess what result would obtain if $y \propto F^4$.)

Now that we have discussed how various non-linearities in response are manifested by the outputs of two sine waves, we now summarize how non-linearity manifested in the ear:

When a sound, having a mixture of two components, with frequencies f_1 and f_2, is incident upon the ear, tones having pitches corresponding to

$$\begin{aligned} f_{c1} &= f_2 - f_1 \\ f_{c2} &= 2f_1 - f_2 \\ f_{c3} &= 3f_1 - 2f_2 \end{aligned} \tag{11.12}$$

can be heard—in addition to the tones corresponding to the frequencies f_1 and f_2. These are the frequencies of the **combination tones**. Their existence indicates that the response of the ear is non-linear response. See Fig. 11.8, wherein these frequencies are graphed as a function of f_2, with f_1 kept fixed.

Experiments show that the basilar membrane within the cochlea responds non-linearly to the fluid forces acting upon it. The fluid does respond linearly to the pressure exerted by the footplate at the oval window. Ultimately, the basilar membrane responds non-linearly to the sound pressure p at the outer ear.

Now the sound pressure can be written as the sum of two components, corresponding to the two frequencies, f_1 and f_2:

$$p = p_1 + p_2 \tag{11.13}$$

The above perceived frequencies indicate that the response of the ear to the sound pressure is given by the sum

$$y = ap + bp^2 + cp^3 + ep^5 \tag{11.14}$$

where a, b, c, and e are constants. The first term accounts for the perception of the tones corresponding to f_1 and f_2, the second to $(f_2 - f_1)$, the third to $(2f_1 - f_2)$, and the fourth to

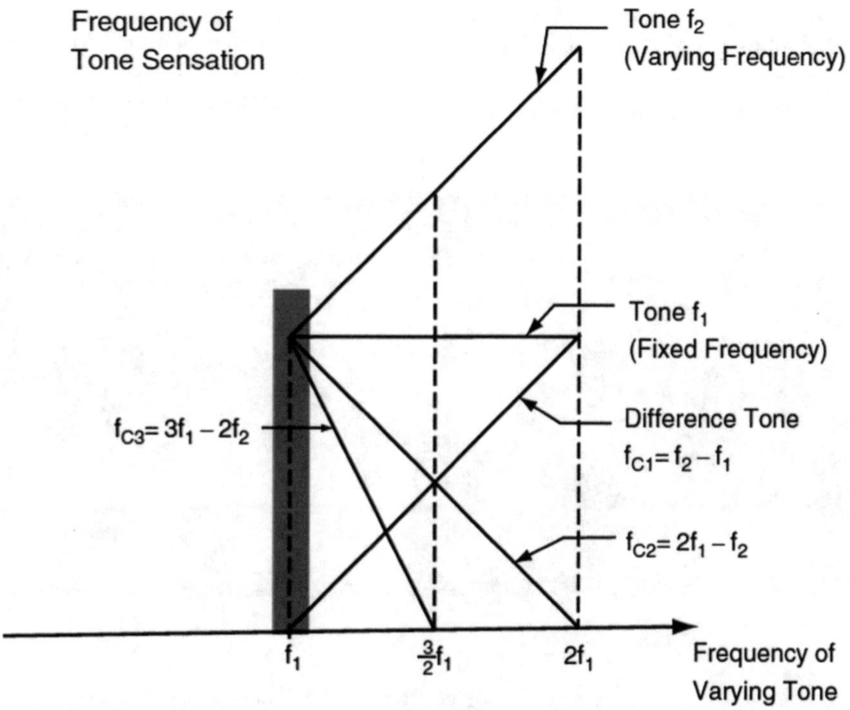

Fig. 11.8 Frequencies of combination tones (source: Roederer, op. cit.)

$(3f_1 - 2f_2)$. A difficult question to answer is why the remaining frequencies, such as $f_2 + f_1$ or $2f_2 + f_1$, which are expected according to the above response relation between y and p, are not perceived.

11.5 The Blue Color of the Sea and Its Connection with Combination Tones

In Chap. 9 we discussed how the scattering of light by the air is responsible for our being able to see the blue sky. Certainly, when we look down into the sea, scattering of the light is responsible for our being able to see its blueness. We might therefore conclude that the preferential scattering of short wavelengths is the most important reason for the blue color of the sea. Evidence to the contrary is indicated by the fact that if you view sunlight from within the sea looking upwards, the light will be blue—in contrast to the red appearance of the setting sun. The primary source of the blue color of the sea is the absorption by water of light in the red region of the visible spectrum. Figure 11.9 exhibits the blueness of water from above the surface as well as from below. The deeper you are in the water, the deeper is the blue of the light because of the increased absorption of red.[12]

Interestingly, it was difficult to understand how water could exhibit strong absorption in this range of wavelengths since there are no corresponding energy level differences among the quantum states of water.[13] There are vibration modes in the infrared region; they are the key to the explanation. It turns out that the vibration of the atoms in a water molecule does not obey Hooke's law precisely. There is a *non-linearity* in the inter-atomic forces. As a result, the spectrum of frequencies includes **combination modes**—analogous to the combination tones that we hear due to the non-linear response of the ear! It has been shown that the strong absorption in the red is due to two modes of vibration—commonly labeled $\nu 1$ and $\nu 3$—that produce a *fourth order combination mode* with frequency $f_c = f_1 + 3f_3$

[12]If you look down into the sea, the light you see is primarily blue scattered light. If you look upwards from the depths of the sea, the effect of preferential scattering *towards* the blue is more than compensated for by the preferential absorption *in* the red. The color is blue, in contrast to the redness of the setting sun.

[13]See Chap. 6 for a review of the connection between absorption and energy level differences.

(a) **(b)**

Fig. 11.9 (**a**) Moraine Lake, Bannf, Canada (**b**) Egypt Orange Spine Unicorn fish in the Red Sea (sources: (**a**) http://en.wikipedia.org/wiki/File:Moraine_Lake-Banff_NP.JPG (**b**) Photo by Sami Salmenkivi, http://seafishes.wordpress.com/category/family/surgeonfish/)

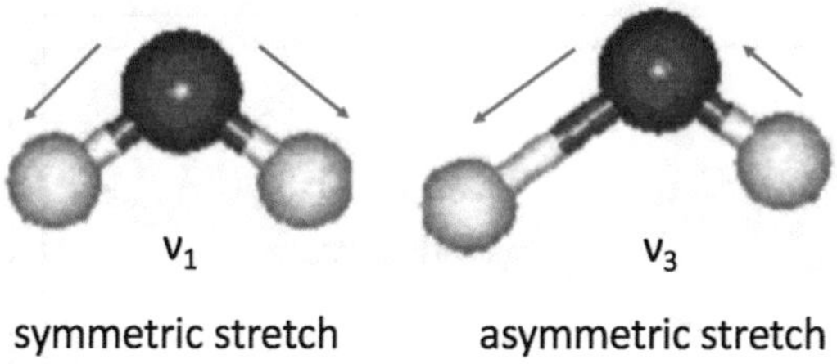

Fig. 11.10 Key Vibration Modes—$\nu1$ and $\nu3$—Responsible for the Blueness of the Sea (source: Martin Chaplin, http://www.lsbu.ac.uk/water/vibrat.html#2)

in the red low wavelength region.[14] These vibration modes are exhibited in Fig. 11.10.

11.6 Duration of a Note and Pitch Discrimination

How long must a single note be played for you to be able to have a clear sense of its pitch? Imagine a pianist playing a series of notes extremely rapidly—say at thirty-two notes a second. The duration of a single note is a mere 1/32 of a second. In case you haven't done so, I suggest that you play a series notes on the piano by striking the keys sharply. See what happens as you move downwards to ever lower notes. I am sure that you will find that it gets very difficult to sense the pitch when you play the very lowest of notes. Why is this so?

The answer to this question has to do frequency. Imagine if the frequency is but 20-Hz. Then with a duration of $\tau = 1/32$ second, there will be less than one cycle of the sine wave—in fact, $f\tau = 20/32 = 5/8$ cycles. It is unreasonable that one can have any sense of pitch in this case. We would expect that the duration must be such that the number of cycles N is at least on the order of two or three.

In general, the number of oscillations N of a sine wave of frequency f over a time interval τ is given by

$$\boxed{N = f\tau} \tag{11.15}$$

For example, if the frequency is 440-Hz and the time interval is 1/100th of a second, the number of oscillations is $N = f\tau = 440 \times 0.01 = 4.4$.

For a given time interval, the lower the frequency, the poorer is our sense of pitch. As a consequence, it is often difficult to perceive clearly the pitch of the notes at the far lower end of the piano. We can express the minimum duration as

$$\tau_m = \frac{N}{f} = NT \tag{11.16}$$

where T is the period.

We see how the necessary duration is inversely proportional to the frequency. We see a comparison of the Fig. 11.11 experimental results of Matti Karjalainen with the expectation—the dotted curve—according to this formula with N=2. Note that the horizontal axis is actually the logarithm of the frequency.[15] The rise in the minimum duration of time τ_m for very high frequencies is not explainable on the basis of our simplified approach. Note that for any given frequency, the time duration is simply proportional to the number of oscillations. Therefore, while increasing N will raise the curve, we wouldn't obtain a good fit for any value of N. For example, doubling N to $N = 2$ will lead to a good fit at 0.2-kHz but would make τ_m too great for a frequency of 0.1-kHz and too small for a frequency of 0.5-kHz.

We can apply the above to our ability study to clearly perceive the pitch of a sequence of notes, say an **A-minor triad**, A, C, E, with the A having a frequency of 440-Hz. One test would be to see how easily this triad can be distinguished from an **A-major triad**, A, C#, E, if for both

[14] See Charles L. Braun and Sergei N. Smirnov, J. Chem. Edu., 1993, 70(8), 612. For a more simplified discussion, see http://www.lsbu.ac.uk/water/vibrat.html#2.

[15] The expected curve is then $\tau \propto 10^{-b \log f}$, where b is a constant. We can see the exponential behavior of the dotted curve.

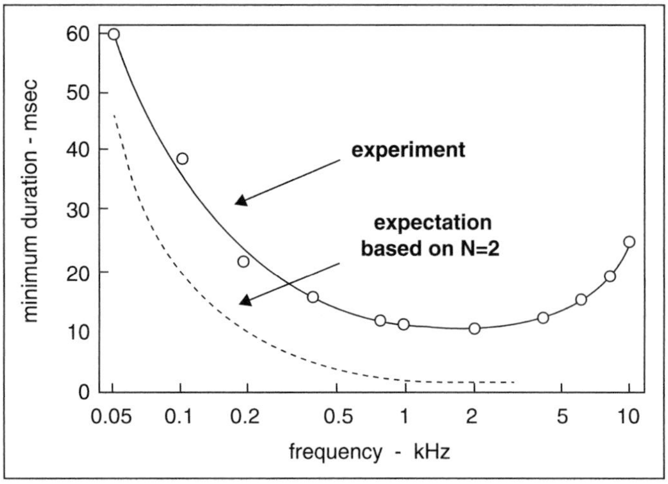

Fig. 11.11 Minimum Duration τ in msec to Discriminate a Pitch vs. Frequency on a log scale. Source: Courtesy of Jouni Hiltunen, http://www.acoustics.hut.fi/teaching/S-89.3320/KA6b.pdf

cases, the notes are played in a sequence of notes each of the same duration, one immediately after the other. In Chap. 12 we will learn that the frequencies for these sequences are, respectively: {440-Hz, 528-Hz, 660-Hz} and {440-Hz, 550-Hz, 660-Hz}. Thus, the two middle notes are 22-Hz apart, which is not too small a value. [22-Hz is $\sim$ 4% of 52-Hz.] If our goal is merely to distinguish between the two triads, we might be content to estimate the time needed to be greater than (1/11-Hz) $\sim$ 0.1-s. The author tested himself and found that indeed a time interval of about 0.1-s sufficed but 0.05-s was too short a time interval.[16]

Note The above results have a great impact on the playing of a double bass. Its lowest frequency is 41-Hz for a four-string bass. As a consequence, the perception of a sense of pitch is extremely difficult in the performance of pizzicatos, which have short durations. Tuning a double bass requires extreme attentiveness.

11.7 Fusion of Harmonics—A Marvel of Auditory Processing

We take for granted that when a musical instrument plays a note we will hear the sound of but one source of sound. We have already learned in Chap. 2 that the frequency spectrum of the periodic wave of a musical instrument is a harmonic series, with the frequency of the wave. We have also learned that the timbre of an instrument is partly determined by the relative amplitudes and phases of the Fourier components associated with the instrument. On the other hand, more than three decades ago, I began to use a device made by the PASCO corporation for producing an electronic signal consisting of a periodic wave having up to nine harmonics, with a fundamental of 440-Hz. Such a device is called a **synthesizer**. The electronic signal was fed into an amplifier, which was connected to a loudspeaker. All listeners reported that they could hear the individual harmonics in the sound. Each harmonic was audibly separated, as if the harmonic had come from a separate source of sound. What is the difference between the wave produced by the Pasco synthesizer and the sound

[16] I decided to test my 12-year-old grandson, Elias Seidman, who has **absolute pitch**. That is, he has the ability to identify a note without the need to be given a base note, such as the note "A" with a frequency of 440-Hz (referred to as "A-440"). Note that the A-440 pitch is engrained in Elias' brain just as a specific red color is engrained in a person's brain. [I myself have **relative pitch**, which allows me to identify a note once given a base note such as the "A-440"]. Moreover, Elias' notes correspond to those on a piano, so that they are in **equal temperament**, a subject discussed in Chap. 12. Elias was able to distinguish between a **C-minor triad** and a **C-major triad** and between an **A-minor triad** and an **A-major triad**, with the notes having a duration of only 0.05=1/20th second.

wave produced by a musical instrument? There is no visible characteristic on an oscilloscope trace that indicates a difference. Perhaps there is a significant difference that is too small to see on the trace with one's eyes.

We will discuss in this section the fact that the brain processes the sound so that we do not hear the individual harmonics. This process is referred to as **fusion of harmonics.**[17,18]

Fusion in the Taste of Food

It might be unclear to some readers what I mean by fusion of a mixture of harmonics. We can get some idea of fusion by considering the taste of a homogeneous dish of food. I recall many years ago, finding Lobster Cantonese an extremely delicious dish. When I first attempted to prepare the dish myself, I was amazed to learn that the essential ingredients were lobster and garlic. How well I knew the taste of garlic and yet how surprised I was that garlic was an essential ingredient in the recipe. Somehow, the blend of garlic and lobster produced a taste all its own—that of Lobster Cantonese—with the flavor of neither ingredient standing out. And so it seems to be with most superb dishes—as long as they are prepared properly. As another example, we can consider curries. Most are such that the ingredients are individually recognizable; there are fortunately some that have a wonderful homogenized flavor all their own. And so it seems to be with the fusion of a mixture of harmonics from a musical instrument!

It is interesting to consider and to try to perceive what the world of music would be like in the absence of fusion: The sound of a musical instrument would be heard as an ensemble of harmonics that would be superimposed with those of other musical instruments. We would lose our ability to separate out the sounds of the ensemble of instruments. Vibratos might lose their sweetness of tone. And so on.

Fusion of Harmonics is responsible for the rich beauty of musical instruments.

Here are some important interesting questions to be investigated:

1. What accounts for the ubiquitous fusion of the sound produced by musical instruments?
2. What are the conditions under which a sound wave with a superposition of harmonics will be fused, will be perceived as having a single source? Factors that seem to be important include:
   (a) the frequency of the fundamental—studies indicate that the lower the fundamental frequency, the greater is the degree of fusion;
   (b) the number of harmonics present and their relative amplitude;
   (c) the presence of "proportional modulation," sometimes referred to as "parallel

---

[17] I have been greatly helped in my attempt to weed out the known understandings of fusion by two audio-psychologists: Alan Bregman of McGill University, Brian Roberts of Aston University (Birmingham, England) and Oliver Knill of Harvard University. Alan Bregman is the author of a book entitled *Auditory Scene Analysis* [MIT Press, Cambridge, 1994], in which he discusses how the brain processes an ensemble of sound inputs and organizes them according to sources. In particular, he explains how the brain is able to focus on one source of sound and ignore or become almost oblivious to other concurrent sources. As a result we are able to hear one person speak in the midst of a dense crowd at a party.

[18] For a resource of introductory material and references on this subject, see: (1-2-2011): http://jjensen.org/VirtualPitch.html#use.

A more general concept than fusion of harmonics is the concept of **virtual pitch**. It takes into account the tendency of the brain to choose a pitch to be perceived even if the frequency spectrum is not a perfect harmonic series and there is ambiguity of the pitch. The concept of virtual pitch is attributed to **Ernst Terhardt**. See his publications: **Pitch, consonance, and harmony**, Journal of the Acoustical Society of America, **55** #5 1974. p.1061–1069, and **Calculating Virtual Pitch**, Hearing Research **1** 1979. p.155–182. Below are references to fascinating illusory responses to sequences of complex sounds:

(1) **Shepard's Staircase**, wherein a sound seems to be ever decreasing in pitch but is actually cycling around like M.C. Escher's staircase. For an incredibly hilarious representation of this illusion, see (1-21-2011): http://www.flixxy.com/escher.htm. To listen to Shepard's staircase, see (1-21-2011):

http://www.cycleback.com/sonicbarber.html

Reference: **Shepard, R.N.** (1964). **Circularity in judgments of relative pitch**, J. Acoust. Soc. Am., **36**, 2346–2353.

(2) **Diana Deutsch**(1-21-2011):

http://www.philomel.com/musical_illusions/octave.php

Included are a number of sound files that allow you to listen to illusions.

modulation." Consider frequency modulation, which is characterized by two parameters. There is a rate at which the frequency is modulated—label it f_m. Next, there is an amplitude of variation of the frequency—label it Δf. Proportional frequency modulation would involve each harmonic being modulated with the same frequency of modulation f_m but with a variation f_a in proportion to the harmonic number n.

Here are some possibly relevant sources of frequency modulation:

Acoustic stringed instrument

A major source of frequency modulation is **vibrato**. A small periodic variation in the length of string that is free to vibrate will lead to proportional variation in each of the harmonics. Here is another possible source of frequency modulation, small as it may be: Normally we think of the vibrating string as having two fixed ends. However, the transmission of sound waves involves the string moving the bridge. Therefore, the string is not absolutely fixed at the bridge. Therefore, the string's length and its tension are modulated.

Wind instrument such as a flute

Here too—there is a modulation of the frequency due to the vibration of the mouth of the musician.

3. Are there strong variations in the auditory processing of people such that sounds that are fused for some people are not fused for others. It is reported that some people can sometimes distinctly hear the individual harmonics produced by a musical instrument.

11.8 Additional Psychoacoustic Phenomena

We summarize below additional psychoacoustic phenomena not covered in this text:

1. The variation of pitch with loudness: Our sense of pitch varies with loudness. This phenomenon reflects a dramatic difference between a **subjective perception** and an **objective input**. We see the results of experiments of this phenomenon in Fig. 11.12.[19] The central features of the graphs are the following: At about 2-kHz, pitch doesn't depend upon loudness. Below 2-kHz, the pitch decreases with increasing loudness, while above 2-kHz it decreases with increasing loudness.

   Note that when musicians tune their instruments for the purposes of having a shared tuning, they share their notes at a low sound level.
2. **just noticeable difference in frequency** as a function of frequency: How different can two sounds be with respect to frequency and still be distinguishable?
3. **just noticeable difference of loudness** as a function of frequency: How different in loudness can two sounds be and still be distinguishable?
4. **second order beats**, also referred to as **mistuned consonances** result in a sense of beating between the corresponding harmonics of the two tones.
5. **masking**: One sufficiently intense tone of a certain frequency will mask a second tone having a different frequency and much lower intensity.
6. A psychoacoustic basis for **consonance**.

The reader is encouraged to read other resources that describe the incredibly rich experiences connected with both psychoacoustics and auditory processing. A recent outstanding book at a layman's level is *This is Your Brain on Music*, by Daniel Levitin [Penguin Group, New York, 2006]. Another book at a higher level is *Tuning, Timbre, Spectrum, Scale*, by William A. Sethares [Springer-Verlag, London, 2nd edition]. To appreciate the latter book the reader should first study Chap. 12 on musical scales. And finally we mention *Musicophilia: Tales of Music and the*

[19] The following website has sound clips that allow you to determine how the pitch you hear changes with increasing intensity. (1-21-2011): http://www.santafevisions.com/csf/demos/audio/412_dependence_pitch_intensity.htm.

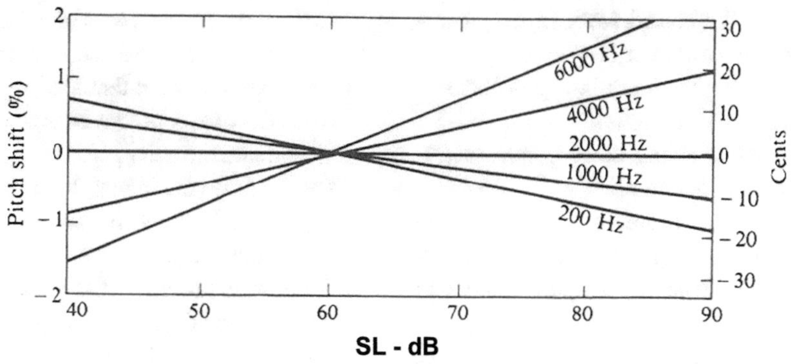

Fig. 11.12 Change in Pitch with Respect to Sound Level (Based upon *Hearing Research*, vol. 1, p. 162, (1979), "Calculating Virtual Pitch," Ernst Terhardt, with permission from Elsevier)

Brain, by Oliver Sachs [Alfred A. Knopf, New York, 2007].

11.9 Terms

- aural harmonics
- combination tone
- difference tone
- equal loudness curve
- general force
- harmonic distortion
- linear response
- masking
- non-linear response
- phon

11.10 Important Equations

mathematical relation between the sone and the phon:

$$s = 2^{\frac{\phi - 40}{10}} \tag{11.17}$$

combination tones:

$$(f_2 - f_1), (2f_1 - f_2), (3f_1 - 2f_2)$$

11.11 Problems for Chap. 11

Five **pure tones** (which we will call A, B, C, D, and E) are sounded with the physical characteristics shown in table above. Use the equal loudness curve in the text to answer the following three questions.

Tone	Frequency (Hz)	Intensity (W/m^2)
A	1000	1.0×10^{-10}
B	100	1.0×10^{-10}
C	100	1.5×10^{-7}
D	10,000	1.5×10^{-8}
E	15,000	9×10^{-8}

1. For a normal young ear, which tone is probably the loudest?
   (a) A
   (b) B
   (c) C
   (d) D
   (e) E
2. For a sixty-year-old person, which tone or tones are probably inaudible?
   (a) B
   (b) E
   (c) B and E
   (d) A, B, and E
   (e) A, B, D, and E
3. For a normal young ear, which tone or tones sound equal in loudness to tone C?
   (a) B
   (b) D
   (c) E
   (d) D and E
4. (a) What is the **phon level** of a 60-dB sound at 100-Hz?

(b) What is the **sound level** in dB of a 10,000-Hz sound of 20-phons?
(c) What is the loudness in sones corresponding to 80-phons?
(d) How many phons corresponds to a loudness of 1/4 sone?

5. Repeat Sample Problem 11.2 with the change that the frequency is 100-Hz and the two phon levels are 20-phons and 100-phons.
6. Consider a sound of frequency of 400-Hz and sound level of 40-dB.
   (a) What is the number of phons?
   (b) Find the loudness in sones.
   (c) Suppose that we want to double the loudness in sones of the sound. Find the required number of phons and the sound level.
7. Suppose that we have one-million bees, each producing a sound of frequency 200-Hz and a sound level of 15-dB. The determination of the loudness of the total sound produced is the object of this problem.

   See Sample Problem 11.3.
   (a) Find the corresponding phon level of the sound from one bee.
   (b) Find the corresponding loudness in sones of the sound from one bee.
   (c) Find the corresponding sound level in **decibels** and the loudness in **phons** of the sound of one-million bees.
   (d) Find the corresponding loudness in sones of the sound of one-million bees.
8. Again referring to Sample Problem 11.3, note that when the two sources with about the same frequency have one sone each, the resulting sone level was about 1.1 sones, which is less than the two sones resulting had the frequencies not been close. Yet, we should note that this value of 1.1 depended upon the frequency being 1000-Hz, for which a change in SL equals a change in phons.
   (a) Explain why, in order for the change in sones to exceed two, the frequency must be such that a change in SL of 3-dB amounts to a change of at least 10 phons.
   (b) Study the equal loudness curves in Fig. 11.2 and see whether there is any such frequency.
9. Here is a problem that professionals who produce musical recordings must contend with. It is also an issue that affects all music we hear, whether live or from recordings.

   For simplicity, suppose that we mix two pure tones, one at 200-Hz and the other at 1000-Hz. We want them to have the same loudness. We mix at a level that the 1000-Hz tone has a level of 20 phons.
   (a) Determine the sound level of the 200-Hz tone?
   (b) Now suppose that the **intensity** of both tones is increased by a factor of one-hundred. Determine the resulting sound level and the phon level of each tone, noting that the latter is now unequal.
10. Most people would regard the moth as being quite primitive and helpless against attacks by a predator. However, it has been known for over 50 years that moths have auditory perception using eardrums and a few sensory cells that can detect an impending attack by predator bats. Recently,[20] it was reported that a moth's eardrum is not static but adjusts itself to the changing sound from bats. Bats use ultrasound to locate their prey and, of course, to avoid obstacles in their path. The range of frequencies of a bat during general

[20] http://www.sciencedaily.com/releases/2006/12/061218122629.htm.

flight is usually in the range of 20 to 40 kHz. This is a range of frequencies for which the moth's eardrum is most responsive in its resting state.

However, there are two interesting changes in bat and moth behavior in the course of a predatory attack.

Explain the physical principles that account for these changes.

(a) When homing in on prey, a bat's frequency is increased to a much higher range, up to about 80-kHz.

(b) When a bat is approaching a moth, the increased sound intensity leads the moth's eardrum to become stiffer.

11. (a) What are the frequencies of the significant **combination tones** produced by two tones having frequencies 500-Hz and 750-Hz?

(b) What is the hypothesized physical basis for the perception of combination tones?

Tuning, Intonation, and Temperament—Choosing Frequencies for Musical Notes

12

Music is defined by a pattern of notes distributed over time with each note being associated with a frequency. However, we are left with the decision as to what these frequencies should be. This chapter focuses on this decision. When an ensemble of musicians, whether small or as large as a symphony orchestra, gather to perform a piece of music, the musicians must tune up their instruments, that is, make sure that the frequencies they produce for each and every note are essentially equal. In the absence of a high level of ensemble tuning, the music produced will not normally be acceptable. There are many tunings that have been used over the centuries and throughout the world. We will focus on Western tunings. The most fascinating observation is that mathematics demonstrates to us that maximal resonance among a set of notes cannot be achieved.

While preparing to focus on the introduction to Chap. 12 for this second edition, I gave great thought to my attempt to perfect my playing the violin part of a magnificent aria in Bach's oratorio the "St Matthew Passion": **Erbarme dich mich, mein Gott**. Achieving a high level of **intonation** is paramount for achieving a beautiful tone. However, before playing a piece of music the strings of my violin have to be tuned to utmost perfection. The reason is that my fingers cannot compensate for the loss of resonances that result from strings that have not been properly tuned.

Imagine yourself seated in a concert hall, anticipating the beginning of a symphony orchestra performance. The musicians are all seated. The concertmaster rises and calls to the oboist to sound the "A," which will be the standard pitch that all others will use to tune their instruments. In advance the oboist has tuned the oboe to the standard frequency chosen by the orchestra, which is usually a frequency of 440-Hz. The winds and brass and tympani tune their instruments accordingly.

Next, the strings tune their A-strings to the oboe's "A." Following that, the strings tune their other 3 strings accordingly. For example, the violins tune the "E," which is a musical "fifth" above the "A," played simultaneously with the "A." The violinist strives for the maximal beauty of a resonance between the two strings. Tests reveal that the ratio of the two frequencies is very close (within a fraction of a percent) to $f_E/f_A = 3 : 2$. The violinist continues in a similar manner with the "D" a fifth below the "A" and then the "G" a fifth below the "D." The frequency ratios—high to low string—again will be close to 3:2. The harpist has already tuned the harp's 47 strings to the standard frequency for the A, but in advance and while on stage so as to avoid changes that might result from moving the harp. Upon hearing the oboe, the harpist might have to quickly make fine adjustments.[1]

[1] A harpist acquaintance of mine, Judith Ross, told me the following: "Harpists are always nervous about the tuning. I tune a little higher than 440, even though the oboe blows 440, because the violins always seem to tune or go sharp.

L. Gunther, *The Physics of Music and Color*,
https://doi.org/10.1007/978-3-030-19219-8_12

If a piano is to be played, its 230 strings (!) would have been previously tuned to the standard "A" before the time of the performance. The pianist will not have an opportunity to make any changes during the concert.

Once tuned, the musicians will perform the concert by producing strings of notes with frequencies that are limited by their respective instruments. Of course, musicians will frequently have to make adjustments of the tuning of their instruments during a concert due to changes that are wont to take place—such as temperature changes, changes in the tension of a string that might not have been stabilized from its new state, and so on. The string players will have liberty to choose notes that involve shortening the length of string that is free to vibrate by varying the placement of their fingers on the strings. Winds and brass can choose notes by the covering holes on the instrument or changing the length of the pipe in the case of a trombone. Other small changes in frequency can be made by the winds and brass by changing the way the mouth blows into the instrument or covering the end of the instrument in the case of the brass. The harpist and pianist cannot make such adjustments.

Those notes that cannot be changed significantly are referred to as **fix tuned**. A piano is an entirely fix tuned instrument. The strings, winds, and brass are only partly fix tuned.

The whole process of tuning an orchestra is a grand display of majesty, reflecting the goal and ability of a group of people to get together to produce an extremely well-organized act of cooperation leading to the heavenly sound of the music of a symphony orchestra.

The focus of this chapter is to study the complex aspects of the choice of frequencies that are played once given a standard "A." This choice has two components: First is the **intended frequency**, which we refer to as **temperament** or **tuning**. Second is the **actual frequency** in a performance. The latter is best referred to as **intonation** and can reflect either intended choice, where choice is possible, or a mistaken, unintended outcome. "Bad intonation" refers to a disagreeable resulting pitch. Above all, in this chapter we will learn about the central role of **numerology** in the process of choosing frequencies and of the fact that a compromise cannot be avoided in an attempt to maximize consonance. Total consonance in its usual meaning can be shown to be impossible mathematically!

Since music is sound and sound consists of waves, what are the waves that music is made of? The focus in this text is on music that has a definite pitch. And, as we have seen, pitch is, for the most part (but not entirely) determined by the frequency of a periodic wave.[2] Given that the audible range of frequencies spans $\sim$ 20 Hz to $\sim$ 20, 000 Hz, a musical composition could call for musical notes whose fundamental frequencies span this entire range, *continuously*. Instead, cultures have produced musical compositions which make use of certain *discrete* sets of frequencies. These sets are called **musical scales**.

To some extent, the choice of frequencies is analogous to the set of colors of paints that a painter places on his/her palette, putting aside the fact that a painter uses the set of colors to produce a continuum of other colors by mixing the base set in various proportions. See Fig. 12.1, in which we exhibit this analogy with a set of frequencies corresponding to a **pentatonic scale** (to be discussed below) laid out on a palette.

The **pentatonic scale** is found throughout the world. The two other most common musical scales are the **diatonic scale** and the **chromatic scale**, which are used in the Western world. We will discuss only these three. Also, we will use the symbols and terminology of Western music. The key question that any serious musician must deal with is:

And I'm used to adjusting as I go. When I play in the pit for shows, I'm constantly re-tuning during the dialogues. I've even replaced broken strings during dialogues. In orchestral concerts if I hear something wacky, I try to tune the offending string(s) as inconspicuously as possible. Stravinsky is believed to have said that 'Harpists spend 90% of their time tuning their harps and 10% playing out of tune.' " Here is a source of this quote (1-23-2011): http://en.wikiquote.org/wiki/Igor_Stravinsky.

---

[2]Note that I refer here to a **'periodic wave'**, not a sinusoidal wave that is associated with a **'pure tone'**.

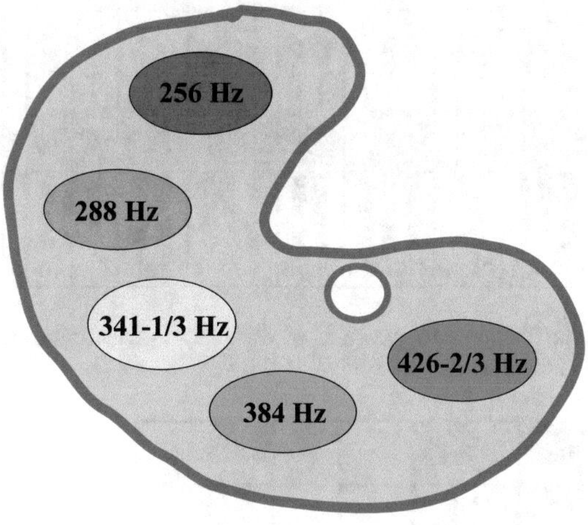

Fig. 12.1 Frequency palette

What should be the frequencies of the musical notes?

The goal of this chapter is to study the bases for the choices in tuning or the choice of the frequencies. We will begin with a discussion of musical scales, which are the backbone of what is referred to as tonal Western music. We then briefly discuss Pythagorean tuning, which is one of the oldest mathematically defined tunings in Western music. We then move on to Just tuning and discuss its drawbacks. Finally, we discuss the most widely used tempered tuning—Equal Temperament. In problem 14 of this chapter, we discuss Werkmeister I(III) temperament, which was one of a number of temperaments that were popular in the Baroque era (ca. first half of the eighteenth century).

12.1 Musical Scales

Every scale has a **key note**, which is the first note of the scale: A musical composition tends to be drawn to that note so strongly that almost invariably, the last note of a Western composition is the key note. The composition tends to feel incomplete and produces a certain tension unless this is so.[3] To be specific, we will assume that we are dealing with a scale that has "**C**" as its key note. We say that the "scale is in the **key of C**".

Most fundamental in determining the discrete set of frequencies chosen for a scale is the **octave**: Two frequencies which are in a 2:1 ratio are said to be an **octave** apart. They sound much alike. Some people cannot tell them apart. Once we choose a certain frequency for the so-called **middle-C** on the piano, we can generate an infinite set of other frequencies that are octaves apart. Consequently, the entire spectrum of notes consists of a series of identical octaves of notes. We will use the symbol C_4 to denote middle-C of a **piano keyboard** (Fig. 12.2). The frequencies indicated are those associated with the A above middle-C set at 440-Hz and the other notes tuned with **Equal Temperament**, which will be discussed in detail in this chapter.

Suppose, for simplicity, we choose $f_{C_4} = 250$ Hz. Then an octave above C_4 is C_5, with $f_{C_5} = 2 \times 250 = 500$ Hz. Continuing on, we obtain

$$f_{C_6} = 2 \times f_{C_5} = 4 \times f_{C_4} = 1,000 \text{ Hz}$$

$$f_{C_7} = 2,000 \text{ Hz}$$

$$f_{C_8} = 4,000 \text{ Hz}$$

$$\ldots$$

Moving downward, we obtain $f_{C_3} = (1/2) f_{C_4} = 125$ Hz, ...etc. The question is: How should we fill in the notes with discrete frequencies between a C and its octave above, C_5?

Note *Once this range is filled in, all other notes are determined by octave relationships, as above.* In this connection, we should note that the term **octave** has an additional meaning: It is also used to refer to the *set of notes* spanning from a given note to its octave note above. Thus, we may also refer to the **octave of notes** ranging from C_4 to C_5 or from C_5 to C_6.

Moving from one note to another is referred to as taking **steps**—e.g., making a *semitone step*.

Considering all the fuss we will be making about tuning and intonation in theory, how do instrumentalists *actually* tend to choose their frequencies? There has been much study of this

[3]In fact, some composers *intentionally* end on a note other than the key note so as to leave the listener in a state of tension!

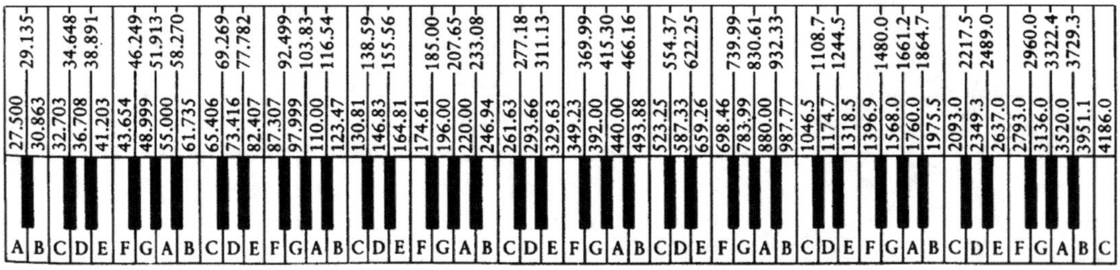

Fig. 12.2 Piano keyboard with a corresponding **musical staff**. Shown to the right of each key is the frequency according to equal tempered tuning, which will be discussed later on in this chapter

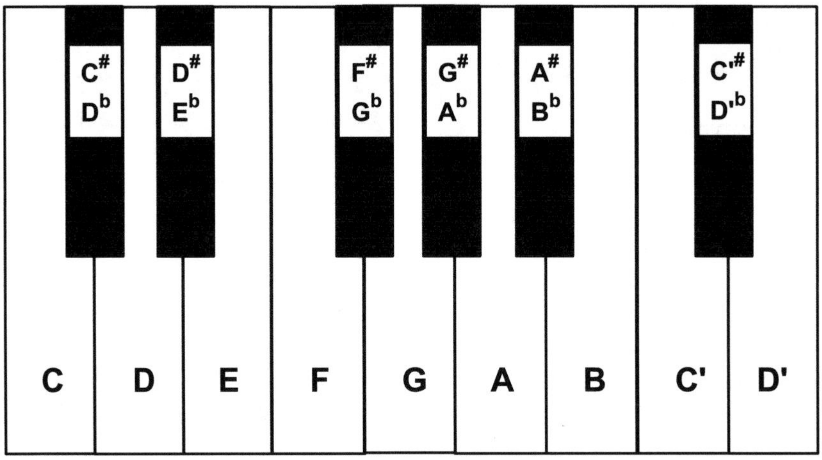

Fig. 12.3 Piano keyboard

last question; unfortunately, the question is beyond the scope of the text.[4]

12.2 The Major Diatonic Scale

The notes of the diatonic scale in the key of C are **C, D, E, F, G, A, B, C'**. They are represented by the white keys on a piano keyboard. See Fig. 12.3.

*The **musical interval** expresses the relationship between the pitches and hence frequencies between a pair of notes.* Musical intervals have the following names:

- C to C': **Octave**
  so-called because there are eight notes in the diatonic scale.
- Intervals between neighboring notes in the diatonic scale are either **semitones** or **whole tones**.
- A semitone is also called a **minor second.**
  Examples: E-F and B-C
- A whole tone is also called a **major second.**
  Examples: C-D, D-E, G-A, and A-B
- **minor third** = 1-1/2 whole tones = 3 semitones
  Examples: D-F and A-C
- **major third** = 2 whole tones = 4 semitones
  Examples: C-E and G-B
- **fourth** = 2-1/2 whole tones = 5 semitones
  Examples: C-F, D-G, and A-D
- **fifth** = 3-1/2 whole tones = 7 semitones
  Examples: C-G, G-D, and D-A
- **minor sixth** = 4 whole tones = 8 semitones
  Examples: B-G and E-C

[4]For further details, see the classic text on the psychology of music: Seashore, Carl Emil, 1866–1949. Psychology of music, by **Carl E. Seashore**.New York, Dover Publications [1967] ML3830.S32 P8 1967.

- **major sixth** = 4-1/2 whole tones = 9 semitones
  Examples: C-A and G-E
- **seventh** = 5-1/2 whole tones = 11 semitones
  Examples: C-B

Thus, in relation to the base note of C, the notes of the diatonic scale in C major are

- D: 2 semitones or one whole tone ... **major second**
- E: 4 semitones or 2 whole tones ... **major third**
- F: 5 semitones or 2-1/2 whole tones ... **fourth**
- G: 7 semitones or 3-1/2 whole tones ... **fifth**
- A: 9 semitones or 4-1/2 whole tones ... **major sixth**
- B: 11 semitones or 5-1/2 whole tones ... **seventh**
- C: 12 semitones or 6 whole tones ... **octave**

The *essential question* now, as before, is the following: *How should the frequencies associated with the notes in the* ***diatonic scale*** *be chosen?* This choice is referred to as **tuning**. We will discuss in detail the three most important tunings, **Pythagorean Tuning**, **Just Tuning**, and **Equal Tempered Tuning**.

The fact that a choice is open to us and must be made might seem foreign to someone who regards music as having a certain natural state of existence. Let it be known that even primitive peoples were aware of the necessity of making a choice. Reread the tale of the Huang Chung in Chap. 1, as a reminder. Most people are mere "consumers" of the music performed by musicians and aren't aware of the details that musicians have to dabble with. Musicians, in turn, have to rely upon the still more diligent studies of others.

The most important fact that must be recognized is that

musical intervals correspond to frequency ratios.

This fact can be understood from the following exemplary situation: Suppose that we choose a certain frequency for f_G to be the fifth above the chosen frequency $f_C = 250$ Hz, say 375 Hz, as in the case of the so-called **Just tuning**. Then the pair of octaves above these two notes, C' and G', must also be a fifth apart.

Now, since $f_{G'} = 2f_G = 750$ Hz and $f_{C'} = 2f_C = 500$ Hz, $f_{G'}/f_{C'} = 2f_G/2f_C$, or

$$\frac{f_{G'}}{f_{C'}} = \frac{f_G}{f_C} = \frac{3}{2}$$

The ratio 3/2 defines the ***Just*** *interval for a fifth.* Alternatively, consider the sequence of notes, C-E-G. C-E is a **major**3rd, while E-G is a minor 3rd. Together, they add up to the musical interval of a fifth, corresponding to C-G. We write

major third + minor third = fifth

This fact is exhibited in Fig. 12.4 below.

This equation is reflected by the following mathematical equation:

$$\frac{f_E}{f_C} \cdot \frac{f_G}{f_E} = \frac{f_G}{f_C} \tag{12.1}$$

We thus see that it is

1. the *ratio* of frequencies that defines the musical interval, and
2. *adding intervals* amounts to *multiplying by frequency ratios*.

We have arrived at the following stage in our study of tuning: We first choose the frequency of one note—say middle C. All octaves are determined by the unchangeable ratio 2:1. Then, since musical intervals correspond to frequency ratios, our task is to decide what the frequency ratios should be corresponding to the musical intervals introduced previously.

12.3 Comments Regarding Western Music

Before we discuss the three most important tunings, we would like to note some important aspects of Western music. We have indicated

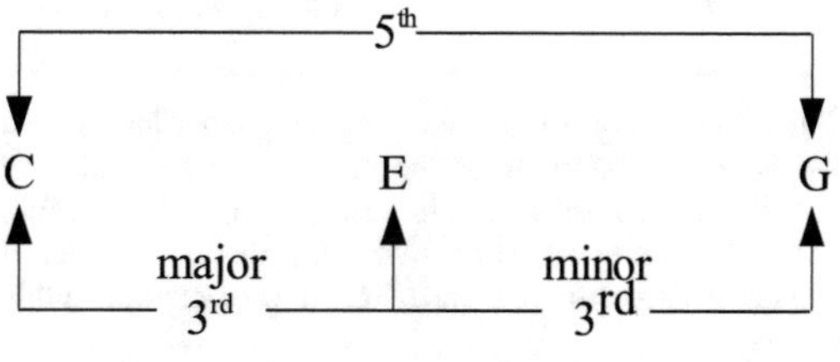

Fig. 12.4 Adding of the major 3rd and the minor 3rd to a 5th

that compositions of Western classical music are drawn strongly towards the **key note**: When we are away from the key note, we feel a pull towards it. Interestingly, the pull is strongest when we are close to a key note. It is true that other notes are pulling too: Most strongly is the fifth above the key note. (The G in the key of C.) Next in strength are the major third (the E) and the minor third (the E-flat). [Those knowledgeable about musical theory will note that which is the stronger of these last two depends upon whether the key is a major one or a minor one.] Next in strength is the fourth-above the C (in this case, the F). [This author must grant that some people feel that the fourth above is stronger than the major and minor thirds.] And finally, we have the major and minor sixths.

The pulling effect is associated only partly with our sense of **consonance** between the key note and each of the above notes—its opposite being **dissonance**. I suggest that this is so in spite of the fact that consonance is associated with a satisfying sound when two notes are played together: The key note is always stored in the memory bank of our brain and other notes are compared with it, whether or not the key note is sounded.

This desire for consonance plays a role in determining the tuning. The relevant question then is: What properties of a musical interval tend to produce a sense of consonance? Often it is said that a sense of consonance guides the tuner. But what characteristics are actually associated with consonance?[5] One definite characteristic traditionally is a sense of the richness of tone that is associated with resonance. To see how resonance is relevant, we will consider the tuning of a viola.

First the A string (which is the A above middle C) is tuned to some standard or reference frequency. These days, that frequency is taken to be a value between 440 and 444 Hz. In Bach's time, it may have been as low as 415 Hz! This frequency seems to be steadily increasing, reflecting an ever increasing preference for bright sounds.

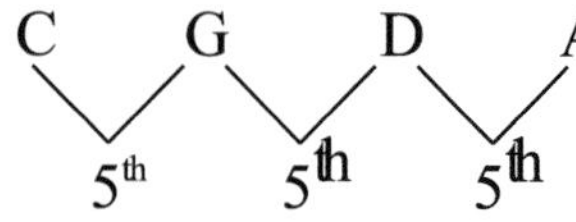

Fig. 12.5 Steps of 5ths

Next, the remaining strings are tuned so as to be fifths apart, as indicated in Fig. 12.5.

Now, suppose that the D is set at a frequency that is exactly 2/3 that of A, as in the case of Pythagorean or Just tuning. For concreteness, let $f_A = 441$ Hz. Then $f_D = 294$ Hz.

When the D-string is excited, the harmonics of 294 Hz are generally present. These are: 294 Hz, $2 \times 294 = 588$ Hz, $3 \times 294 = 882$ Hz, ...

When the A-string is excited, the harmonics of 441 Hz are generally present. These are: 441 Hz, 2×441 Hz = 882 Hz, ...

Note the match between the third harmonic of the D and the second harmonic of the A. A consequence is resonance between the vibrating strings. If the D and A are played together, this resonance enriches the tone quality. [A fascinating question is the extent to which there is a resonance taking place in the ear and possibly in the brain.[6]] It is interesting to note that it is much more difficult to adjust by ear two **pure tones** to a 3:2 ratio of frequencies. There is no sharp maximum of consonance at this ratio. Highly accurate tuning can, however, be accomplished by listening to the *combination tones* produced in the hearing process. [See Chap. 11.]

Generally, it is found that pairs of notes sound most consonant when their frequencies are in a ratio of small integers, an observation that leads us to **Just tuning**, which will be treated in Sect. 12.5. However, we will first study the simpler Pythagorean intonation.

[5] Of course, if the tuner has perfect pitch (that is, has an internal sense of what the pitch of an isolated note is without the need of a reference sound such as tuning fork), then consonance may not be a factor. We should want to understand better the nature of perfect pitch in this context.

[6] See The Physics of Musical Sounds by C.A. Taylor (The English Universities Press, London, 1965), Introduction to the Physics and Psychophysics of Music by Juan G. Roederer (The English Universities Press, London, 1973), and On the Sensations of Tone, by **Hermann L.F. von Helmholtz** (originally published in German in 1877, reissued in English by Dover, N.Y., 1954).

12.4 Pythagorean Tuning and the Pentatonic Scale

Pythagorean tuning is based upon assigning a frequency ratio of 3 : 2 to the fifth and treating all the remaining intervals as subservient to the fifth, as follows. We start by choosing the frequency f_{C_4}. Then, we determine the frequencies for the notes that arise as we ascend and descend in a series of fifths:

NOTE

F_3 →	C_4 →	G_4 →	D_5 →	A_5 →	E_6 →	B_6
2/3	1	3/2	$(3/2)^2$	$(3/2)^3$	$(3/2)^4$	$(3/2)^5$

Note how in descending down by a fifth from C_4 to F_3, we multiply by 2/3. The reason is that in doing so we must divide by 3/2, which is equivalent to multiplying by 2/3.

We will discuss only the *five* notes C, D, F, G, and A. These notes form a **pentatonic scale**, which is the essential scale in many parts of the non-Western world.[7] The process of determining the remaining frequencies that are needed to form the diatonic and chromatic scales is a mere continuation of the process we will demonstrate below.

What we want are the frequencies in the octave starting with C_4. They are determined by a process called **reducing to the octave** the above frequencies, as follows: $f_{F_4} = 2f_{F_3}$, $f_{D_4} = (1/2)f_{D_5}$, $f_{A_4} = (1/2)f_{A_5}$.

Thus, using $2(2/3) = 4/3$, $(1/2)(3/2)^2 = 9/8$, and $(1/2)(3/2)^3 = 27/16$, we obtain the pentatonic scale in **Pythagorean tuning** (see Fig. 12.6).

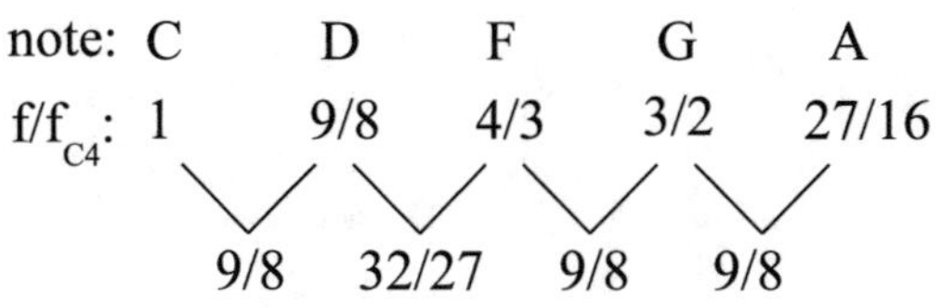

Fig. 12.6 Construction of the Pythagorean scale

[7]An easy way to hear this scale is to play the five black keys of the piano in order, starting with whatever first note you wish.

On the last line, we have indicated the intervals between neighboring notes. For example, for the interval between D and F we have:

$$\frac{f_F}{f_D} = \frac{\frac{f_F}{f_C}}{\frac{f_D}{f_C}} = \frac{\frac{4}{3}}{\frac{9}{8}} = \frac{4}{3} \times \frac{8}{9} = \frac{32}{27} \tag{12.2}$$

We note the results for the following **Pythagorean intervals** in Pythagorean tuning:

whole tone (C-D) :	9/8
minor third (D-F) :	32/27
major third (F-A) :	81/64
fourth (C-F) :	4/3
major sixth (C-A) :	27/16

We note that the two intervals of a third (the minor third and the major third) each involve a ratio of rather large integers. As such, these intervals do not sound very consonant in the traditional sense.

12.5 Just Tuning and the Just Scale

Just tuning sets a *priority* on ratios of small integers for the frequency ratios of two musical intervals, the Fifth—3/2, and Major third—5/4.

As we pointed out above, these two musical intervals are the central ones in Western classical music. We will now see how this choice for the two intervals is used to *generate* the Just notes.

As in the case of generating the Pythagorean scale, we start by choosing the frequency f_C of the key note C. From thereon, we are interested only in the ratio of the frequency of interest to this frequency. Let us lay out the notes of the diatonic scale with the following notation. We obtain the E and G straightforwardly, since C-E is a major third and C-G is a fifth. See Fig. 12.7.

We can obtain A by descending from E to A by a fifth and then reducing to the octave by ascending by an octave. This involves the product

$$2\left(\frac{2}{3}\right)\left(\frac{5}{4}\right) = 2\left(\frac{5}{6}\right) = \frac{5}{3}.$$

See Fig. 12.8.

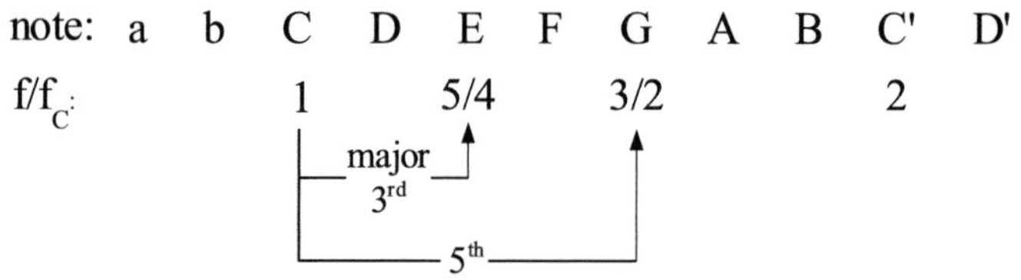

Fig. 12.7 Obtaining G from C in Just tuning

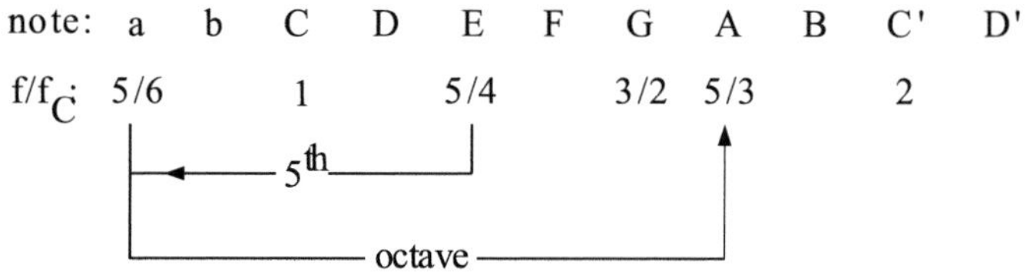

Fig. 12.8 Obtaining A from C in Just tuning

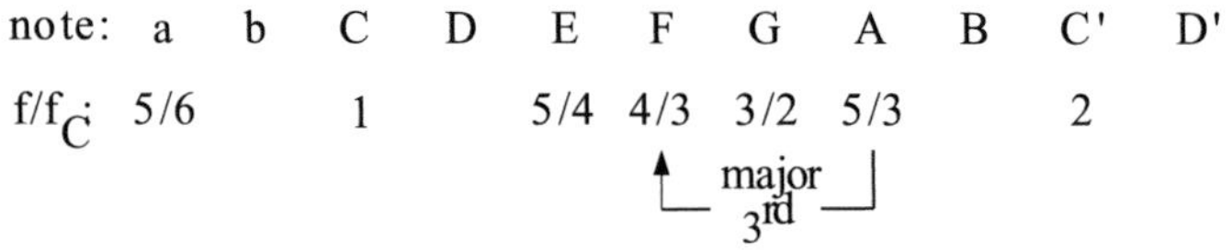

Fig. 12.9 Obtaining F from C in Just tuning

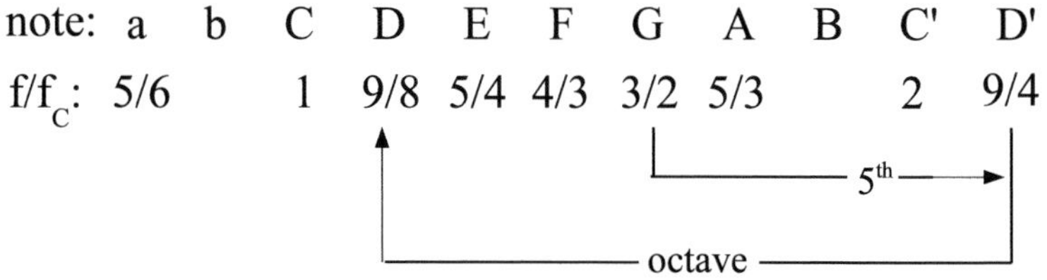

Fig. 12.10 Obtaining D from C in Just tuning

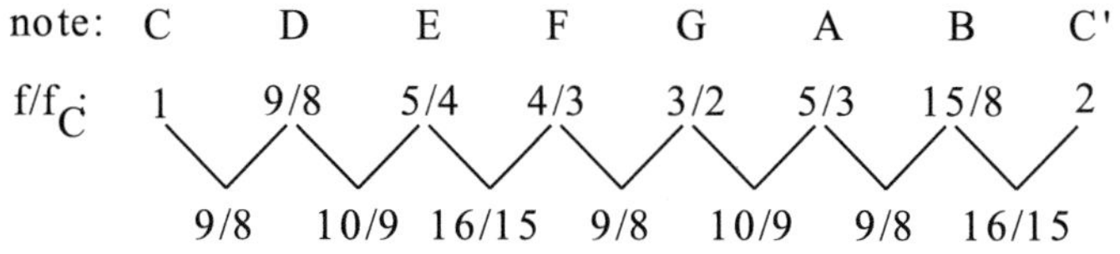

Fig. 12.11 Full Just diatonic scale

We can obtain the F by descending from A by a major third, thus obtaining the ratio $(4/5)(5/3) = 4/3$. See Fig. 12.9.

We can obtain the D by ascending from G by a fifth and then descending by an octave. The product is $(1/2)(3/2)(3/2) = (1/2)(9/4) = 9/8$. See Fig. 12.10.

Finally, we can obtain the B by ascending from E by a fifth. The product involved is $(3/2)(5/4) = 15/8$. The full Just diatonic scale is shown in Fig. 12.11.

Intervals between neighboring notes are indicated on the last line. They are obtained by taking ratios of the corresponding pair of ratios. For example,

$$\frac{f_F}{f_E} = \frac{\frac{f_F}{f_C}}{\frac{f_E}{f_C}} = \frac{\frac{4}{3}}{\frac{5}{4}} = \frac{4}{3} \times \frac{4}{5} = \frac{16}{15} \tag{12.3}$$

The basic music intervals in **Just tuning** are listed below:

semitone	16/15	fifth	3/2
whole tone	10/9 and 9/8	minor sixth	8/5
minor third	6/5	major sixth	5/3
major third	5/4	seventh	15/8
fourth	4/3		

We obtained the minor third, 6/5, from the interval E-G as follows:

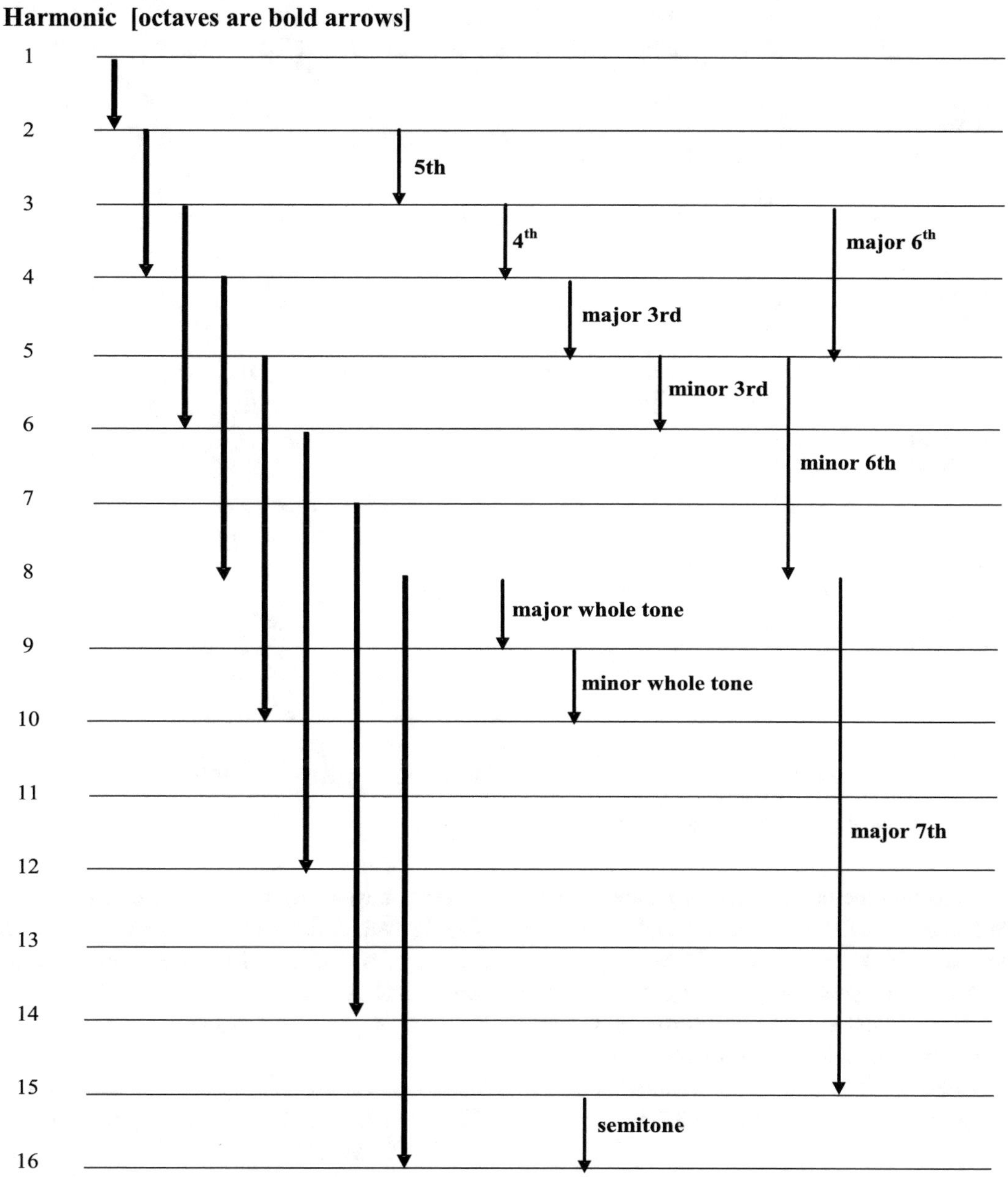

Fig. 12.12 Chart of important Just intervals

$$\frac{f_G}{f_E} = \frac{\frac{f_G}{f_C}}{\frac{f_E}{f_C}} = \frac{\frac{3}{2}}{\frac{5}{4}} = \frac{3}{2} \times \frac{4}{5} = \frac{6}{5} \qquad (12.4)$$

We obtained the minor sixth, 8/5, from the interval A-F', in a similar way.

Fig. 12.12 is a chart that lays out the important intervals in Just tuning in a clear way.

12.6 The Just Chromatic Scale

The **Chromatic Scale** includes all the so-called **sharps** and **flats** of the notes of the diatonic scale in the key of C. These are, respectively, semitones above and below the corresponding notes. Thus $C^{\#}$ is a semitone above C, B^{b} is a semitone below B.

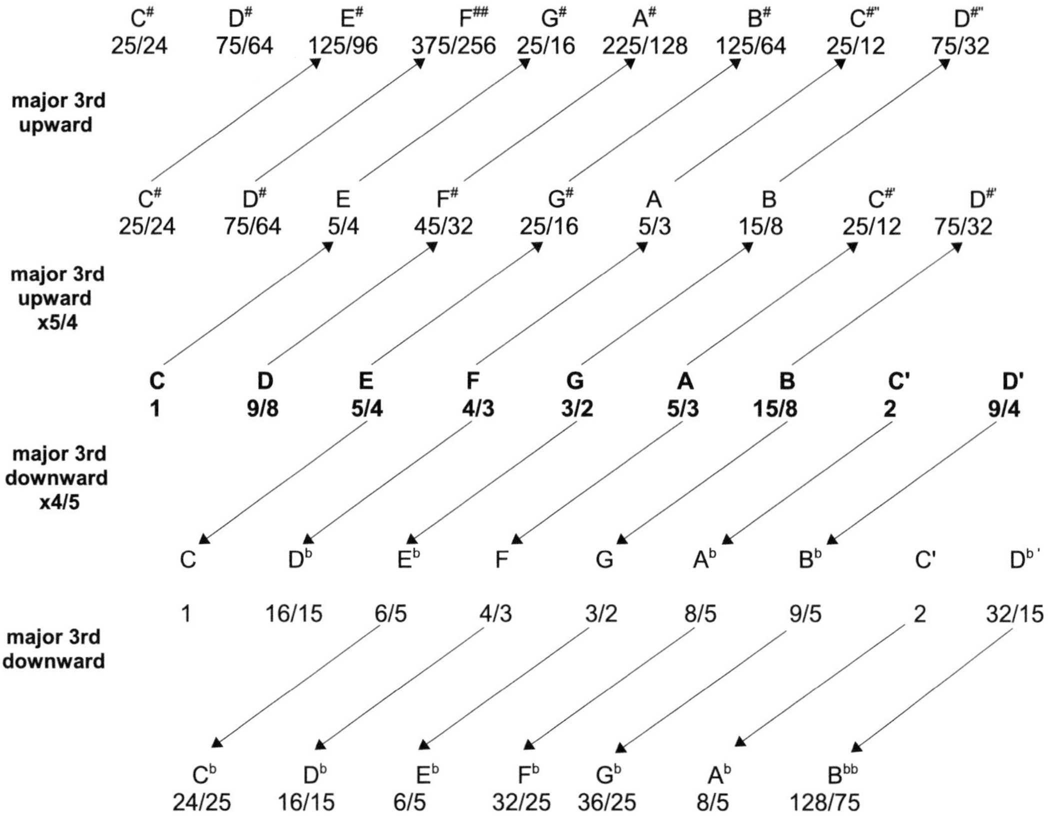

Fig. 12.13 Building the Just chromatic scale

To obtain the notes with **sharps** and **flats** involves a procedure of ascending and descending by major thirds, as shown on the next page.

Note that $f_{G^\#} \neq f_{A^b}$, $f_{A^\#} \neq f_{B^b}$. Such pairs of notes are called **enharmonic equivalents**. On a piano, they are not distinguishable since they are produced by the same key. The interval between the above pairs is $125/128 = 0.9755$:

$$\frac{f_{G^\#}}{f_{A^b}} = \frac{25/16}{8/5} = \frac{125}{128}$$

$$\frac{f_{A^\#}}{f_{B^b}} = \frac{225/128}{9/5} = \frac{125}{128}$$

We obtain the same ratio for the enharmonic equivalents, f_E/f_{F^b} and $f_{B^{bb}}/f_A$.

The detailed construction of the Just chromatic scale is laid out in Fig. 12.13. It is interesting to note the inclusion of double sharps and double flats, which are necessary for some scales.

Many musicians, who are not constrained in playing with a fixed tuned instrument such as the piano, will distinguish between enharmonic equivalents. As a violinist, this author happens to have a strong tendency to do so.

12.7 Intrinsic Problems with Just Tuning

Just intervals have the beauty of rich consonance. Unfortunately, for fundamental, mathematical reasons, they are impossible to realize fully in a composition, as we will observe in the following three examples.

1. In the key of C, D-F is a minor 3rd and yet has a frequency ratio of $(4/3)/(9/8) = 33/27$ instead of the standard Just minor third of $6/5$. Thus there are two different minor thirds in the scale. The ratio $32/27$ is rather harsh and

defeats the goal of Just tuning. We would hope that this interval could be avoided; unfortunately, mathematical analysis shows this to be impossible.

2. Suppose a violinist tunes the strings with fifths having 3/2 frequency ratios. Recall that the A-string is A_4. Then $f_{E_5} = (3/2)f_{A_4}$ and $f_{D_4} = (2/3)f_{A_4}$.

Now, suppose that the violinist plays a B_4 so as to be a Just interval with respect to E_5, which is a fourth above. Then $f_{E_5}/f_{B_4} = 4/3$ or

$$f_{B_4} = \frac{3}{4} f_{E_5} = \left(\frac{3}{4}\right)\left(\frac{3}{2}\right) f_{A_4} = \left(\frac{9}{8}\right) f_{A_4}$$

We now compute the interval between this B_4 and the D_4 below:

$$\frac{f_{B_4}}{f_{D_4}} = \frac{\frac{9}{8} f_{A_4}}{\frac{2}{3} f_{A_4}} = \frac{27}{16}.$$

This interval is the major sixth in Pythagorean tuning. It differs from the Just major sixth of 5/3 by a small amount, the interval being equal to $(5/3)/(27/16) = 80/81$.

The interval 81/80 is well-known, being referred to as the **syntonic comma**. While it is equivalent to only about one-fifth of a semitone, it is large enough to make the interval 27/16 noticeably dissonant. We will see the syntonic comma a number of times in this chapter since it is ubiquitous in the mathematics of tunings.[8,9] Our result is summarized in Fig. 12.14.

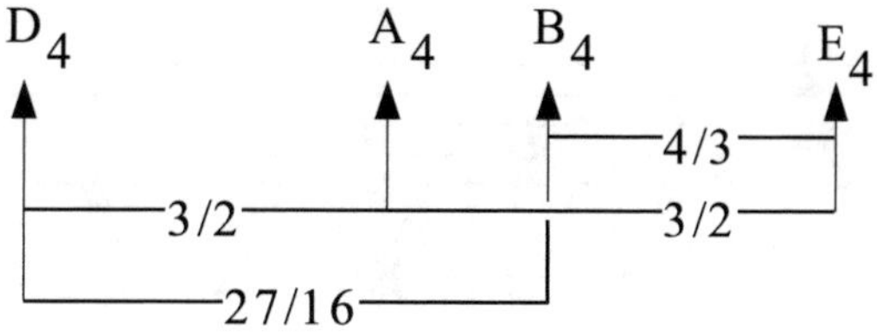

Fig. 12.14 Syntonic comma reflected by the ratio 27/16

Alternatively, one could choose to play the major sixth between the D and the B as a Just major sixth, that is, 5/3. But then, the interval of a major fourth between the B and the E would be $[(3/2)(3/2)]/(5/3) = 27/20$, rather than the Just ratio of 4/3. The ratio between these two numbers is $(27/20)/(4/3) = 81/80$, the syntonic comma.

We see then that it is absolutely impossible to perform pieces fully with just intervals between all pairs of significant notes. Furthermore, the errors are significant.

Now note that we have pieces wherein the violinist has to play these three notes, D, B, & E in rapid sequence. with open D and E strings desired, or worse yet, with the pair DB followed immediately by the pair BE. There would be too little time for the violinist to move the finger for the B to produce two Just intervals. We see that the violinist would have to compromise a bit and to make the two intervals sound acceptable. On the other hand, there is a natural help here: if the violinist plays the B with a **vibrato**,[10] the desired consonances are not badly affected by the lack of Just intervals.

3. As we progress through a piece of music, we might choose to have the interval between every neighboring note be a Just interval. If we do so, we may follow such a path that there is no guarantee that we return to the frequencies of the initial notes. An example of such a problematical sequence of notes is seen in Fig. 12.15.

[8]It is interesting to calculate the ratio 80/81 using a pocket calculator. The ratio consists of a repeated set of integers—called a "repeated decimal." Amazingly, each digit, from 1–9 appears once and only once. The reader should also calculate other ratios, such as 31/81 or 13/81.

[9]The syntonic comma is often mistaken for the **Pythagorean comma**. The latter is the interval that we obtain if we start with a base frequency and increase it by a product of twelve fifths, each with a ratio of 3/2. Thus, if we start with the note F, we would obtain the series of notes: $F \Rightarrow C \Rightarrow G \Rightarrow D \Rightarrow A \Rightarrow E \Rightarrow B \Rightarrow F\# \Rightarrow C\# \Rightarrow G\# \Rightarrow D\# \Rightarrow A\# \Rightarrow E\#$. $E\#$ is enharmonic with F and is very close to being seven octaves above (not below as we might expect) the original note F by a factor of $(3/2)^{12}/2^7 \sim 1.0136\ldots$. This number is the **Pythagorean comma**.

[10]See Sect. 2.15.

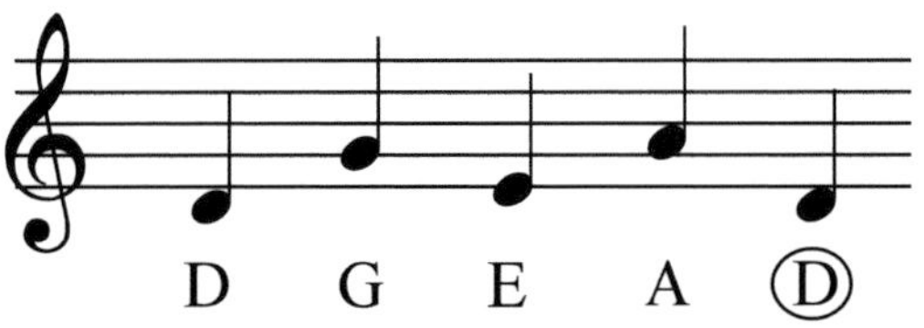

Fig. 12.15 Problematic sequence of notes

Consider the intervals between neighboring notes:

- D to G above: perfect fourth
- G to E below: minor third
- E to A above: perfect fourth
- A to D below: fifth

With Just intervals between neighboring notes, we obtain

$$f_D = f_D \times \frac{4}{3} \times \frac{5}{6} \times \frac{4}{3} \times \frac{2}{3} = \frac{80}{81} f_D \neq f_D$$

If Pythagorean intervals are used, the above problem is avoided; however, their major and minor thirds are less consonant than corresponding Just intervals.

Note that the above sequence, DGEA, is the theme of the first movement of Ralph Vaughan-Williams' Symphony #8. It is an interesting fact that the resulting A, as determined above, has a ratio of 40/27 with respect to the opening note of D instead of 3/2, results in a psychoacoustic tension that is musically exciting.

Homework: Check this last statement by calculating the same product using Pythagorean intervals. Recall that the Pythagorean minor third is the ratio 32/27.

Note I have been told that a Mozart opera was analyzed in the above manner to see whether the final key-note had any shift relative to the opening note. One would expect shifting by 80/81 or 81/80 to occur many times in a long piece. Upward shifts by $81/80$ and downward shifts by $80/81$ would not be expected to cancel if they were random, much as when a drunkard carries on a random walk: Starting from the center of the top of a mesa, we would certainly expect the drunkard to eventually fall off. Amazingly, after the entire length of the Mozart opera, there was no shift at all! This result probably reflects well of the quality of Mozart's music.

4. Let us now examine the Diatonic Major scale in the key of D, using the frequencies that were determined earlier **based upon the key of C**. We need two sharps from the chromatic scale, $F^{\#}$ and $C^{\#}$. The result we obtain is depicted in Fig. 12.16.

   Note that the last semitone is equal to $27/25$, rather than $16/15$. The ratio of the two numbers is $(27/25)/(16/15) = 81/80$, the syntonic comma.

   In order to handle such a change, harpsichords used to be repeatedly tuned, according to the key of the piece being performed and in the manner by which we constructed the scale for the key of C, above. In order to avoid this labor, equal-tempered tuning was introduced. We discuss this tuning in the next section.

12.8 Equal Tempered Tuning

In **Equal Tempered Tuning**, the *octave is divided into twelve equal semitone intervals.* The gain is that the problems of Just tuning are removed. The loss is that resonances and consonances are not as strong as they are in Just tuning. Consonant intervals such as the fifth or the major third are harsher. The greatest gain, perhaps, is that fixed tuned instruments such as the piano do not have to be retuned for each key change. Furthermore, single compositions that have changes of key can be played in a consistent manner on fixed tuned instruments, that is, without biasing one key.[11]

[11] Bach's **Well-Tempered Clavier**, a collection of 48 short pieces, two for each of the possible keys (including only those with at most one sharp or flat), was famous in promoting what is referred to as one of the numerous **well tempered tunings**. According to Anton Kell-

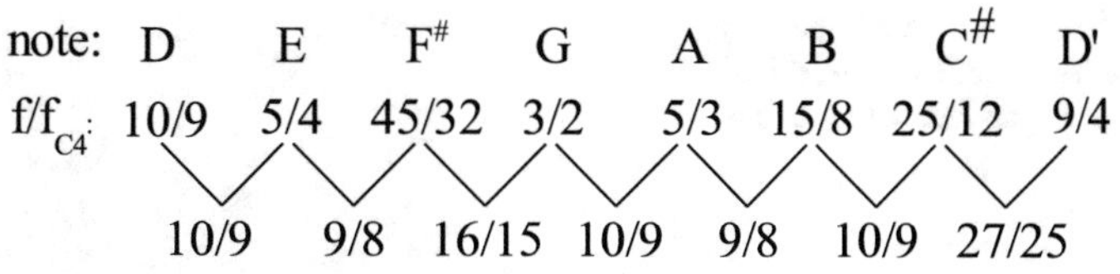

Fig. 12.16 Major D scale

For equal tempered tuning, all we need is to determine the frequency ratio corresponding to a semitone. Label it with the symbol r. Since there are twelve semitones to the octave, we must have

$$r^{12} = 2$$

$$r = 2^{1/12} = 1.05946\ldots$$

For comparison sake, the Just semitone is $16/15 = 1.066\ldots$ and is therefore slightly larger. To compensate for this increased value within the octave, Just tuning has a whole tone interval, $10/9 = 1.1111\ldots$, which is less than the equal tempered whole tone (= two semitones) interval of

$$r^2 = 2^{2/12} = 2^{1/6} = 1.12246\ldots$$

On the other hand, recall that Just tuning uses two different whole tones; the other Just whole tone interval of $9/8 = 1.125$ is *greater* than r^2.

Table 12.1 compares the important intervals in Just (J) and Equal Tempered (ET) tuning.

12.9 The Cents System of Expressing Musical Intervals

To express intervals that are smaller than a semitone in a quantitative way, so as to reflect very small changes of pitch, the equal tempered semitone, $r = 2^{1/12}$, is divided into one hundred (100) units, called **Cents**, as follows:

100 cents = 1 ET semitone
1200 cents = 1 octave

Generally, in the cents system the interval between two frequencies, f_2 and f_1, is given by:

$$\boxed{C \equiv \frac{1200}{\log 2} \log \frac{f_2}{f_1}} \qquad (12.5)$$

where the unit of the parameter C is the number of Cents in the interval.[12]

We have here an explicit, *mathematical expression for the musical interval as a function of the corresponding frequency ratio.*[13]

Discussion and applications of this equation follow in the examples below.

---

ner [http://plaza.ufl.edu/wnb/baroque_temperament.htm#German] Bach's temperament was not at all equal tempered. One possibility is that it was similar to one of the tunings of **Andreas Werkmeister** (1645–1706) [http://en.wikipedia.org/wiki/Werckmeister_temperament]. See the problem on Werkmeister tuning at the end of the chapter.

[12] This equation can be written in a simpler form by making use of the logarithm to the base 2 instead of the current base 10:

$$\boxed{C = \log_2 \frac{f_2}{f_1}}. \qquad (12.6)$$

[13] **"Musical instrument digital interface"** (**MIDI**) is a protocol for allowing various devices, such as musical instruments and computers, to communicate musical scores. It gives a numerical value to all notes based upon the frequency. The value is clearly closely related to our sense of pitch and is given by

$$\boxed{\mathbf{p} = 69 + 12 \log_2 \frac{f}{440}} \qquad (12.7)$$

Note that **p**=69 is associated with A440. The number 69 was chosen so that the C below A440 using ET tuning, so that this C is simply nine semi-tones below A440, will have the value **p**=60. In addition, for each semi-tone in ET **p** changes by unity:

$$\Delta\mathbf{p} = 12 \log_2 \frac{f_2}{f_1} = 12 \log_2 2^{1/12} = 1 \qquad (12.8)$$

A related objective measure of pitch is the **mel scale** defined by the equation

$$\boxed{m = \frac{1000}{\log\ 2} \log\left(\frac{f}{1000} + 1\right)} \qquad (12.9)$$

Note that $m = 1000$ mels at a frequency of 1000 Hz and zero mels at 0 Hz.

Table 12.1 A comparison of some intervals in Just vs. Equal Tempered tuning

Interval	Just (J)	Equal Tempered (ET)
Semitone	$16/15 = 1.066\ldots$	$2^{1/12} = 1.059\ldots$
Whole tone	$9/8 = 1.125$, $10/9 = 1.11\ldots$	$2^{2/12} = 1.122\ldots$
Minor third	$6/5 = 1.2$	$2^{3/12} = 2^{1/4} = 1.189\ldots$
Major third	$5/4 = 1.25$	$2^{4/12} = 2^{1/3} = 1.26\ldots$
Fourth	$4/3 = 1.333\ldots$	$2^{5/12} = 1.335\ldots$
Fifth	$3/2 = 1.5$	$2^{7/12} = 1.498\ldots$
Minor sixth	$8/5 = 1.6$	$2^{8/12} = 2^{2/3} = 1.587\ldots$
Major sixth	$5/3 = 1.666\ldots$	$2^{9/12} = 2^{3/4} = 1.68\ldots$
Octave	2	2

Sample Problem 12.1 Calculate the number of cents in an ET semitone.

Solution We have

$$\frac{f_2}{f_1} = 2^{12} \tag{12.10}$$

$$\mathcal{C} = \frac{1200}{\log 2} \log 2^{12}$$
$$= \frac{1200}{\log 2} \times \frac{1}{12} \times \log 2 = 100 \text{ cents}$$

Sample Problem 12.2 What is the frequency ratio corresponding to one cent?

Solution Since a ratio of $2^{1/12}$ corresponds to 100 cents, we have

$$\frac{f_2}{f_1} = \left(2^{1/12}\right)^{1/100} = 2^{1/1200} \simeq 1.00058$$

As a check we note that

$$\mathcal{C} = \frac{1200}{\log 2} \log 2^{1/1200}$$
$$= \left(\frac{1200}{\log 2}\right)\left(\frac{1}{1200}\right) \log 2 = 1 \text{ cent}$$

Sample Problem 12.3 By how many cents do the Just and equal tempered fifth differ?

Solution For the J-fifth we have

$$\mathcal{C} = \left(\frac{1200}{\log 2}\right) \log(3/2) \cong 702 \text{ cents}$$

while for the ET-fifth (7 semitones) we have precisely $\mathcal{C} = 700$ cents.

Therefore the ET-fifth is 2 cents ($\sim 2/100 = 1/50$ or 2% of a semitone) smaller than the J-fifth.

We now demonstrate how **adding intervals in cents corresponds to multiplying ratios**.

Suppose that we have three frequencies, $f_3 > f_2 > f_1$. We know that the interval $f_1 \rightarrow f_3$ must equal the *sum* of the two intervals $f_1 \rightarrow f_2$ and $f_2 \rightarrow f_3$. This fact is reflected mathematically as follows. Let:

$$\mathcal{C}_{21} = \text{interval for the frequency ratio } \frac{f_2}{f_1}$$
$$\mathcal{C}_{32} = \text{interval for the frequency ratio } \frac{f_3}{f_2}$$
$$\mathcal{C}_{31} = \text{interval for the frequency ratio } \frac{f_3}{f_1} \tag{12.11}$$

so that we can then add them to become

$$\mathcal{C}_{32}+\mathcal{C}_{21} = \frac{1200}{\log 2} \log \frac{f_3}{f_2} + \frac{1200}{\log 2} \log \frac{f_2}{f_1} \tag{12.12}$$

or

$$\mathcal{C}_{32}+\mathcal{C}_{21} = \frac{1200}{\log 2} \left[\log \frac{f_3}{f_2} + \log \frac{f_2}{f1}\right] \tag{12.13}$$

or

$$\mathcal{C}_{32}+\mathcal{C}_{21} = \frac{1200}{\log 2} \log \frac{f_3}{f_1} = \mathcal{C}_{31} \tag{12.14}$$

Table 12.2 Table of musical intervals

NOTE	Pythagorean frequency		Cents		Just frequency		Cents		Equal Temperament frequency		Cents	
C	1		0		1		0		1		0	
	$C^{\#}$	2187/2048		114		25/24		71		$2^{1/12}$		100
	D^{b}	256/243		89		16/15		112		$2^{1/12}$		100
D	9/8		204		9/8		204		$2^{1/6}$		200	
	$D^{\#}$	19683/16384		317		75/64		275		$2^{1/4}$		300
	E^{b}	32/27		294		6/5		316		$2^{1/4}$		300
E	81/64		409		5/4		386		$2^{1/3}$		400	
	$E^{\#}$	177147/131072		522		125/96		457				
	F^{b}	8192/6561		385		32/25		427				
F	4/3		498		4/3		498		$2^{5/12}$		500	
	$F^{\#}$	729/512		612		45/32		590		$2^{1/2}$		600
	G^{b}	1024/729		588		36/25		631		$2^{1/2}$		600
G	3/2		702		3/2		702		$2^{7/12}$		700	
	$G^{\#}$	6561/4096		816		25/16		773		$2^{2/3}$		800
	A^{b}	128/81		792		8/5		814		$2^{2/3}$		800
A	27/16		906		5/3		884		$2^{3/4}$		900	
	$A^{\#}$	59049/32768		1019		225/128		977		$2^{5/6}$		1000
	B^{b}	16/9		996		9/5		1018		$2^{5/6}$		1000
B	243/128		1109		15/8		1088		$2^{11/12}$		1100	
	$B^{\#}$	531441/262144		1228		125/64		1159				
	C^{b}	4096/2187		1086		48/25		1129				
C	2		1200		2		1200		2		1200	

That is,

$$\mathcal{C}_{32} + \mathcal{C}_{21} = \mathcal{C}_{31} \tag{12.15}$$

The cents system is convenient for expressing in a precise quantitative way musical intervals.

Table 12.2 displays a comparison of the three tunings discussed in this chapter with respect to the musical interval from C to various notes in the chromatic scale. Frequency ratios are given as well as the number of cents.

12.10 Debussy's 6-Tone Scale

Given the power of impressionistic music, such as the music of **Claude Debussy** or **Maurice Ravel**, it is worth mentioning Debussy's **6-tone Scale**. These scales give the listener a powerful sense of the mystical. Why this is so is not clear. There are two scales:

$$F \;\; G \;\; A \;\; B \;\; C^{\#} \;\; D^{\#}$$
$$F^{\#} \;\; G^{\#} \;\; A^{\#} \;\; C \;\; D \;\; E$$

Each scale consists of ET whole-tone intervals. They are often played juxtaposed against each other.

12.11 String-Harmonics

Music for strings often makes use of the sound produced by what are casually called "harmonics." To distinguish such notes from the harmonics of a harmonic series that are played in a normal way, they are referred to as "**string-harmonics**." In Fig. 12.17 we see a series of four notes in the violin part of Maurice Ravel's first violin-piano sonata, which he composed when he was 22-years old.[14]

Let us examine the first two notes. The first B in the sequence, referred to as B_5,[15] has a frequency that is 3/2 above the frequency E_5 of

[14] The reader can hear these four notes of the sonata, starting at the time 0:43 on the YouTube recording accessible from the link

https://www.youtube.com/watch?v=C1PA_qQeEdU.

[15] See Fig. 12.2 for the nomenclature of all the strings on a standard piano, which ranges from A_0 through C_8, for a total of seven octaves plus four notes, thus $(7\times12=84)+4=88$ notes.

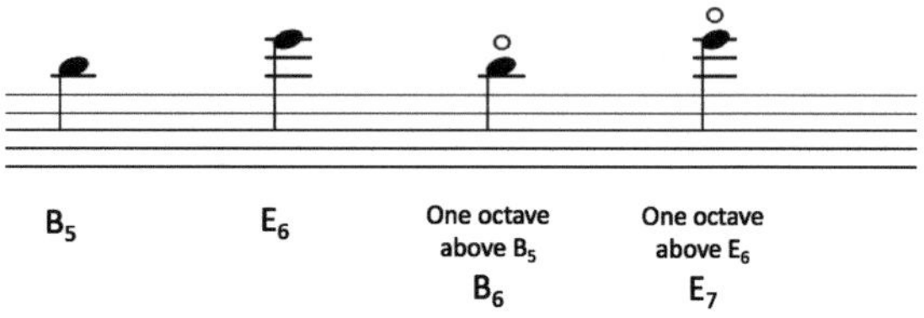

Fig. 12.17 Sequence of four notes in Ravel's Sonata no.1 for violin and piano

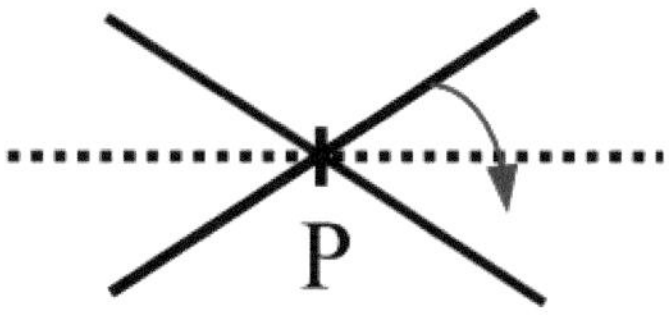

Fig. 12.18 Rotation of the string about a point that is touched lightly

the open E-string, of length ℓ.[16] To play B_5, the finger is pressed hard at a point that is a distance $2\ell/3$ from the bridge. Then the length of string that is free to vibrate will be $2\ell/3$ and lead to the increase in frequency by a factor of 3/2. The note E_6 is an octave above that of the open string and requires the finger to be pressed hard at the midpoint of the string.

We note that there are circles above the next two notes. The circles indicate that the violinist should play these notes as "**natural harmonics**."[17] These two notes have frequencies that are, respectively, one octave above the previous two notes. In order to produce these frequencies in a normal way, the finger could be pressed hard at distances $\ell/3$ and $\ell/4$, respectively, from the bridge. **However**, to produce the harmonics, the finger is **not pressed hard** on the string. Instead, the string is pressed ever so lightly at different specific points, with extremely low pressure. I will refer to such a contact as a **light-touch**. While the string cannot move at such a point in a direction **perpendicular** to the axis of the string at the point that is pressed, the string can **rotate** about such a point, as shown in Fig. 12.18, wherein the string is pressed lightly at the point P. The dotted line represents the axis of the string. A violinist plays the two harmonics by a light-touch of the finger at the positions of B_5 and A_5 along the string.[18] The physical basis will be explained below.

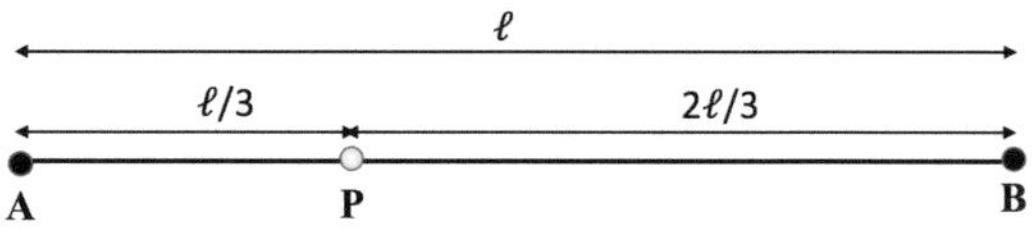

Fig. 12.19 Identification of light touch point to produce B_6

Suppose that a string of length ℓ is pressed lightly at point P in Fig. 12.19, a distance $2\ell/3$ from the bridge at point B, and is bowed between point P and the bridge. Let us first note that the frequencies in the harmonic series of the right side are B_5, B_6, $F_7^{\#}$, B_7, ... Those of the left side are B_6, B_7, $F_8^{\#}$, B_8, ... Next, let us note that the excitation of the **right side** leads to the vibration of the **left side** of point P also. **The two sides are coupled.** As a result, the **fundamental** of the length of string to the right of P cannot be excited and sustained. The only frequencies in the harmonic series of the right side that can be excited and sustained must be shared by frequencies of the harmonic series of the left side. **The two lowest frequencies that are shared are B_6 and B_7**. As an example we see the vibration of the full length of the string for the frequency B_6 in Fig. 12.20. To the left of P we see the first

[16]The word "open" means that there are no fingers placed on the string.

[17]See later for an explanation of this nomenclature.

[18]Because the notes are produced by a light touch on the open string, the harmonics are called **natural harmonics**. Those harmonics that require that a finger be placed strongly at a point on the string, along with a light touch elsewhere on the same string, are referred to as **artificial harmonics**. See http://www.musicalobservations.com/publications/harmonics.html.

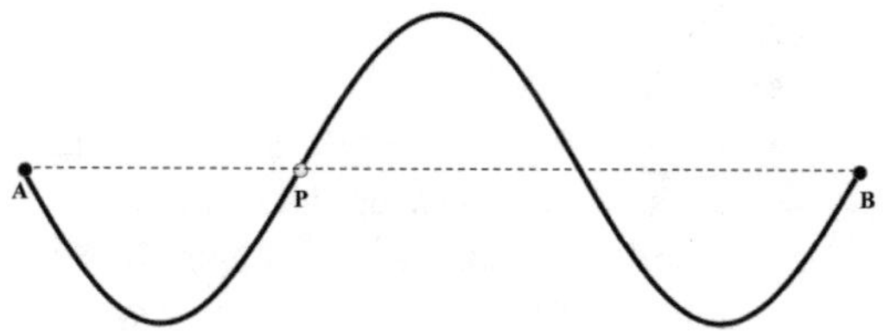

Fig. 12.20 The vibration of the lowest frequency of excitation of the string-harmonic B_6

harmonic while to the right we see the second harmonic.

When the string is bowed between point P and the bridge, the wave traveling along the string contains both these Fourier components, B_6 and B_7, and the sound has the frequency B_6. As is known, the timbre of the bowed string-harmonic B_6 is different from that of the normally played B_6. The reason is that the admixture of harmonics that are excited is different for the normally played note and the string-harmonic. In particular, the spectrum for the string-harmonic is relatively weaker in the high harmonics and the sound "purer" and less harsh.

Now let us turn to the string-harmonic for E_6. Its frequency is an octave above E_6, that is, E_7. E_6 can be produced by pressing the string hard at the midpoint of the string, a distance $\ell/2$ from the bridge. Thus, the frequency E_7 can be produced by pressing hard at a point a distance $\ell/4$ from the bridge. On the other hand, as we will see shortly, the **string-harmonic** can be produced by pressing lightly at a point $3\ell/4$ from the bridge, that is, at a point a distance $\ell/4$ from the end opposite to the bridge, known as the **nut**.

We begin the explanation by noting that if a finger is pressed **hard** at point P a distance $3\ell/4$ from the bridge, the note A_5 will be produced, corresponding to the fundamental of a length of string $3\ell/4$. The harmonic series corresponding to the length of string between the point P and the bridge are A_5, A_6, E_7, A_7, ... Those corresponding to the length of string between the point P and the nut are E_7, E_8, B_8, E_9, ... However, by pressing **lightly** at point P, the two lengths of string to the right and left of point P will be coupled. As a result, the note sounded will have a frequency corresponding to the lowest frequency

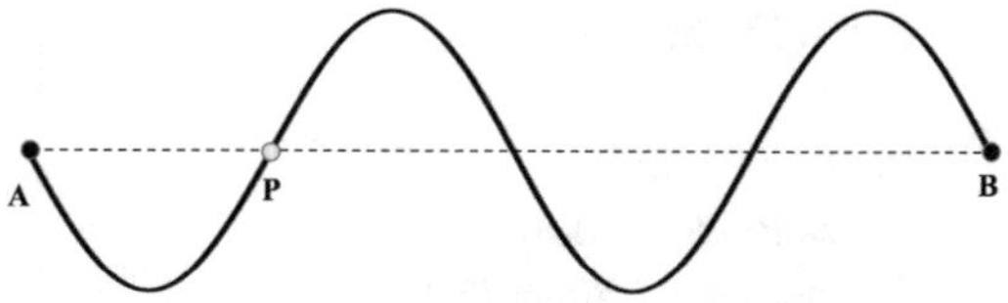

Fig. 12.21 The vibration of the lowest frequency of excitation of the string-harmonic E_7

common to the harmonic series of the lengths $3\ell/4$ and $\ell/4$. That frequency is E_7, which is the third harmonic of A_5 of the right side and the first harmonic E_7 of the left side. We exhibit the lowest excitation of the full length of the string for the string-harmonic E_7 in Fig. 12.21.

In sum, the four notes are played with hard pressure at the positions of B5 and E6, followed by light-touches at the positions of B5 and A5.

12.12 Terms

1. building a musical scale
2. cent
3. cents system of expressing musical intervals
4. chromatic scale
5. consonance
6. diatonic scale
7. dissonance
8. enharmonic equivalents
9. equal tempered tuning
10. fifth
11. fix tuned instrument
12. flat
13. fourth
14. intonation
15. Just temperament
16. Just scale
17. key note
18. light-touch
19. major second
20. major sixth
21. major third
22. minor second
23. minor third
24. minor sixth
25. musical interval
26. musical scale
27. musical staff

28. natural harmonic
29. octave
30. pentatonic scale
31. Pythagorean comma
32. Pythagorean temperament
33. Pythagorean scale
34. semitone
35. seventh
36. sharp
37. six tone scale (of Debussy)
38. string harmonic
39. syntonic comma
40. temperament
41. tuning
42. vibrato
43. well tempered tuning
44. Werkmeister temperament
45. whole tone

12.13 Important Equations

– objective numerical expression of the musical interval between two frequencies, f_1 and f_2

$$\frac{f_2}{f_1}$$

– frequency ratio for musical intervals in equal tempered intonation:

$$frequency\ ratio = 2^{n/12} \tag{12.16}$$

where n is the number of semitones

– equation defining **cents** in relation to frequency ratio

$$C \equiv \frac{1200}{\log 2} \log \frac{f_2}{f_1} \tag{12.17}$$

12.14 Problems for Chap. 12

1. One tone is produced with a frequency of 150 Hz. The tone three octaves higher is produced with a frequency of (choose one).
   (a) 300 Hz
   (b) 450 Hz
   (c) 600 Hz
   (d) 900 Hz
   (e) 1200 Hz
2. (a) How many *semitones* are there in each of the following intervals: major second, minor third, major third, fourth, fifth, minor sixth, major sixth, octave?
   (b) What are the frequency ratios of the above intervals in Equal Tempered tuning?
   (c) Show that the sum of a major sixth and a minor 3rd is an octave by adding up the number of semitones for each interval. Check whether the product of the corresponding frequency ratios is two for Pythagorean tuning, as shown in the table of musical intervals (Table 12.2).
3. (a) What advantages does equal tempered tuning have over Just tuning?
   (b) What advantages does Just tuning have over equal tempered tuning?
4. Suppose the note A above middle-C is tuned at 440 Hz.
   (a) To what frequency should the E above be tuned to produce a Just fifth in the key of C? Note that the 3rd harmonic of the A will equal the 2nd harmonic of the E, so that there will be *resonance*.
   (b) To what frequency should the E above be tuned so as to produce an *equal-tempered* fifth?
   (c) In the latter case, what will be the **beat frequency** between the 3rd harmonic of the A and the 2nd harmonic of the E?
5. We learned that in Just tuning, there are two different frequency ratios for a whole tone. Show that the ratio of these two ratios is a syntonic comma, 81/80.
6. The **tritone** is a sum of a perfect fourth and a minor second and can be regarded as half of a full octave. The interval is famous for its association with horror and gloom. Calculate the corresponding frequency ratio in both Just tuning and equal temperament.

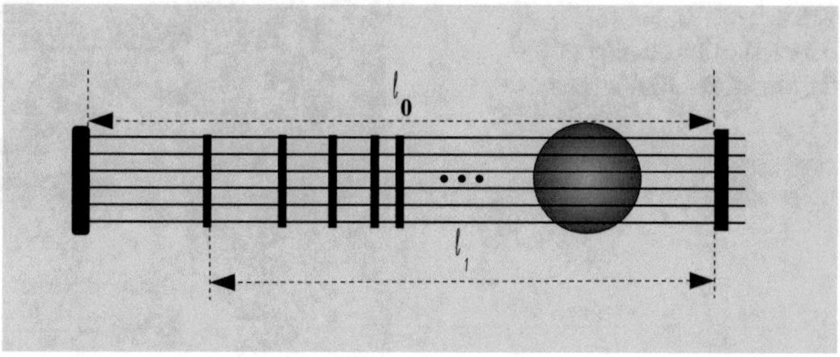

Fig. 12.22 Guitar frets

7. The six strings of a guitar are tuned to the following notes, in order of increasing pitch:

E – A – D – G – B – E"

so that the two E's—E and E"—are two octaves apart.
   (a) What are the intervals (by name) D-G and G-B?
   (b) Suppose all the above intervals are set as JUST intervals. What will be the frequency ratio—$f_{E''}/f_E$? [It will **not** be 4 : 1!]
   (c) By how many cents does $f_{E''}/f_E$ differ from 2400 cents (corresponding to two octaves)?
   (d) Show how the use of equal temperament eliminates this discrepancy.
8. The frets on a guitar are set so as to produce equal tempered frequencies as shown in Fig. 12.22.
Let ℓ_0 be the length of the "open" string (without any fingers on the frets). Let ℓ_n be the length of string that is free to vibrate corresponding to the nth note above that of the open string. The length ℓ_1 is shown in Fig. 12.22. We see in the figure how the spacing between neighboring frets decreases as the frequency is increased.
Show that the length ℓ_n is given by

$$\frac{\ell_n}{\ell_0} = 2^{-n/12} \tag{12.18}$$

This function is *mathematically* similar to the exponential decay in the attenuation of a wave. Note that

$$\log\,\ell_n = \log\,\ell_0 - n\,\frac{\log\,2}{12} \tag{12.19}$$

As a result, a plot of the *logarithm* of the length ℓ vs. n will yield a straight line with a slope given by $-(\log\,2)/12$.
9. One summer, at a day camp of two of my grandchildren, we were sitting outdoors on a lawn that was close to a small airport. Suddenly a propeller driven airplane passed directly overhead. We could hear a distinct tonality to the whirring sound that varied from one note to another that was a major third below.
Use this information to estimate the speed of the airplane.
HINT: **Refer to a sample problem in Sect. 9.11 of Chap. 9.**
10. Suppose that the following musical passage is played with **Just intervals between neighboring notes**:

major second	9/8	minor third	6/5	major third	5/4
fourth	4/3	fifth	3/2		

$$C \to E \to A \to B \to E' \to A \to F \to D \to C$$

   (a) Find the frequency ratios for all the neighboring notes—

$$\frac{f_E}{f_C},\ \frac{f_A}{f_E},\ \frac{f_B}{f_A},\ \frac{f_E}{f_B},\ \ldots$$

   (b) Show that the ratio of the frequency of the last C to the first C in the passage is equal to $80/81$ = inverse of the **syntonic comma**.
   (c) Evaluate $80/81$ using a pocket calculator, exhibiting all decimal places in your answer.
   (d) To how many cents does a syntonic comma correspond?

Table 12.3 Table of intervals in Just tuning in the key of D - J(D)

NOTE	f/f_C in J(C)	Frequency in J(C)	f/f_D in J(C)	Frequency in J(D)
C	1	264		
D	9/8	297	1	297
E	5/4		10/9	
F	4/3			
$F^{\#}$	45/25			
G	3/2			
A	5/3			
B	15/8			
$C'^{\#}$	25/12			
D'	9/4		2	

11. Let J(C) represent Just tuning in the key of C, as displayed in table of musical intervals and in Fig. 12.13. Suppose that we choose the frequency of the C, as $f_C = 264$Hz. This choice corresponds to a frequency of 440Hz for A-440 in J(C). Suppose further that we have tuned a harpsichord in J(C) in order to play a piece in the key of C and then want to play another piece in the key of D.
A central question is: How will the frequencies of the strings compare to what we would need in order to have them tuned to J(D), that is, Just tuning in the key of D? You will use the table below to exhibit the results of your study.
    (a) In order to answer this question, first calculate the actual frequencies in J(C) for the notes in the diatonic scale in the key of D: These notes are D E $F^{\#}$ G A B $C'^{\#}$ D'. Note that in the table, the ratios of the frequencies are exhibited, as taken from Table 12.2. You should fill in the third column with these frequencies of Table 12.3. The frequencies can be determined by multiplying $f_C = 264$Hz by the corresponding ratios in the second column.
    (b) Fill in the fourth column with the ratio of each these frequencies to the frequency $f_D = 297$Hz: 1, 10/9, ...
    (c) You will next obtain the frequencies in J(D) using $f_D = 264\text{Hz}\times(9/8) = 297$Hz. All you have to do is to multiply 297 by the corresponding ratios in the second column—that is, 1, 9/8, ... 15/8, 2. Enter the resulting frequencies in the fifth column.
    The differences between the corresponding frequencies in J(D) and J(C) help us to appreciate the problem in using Just tuning in situations where one wants to play a piece of music within which the key changes.
12. Suppose a car is moving towards you and produces a **Doppler effect** on its tooted horn corresponding to a Just semitone ($f'/f = 16/15$). Determine how fast the car is moving.
13. **Stretch Tuning of Pianos***
In Chap. 2, we learned that the frequency spectrum of a string is not a harmonic series on account of stiffness. In particular, the frequency of the nth mode is given by[19]

$$f_n = n\frac{\sqrt{\frac{\mathcal{T}}{\mu}}}{2\ell}\sqrt{1+\mathcal{B}n^2} \tag{12.21}$$

An important consequence is that the harmonics of a given piano string will not be

[19] In detail, the constant $\mathcal{B}$ is given by

$$\mathcal{B} = \frac{\pi a^2 Y}{\mathcal{T}}\left(\frac{\pi a}{2\ell}\right)^2 \tag{12.20}$$

where a is the radius of the string (assumed to be a solid cylinder), Y is Young's modulus of the string's material, $\mathcal{T}$ is the tension, and ℓ is the length of the string. Readers who have a background in the basic physics of elasticity will recognize that the ratio $(\mathcal{T}/(\pi a^2 Y))$ is the fractional increase $\Delta\ell/\ell$ in the length of the string due to the tension.

consonant with some of the harmonics of other strings, as discussed in Chap. 2 and we will lose resonance. Of course, the use of equal temperament tuning already destroys this resonance *except for* the set of octaves of a given note.

Consider for simplicity the A-440 string and the string an octave above, which we will refer to as A-880. All tunings assign a fundamental frequency of f_1^{440} =440-Hz for the A-440. Moreover, all of the standard tunings discussed in this chapter would assign a frequency of 880-Hz to the A-880. The combination of these two frequencies will produce a consonant sound. On the other hand, the first overtone of the A-440 will be slightly higher than 880-Hz on account of stiffness.

(a) Show that the frequency of the first overtone of A-440 in the presence of stiffness is given by

$$f_2^{440} = 880\frac{\sqrt{1+4\mathcal{B}}}{\sqrt{1+\mathcal{B}}} \qquad (12.22)$$

Note that if A-880 is tuned to this frequency, we would sacrifice the resonance in our hearing a 2:1 ratio of frequencies for resonance between the two strings, A-440 and A-880. Such is the case in **stretch tuning of pianos**.

(b) Calculate this frequency if $\mathcal{B} = 0.008$.

(c) What would be the beat frequency between 880-Hz and this frequency?

(d) Calculate the interval $\mathcal{C}$ in cents for the two frequencies f_2^{440} above and 880-Hz.

Now let us examine the next octave above A-440, namely A-1760. We have two choices to make if we want to achieve resonance between the three strings under consideration thus far: We could tune A-1760 to so as to have a frequency equal to the second overtone of A-440 or we can choose a frequency equal to the first overtone of A-880.

(e) Explain why these two choices lead to different frequencies. You can simplify your response with formulas if you assume that the A-440 and A-880 string have the same value of the constant $\mathcal{B}$.

(f) Suppose that we apply one of the above stretch tunings for each of the eleven notes in the octave lying between A-440 and A-880. Will such a stretch tuning lead to resonances between corresponding harmonics of two different notes in the octave?

14. Recently, because of the development of electronic keyboards and pianos, one can change the frequencies of the notes in a small fraction of a second. As a result, access to temperaments used centuries ago that have a different sound (some might say more consonant sound) is available and is attractive to some musicians. Many expensive keyboards and pianos now provide these temperaments. In this problem, you will learn about one of these temperaments

Werkmeister I(III) Temperament. It is based upon the Pythagorean temperament.

(a) Consider the sequence of notes starting with C and increasing by fifths: C, G, D, A, E, B, $F^{\#}$, $C^{\#}$, $G^{\#}$. Thus, $G^{\#}$ is eight fifths above C.

Find the frequency $f_{G^{\#}}$ in proportion to C in Pythagorean tuning by multiplying by factors of (3/2) and reducing to the octave.

(b) Note that the interval C-E (a major third), is a series of four fifths followed by a reduction down by two octaves. The Pythagorean ratio is $(3/2)^4/2^2 = 81/64$ and differs significantly from the more consonant (5/4) of Just tuning.

Show that the ratio of these two ratios is the **syntonic comma** (81/80)

(c) **Werkmeister I(III)** temperament was introduced by Andreas Werkmeister. It **"well-tempers"** Pythagorean temperament most of all in order to produce a Just major third. To do so,

Werkmeister makes a compromise. Instead of producing the Pythagorean scale by multiplying by a sequence of ratios (3/2), he multiplies an additional ratio of $(80/81)^{1/4}$ for each fifth. The ratio $(81/80)^{1/4}$ is referred to as a **quarter comma**; thus, each multiplication by (3/2) is accompanied by a **reduction** by a quarter comma.
Show that the result is that the interval between C and E is a Just major third. Note that the intervals between C and notes other than E are no longer Just. In particular, the interval of fifth between C and G is not equal to the "sacred value" of 3/2 of Just Temperament. Rather, it is reduced by a quarter comma, or $(80/81)^{1/4}$.

(d) Evaluate the correction factor $(80/81)^{1/4}$ and express it in cents.

(e) Consider the sequence of notes starting with C and decreasing by four intervals of a fifth each: C, F, B^b, E^b, A^b.
Find the frequency f_{A^b} in proportion to C in Pythagorean tuning by dividing by factors of (3/2) and reducing to the octave. Note first that the resulting f_{A^b} is a Just interval above C. Note, too, that $f_{A^b} \neq f_{G^\#}$. [You can show that this inequality persists with Werkmeister I(III) temperament.]

15. Flutists often have a tendency to sway when they play. We might wonder whether this motion can have a serious effect due to the Doppler effect. To answer this question, we need to know how much of a shift in frequency is tolerable. Tests reveal that a good listener can detect a frequency shift of about 3.5 Hz for an A440. This is the **just noticeable difference of pitch** vs. frequency. [Experienced ears can do better.]
Determine the velocity that will lead to a Doppler shift of 3.5 Hz.

16. The **open A-string** of a violin has the note A_4 and is usually tuned to a frequency of 440-Hz. Suppose that the following harmonics are played using a light-touch:
(a) As above, with finger is positioned a distance $2\ell/3$ from the bridge.
What is the frequency sounded in standard notation?
(b) As above, with finger is positioned a distance $3\ell/4$ from the bridge.
What is the frequency sounded in standard notation?
(c) As above, with finger is positioned a distance $4\ell/5$ from the bridge.
What is the frequency sounded in standard notation?
(d) What are the frequencies of the three notes sounded if the finger presses hard at the above respective positions?

The Eye

13

The human eye has the purpose of converting the light that enters it into nerve impulses that are sent to the brain. Up to the stage of the retina, it functions much like a camera: It has an iris, a lens, and a retina that is like the photographic film of old cameras or the array of electronic sensors of the more recent digital cameras. In this chapter, we review these features, with some detail and apply what we learned about lenses in Sect. 9.10 to the focusing of light by the eye lens onto the retina.

We see a schematic comparison of the eye and a camera in Fig. 13.1 taken from Scientific American article by **George Wald**, in the August, 1950 issue. The principal structures of the eye are the:

- **cornea**—it acts like a primary lens
- **iris**—it provides a variable aperture
- **lens**—it provides a variable focal length
- **retina**—it acts like camera film in responding to light by producing nerve impulses which are sent to the brain down the **optic nerve**.

The eye serves to provide:

1. a two-dimensional representation of a scene, including both **light intensity** and a **sense of color** as a function of position in space;
2. a sense of distance of light sources from our eyes—that is, **depth perception**.

13.1 The Cornea and Lens

To appreciate fully this chapter, you should have a good understanding of the material on lenses in Chap. 9. The lens and cornea act as a **compound lens** system with variable effective focal length f. The effective image distance of the eye-lens system, d_{ie}, is *fixed* because the positions of the lens and retina are fixed. This effective distance is about 24 mm. According to the lens equation (see Eq. (13.1) below), the farther an object is, that is, the larger the object distance d_o is, the larger must be the focal length. At infinite object distance, the effective focal length f will be a maximum and equal to the image distance d_{ie}, that is about 24 mm. Conversely, the closer an object is, the smaller must be the focal length. Surprisingly, we will see that the normal requirement of being able to see objects, whose distance ranges from as close as a foot or so all the way to infinity, does not require a very large variation in focal length. As usual, the focal length, object distance d_o, and image distance d_{ie} must satisfy the lens equation, which we write as

$$\frac{1}{f} = \frac{1}{d_o} + \frac{1}{d_{ie}} \tag{13.1}$$

In order to increase the **focal length**, **ciliary muscles**, which act on the lens, must relax (not tighten) so as to flatten the lens. The process of varying the focal length is known as

L. Gunther, *The Physics of Music and Color*,
https://doi.org/10.1007/978-3-030-19219-8_13

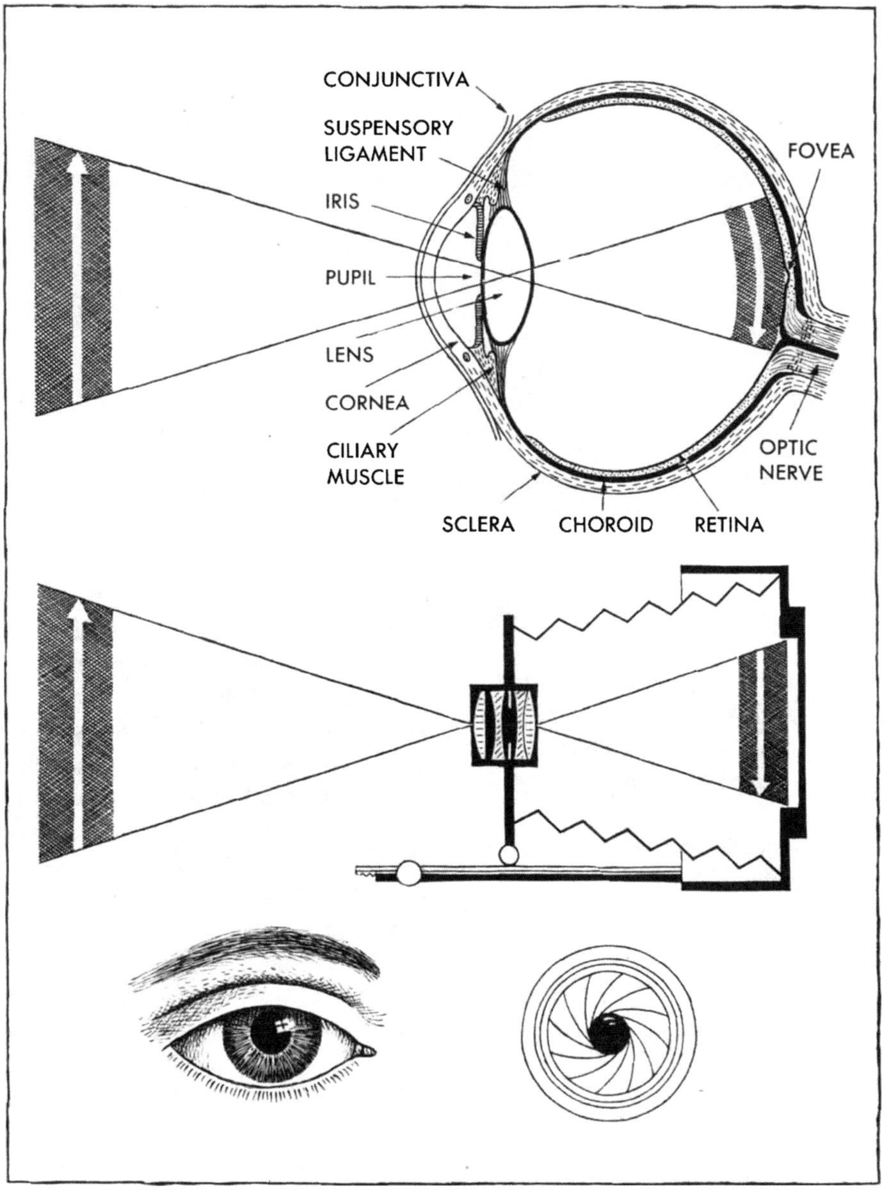

Fig. 13.1 The camera and eye compared (source: **George Wald**, *Camera and Eye*. Reproduced with permission. © (1950) Scientific American, Inc. All rights reserved)

accommodation. Accommodation thus provides for a range of focal lengths, from some minimum, **min f** to some maximum, **max f**.

A **myopic** (or **near-sighted**) eye has a difficulty accommodating to great distances. It has a maximum focal length, $max\ f$, which is less than d_{ie}. As a result, the maximum object distance, $max\ d_o$ that can be brought into focus is not at infinity, as one would wish. Using the lens formula we obtain:

$$\frac{1}{max\ d_o} = \frac{1}{max\ f} - \frac{1}{d_{ie}} \tag{13.2}$$

To see near objects, the focal length of the eye must be decreased so that the lens must have a larger "bulging." The ciliary muscles are tightened to reduce f. A **hyperopic (far-sighted)** eye has a minimum focal length, $min\ f$, which is so large that the closest an object can be to the eye, and still be brought into focus on the retina, is too great for the person's needs. That closest distance, $min\ d_0 \equiv d_{np}$, is called the **near point**. We have:

$$\frac{1}{d_{np}} = \frac{1}{min\ f} - \frac{1}{d_{ie}} \qquad (13.3)$$

Sample Problem 13.1 An eye has a retina that is 24.0mm from the eye lens. It focuses an object that is 2.00m away and has a height of 40cm.

Find the focal length of the eye lens and the height of the image on the retina.

Solution We have

$$\frac{1}{d_o} = \frac{1}{f} - \frac{1}{d_{ie}} \qquad (13.4)$$

with $d_o = 200\text{cm}$, $d_{ie} = 24\text{mm}$, and $h_o = 40\text{cm}$.

We will express all lengths in centimeters. Then

$$\frac{1}{f} = \frac{1}{200\text{cm}} + \frac{1}{2.4\text{cm}} = 0.419\text{cm}^{-1} \qquad (13.5)$$

By inverting the right-hand side we obtain $f = 2.39\text{cm}$.

Note that f is extremely close to the image distance because of the large object distance. Also, relatively large changes in object distances do not require significant changes in the focal length.

Now we turn to the image height. We recall that

$$h_i = Mh_o \qquad (13.6)$$

We don't need to know the focal length in order to determine the magnification. It is determined by the object distance and the image distance.

Thus

$$h_i = Mh_o = \left|\frac{d_i}{d_o}\right| h_o = \frac{2.4\text{cm}}{200\text{cm}} 40\text{cm} = 0.48\text{cm} = 4.8\text{mm} \qquad (13.7)$$

Note It is useful to appreciate the slight ambiguity of the near point: As a **home exercise**, estimate the near point of each of your eyes separately, as well as when used together. I suggest that you hold text in front of you having various font sizes. How different are the results?

As one ages, $min\ f$ tends to increase because of increasing stiffness in the lens. This stiffening involves a process of crystallization of the lens material.

Note For the *myopic* eye, $min\ f < max\ f < d_{ie}$, while for the *hyperopic* eye, $min\ f$ **is too large for practical purposes**.

For an excellent applet that explains myopia and hyperopia, go to the website (accessible on 6/13/06) http://webphysics.davidson.edu/physlet_resources/dav_optics/Examples/eye_demo.html.

The lens is an excellent filter in the ultraviolet (UV) region of the spectrum. However, filtering extends into the violet end of the visible region. Thus, the fraction of incident light that is transmitted drops from close to 100% for 700 nm $> \lambda >$ 500 nm, to 15% for $\lambda =$ 400 nm, to 0.1% for $\lambda =$ 365 nm. (See Chap. 14 for further discussion on the **transmittance**, also called the **transmission coefficient**.) Thus, people whose lens has been removed because of cataracts experience a new world of vision in the UV.

13.2 The Iris

Experts claim that the human eye can see a steady light intensity, without suffering pain or eye damage, that spans about twelve orders of magnitude. [This range compares well with the span of twelve orders of magnitude of the ear.]

To help the eye deal with such a huge range, the iris varies the aperture area from $0.02\ cm^2$ to $0.3\ cm^2$. This function of the iris is analogous to the role of the ossicles of the ear in protecting the ear from excessively loud sounds by restraining their mobility.

13.3 The "Humourous" Liquids of the Eye

The volume of the eye between the cornea and the lens is filled with a water-like liquid called **aqueous humour**. One of its purposes is to nourish the eye in place of blood, which would be opaque to light. Fluid is constantly flowing into and out of the chamber containing aqueous humour. Unfortunately, the pressure in the fluid can become excessive, resulting in the dangerous disease known as **glaucoma**.

The volume within the eye between the lens and the retina is filled with a gelatinous material called **vitreous humour**. Its consistency is close to that of egg white. Both humours serve to keep the eye's size and shape firm. Their index of refraction is about 1.3, in comparison with the index of refraction 1.4 of the lens. There is therefore refraction at the two **lens–humour interfaces**.

Clearly, both humours should remain as clear as possible so as not to block light rays heading for the retina. Fortunately, because many light rays emanating from a point source of light enter the eye on their way to being focused on a corresponding position on the retina, strands of material floating within the vitreous humour do not necessarily severely disturb the image produced on the retina.[1]

13.4 The Retina

A photograph of the retina is shown in Fig. 13.2 as taken using an **ophthalmoscope**.

The retina contains a layer of two types of light sensitive cells together referred to as **light receptors**. These are the **rods** and the **cones**, so-called because of their respective shapes. In Fig. 13.3 we see a beautiful electron microscope image of rods and cones. They are capable of responding to the absorption of a single photon of light by producing a nerve impulse that is sent down a nerve fiber. Typically, though, a few photons are necessary to produce such excitation.

Near the middle of Fig. 13.2, we see a yellow region about 2×1.5 mm across, within which the optic nerve fibers pass through the retina, so that there are no rods or cones. This region produces a blind spot in our vision, which we usually don't notice. Towards to left of the yellow region is a dark orange region called the **macula**, with its even darker **fovea** at the center. The fovea is about one-millimeter in diameter.

Figure 13.4 is a schematic drawing of a section of the retina. Note the amazing fact that the nerve fibers leading to the optic nerve leave the retina within the vitreous humour, and thus lie in the path of incoming light. In contrast to invertebrates, vertebrates have this so-called **inverted retina**.[2]

The following are some of the important characteristics of rods and cones:

Rods:

1. There exists one type of rod, with a peak frequency sensitivity at around 500 nm (greenish). As we will see in Chap. 15 on color vision, there being one type is connected with their not being used for color vision.

[1] In order to appreciate this fact: If you have an opportunity to be present when slides are being projected onto a screen by a slide projector, place some fingers over the lens and notice that your fingers do not cast a shadow on the screen; instead, the image on the screen is simply dimmed non-uniformly.

[2] For a much more detailed drawing, see *Eye and Brain*, by **R. L. Gregory**, Princeton University Press; 5th edition (December 15, 1997).

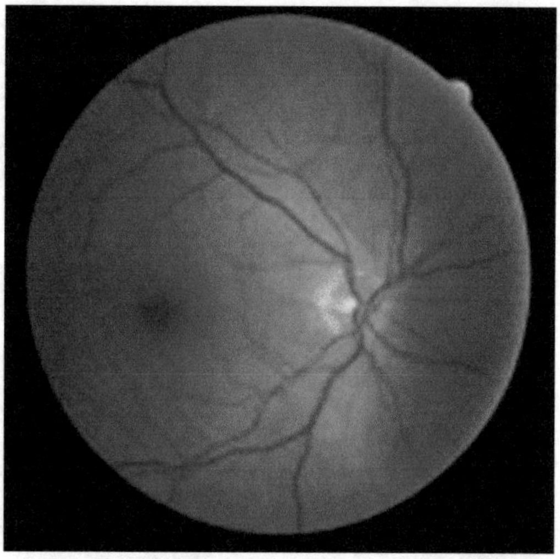

Fig. 13.2 The human retina (source: "Phenx Toolkit," funded by the National Human Genome Research Institute, https://www.phenxtoolkit.org/)

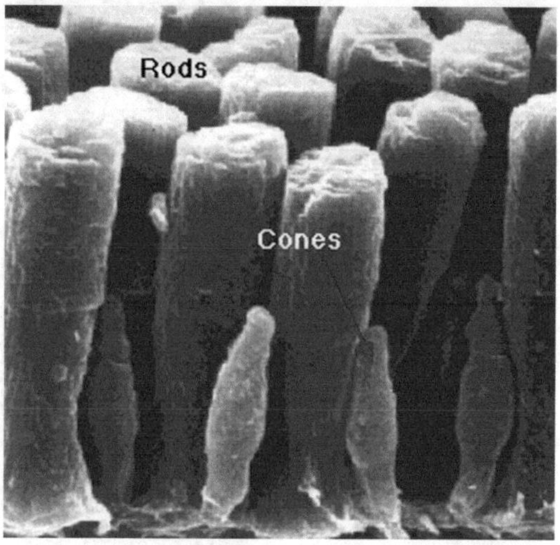

Fig. 13.3 Electron microscope image of rods and cones (source: http://users.rcn.com/jkimball.ma.ultranet/BiologyPages/V/Vision.html, courtesy of David Copenhagen; produced by David Copenhagen, Scott Mittman, and Maria Maglio)

2. The rods are about 1000 times more sensitive than the cones. This property is connected with their providing for **night vision**, referred to technically as **scotopic vision**. A June 2015 article in the highly selective journal NATURE reported experimental evidence that a human rod can detect as few as three photons that are incident on the rod![3]
3. Rods have a relatively long recovery time—about 25 min. That is, once they have been excited and have transmitted a nerve impulse, they require about 25 min in order to return to a receptive state. This property is connected with the observation that light temporarily **bleaches** rods from a purple color to a transparent state indicative of their inability to absorb light. The light sensitive material in rods is called **rhodopsin** or **visual purple** and was extensively studied by **George Wald**, who received the Nobel Prize for his work.
4. About 120-million in number, rods are distributed mostly peripherally to the centrally

[3]https://www.nature.com/news/quantum-technology-probes-ultimate-limits-of-vision-1.17731.

Fig. 13.4 Rods and cones in a section of the retina (source: GREGORY, Richard; EYE AND BRAIN. ©1990 Princeton University Press, Reprinted by permission of Princeton University Press)

located **fovea**, which has a diameter of about 1 mm. (See Fig. 13.5, below.)

Cones:

1. The cones, concentrated in the macula, provide for color vision by the existence of three types. They have peaks in their sensitivity as a function of frequency at the following respective wavelengths: $\lambda \sim 440$ nm (blue), $\lambda \sim 520$ nm (green), and $\lambda \sim 570$ nm (orange). The three types are often called **blue cones**, **green cones**, and **red cones**, respectively. **Genes** for each type of cone have recently been isolated at Johns Hopkins University. How the three provide for color vision will be described later, in Chap. 15.
2. About 6 million in number, cones are concentrated in the region of the fovea. It is estimated that about 64% are red cones, 32% are green cones, and 2% are blue cones.
3. Cones provide for **day vision** or **photopic vision**.
4. As with all nerves, when a cone is excited and emits a nerve impulse, it takes time for it to be able to respond again. This time is referred to as the **recovery time**. Cones have a relatively short recovery time. Recovery time

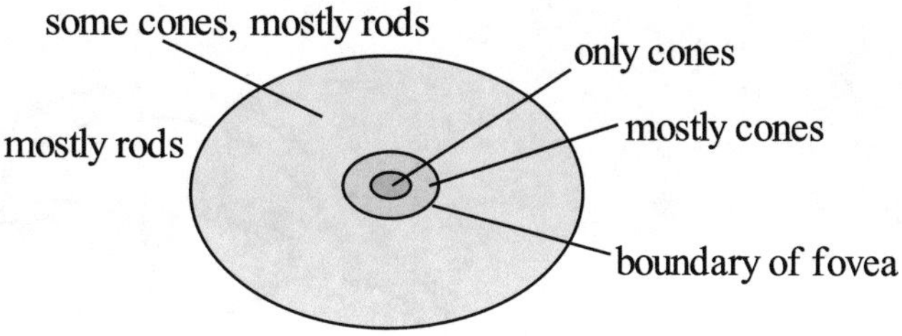

Fig. 13.5 Rough sketch of the distribution of rods and cones in the retina

Fig. 13.6 Complement of the US flag

is responsible for **after images**. Below is a reproduction of a famous painting of the flag of the United States. The reader should stare at the center of the flag for about 30 s, with minimal eye motion. Then the reader should look to the side at a blank part of the page. The original red, white, and blue colors will appear. They are the **complementary colors** discussed in Chap. 15 to the colors in Fig. 13.6 This phenomenon results from fatigue of the cones corresponding to the color of a region.

In Fig. 13.7 we see the distribution along a line running from the top to the bottom of the eye that runs through the fovea and the blind spot at the exit to the optic nerve. From Figs. 13.5 and 13.7 it is clear that the rods provide for excellent peripheral but poor head-on **scotopic** (*night*) vision. On the other hand, the cones provide for excellent head-on but poor peripheral **photopic** (*day*) vision. See Fig. 13.7.

13.5 Dark Adaptation

Suppose that you are in a well lit room and then suddenly turning off the lights. Most of us are familiar with the experience that we don't immediately see well. Our eyes need some time to adjust and increase their sensitivity to low light intensities. This adjustment is called **dark adaptation**. A minor reason is the need for our pupils to dilate—but that response is relatively fast—about 10–20 s.[4] Full adaptation, that is maximum sensitivity, is achieved by the cones in about seven minutes, that by the rods in about an hour. The reason is the fatigue of the rods and cones, which will be discussed in Chap. 15.

13.6 Depth Perception

Depth perception is apparently achieved by two primary means, both depending upon the use of two eyes:

1. **convergence**: In order to produce the same field of vision on the two retinas, the two eyes must be turned through different angles.

[4]See Wikipedia (1-28-2011): http://en.wikipedia.org/wiki/Pupillary_light_reflex and M.H.Pirenne, *Vision and the Eye* [Associated Book Publishers, London, 1987].

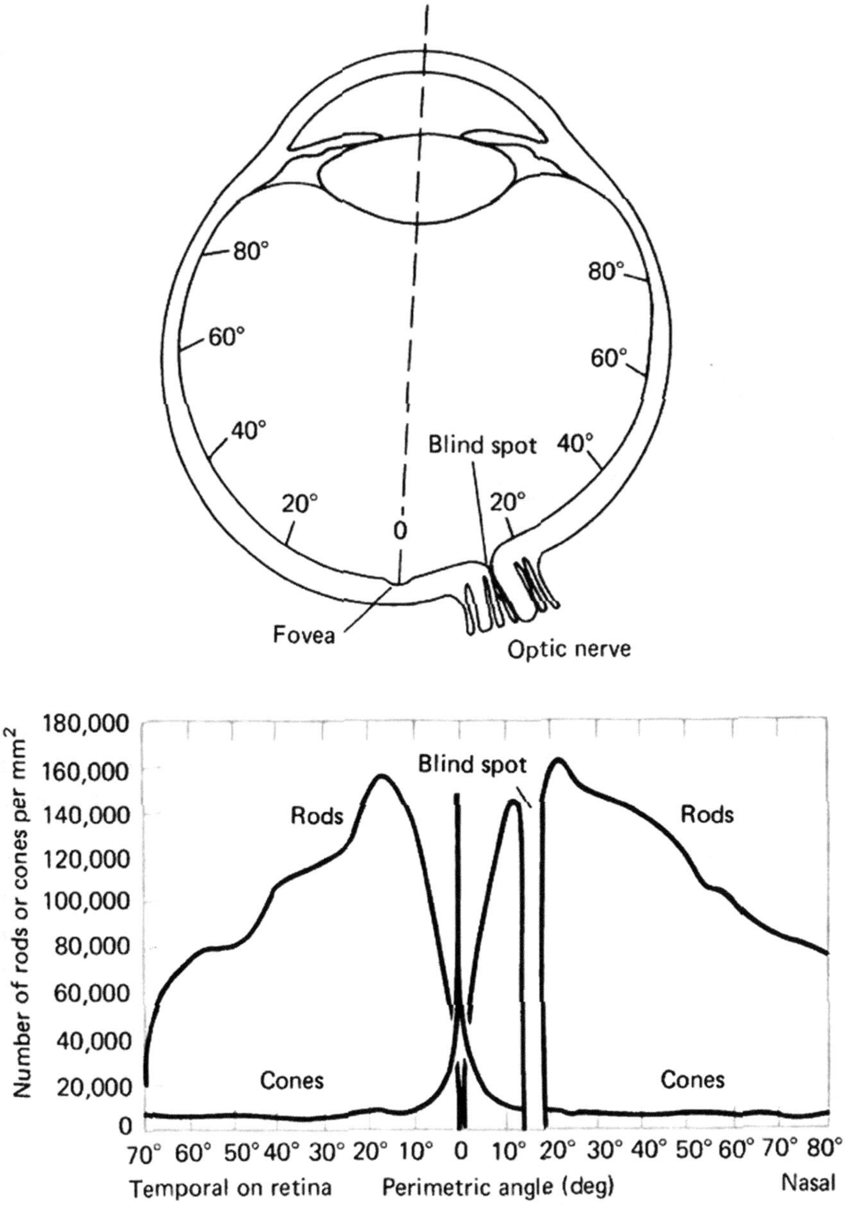

Fig. 13.7 Distribution of rods and cones
(source: M.H.Pirenne, *Vision and the Eye*, (Associated Book Publishers, London, 1987))

See Fig. 13.8. That difference decreases with increasing distance of the object.

2. **disparity of eye position**: Because of the physical separation of the two eyes (by about 6.3 *cm* according to R. L. Gregory's *Eye and Brain*, the two retinas necessarily end up receiving images which are slightly shifted with respect to one another. This shift is interpreted by the brain in terms of the distance to the light source. **Stereoscopes** use this phenomenon to produce a sense of depth: The light coming from a given two-dimensional image—as on a photo—is split into two identical beams of light, with one beam incident upon one eye, the other upon the second eye. The beams are such that

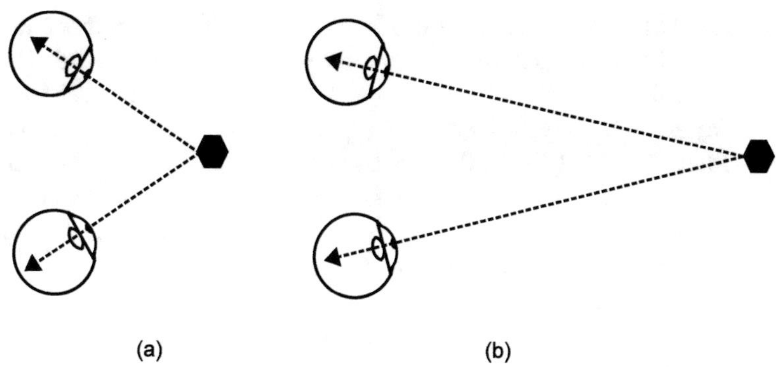

Fig. 13.8 Depth perception via convergence of eyes

they are incident with different positions with respect to the two eyes, thereby reproducing the disparity of eye position of a real 3D scene.

13.7 Terms

- accommodation
- aqueous humour
- ciliary muscles
- compound lens
- cones – blue, green, and red
- convergence & disparity
- cornea
- day vision
- depth perception
- fovea
- hyperopia (far-sightedness)
- iris
- lens
- light receptors
- myopia (near-sightedness)
- near point
- night vision
- optic nerve
- retina
- rods
- transmittance
- vitreous humour

13.8 Problems for Chap. 13

1. (a) Roughly what is the value of the effective focal length of the eye?
   (b) What is the light-sensitive layer in the eye called?
   (c) What is myopia and how is it corrected?
   (d) What is hyperopia and how is it corrected?
   (e) What is the area of the retina called that has the greatest visual acuity?
   (f) What function does the crystalline lens of the eye have?
   (g) For what object distances is the eye focused when relaxed—*near* or *far*?
   (h) What is the **near point**?
   (i) Why does the eye have a blind spot?
   (j) What are rods and cones? Which are more numerous?
   (k) Which type of receptor cell does the fovea contain?
   (l) Which is more sensitive to light, photopic or scotopic vision? By how much more?
   (m) List two methods of binocular depth perception.
2. The greatest amount of refraction in the lens system of the eye occurs when the light
   (a) passes through the middle of the lens.
   (b) passes from the vitreous humor to the aqueous humor.
   (c) passes from air into the vitreous humor.
   (d) enters the cornea.
3. The old saying "at night, all cats appear gray" has the following scientific basis:
   (a) At low light levels, all cones are stimulated, so every object appears white or off-white.

   (b) At low light levels, the cones are not stimulated, and rods cannot distinguish colors.
   (c) At night, only the fovea reacts to light.
   (d) In low light levels, the lens transmits only blue light, which we interpret as gray.
   (e) At night, the optic nerve only reacts to light and dark.
4. Can an eye be both near-sighted and far-sighted? Explain.
5. Explain why in *dim light*, objects appear more distinct if they lie off to the side of the field of vision.
6. Why do you suppose the eye's lens must have a *bulging much greater* than it would need if it were suspended in air, in order to provide its focal length?
7. Suppose that the image distance from the effective lens to the retina is 2.25cm.
   (a) Find the focal length of the eye lens when viewing an extremely distant object.
   (b) Suppose that the eye now views an object a distance 30cm away from the eye. Do you expect the focal length to increase or decrease? Calculate the focal length .

   NOTE: If a screen is held beyond a lens, the object distance is where an image will have maximum resolution. However, if the screen is held a little before the object distance or a little beyond the image will merely have lower resolution but might be acceptable nevertheless. Below, we will assume that maximal resolution is necessary.
8. Here is a problem on **accommodation**. Suppose that an eye can bring into focus an object that has a range of object distances from 40-cm to 200-cm. Assuming that $d_{ie} = 24$ mm, **calculate** this eye's range of focal lengths.
9. Consider an object with a height of 2m. Estimate the minimum distance it can have from the eye so that its image can be cast entirely on the fovea.
10. The author had his eyes checked on December 8, 2005. The prescription for his left eye read -0.25 for far vision and $-0.25+2.25 = 2.00$ for near vision. These numbers refer to what is known as the **diopter** value. The diopter value **D** is a measure of the strength of a lens. Mathematically

$$D \equiv \frac{1}{f} \tag{13.8}$$

where f is the focal length *expressed in meters*. Thus, for far vision I need a diverging lens with a focal length

$$f_{\text{lens}} = \frac{1}{D} = \frac{1}{-0.25} = -4.00\text{m} \tag{13.9}$$

Let us now again assume that the fixed image distance in my eye is 24mm.

   (a) Calculate the focal length f_{far} needed by my eye to bring into focus objects at infinity **without** any corrective help.
   (b) The focal length with my corrective lens is determined by the maximum focal length of my eye $max\ f_{\text{eye}}$ and f_{far} as follows. If we neglect the fact that the corrective lens is held outside the eyeball and assume as an approximation that it is coincident with the center of the eye-lens, it can be shown that the effective focal length of the combination of the two lenses is obtained by adding the inverses of the respective focal lengths:

$$\frac{1}{f_{\text{far}}} = \frac{1}{f_{\text{lens}}} + \frac{1}{max\ f_{\text{eye}}} \tag{13.10}$$

   Expressed in terms of diopters, we have direct addition:

$$D_{\text{far}} = D_{\text{lens}} + min\ D_{\text{eye}} \tag{13.11}$$

   Calculate D_{far} and $min\ D_{\text{eye}}$.
   (c) Calculate $min\ f_{\text{eye}}$.
   (d) The maximum focal length of my eye, used for near vision can be

determined by the desired near point, which we choose here to be d_{np} = 25cm, and the focal length of the corrective lens. First calculate the needed focal length without correction:

$$\frac{1}{f_{\text{near}}} = \frac{1}{d_{np}} + \frac{1}{24\text{mm}} \quad (13.12)$$

(e) Next calculate

$$D_{\text{near}} = D_{\text{lens}} + max\ D_{\text{eye}} \quad (13.13)$$

(f) Finally calculate $min\ f_{\text{eye}}$.
It is useful to mark off on a line the positions of the eye-lens, f_{far}, f_{near}, $max\ f_{\text{eye}}$, and $min\ f_{\text{eye}}$.

11. A tree of height 5 m is a distance 10 m away from you. How high is the image of the tree on your retina if your lens-to-retina distance is 24 mm?
12. Suppose we treat the human eye as having a uniform index of refraction of 1.37. What fraction of light intensity, incident directly at the eye, is *transmitted* into the eye from air?
13. The **effective focal length** of a particular person's eye is 22 mm when the lens is relaxed. The effective lens-to-retina distance is 24 mm.
    (a) Is the person near-sighted or far-sighted?
    (b) To bring distant objects into focus, a diverging lens is used.
    Assuming that the **corrective lens** is *adjacent* to the eye: Determine the focal length of the lens so that it will produce a **virtual image** of an infinitely distant object, such that the virtual image will serve as an object which the eye will focus on the retina in the relaxed-lens state.
14. It has been reported that the density of rods on the retina can be as high as 160,000 per mm^2. On this basis, assuming that the rods are touching each other, **estimate the diameter** of a single rod. To do so, imagine the rods distributed as squares on a checkerboard, each square having a side d, which will represent the diameter. Then write down an expression relating d to the density. How many squares, each of area d^2 would there be in an area of 1 mm^2?

14 Characterizing Light Sources, Color Filters, and Pigments

14.1 Characterization of a Light Beam

If we are to be able to understand the way the eye transmits nerve impulses, we need to be able to characterize accurately the light that is incident upon the eye. Obviously, the light generally casts an image with great variation in detail, with respect to both color and intensity. In this chapter, we will restrict ourselves to an image that consists of a field of **uniform color and intensity**. Such an image can be produced by having a light beam be cast upon and then reflected by a white screen. Therefore, we will focus our attention upon the characterization of a light beam.

A complete characterization of a light beam and hence its source would include a specification of the wave pattern, that is, the magnitude of the electric field, as it varies in space and time, as well as the state of polarization. In this chapter, we will ignore the state of polarization of the beam. The reason is that our goal is to relate the physical characteristics of a beam with color perception. And, polarization plays little role in this regard.

Note Before we begin, some important comments are in order. What does it mean to say that I see the color **blue** and that you agree with me? Do we experience the same sensations? Does the color blue look the same to you as it does to me? Some thought reveals that no one can tell what another person's actual sensation and experience is in connection with the color blue. We could say the same thing about any sensation or feeling that we have. For example, I know what I myself experience when I say that I feel cold or I feel sad. But I cannot tell what any other person's experience is like when they refer to these experiences with the same words. If someone says that they feel cold, I might observe behaviors that are commonly displayed by people who feel cold—shivering, for example. Observing such behaviors supports and/or gives me confirmation that a person is experiencing a cold feeling and leads me to assume that their experience is the same as mine.

If I cannot tell what another person's experience is like, based upon how they describe their experience in standard verbal terms, what can I be essentially sure of? One possibility is the following: People have a multitude of experiences. These experiences are **mapped** onto a language of words. *Ideally*, this mapping would be a **one-to-one correspondence**, as with the object and image of a lens. The reader is encouraged to read Appendix J wherein the roll of **mappings** is discussed in many contexts—including its important role in the discipline of physics. The complexity of assigning words to colors has also been studied by philosophers. See, for example, the article entitled "The Essence

L. Gunther, *The Physics of Music and Color*,
https://doi.org/10.1007/978-3-030-19219-8_14

of Color, According to Wittgenstein": http://wittgensteinrepository.org/agora-alws/article/view/2704/3132.

People argue incessantly within families, within relationships of all sorts, in the interpretation of laws and certainly in courtrooms about the meaning of verbal language. Unfortunately people speaking the same language are often not capable of clearly communicating their experiences with each other.

Legend has it that mankind suffered from the curse of God during the construction of the ***Tower of Babel*** *by the creation of many languages, thereby destroying communication between people speaking different languages so that people couldn't work together during the construction.*

14.2 Spectral Intensity

For the purposes of characterizing the color and brightness alone, a spectral analysis is sufficient. That is, all we have to know is the intensity of all the Fourier (monochromatic) components in the visible region of the electromagnetic (EM) spectrum. A specification of how the intensity is distributed over various wavelengths λ is called the **spectral intensity with respect to the wavelength**, and will be symbolized by $I(\lambda)$. Back in Chap. 4 we introduced the **spectral intensity with respect to the frequency**, $\mathbf{I}(f)$. Unless otherwise stated in this chapter, we will assume that the spectral intensity refers to $I(\lambda)$. Because measurements of the spectra of electromagnetic waves—like light—typically use devices that are directly related to wavelength, we deal with **spectral intensities with respect to wavelength**, symbolized by $I(\lambda)$.

Note We pointed out that a spectral intensity is not an intensity but rather an **intensity density**. [The reader is encouraged to review this subject in Sect. 4.7.] The spectral intensity with respect to the wavelength can be expressed in units of [W/m^2 per nm].

In practice, the spectral intensity of a light beam can be determined by passing the beam through a prism or reflecting the beam from a diffraction grating. In the figure below, a light beam is shown passing *through* a diffraction grating. Equations (14.1) and (14.2) provide us with a relationship between the angle θ and the wavelength. Therefore, the spectral intensity of the light is measured by a light meter set at various angles θ.

In order to simplify our discussion, we will be concrete by assuming that the line spacing of the grating is **10,000 lines per centimeter** (see Chap. 9 regarding gratings). The line spacing d is equal to $(10,000)^{-1}$cm= 1000-nm.[1] The first order diffraction pattern has the angle θ related to the wavelength via the equation:

$$\boxed{\sin\theta = \frac{\lambda}{d} = \frac{\lambda[\text{nm}]}{1000\text{-nm}}} \qquad (14.1)$$

Alternatively, we could write

$$\lambda\ [\text{nm}] = 1000\text{-nm} \cdot \sin\theta \qquad (14.2)$$

The symbol $\lambda[nm]$ means that the use of Eqs. (14.1) or (14.2) requires that the wavelength be expressed in nanometers (nm).

We will restrict the wavelength λ to the range 400-nm to 700-nm (the approximate visible range). Then $\sin\theta$ ranges from 0.40 to 0.70 and, correspondingly, the angle θ ranges from 23.6° to 44.4°.

In Fig. 14.1, the light meter is set at an angle of 40°. It therefore measures the intensity of the spectral component having a wavelength $\lambda \approx (\sin 40°)(1000) = 643$-nm.

14.2.1 Measurement of Spectral Intensity

The grating device is an example of what is called a **spectral photometer**, as would be a device that uses a prism. No spectral photometer can measure intensity at a specific wavelength. If we examine Fig. 14.1, we see that the light meter has an **aperture** that allows light to enter only over a range of angles corresponding to a range of wavelengths centered on the wavelength

[1] Note that 1-cm=10^{-2}m and one meter=10^9nm.

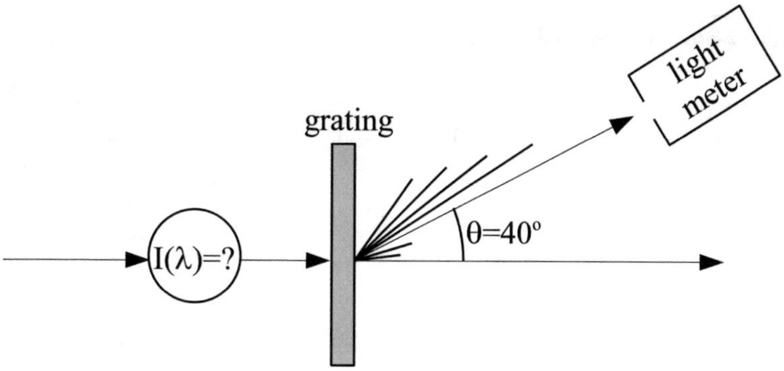

Fig. 14.1 Diffraction by a grating

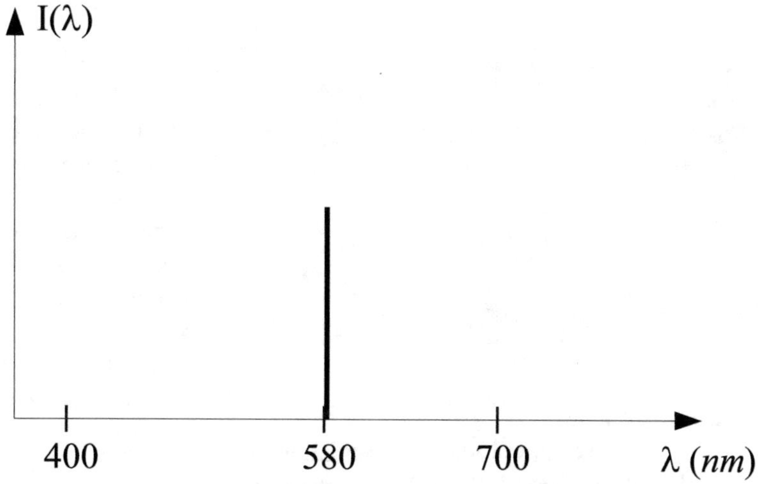

Fig. 14.2 Ideal monochromatic yellow spectrum

corresponding to 40^0. If we let $\Delta\lambda$ be this range, we see that if the meter registers an *intensity I*, the *spectral intensity* is given by[2]

$$\boxed{I(\lambda) = \frac{I}{\Delta\lambda}} \qquad (14.3)$$

14.2.2 Examples of Spectral Intensities

Examples of some spectral intensities are exhibited below:

1. **Monochromatic light**, as the term is loosely defined, is represented by a spike in the spectral intensity. See Fig. 14.2, wherein we see monochromatic light at a wavelength of 580-nm. Absolutely monochromatic light is a non-existent ideal.
2. **Neon laser light**—see Fig. 14.3. The spectral intensity of the laser light has a peak at 632.8-nm. The **bandwidth** $\Delta\lambda$, shown in the figure, gives us a measure of how monochromatic the light is. It is usually defined as the difference between the two wavelengths for which $I(\lambda)$ is equal to one-half the maximum value of $I(\lambda)$. Laser light is very highly monochromatic in that $\Delta\lambda$ is a minuscule fraction of the wavelength λ_{max} at which $I(\lambda)$ is a maximum. For He-Ne laser light, $\Delta\lambda$=0.002-nm. Clearly $\Delta\lambda$ is highly exaggerated in Fig. 14.3.

In Sect. 4.9.1 we pointed out that the response of a **simple harmonic oscillator** to an external oscillating force has a peak at the resonant frequency (of the oscillator) whose width is inversely proportional to the rate of attenuation of the oscillator.

[2]Cf. mass density=mass/volume.

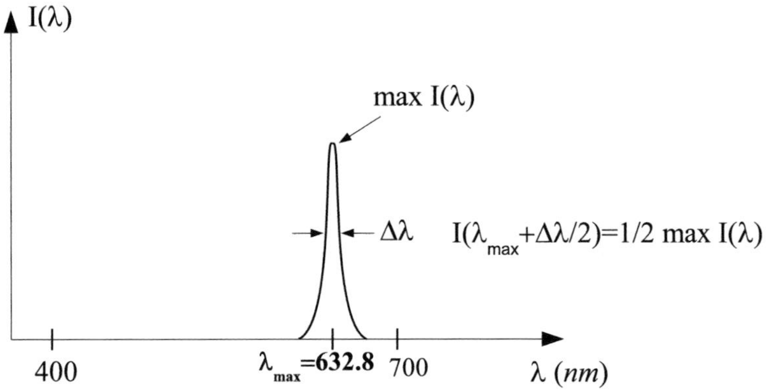

Fig. 14.3 Quasi-monochromatic red

Table 14.1 Range of wavelengths vs. color

λ (*nm*)	Spectral color
400–420	Violet
420–455	Indigo
455–490	Blue
490–575	Green
575–585	Yellow
585–650	Orange
650–720	Red

3. **Spectral Colors**

   The color sensations produced by **monochromatic light** span the full continuous range of colors of the rainbow. These color sensations are called **spectral colors**. For simplicity, the spectral colors are often grouped according to certain ranges of wavelengths. For example, the term **spectral red** refers to a monochromatic light that is red in color. The groups are listed in the table below, according to ranges of wavelengths (Table 14.1).

   According to this table, neon laser light would be labeled **orange**. Boundaries between color-labeled regions are not unanimously agreed upon!

   It is important to note that certain color sensations cannot be produced by monochromatic light. They include: *white*, *magenta*, *cyan (turquoise)*, *brown*, and *gray*. These will be further discussed in Chap. 15.

4. In Fig. 14.4 below is an example of a *broad* spectral intensity that produces a distinct bluish color sensation. Note that quite a bit of monochromatic light in the red region is present.
5. The pale blue color in Fig. 14.5 below requires a much broader spectral intensity than the previous one in Fig. 14.4.
6. A spectral intensity that is constant, that is, the same, for all visible wavelengths is referred to as **pure white** and is shown in Fig. 14.6. Another term for this spectrum is **equal energy spectrum**, for obvious reasons. Such a spectrum will appear white. For simplicity, Fig. 14.6 has a range of wavelengths only from 400-nm to 400-nm. As we see in the transmission spectra of the filters, specifications normally extend beyond these limits.

   As we move from a spectral blue to a broad spectrum blue to a pale blue, and on to pure white, the color sensation becomes less distinct. We say that the color becomes less **saturated**. A spectral color has maximum saturation. A pale blue is a low saturated blue, with "blue" referred to as the **hue**. It can be simply produced by adding white light to a saturated blue light. Similarly, **pink** is low saturated red.

   So far, we have characterized our psychological perception of a light source according to *hue* and *saturation*. Hue has so far been associated with a peak frequency in the spectrum. The **third** characteristic of a light source is its **brightness**, which is a **subjective** characteristic related to the overall intensity of the light source. The range of light intensities that can be seen without pain is from $\sim 10^{-10}$-W/m^2 to $\sim$ 100-W/m^2 and therefore spans *twelve orders of magnitude*, as in the case of sound.

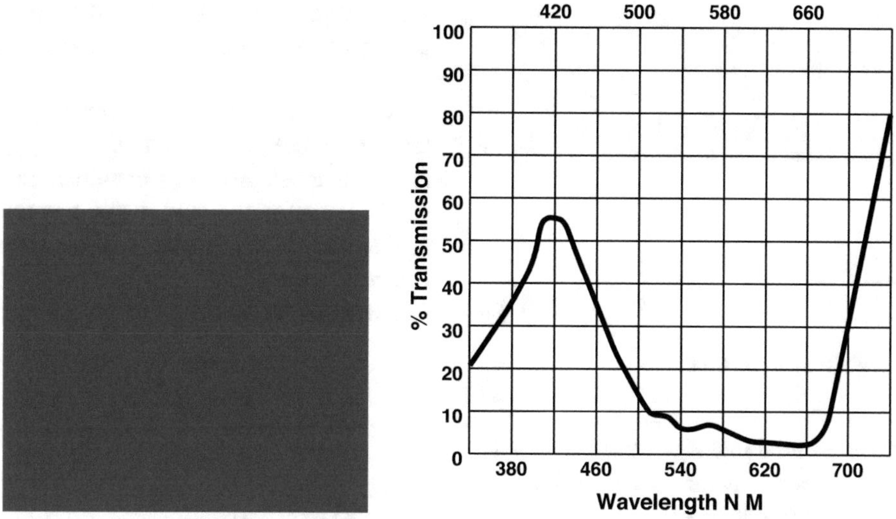

Fig. 14.4 Double blue slide and corresponding spectral intensity (source: Courtesy of Rosco, Inc, Stamford, CT)

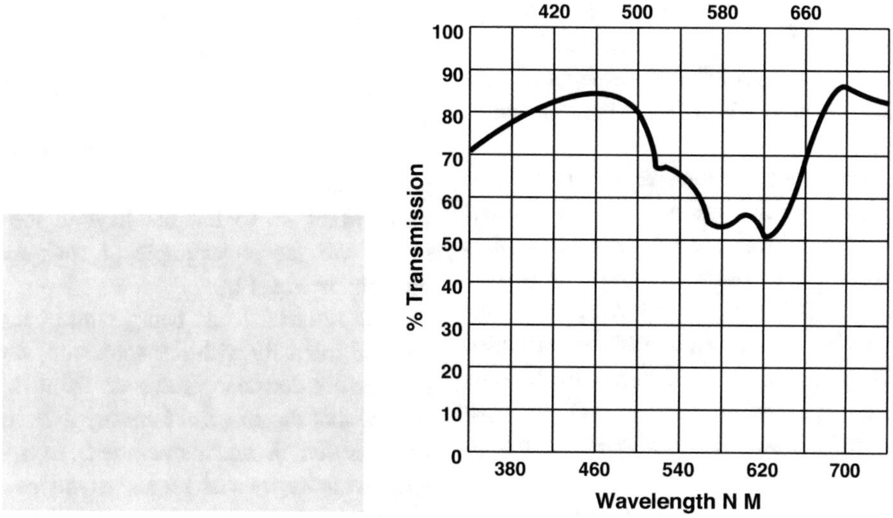

Fig. 14.5 Mist blue slide and corresponding spectral intensity (source: Courtesy of Rosco, Inc, Stamford, CT)

The following table summarizes the *psychological* perception characteristics with the *physical* characteristics of light and sound:

	Psychological	Physical
Light	Hue	Central frequency
	Saturation	Bandwidth or white admixture
	Brightness	Intensity
Sound	Pitch	Fundamental frequency
	Timbre	Admixture of overtones
	Loudness	Intensity

14.2.3 Determining an Intensity from the Spectral Intensity

Analogous to the significance of the spectral intensity with respect to the frequency, there is no intensity associated with the value of the spectral intensity at a given value of the wavelength. If we plot the spectral intensity with respect to the wavelength as a function of the wavelength, the total intensity associated with a range of wavelengths is equal to the **area under the**

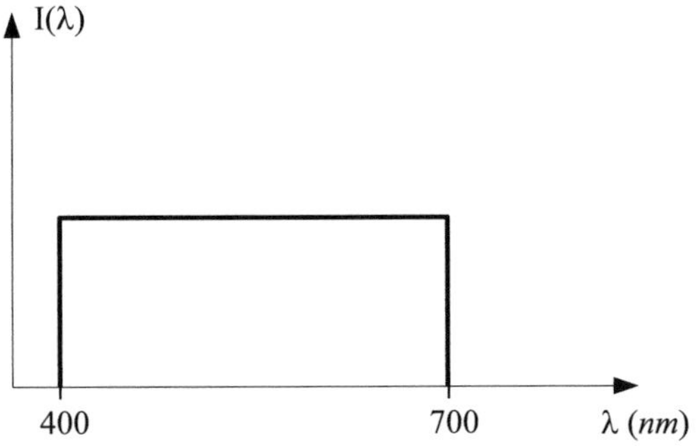

Fig. 14.6 White light spectrum

curve of the spectral intensity between these two wavelengths. The following sample problem will explain clearly what we mean.

Sample Problem 14.1 In Fig. 14.7 we see a plot of a particular spectral intensity. Note that the unit of spectral intensity in the plot is 1-W/m^2 per nm. The wavelength difference between ticks on the plot is 100-nm. What is the total intensity?

Solution We see two cross-hatched rectangles. From 400-nm to 500-nm we have a spectral intensity of 3-mW/m^2 per nm, so that the area of this rectangle is $100\times\ 3 = 300$ units, corresponding to 300-mW/m^2=0.3-W/m^2. With the same analysis we obtain an intensity contribution of 0.2-W/m^2 in the range 500-nm to 700-nm, wherein the spectral intensity is 1-mW/m^2 per nm. The total intensity is therefore 0.5-Wm^2.

14.3 Comments on the Two Spectral Intensities

The spectral intensity with respect to the frequency, $\mathbf{I}(f)$, is used when dealing waves that are analyzed by devices that measure frequency. Examples are electronic signals produced by a microphone that respond to sound waves or the output of an antenna that picks up radio, TV, or radar waves. The spectral intensity with respect to the wavelength, I(λ), is used in the case of devices that measure directly the wavelength of a wavelength; for example, a prism or grating for light waves. See Sect. 7.4.1 that discusses the grating and Sect. 9.9.1 that discusses the prism.

Note The frequency and wavelength are directly related through the relation $\lambda = v/f$, where v is the wave velocity. As a consequence, the two spectral intensities are in one-to-one correspondence: That is, either one determines the other. It can be shown that the two are related by the following equation:

$$\boxed{\mathbf{I}(f) = \frac{\lambda^2}{v}\mathrm{I}(\lambda)} \tag{14.4}$$

14.4 White Noise

A very important spectrum of sound and of electronic signals is referred to as **white noise.**[3] It is so designated because the spectral intensity with respect to frequency, namely I(f), is constant. Typical so-called *white noise* has a constant spectrum up to a high **cutoff frequency** beyond what is relevant for the signal.[4] For example, it might be constant up to but not beyond the audible range. We see an example of such a spectral intensity in Fig. 14.8.

It is referred to as being white because the spectral intensity with respect to the wavelength for light is constant. However, from Eq. (14.4) we see that *the spectral intensity with respect to the wavelength that corresponds to a constant spectral intensity with respect to the frequency* is given by

$$\mathrm{I}(\lambda) = \frac{\mathrm{v}}{\lambda^2}\mathbf{I}(f) \propto \frac{1}{\lambda^2} \tag{14.5}$$

The color of the corresponding light would be a pale blue, not white.

Another important type of noise of a sound or of electronic equipment is known as **pink noise**. Its spectral intensity is inversely proportional to the frequency and is shown in Fig. 14.9. We have

[3]Noisy sound is often used in the waiting room of an MD or a psychotherapist to maintain the privacy of patients.

[4]It isn't possible to produce a signal that is constant for all frequencies ranging from zero to infinity. There will always be a cutoff.

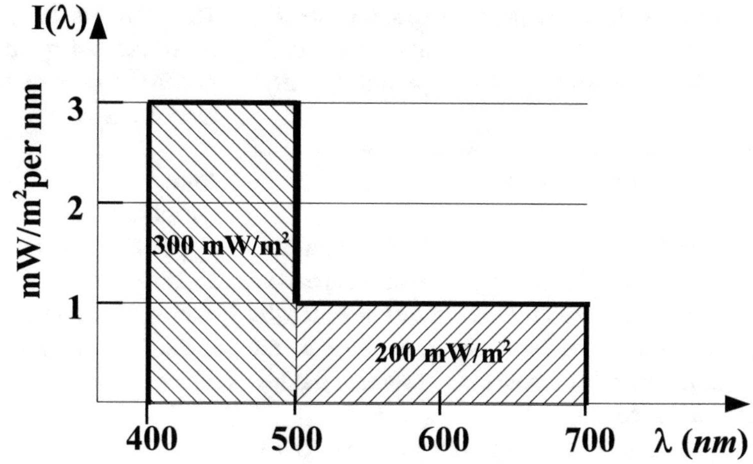

Fig. 14.7 Spectral intensity

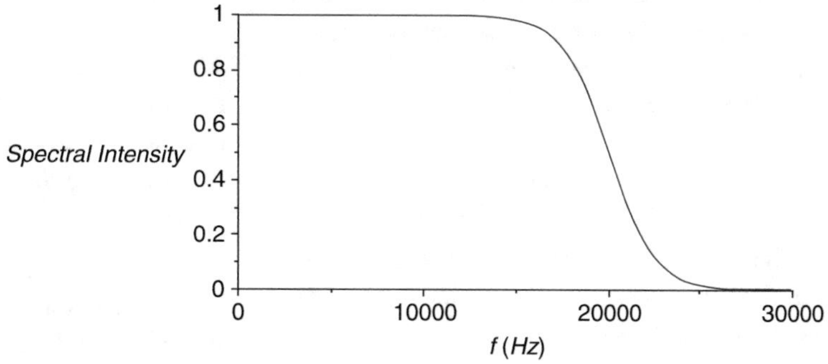

Fig. 14.8 An example of a spectral intensity of white noise

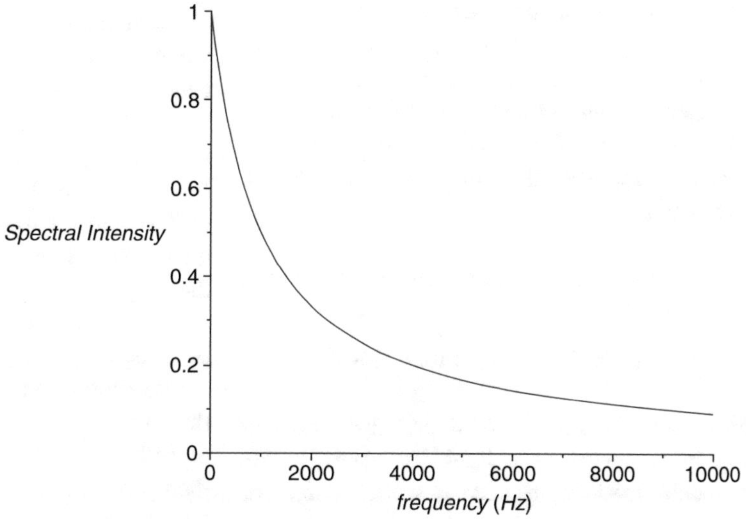

Fig. 14.9 Spectral intensity of pink noise

$$\mathbf{I}(f) \propto \frac{1}{f} \tag{14.6}$$

In electronics the noise is often referred to as "**one-over-f noise**."

Questions:

1. Why do you suppose this noise is referred to as "pink noise"?

2. What is the corresponding spectral intensity $I(\lambda)$ with respect to wavelength? What is the color corresponding to this spectral intensity?

14.5 Color Filters

A color filter is transparent to light, with a fraction of the intensity transmitted that is dependent upon the wavelength. Color filters function by a process of **selective absorption**—that is, the more a spectral component is absorbed, the less is transmitted.

If a light beam of spectral intensity $I(\lambda)$ is passed through a color filter, the outgoing beam has a spectral intensity $I'(\lambda)$ that is given by the product of the **transmittance** $T(\lambda)$ and the incoming spectral intensity:

$$\boxed{I'(\lambda) = T(\lambda)I(\lambda)} \tag{14.7}$$

The transmittance can be expressed as a fraction (less than one) or a percentage. The process is exhibited in Fig. 14.10.

Note If the incoming intensity is pure white, the outgoing beam will have a spectral intensity that is proportional to the transmittance. Thus, to produce the blue and pale blue colors in the previous figures requires filters with transmittances that are proportional to the respective spectral intensities.

An example of the transmittance of a red filter is shown in Fig. 14.11.

Homework: Sketch $T(\lambda)$ for a green filter.

Note We do not have filters that transmit highly monochromatic light. To obtain highly monochromatic light from a white source (or non-chromatic source) a prism or a diffraction grating is used. In this capacity, these devices are referred to as **monochrometers**.

Note One might ask, "What happens to the incident light that is not transmitted through a filter?" The answer is that the remaining part of the light is represented by reflected light as well as light that is absorbed by the material out of which the filter is made.

Note

Question: Does the value of the transmittance at a specific value of the wavelength have a direct significance experimentally? The answer is affirmative: The transmittance at a given wavelength is the fraction of incident **monochromatic light at that wavelength** that is transmitted.

Stacking Filters (Filters in Series)

When two filters are laid one on top of the other, in series, we say that they are **stacked**. The transmittance of a **series of two filters** is obtained by *multiplying* the two respective transmittances.

Thus,

$$\boxed{T(\lambda) = T_1(\lambda)T_2(\lambda)} \tag{14.8}$$

Note that the *order* in which the filters are stacked is not relevant since $T_1(\lambda)T_2(\lambda) = T_2(\lambda)T_1(\lambda)$.

To illustrate an application of this formula, we will examine a process whereby an incoming beam of white light has its spectral intensity increasingly narrowed (hence made increasingly saturated) by a series of *two identical* filters, so that $T_1 = T_2$ and $T(\lambda) = T_1(\lambda)^2$. See Fig. 14.12.

The length and color of the arrows in the figure are meant to show that on passing through a filter, the outgoing light has a reduced intensity and an increased saturation. This effect can be achieved by using only **one** filter as follows:

We simply have the light that passes through the single filter reflect backwards from a mirror or reflect diffusely from a white surface and then pass through the filter a second time. [The term **white** here means that all frequencies are reflected to an equal extent.] See Fig. 14.13.

Let us consider a simple **numerical** example to illustrate the increase in saturation by a **series of two identical filters**. Suppose that $T_1(\lambda)$ and $T_2(\lambda)$ are given by the following:

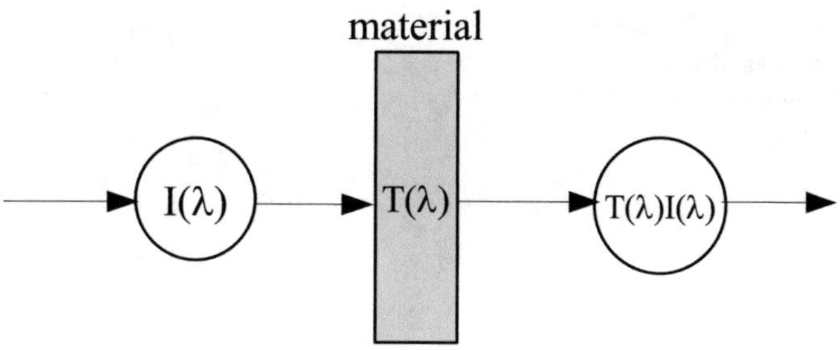

Fig. 14.10 Transmittance of a material

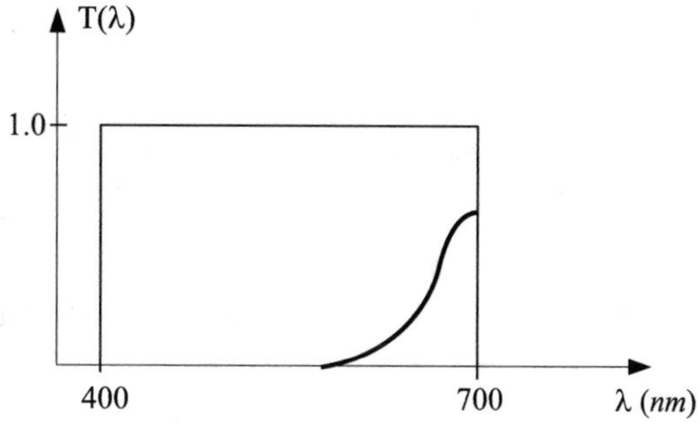

Fig. 14.11 Transmittance of a red filter

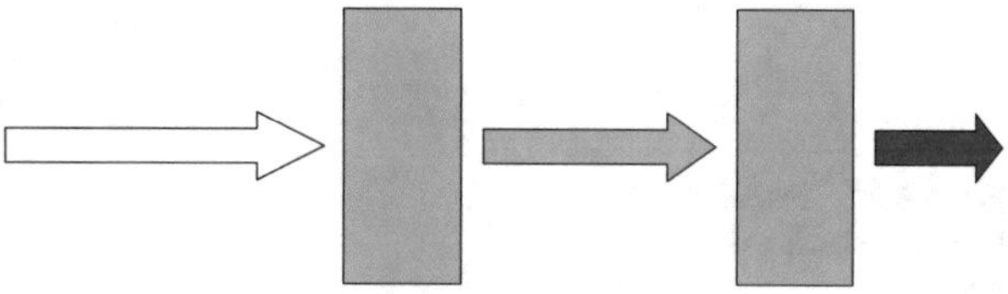

Fig. 14.12 Two identical stacked filters

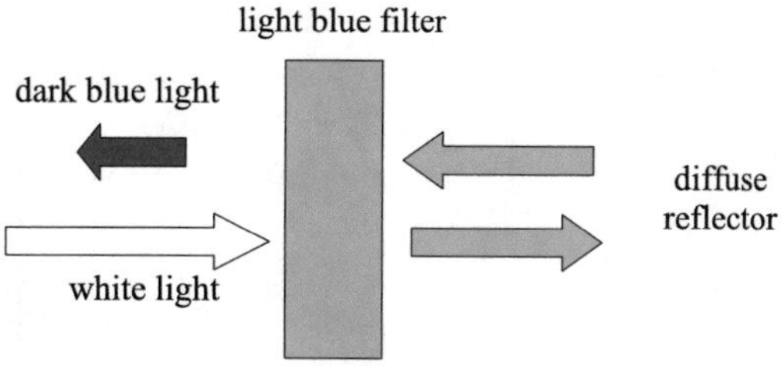

Fig. 14.13 Light through a filter, reflected back through the filter

$$T_1(\lambda) = T_2(\lambda) = T(\lambda)$$
$$= \begin{cases} 0.8 \text{ for } 500 \text{ nm} < \lambda < 600 \text{ nm} \\ 0.3 \text{ otherwise} \end{cases}$$

This transmittance is shown in Fig. 14.14.

$$T(\lambda)^2 = \begin{cases} 0.64 \text{ for } 500 \text{ nm} < \lambda < 600 \text{ nm} \\ 0.09 \text{ otherwise} \end{cases}$$

The resulting net transmittance is shown in Fig. 14.15.

We see that the resulting transmittance is more sharply peaked. A series of identical filters or a single thick filter can be used to make a white beam more monochromatic. However, note that *the price paid is reduced intensity*.

Example Two: We consider a series of two filters, one is a red filter and the other a blue one. We see that wherever $T_{\lambda_R}(\lambda)$ is non-zero, $T_{\lambda_B}(\lambda)$ is zero, and conversely. Thus, the net transmittance vanishes for all visible wavelengths. *That is, the two filters remove all the light!*. See Fig. 14.16.

To further simplify our discussion we will use the following symbols to refer to various general colors:

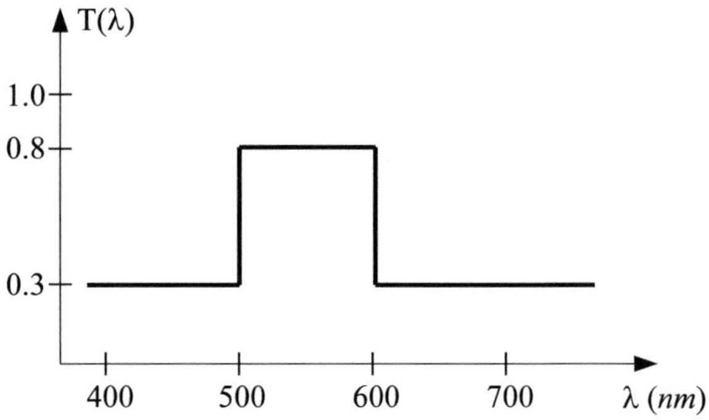

Fig. 14.14 Graphical representation of the transmittance in the above example

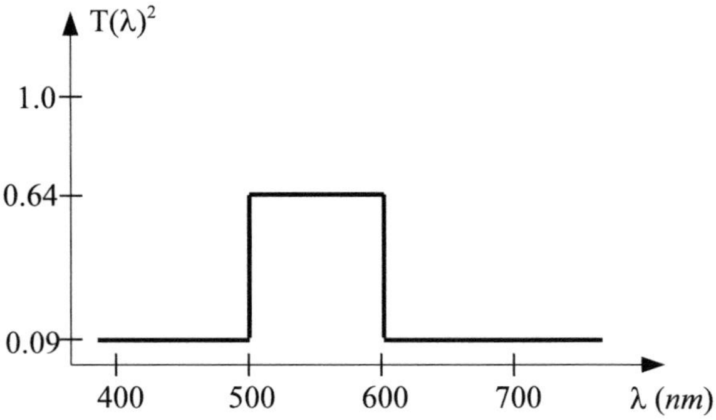

Fig. 14.15 Net transmittance for the sequence of two identical filters

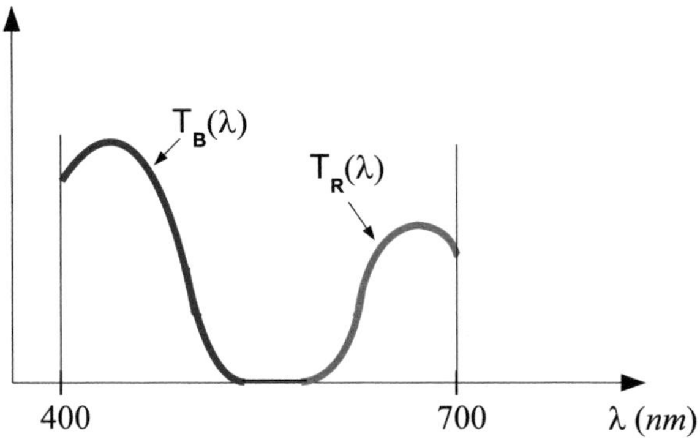

Fig. 14.16 Two filters together filtering out all light

W: white	**G**: green	**M**: magenta	**BN**: brown
R: red	**B**: blue	**C**: cyan	**BK**: black
O: orange	**V**: violet	**P**: purple	**Y**: yellow

In order to indicate the effect that a filter has on a beam of light, we will use the symbols in Fig. 14.17 below. Their significance should be self-explanatory.

We can use a prism to determine the transmittance of a filter as shown below for a *typical yellow filter*. See Fig. 14.18. First, let us recall that yellow is a spectral color. As such, we might expect to see one outgoing yellow beam. Instead, we would see two outgoing beams, green and red.

This analysis reveals that the filter does not transmit spectral yellow. Instead, it transmits the two colors, red and green. We will label such a filter with an asterisked Y: **Y***, in order to distinguish it from a spectral yellow filter. From the above we learn that the eye sees yellow when a mixture of red and green light is incident on the eye. We will explain this phenomenon in Chap. 15, where we will discuss a theory of Color Vision.

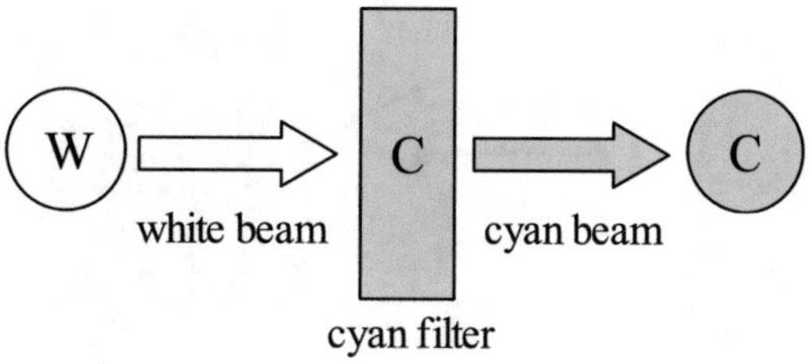

Fig. 14.17 Cyan filter

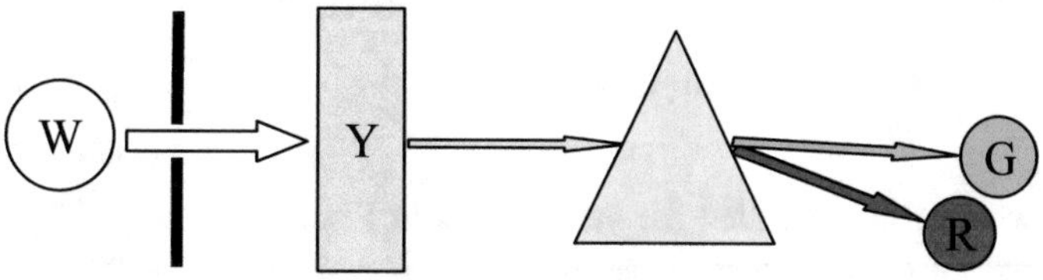

Fig. 14.18 Yellow beam and a prism

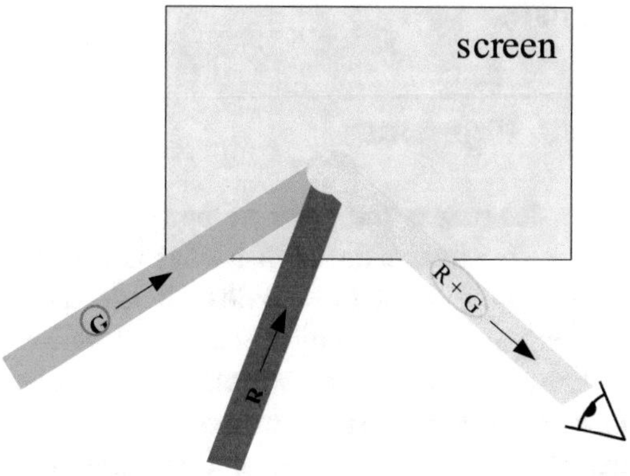

Fig. 14.19 Addition of red and green to give yellow: $R \bigoplus G \equiv Y$

The most significant conclusion of these observations, a conclusion that has widespread validity, is that there is

NO one-to-one correspondence between the multitude of spectral intensities I(λ) and multitude of color sensations.

In fact, every color will, can be produced by an infinite number of spectral intensities.

Now suppose that we cast a circular beam of light onto a white screen that scatters light diffusely. Such an image is called a **color patch**. If we cast two color beams onto the screen, and have the patches overlap, our eyes will see the light from both beams that has scattered diffusely from the screen into our eyes. Our eyes will then receive the *sum* of the two light intensities. In particular, overlapping patches from a red filter and a green filter will look like a *yellow patch*! See Fig. 14.19.

The sum of red light and green light appears yellow as with the production of yellow with the yellow filter Y*. The process reflects a synthesis, from a mixture of spectral green and spectral red, of a color that is *essentially* indistinguishable[5] from the sensation produced by a spectral yellow. This synthesis is the reverse of the analysis carried out by the prism on the light transmitted by the yellow filter.

[5]The word *essentially* is inserted here because light from a single mono-chromatic source will always be more saturated than light from a mixture of monochromatic sources, as we will learn in Chap. 15.

Most people are surprised at the above observation of mixing red and green to produce yellow. They expect the same result as is produced by mixing red and green **paints**. In that case, **red** added to **green** equals a brown. In the next section we will discuss the behavior of **pigments**, which are the basic ingredients of paints.

14.6 Reflectance

If a beam of light with spectral intensity I(λ) is reflected off a surface, the spectral intensity of the reflected light is given by

$$\boxed{\mathrm{I}'(\lambda) = \mathrm{R}(\lambda)\mathrm{I}(\lambda)} \tag{14.9}$$

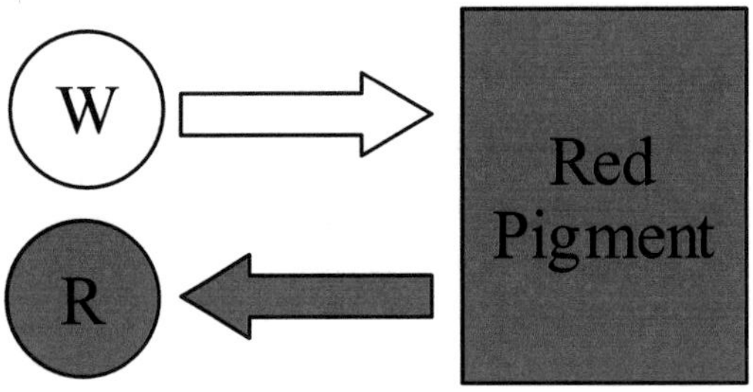

Fig. 14.20 Action of red pigment

where R(λ) is called the **reflectance of light**. It ranges from zero to unity. However, it is often expressed as a percentage, therefore ranging from 0 to 100%.

14.7 Pigments

What determines the color of opaque objects? Clearly, the transmittance of a filter is mirrored by the ability of the filter to **absorb selectively**. Similarly, an opaque object absorbs selectively and has a corresponding wavelength-dependent *reflectance*. **Pigments** are materials that are extremely selectively absorptive. They can color a material when present in very dilute concentrations and are most commonly used in the dyeing of cloth, paints, the color filters described in this chapter, and food coloring.[6] The basis for their strong selective absorption is typically a specific molecule. For example, the pigment **lapis lazuli**, well known and cherished in ancient times, produces a deep ultramarine color due to a sodium-silicate molecule (Na8-10Al6Si6O24S2-4). Here are descriptions as to how pigments function, with a simplified account:

1. RED pigment absorbs all but red and reflects red. See Fig. 14.20
2. GREEN pigment absorbs all but green and reflects green.
3. BLUE pigment absorbs all but blue and reflects blue.
4. Y* pigment absorbs B, and reflects R and G.
5. Y* pigment mixed with B pigment absorbs B, R, and G—so absorbs all!

Thus, Y* pigment mixed with blue pigment should appear black according to our simple description. Most often, such a mixture appears green.

Oil paints behave like the pigments described above. How they are prepared so as to have the desired absorption-reflection characteristics is fascinating.[7]

In Fig. 14.21 we see a display of powdered pigments.[8]

14.8 Summary Comments on Filters and Pigments

1. Both filters and pigments function by a process of *selective absorption*.
2. With a filter, we observe *transmitted light*.
3. With pigment, we observe *reflected light*.
4. When the color patches of light beams are *overlapped*, spectral intensities are added and we have **additive mixing**.
5. When color filters are *stacked* or pigments are *mixed*, spectral absorptions are added and we have **subtractive mixing**.

Figure 14.22 below illustrates very nicely the difference between the two processes, addition and subtraction. Results for both types of mixing of the respective primaries are summarized in Fig. 14.23.

14.9 Terms

– additive mixing
– bandwidth of a spectrum
– color filter

[6]For a fascinating discussion of food colorings, see https://www.acs.org/content/acs/en/education/resources/highschool/chemmatters/.

[7]The following are suggested for further reading: *Light and Color in Nature and Art*, by S. Williamson and H. Cummins (J. Wiley and Sons, N.Y., 1983), *Light and Color*, by R. D. Overheim and D. L. Wagner (J. Wiley and Sons, N.Y., 1982), and *Seeing the Light*, by D. Falk, D. Brill, and D. Stork (Harper and Row, N.Y., 1986).

[8]Source: https://upload.wikimedia.org/wikipedia/commons/f/fb/Indian_pigments.jpg.

Fig. 14.21 Powders of concentrated pigments

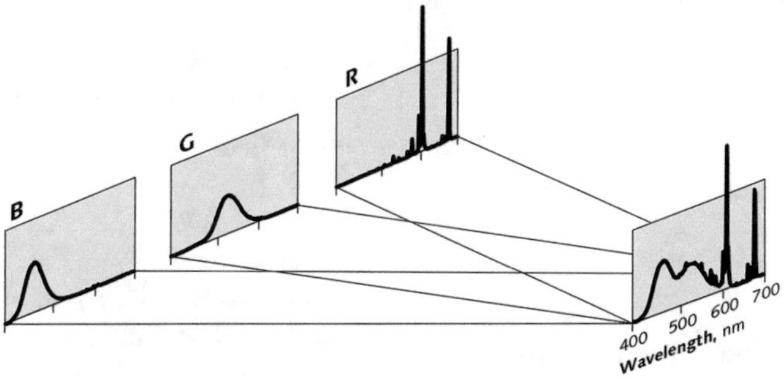

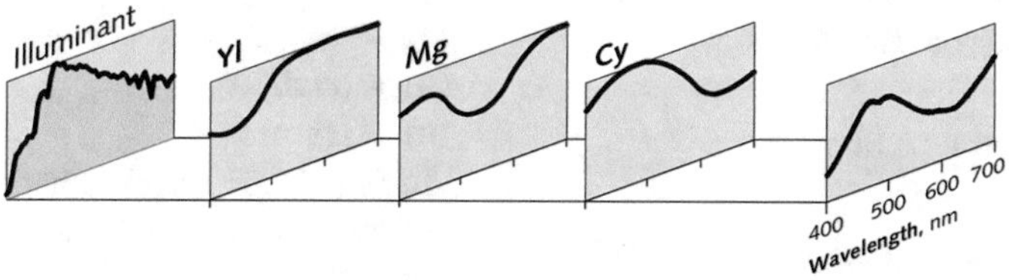

Fig. 14.22 Comparison of addition and subtraction of spectral intensities
(source: Charles Poynton and Garrett Johnson, "Color science and color appearance models for CG, HDTV, and D-Cinema"; ACM Digital Library, ©2004 http://dl.acm.org/citation.cfm?id=1103903, reprinted with permission)

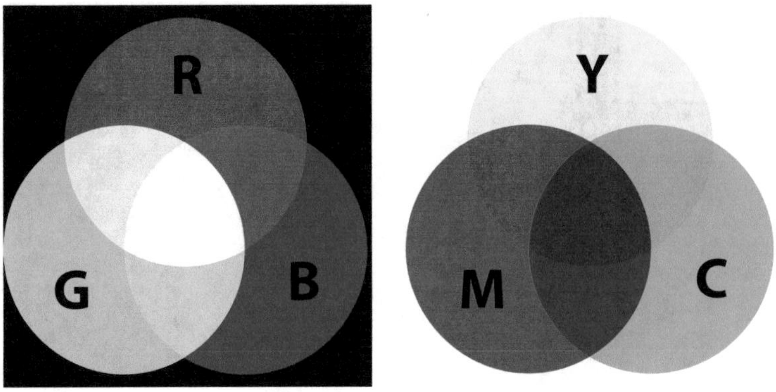

Fig. 14.23 Additive vs. subtractive mixing of colors (sources: **Addition**—http://upload.wikimedia.org/wikipedia/commons/thumb/c/c2/AdditiveColor.svg/1000px-AdditiveColor.svg.png **Subtraction**—http://upload.wikimedia.org/wikipedia/commons/thumb/1/19/SubtractiveColor.svg/1000px-SubtractiveColor.svg.png)

- color
- bandwidth of a spectrum
- color patch
- equal energy spectrum
- hue
- monochrometer
- monochromatic light
- pigment
- saturation
- spectral color
- spectral intensity with respect to wavelength
- subtractive mixing
- transmittance

14.10 Important Equations

Intensity transmitted by a filter:

$$I'(\lambda) = T(\lambda)I(\lambda) \tag{14.10}$$

Transmittance of two stacked filters:

$$T(\lambda) = T_1(\lambda)T_2(\lambda) \tag{14.11}$$

14.11 Problems for Chap. 14

1. What are the physical characteristics of a light source that are, respectively, related to the following:
   (a) hue
   (b) saturation
   (c) brightness
2. When white light is passed through two identical stacked filters having a strong degree of monochromaticity, which of the three attributes of the outgoing beam *change considerably in comparison* with passage through *one* of the filters?
   (a) hue
   (b) saturation
   (c) brightness
3. In Fig. 14.24 are sketched the transmittances of two filters, with $T_1(\lambda)$, and $T_2(\lambda)$, respectively.
   (a) What colors do these filters individually produce?
   (b) If the filters are stacked, sketch the resulting transmittance. What is the corresponding color?

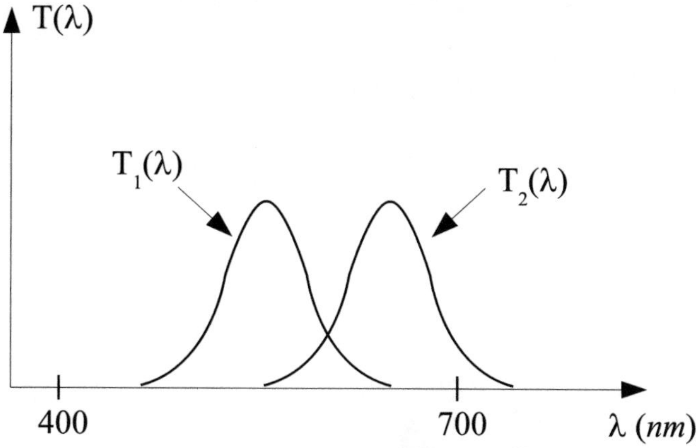

Fig. 14.24 Two filters with overlapping maxima

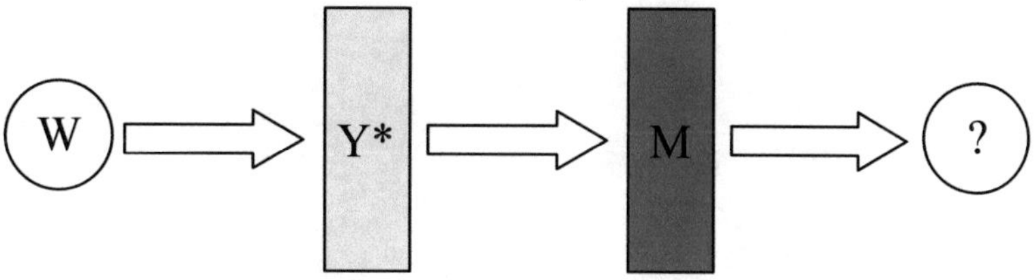

Fig. 14.25 White light passes through a yellow filter followed by a magenta filter

(c) Suppose a white beam is passed through each of the filters and the beams are projected onto a screen so as to produce two overlapping patches. What is the color of the patch? Sketch the spectral intensity of the reflected light.

4. (a) In mixing blue and yellow (Y*) lights, one can produce ________ light.

(b) In mixing blue and yellow (Y*) pigments, one can produce ________ pigment.

(c) Explain why your answers to parts (a) and (b) are different.

5. Complete Fig. 14.25.

Theory of Color Vision

15

In Chap. 14, we saw that one can characterize a light source in terms of its **spectral intensity with respect to the wavelength** I(λ), which is an objective function. We now turn to the question as to the relationship between the **objective** characteristics of a light source and the **subjective** perception of the light. We have already identified three attributes of one's visual perception—**hue**, **saturation**, and **brightness**. The first two are the attributes that, together, are referred to as the **color**, or alternatively, the **chromaticity**. The technical term for the third attribute, brightness, is referred to as the **luminance**, which has the symbol **Y**. All three are numerical parameters and are therefore **objective parameters**.

Note We will learn how to calculate these three parameters from the spectral intensity with respect to the wavelength I(λ). Then we will discuss how this relationship can be understood in terms of the physiological behavior of the visual apparatus, the **rods of the retina**.

It is noteworthy that music focuses on TIME while color focuses on SPACE.[1] In the case of sound and music, we know that the amplitude of an incident sound wave as a function of time determines the sound that a person hears. The amplitude can be analyzed so as to determine the **spectral intensity with respect to the frequency**, I(f). Since frequency varies continuously, there are an infinite number of frequencies. If the spectrum contains only a **harmonic series** of frequencies, the amplitude will be periodic and the person will hear a specific **pitch**, corresponding to the **fundamental frequency** of the harmonic series.

In the case of color, we will focus on a uniform beam of light that is projected onto a limited area of a surface so as to produce a **color patch**.[2]

In addition to the elementary book, *Light and Color* by **Overheim and Wagner** (op, cit. in Chapter 13), the reader is referred to the advanced texts: **T. N. Cornsweet**'s *Visual Perception* (Academic Press, N.Y., 1970) and **Y. Le Grand's** *Light, Color, and Vision* (Dover, N.Y., 1957).

[1] A few decades ago I was privy to a conversation between my violin teacher's wife and her close friend—the first a pianist, the second a painter. They were discussing the differences between their personal artistic experiences. My pianist friend pointed out that *a performance takes place in time, having a beginning and an end in time.* An audio recording cannot fully capture the depth of the experience she had. She expressed her envy of the painter. While the pianist will have only with great difficulty a deep pleasure contemplating her past experience, the painter produces *a painting in space that is confined only by its spatial boundary*, so that she has all time in the world to contemplate and treasure her piece of art.

[2] Dealing with variations of spectral intensity on a surface leads to complexity of the excitation of light sensors on an area of the retina; this interesting and very important interaction will not be dealt with in this book. Of course, quite important variations are associated with paintings.

L. Gunther, *The Physics of Music and Color*,
https://doi.org/10.1007/978-3-030-19219-8_15

MUSIC	COLOR
Spectral intensity I(f)	Spectral intensity I(λ)
A **pitch** will be heard only if the sound wave is periodic over a significant **TIME** interval	All spectral intensities I(λ) on a 2-dimensional patch in **SPACE** will produce a **color**
phon measure of loudness ϕ	**luminance** measure of brightness **Y**
sone measure of loudness **s**	**lightness** measure of brightness, **L***

The light can be analyzed so as to determine its spectral intensity with respect to the wavelength, I(λ). While the wavelength spans a continuum from zero to infinity, we have learned that the range of light (visible electromagnetic waves) is from about 400-nm to about 700-nm. There are an infinite number of wavelengths in this range. [See the discussion of spectral intensities.] Furthermore, there are an infinite number of graphs of the spectral intensity over this range.[3]

We will begin by discussing color vision as experienced by people with **normal vision**. [Later in the chapter we will discuss color blindness.] A spectral intensity will be shown to be associated with a pair of numbers referred to as **color coördinates**. Such a pair can be placed at a point in a plane. While there are an infinite number ($\aleph_0^{\mathcal{C}}$ to be precise) of spectral intensities, there are only an infinite number ($\mathcal{C}$ to be precise) of pairs of numbers—or, correspondingly, points in a plane. See the footnote on this page for a discussion of different **orders of infinity**.

Note Suppose two people view the same color patch. Because of the associations they have made from childhood between the color they perceive and the word they use to refer to their perceived color, they will use the same word to refer to the observed color. Yet, neither can know whether the other person is experiencing the same color sensation. However, if they view *two* patches of color, to a great degree they will tend to agree as to whether or not the two patches appear to have the same color.

A reliable specification of color is essential for the communication of color among interested individuals and for the reproduction of color by mixing various color sources. Examples are paints, color monitors, color printers, and fabrics. In fact, it is possible for someone to "discover" a color among the infinite number of possible colors, a color that happens to be extremely pleasing or exciting to most people.[4] Such a color can then be patented![5]

15.1 The Three Primary Colors Theory of Color Vision

The observation that any color might be produced by adding together **three primary colors** is well known.[6] This idea is the basis for color television

[3]In the late 1800s, the mathematician Georg Cantor introduced the concept of different orders of infinity and provided a clearly defined method of comparing these levels. The lowest order of infinity is the number of integers, given the symbol $\aleph_0$. The next order of infinities is the set of real numbers, given the symbol $\mathcal{C}$. It can be shown that $\mathcal{C} = 10^{\aleph_0}$. As surprising as it may seem, Cantor was able to show using his method of comparing infinities that $\mathcal{C}$ is also the number of points along a line, a finite line segment, as well as the number of points in an *area*, as well as in a volume! This infinity is the infinite number of different chromaticities. The number of distinct spectral intensities is the number of ways you can draw a continuous graph along an axis. This number is an even higher infinity than $\mathcal{C}$ and can be shown to be equal to $\aleph_0^{\mathcal{C}}$. See the Wikipedia article (1-8-2011): http://en.wikipedia.org/wiki/Georg_Cantor.

[4]It should be recognized that any particular individual is limited in their ability to discriminate one color from another. There are **Just Noticeable Differences in Color** in analogy with **Just Noticeable Differences in Frequency** Therefore, a particular individual can discriminate among only a finite number of distinct colors.

[5]See the interesting article on this complex subject on the website https://lizerbramlaw.com/2012/10/30/color-as-a-trademark/.

[6]We will discuss the modern developments in the field of color. The history of the science of color vision started with Isaac Newton, who in the 1600s proposed that there are seven primaries that can be mixed in appropriate

and was used by the **pointillist** painters, such as Seurat, to produce a rich color painting by using dots of three colors alone. We will begin this chapter by discussing a very simplified version of the three-primary theory of color vision. This version recognizes only six hues. Later, we will refine it with the **chromaticity diagram**, which reflects much more accurately the true nature of color vision, treating the full range of hues and degrees of saturation. It turns out that the ideal three-primary theory is quite good in reflecting human color vision. Finally, we will see how the three-primary theory is connected with the existence of three types of cones in the retina, the **red cone**, the **green cone**, and the **blue cone**. We will also see how these three cones can provide us with the three basic attributes of our visual sense: *hue*, *saturation*, and *brightness* through their physiological behavior. In 1986, **genes** for these three cones were identified at Stanford University by Jeremy Nathans, thus establishing a physiological basis for the primary color theory.

According to the ideal **Three-Primary Theory of Color Perception**, there exists a set of three colors—referred to as **primaries**—which have the following property:

Any color sensation can be produced by an appropriate mixture of spectral intensities corresponding to the three primaries.

We will henceforth label these spectral sources as R, G, and B. We must keep in mind that the **choice of these primaries is not unique**. There might be more than one set of three sources that can serve as primaries. The names and symbols we are using merely indicate that the choices of primaries that work well tend to be in their respective regions of color, that is, red, green, and blue. Finally, a primary can have a **spectral color**—that is, be produced by monochromatic light. We call such primaries **"spectral primaries."**

15.2 Metamers

It is found that an infinite number of different spectral intensities can produce the same color sensation. Any two spectral intensities producing the same color sensation are said to be **metamers**. Alternatively, two spectral intensities that have the same pair of color coördinates r and g are **metamers**. An example of such a pair of metamers is shown in Fig. 15.1.

We see two distinctly different spectral intensities at the top. Below we see the match of the admixture of primaries (the two sets of three colored vertical bars) and the match in the color of the discs.[7]

As we will see, the fact that *two* numbers are sufficient to label a metamer is connected with the fact that **chromaticity** is specified by two characteristics—hue and saturation. In fact, they are specified by a pair of numbers. We have an infinite number of spectral intensities as well as an infinite number of pairs of numbers that specify the chromaticity. Mathematicians would say that the infinite set of all possible spectral intensities are **mapped** onto an infinite set of pairs of numbers.

proportions to produce any color sensation. The basis of the proposal was Newton's studies of the decomposition of white light by a prism into its rainbow of colors. He identified the color of an object as an attribute of the response of the eye to various wavelengths of light that are reflected off the object as opposed to the idea that the color "resides" in the object itself. The proposal that there are three primaries is due to Thomas Young (1807). Many other scientists helped develop the basic principles of color mixing—in particular, James C. Maxwell, who provided a theoretical basis for electromagnetic waves, as discussed in Chap. 5. In 1860, Maxwell produced the first, albeit crude, set of **color matching functions**, which will be discussed in detail in this chapter. For an excellent history of studies of color vision, see Deane Judd in the publication: NATIONAL BUREAU OF STANDARDS: VOL. 55, p. 1313, (1966).

[7]The figure was produced using the applet on the following wonderful website: http://www.cs.brown.edu/exploratories/freeSoftware/repository/edu/brown/cs/exploratories/applets/spectrum/metamers_guide.html. The applet enables you to play around with pairs of spectral intensities, each independently and see their respective color patches. You can then produce metamers galore.

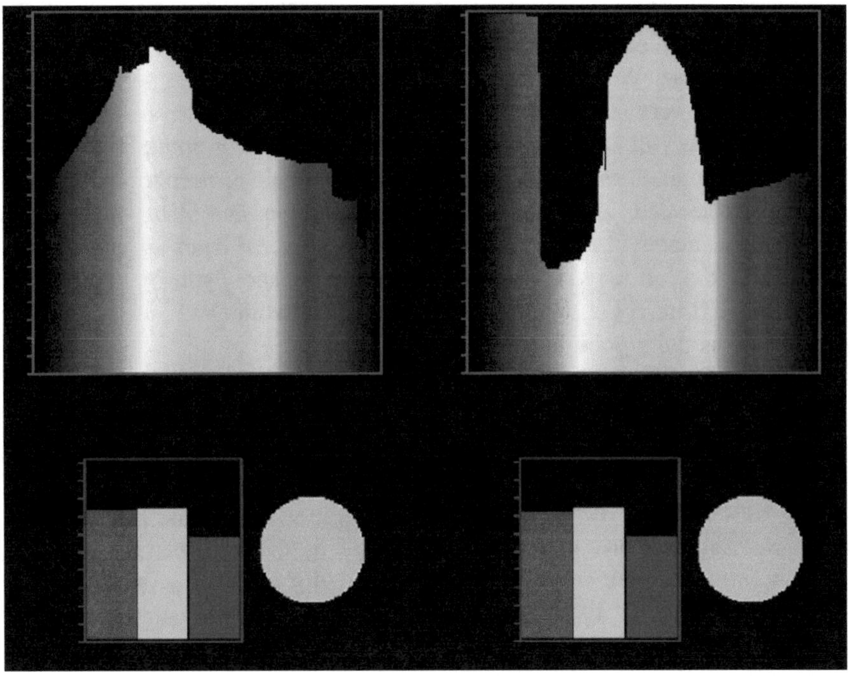

Fig. 15.1 Metamers—two spectral intensities that produce the same color sensation

15.3 Simplification with Just Six Hues

Suppose that we choose three sources of light that we hope to serve as **primaries**. Ideally, we would be able to add these sources, with appropriate total intensities, so as to produce the same color as a given color patch. We say that the sum of the spectral intensities matches the given color. Let the intensities of the primaries that produce a match be

$$\boxed{I_R \quad I_G \quad I_B}$$

Then

$$\mathrm{I}(\lambda) = I_R + I_G + I_B \qquad (15.1)$$

To introduce the ideas of the Three-Primary theory, we will neglect saturation and consider only **SIX HUEs**: R, G, B, Y, C, and M. [See Chap. 14.] The three non-primary hues can be produced from mixtures of the primaries as follows:

Let the symbol $\oplus$ represent the addition of two light sources. Thus, for example,

Yellow = Red + Green
Alternatively we will write $Y = R \oplus G$

Cyan = Blue + Green
Alternatively we will write $C = B \oplus G$

Magenta = Blue + Red Alternatively we will write $M = B \oplus R$

White = Blue + Green + Red Alternatively we will write $W = B \oplus G \oplus R$

R, G, and B are referred to as **additive primaries**, in that they can produce the remaining basic hues, Y, C, and M, along with white (W).

Within the framework of this simplified set of hues,

- a C-filter filters out R
- a Y-filter filters out B
- an M-filter filters out G

Suppose that we pass white light through these filters, C, Y, and M. Furthermore, let the symbol $\ominus$ represent the removal of the component that follows the symbol from the component that precedes it. Then we can write:

$$C = W \ominus R$$
$$Y = W \ominus B$$
$$M = W \ominus G$$

When filters are **stacked** in **series**, C, Y, and M filters act as **subtractive primaries**, in that they can produce the remaining basic hues, R, G, and B, along with black (BK): Let us introduce the symbol $\odot$ to indicate that two filters are in series. Thus, $C \odot Y$ represents C and Y in series. Then $C \odot Y = W \odot R \odot B = G$.

Homework: Explain why $B \odot Y = BK$, $C \odot M = B$, $Y \odot M = R$, and $C \odot Y \odot M = BK$.

The three subtractive primaries form **complementary color pairs** with the additive primaries, in that the addition of a pair of such colors produces WHITE:

$$\begin{aligned} B \oplus Y &= W \\ G \oplus M &= W \\ R \oplus C &= W \end{aligned} \tag{15.2}$$

Next, we want to move on to consider the full set of colors perceived with normal vision. This set is infinite in number.

15.4 Exploration of Color Mixing with a Computer

It is quite an experience to see how a mixture of primaries can produce a vast set of colors. You can do so in a simple way by using **Paintbrush**, which is available with either a **PC** or a **MAC**. Typically this program comes with any purchased PC. In the case of a MAC, you can obtain a free download from the apple.com website.

Your first step is to gain access to a window that displays a box whose color is determined by a set of the three numbers corresponding to the admixture of the primaries of your monitor. I suggest that you choose the setting that provides 8-bit **Truecolor**. The number eight means that there are $2^8 = 256$ possible values for the intensity of each primary, the values being given by 0, 1, 2, 3, ...255. The total number of possibilities is then $256 \times 256 \times 256 = 16,777,216$. Here is how you can access the **palette of colors**: In the **PC** version, you should click on colors/edit colors. In the MAC version you should click on **Tools/Font/Colors**. I will refer to the **MAC** version in what follows. In Fig. 15.2 you can see two of the windows that should appear.

At the very top you will see a small color wheel to the far left, followed by three color sliders.

Choose RGB sliders. Here you can choose the sets of three numbers that are associated with the color in a computer. I will use the bold letters **R**, **G**, **B** here to refer to these values. *Be forewarned that the unbolded letters R, G, and B will refer to another set of numbers used later on in this chapter—the so-called* ***tristimulus values****. The two sets of three numbers are both used to specify the color but are not the same! Unfortunately, you will find both sets of symbols used interchangeably in the literature, so that you will have to make sure that you are certain about which set the symbols represent.* Another important fact to keep in mind is that the actual color you will see on a monitor or produced by a printer by a given set of monitor color coördinates **R**, **G**, **B** will vary considerably. As a result there is incredibly extensive literature on the problems of matching colors.

Experiment by moving the sliders and observing the changing values **RGB** and the corresponding color in the box at the top. Next, choose a set RGB and switch to the **color wheel**. In this window, you will see a circle on the left filled with colors. Note the small "pointer circle" someplace within the colored circle. It characterizes the color. **Color** refers here to **hue** and **saturation**. The position of the pointer circle is characterized by its distance from the center of the color circle and by its angle with respect to a direction to the right. The two numbers are the **polar**

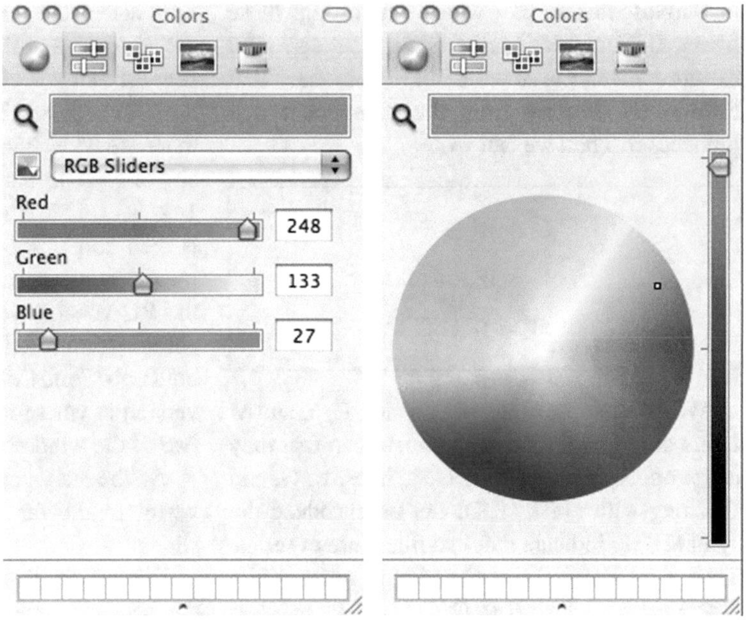

Fig. 15.2 Color mixing with paintbrush in a Mac computer—RGB sliders on the left, color wheel and intensity slider on the right

Table 15.1 Preucil circle of colors

Ordering	Hue region	Formula
$R \geq G \geq B$	Red-Yellow	$h_{\text{Preucil circle}} = 60° \cdot \frac{G-B}{R-B}$
$G > R \geq B$	Yellow-Green	$h_{\text{Preucil circle}} = 60° \cdot \left(2 - \frac{R-B}{G-B}\right)$
$G \geq B > R$	Green-Cyan	$h_{\text{Preucil circle}} = 60° \cdot \left(2 + \frac{B-R}{G-R}\right)$
$B > G > R$	Cyan-Blue	$h_{\text{Preucil circle}} = 60° \cdot \left(4 - \frac{G-R}{B-R}\right)$
$B > R \geq G$	Blue-Magenta	$h_{\text{Preucil circle}} = 60° \cdot \left(4 + \frac{R-G}{B-G}\right)$
$R \geq B > G$	Magenta-Red	$h_{\text{Preucil circle}} = 60° \cdot \left(6 - \frac{B-G}{R-G}\right)$

coördinates of its position. The angle determines the hue, ranging from red to magenta, as the angle moves around from 0° to 360°. Thus RED is at 0°, GREEN is at 120°, and BLUE is at 240°.

The angle is determined by the **RGB** values according to the formula introduced by Frank Preucil and produces the **Preucil Circle.**[8] Table 15.1 provides the equations needed to calculate the angle.

Next you can control the degree of **saturation**: By moving the pointer circle radially inward you will see how the color becomes more pale. Red becomes a pink. Blue becomes pale blue, and so on. Lowering the saturation decreases your ability to distinguish among various hues. Ultimately, you will reach white at the center.

By moving the vertical slider at the right up and down you can change the overall intensity while maintaining the color. Doing so will not change the hue. However, as you lower the intensity all the **RGB** coördinates will be reduced. The question is whether they are reduced in the same proportion. Check this out by switching back and forth between the circle window and the RGB slider window. A good place to start is with RGB set initially to full brightness white: **R=G=B=255**. As you reduce the intensity, you should find the **RGB** values reduced in proportion, so that they are always equal.

In Fig. 15.2 we see the two windows with an orange patch having **RGB** coördinates {248, 133, 27}, corresponding to 243 units of red primary, 128 units of green primary, and 2 units of blue primary. The actual colors seen will vary from monitor to monitor because of

[8]Frank Preucil, *Color Hue and Ink Transfer É Their Relation to Perfect Reproduction*, TAGA Proceedings, p 102–110 (1953).

the varying mapping systems between computer color coördinates and intensities. If you now move the intensity slider downward towards decreasing intensity, you would find that the orange looks brown. Thus we learn that brown is not a specific color in the sense that we are using the term; rather, it is the color orange with low intensity. I am reminded of how a slice of bread that starts out with a very pale yellow color turns brown as it is toasted. Toasting increases the level of absorption of light so that the intensity of reflected light decreases. When the toast is fully burned, it absorbs a great fraction of the light and looks black.

15.5 Introduction to the Chromaticity Diagram

In this section we will begin our study as to how one can determine the admixture of primaries that will produce a match with the color of a given spectral intensity $I(\lambda)$. In parallel, we will be determining the **color coördinates** of the color.

Recall that our perception of a light source has three distinct characteristics, hue, saturation, and brightness. Hue and saturation together constitute the **color** of the light—technically referred to as the **chromaticity** of the light source. Thus,

$$\boxed{\text{chromaticity} = \text{hue} + \text{saturation}} \tag{15.3}$$

In Sect. 15.3, we ignored brightness and saturation and recognized only six hues. Here we still neglect brightness but consider the full range of chromaticities.

The results of tests on people with normal vision—those who are not classified as color blind—are summarized in what is known as a **Chromaticity Diagram**. It reflects the observation that two numbers alone are sufficient to specify the chromaticity of a light source:

For any spectral intensity $I(\lambda)$, one can calculate the values of these two numbers, referred to as **color coördinates**. They are given the symbols r and g.

We can easily appreciate why the chromaticity diagram can be of great interest and use to the student of color vision, for purely academic reasons as well as for medical reasons. However, the diagram is also invaluable to artists, to stage designers, as well as to designers of cloth for clothing. Imagine how useful it is to be able to refer to any color precisely by telephone or mail by merely specifying two numbers, say $\{0.2023, 0.4285\}$.

However—this is the central fortunate observation—experiments have shown that when people with normal vision are asked to look at two color patches of light and asked whether they are the same or not, there is a strong agreement in their responses. If the colors of a pair of color sources appear different for one observer they appear different for the other observer, and conversely. Because of this observation, in the late 1920s and early 1930s, pioneers in color specification carried out extensive experiments to quantify color perception. The foremost were W. David Wright and John Guild.[9]

15.5.1 A Crude Chromaticity Diagram—Color Coördinates

In this subsection, we will introduce the essence of color coördinates and a chromaticity diagram. Let us suppose that **three** primaries can be mixed to match the visual sensation of any given spectral intensity $I(\lambda)$. The mixture is specified by the intensities of each of the three primaries, namely I_R, I_G, and I_B.

Note At this stage, we don't have a prescription as to how these three intensities are determined by the spectral intensity. We will provide a prescription later on in this chapter.

---

[9] **Wright, William David** (1928). "A re-determination of the trichromatic coefficients of the spectral colours." Transactions of the Optical Society 30: 141–164. **Guild, John** (1931). "The colorimetric properties of the spectrum." Philosophical Transactions of the Royal Society of London (Philosophical Transactions of the Royal Society of London. Series A, Containing Papers of a Mathematical or Physical Character, Vol. 230) A230: 149–187.

We will now see how these three functions can determine the three parameters of color perception: hue, saturation, and brightness.

- The *total* intensity, $I = I_R + I_G + I_B$, associated with I(λ) is a measure of *brightness*.
- The three *fractions*

$$\boxed{\begin{aligned} r' &= \frac{I_R}{I_R + I_G + I_B} \\ g' &= \frac{I_G}{I_R + I_G + I_B} \\ b' &= \frac{I_B}{I_R + I_G + I_B} \end{aligned}} \tag{15.4}$$

can characterize the *hue* and *saturation*, that is, the chromaticity.

It is easy to check from Eq. (15.4) that we have $r' + g' + b' = 1$. Therefore, only **two** of these three fractions—that is two out of the three parameters (r', g', and b')—are independent. For example, if you know r' and g', b' is determined to be (1-r'-g'). The convention is to specify r' and g'. They are referred to as the **color coördinates** and determine the two characteristics, **hue** and **saturation**.

Expressed succinctly, associated with any spectral intensity I(λ) is a pair of numbers, the color coördinates (r', g'), which characterize fully the color sensation produced by the spectral intensity.

Suppose that the three intensities have the values of 1, 2, and 4, respectively, in some arbitrary units. The choice is irrelevant as far as chromaticity is concerned since only the **ratios** are relevant.[10] The total intensity is $1+2+4=7$ units and determines the brightness.

The three fractions are $r' = 1/7$, $g' = 2/7$, and $b' = 4/7$, which add up to unity. The first two fractions, here $1/7$ and $2/7$, are used to specify the chromaticity.

Since two numbers specify a point in a plane, we can specify the chromaticity by a point in a plane. Furthermore, since the sum of the two numbers cannot exceed one, it can be shown that the point must lie within or on the boundary of the triangle shown in Fig. 15.3.

Note that at the corners of the triangle, we have the corresponding pure, fully saturated primaries, R, G, and B, respectively.

It happens that in order to match **white**, one needs to add a mixture of primaries such that the intensity of the R-primary is much greater than that of the other two primaries, G and B. As a result, white (W) will lie very close to the R-corner at $(1, 0)$, as shown in Fig. 15.3. The chromaticity diagram then has the undesirable feature of having the region of rapid variation of hues strongly concentrated in this corner, with much less variation occurring elsewhere. This feature can be removed in the actual chromaticity diagram by using a different unit of intensity for each primary. Such a change in the diagram will be discussed in the next section.

Ideally, monochromatic colors would lie along the two line segments, joining "B" to "G" and "G" to "R," respectively. The line segment from "B" to "R" is called the **line of purples. None of these purple colors is represented by monochromatic light**. In more common terms: these colors are not part of the rainbow of colors. **Magenta** is such an example.

Unfortunately, experiments reveal that it is impossible to match all colors with a sum of three primaries. Any such addition cannot be fully saturated. We produce a large fraction of colors by using monochromatic colors as primaries—referred to as spectral primaries. However, the only colors that can be perfectly matched will be the spectral primaries.

In the following section we will describe in detail how one calculates the color coördinates with respect to a given set of three monochromatic primaries that correspond to any given

[10]If a given spectral intensity is increased uniformly for all wavelengths by the same factor, it has been found from studies of human subjects that the color coördinates do not change. This observation amounts to saying that color is independent of brightness or, analogously, that pitch is independent of loudness.

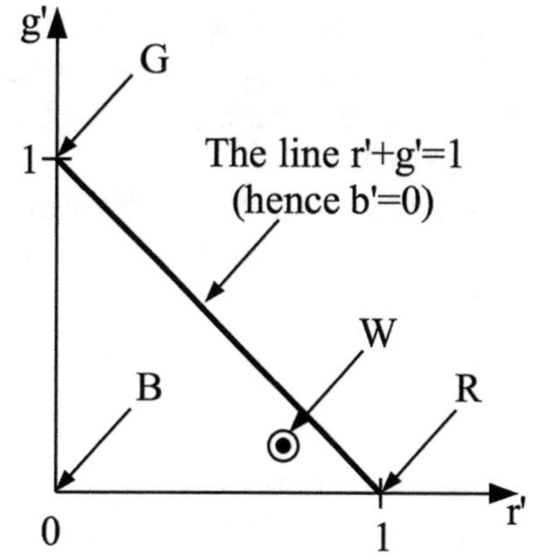

Fig. 15.3 The color coördinates are restricted to the triangle in the diagram

spectral intensity. We will also describe the resulting chromaticity diagram and the information it contains more fully.

15.6 A Standard Chromaticity Diagram

The chromaticity diagram we will now discuss is a detailed standard version of the diagram described in the previous section and involves the determination of the color coördinates corresponding to any spectral intensity I(λ). It is based upon the experimental results obtained by Wright and Guild using the following three monochromatic primaries:

$$\lambda_B = 435.8\text{-nm}, \quad \lambda_G = 546.1\text{-nm}, \quad \lambda_R = 700\text{-nm} \tag{15.5}$$

With apologies for repetition:

1. The choice of the three primary sources is arbitrary, yet all choices leave a set of unmatchable chromaticities. In fact, we will see that the above choice of primaries leaves a large range of green chromaticities unmatchable.
2. The fraction of chromaticities that are matchable can be increased by using three monochromatic primaries, that is, **spectral primaries**.

Interestingly, we will see that

No given set of three monochromatic primaries can be mixed so as to match any monochromatic colors but themselves.

Here is why the above wavelengths were chosen by Wright and Guild: The wavelengths of the first and second primaries—the green and the blue—are quite specific: They correspond to two highly monochromatic spectral lines produced by the excitation of mercury vapor using an electrical discharge. On the other hand, at the time of the experiments, an intense monochromatic source of a spectral line in the red region was not available. Fortunately, it happens that color vision is not extremely discriminating in the red region. Thus, an intense source of a red primary, albeit not with too small a spread of wavelengths, could be obtained from an intense light source passing through a red filter with a peak transmittance near 700-nm. While this primary was not highly monochromatic, the primary is **labeled** with the wavelength value of 700-nm.

For any given spectral intensity I(λ), there is a method—to be discussed later in this section—for calculating the corresponding **color coördinates**, also called **chromaticity coördinates**. These two numbers specify the **chromaticity** of that spectral intensity. Chromaticity is the technical term for **color** and provides us with a clear-cut distinction from **brightness**. Brightness is a **subjective** term that is related to the technical **objective** term **luminance**, which will be discussed later in this chapter. The set of all possible color coördinates corresponds to the entire set of possible colors that a person with normal vision can perceive.

We can regard the relationship between the spectral intensity and the color coördinates as a mapping between the two.

Note

- All spectral intensities that yield the same set of color coördinates appear to have the same color and are therefore **metamers**.
- Had a different set of primaries been used, the color coördinates corresponding to a given spectral intensity would change. However, if color coördinates are the same for two metamers using one choice of primaries, the color coördinates will be the same for any other choice of primaries. The reason is that the color coördinates for one choice of primaries are in a one-to-one correspondence with the color coördinates with any other choice of primaries. How the two sets are related will be discussed qualitatively later in this chapter.
- A set of primaries need not all be monochromatic. We will see that choosing *monochromatic primaries* is highly beneficial, as will be explained later in this chapter. In Appendix M we show in mathematical detail how you can calculate one set of coördinates from another set.

15.6.1 The Calculation of Color Coördinates

We will now show how one calculates the color coördinates for a given spectral intensity. The method is based upon a **table of color matching functions** shown in Table 15.2. We will sometimes use the acronym **TCMF** for such a table. Here, we display the **color matching functions**, $\overline{r}(\lambda)$, $\overline{g}(\lambda)$, and $\overline{b}(\lambda)$ for the Wright-Guild primaries $\lambda_B = 435.8$ nm, $\lambda_G = 546.1$ nm, and $\lambda_R = 700.0$ nm.

Let us study the details of the table.

- In the first column we have a discrete set of wavelengths. The wavelengths are separated by 10-nm. The actual wavelengths are distributed continuously. The spectral intensity should be plotted as a continuous curve. Instead, this table allows us to represent the spectral intensity by a **histogram approximation**. There exist TCMFs with finer spacings, such as 5-nm. I have used a 10-nm spacing to save space in this book and because it suffices to demonstrate the essential characteristics.
- In the next three columns we display the color matching functions $\{\overline{r}(\lambda), \overline{g}(\lambda), \overline{b}(\lambda)\}$ for each of the three primaries. We will later see what they represent and how they can be determined.
- In the fifth column we display the spectral intensity I(λ). At the bottom of the fifth column is the sum of the column of values of the spectral intensity for the wavelengths multiplied by the spacing in the table, Δλ=10-nm. The sum is an approximate value of the area under the graph of the spectral intensity.[11]
- In the final three columns we display the products of the color matching functions and the spectral intensity.
- **Tristimulus Values**

  The sums of the last three columns are referred to as the **tristimulus values**, R, G, and B, corresponding to the spectral intensity.[12]
- Notice that the sums of each of the second, third, and fourth columns are equal to 1.89. That is,

$$\text{Sum of } \overline{r}(\lambda) = \text{Sum of } \overline{g}(\lambda) = \text{Sum of } \overline{b}(\lambda) \tag{15.6}$$

The number 1.89 is not significant here; what is significant is that they are equal. Now recall that if a spectral intensity is constant, the color is referred to as**Equal Energy White**,

[11] See Sect. 14.2.3.

[12] For those who know the calculus: The sums should actually be integrals. E.g., $R = \int_{400\text{-nm}}^{700\text{-nm}} \overline{r}(\lambda)\mathrm{I}\lambda) d\lambda \approx$ Sum $[\overline{r}(\lambda)\mathrm{I}(\lambda)\Delta\lambda]$. If the spacing is halved, the number of entries will be doubled, but Δλ is reduced to 5-nm to compensate.

Table 15.2 Table of color matching functions for Wright-Guild primaries

λ-nm	$\bar{r}(\lambda)$	$\bar{g}(\lambda)$	$\bar{b}(\lambda)$	I(λ)	$\bar{r}(\lambda)$I(λ)	$\bar{g}(\lambda)$I(λ)	$\bar{b}(\lambda)$I(λ)
400	0.0003	−0.00014	0.01214				
410	0.00084	−0.00041	0.03707				
420	0.00211	−0.0011	0.11541				
430	0.00218	−0.00119	0.24169				
440	−0.00261	0.00149	0.31228				
450	−0.01213	0.00678	0.3167				
460	−0.02608	0.01485	0.29821				
470	−0.03933	0.02538	0.22991				
480	−0.04939	0.03914	0.14494				
490	−0.05814	0.05689	0.08257				
500	−0.07173	0.08536	0.04776				
510	−0.08901	0.1286	0.02698				
515	−0.09398	0.153839	0.018589				
520	−0.09264	0.17468	0.01221				
530	−0.07101	0.20317	0.00549				
540	−0.03152	0.21466	0.00146				
550	0.02279	0.21178	−0.00058				
560	0.0906	0.19702	−0.0013				
570	0.16768	0.17087	−0.00135				
580	0.24526	0.1361	−0.00108				
590	0.30928	0.09754	0.00079				
600	0.34429	0.06246	−0.00049				
610	0.33971	0.03557	−0.0003				
620	0.29708	0.01828	−0.00015				
630	0.22677	0.00833	−0.00008				
640	0.15968	0.00334	−0.00003				
650	0.10167	0.00116	−0.00001				
660	0.05932	0.00037	0				
670	0.03149	0.00011	0				
680	0.01687	0.00003	0				
690	0.00819	0	0				
700	0.0041	0	0				
	Sum=1.89	**Sum=1.89**	**Sum=1.89**	I=Sum I(λ) · Δλ)	R=Sum	G=Sum	B=Sum

represented by the letter **E**. Correspondingly, we see that the tristimulus values R, G, and B are equal.

- In Chap. 4 we discussed **white noise**, which corresponds to a **spectral intensity with respect to frequency** that is constant; having a constant spectral intensity gave white noise its name. The reader is encouraged to review Sect. 14.3.
- **Color Coördinates**

  We define the color coördinates, as follows:

$$\boxed{\begin{aligned} r &= \frac{\text{R}}{\text{R+G+B}} \\ g &= \frac{\text{G}}{\text{R+G+B}} \\ b &= \frac{\text{B}}{\text{R+G+B}} \end{aligned}} \tag{15.7}$$

Note that $r+g+b = 1$ so that $b = 1-r-g$; thus, r and g determine b. The two values, r and g serve as a coördinate system for identifying colors. Each point in the $r - g$

plane represents a color. The only points that represent a matchable color lie in a triangle defined by the points {0,0}, {1,0}, and {0,1}.

For **equal energy white**, I(λ) = constant, and the tristimulus values R, G, and B are equal. Then, $r = g = b = 1/3$.

- Note that if the spectral intensity were multiplied by whatever factor, all the entries in the last three columns would be multiplied by the same factor and so would the three tristimulus values. According to Eq. (15.7), the color coördinates will not change. The multiplication corresponds to a change in brightness but not in color.
- Similarly, if all the color matching functions were multiplied by the whatever factor, the three tristimulus values would be multiplied by the same factor and the color coördinates would not change.

Since this is so, what is the basis for the choice of the particular set of color matching functions if all we are interested in are the color coördinates corresponding to a given spectral intensity?

15.6.2 Color Coördinates of Butter

If Equal Energy White light [**E**] is incident upon butter, the outgoing spectral intensity will be proportional to the reflectance of butter, R(λ), which is displayed in Fig. 15.4. Since we are interested here only in the color coördinates of butter we can set spectral intensity of the incident light, I(λ), equal to any constant. We will choose I(λ) =1. From Eq. (14.9) we obtain the spectral intensity of the reflected light:

$$\boxed{\mathrm{I}'(\lambda) = \mathrm{R}(\lambda)} \tag{15.8}$$

Let us see how to use the TCMF table to obtain the color coördinates of butter. In the TCMF we will set I(λ) =R(λ).

Study Table 15.3. In the column labelled R(λ) is the **reflectance of light** by butter expressed as a percentage. The values of R(λ) can be read off Fig. 15.4. I have increased the wavelength spacing from 10-nm to 20-nm for simplification, at the expense of accuracy. The three columns to the far right have obvious labels: For example, $\overline{r}(\lambda)$I(λ) is the product of the 2nd column and the 5th column. On the bottom line are the sums of the numbers in the respective columns. In symbolic form we have

$$\begin{aligned} \mathrm{R} &= SUM\ over\ \lambda\ of\ the\ products\ \overline{r}(\lambda)\ \mathrm{R}(\lambda) \\ \mathrm{G} &= SUM\ over\ \lambda\ of\ the\ products\ \overline{g}(\lambda)\ \mathrm{R}(\lambda) \\ \mathrm{B} &= SUM\ over\ \lambda\ of\ the\ products\ \overline{b}(\lambda)\ \mathrm{R}(\lambda) \end{aligned} \tag{15.9}$$

We see in the table that the tristimulus values are R=82.112, G=57.852, and B=33.836. We then obtain for the color coördinates

$$\begin{aligned} r &= \frac{\mathrm{R}}{\mathrm{R+G+B}} = \frac{82.112}{82.112+57.852+33.836} = 0.472 \\ g &= \frac{\mathrm{G}}{\mathrm{R+G+B}} = \frac{57.852}{82.112+57.852+33.836} = 0.333 \\ b &= \frac{\mathrm{B}}{\mathrm{R+G+B}} = \frac{33.836}{82.112+57.852+33.836} = 0.195 \end{aligned} \tag{15.10}$$

15.6.3 Chromaticity Diagram

In Fig. 15.5 we display the chromaticity diagram for the WG primaries, λ_B = B=435.8-nm, λ_G = 546.1-nm, and λ_R = 700-nm. All spectral intensities are represented within the colored boundary, with their corresponding colors.

- **The Upper Perimeter**
  Along the perimeter of the curved part of the boundary we see marked off the spectral colors, having wavelengths ranging from 380-nm to 700-nm.[13] The W-G primaries lie at the corners of the triangle. The only spectral colors with wavelengths that can be matched with the W-G primaries are the primaries themselves. The complete boundary is sometimes referred

[13] In the table, I have cut off the range of wavelengths below 400-nm in order to coincide with Chap. 14.

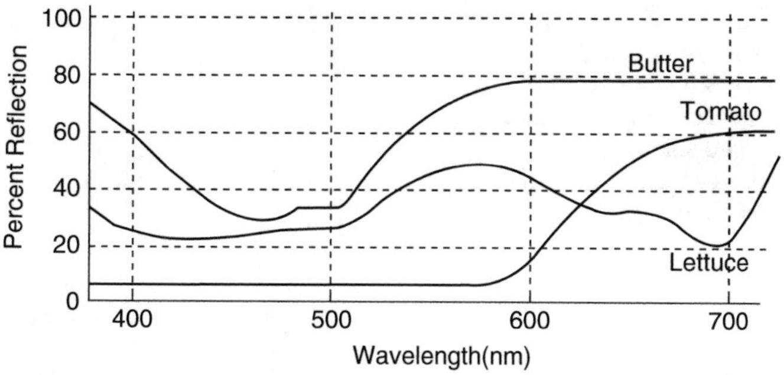

Fig. 15.4 Reflectance of butter, tomato, and lettuce Source: After Williamson and Cummins, who cite General Electric Company Publication TP119

Table 15.3 Analysis of butter (source: based upon Williamson and Cummins, op. cit., section 3-5)

λ(nm)	$\overline{r}(\lambda)$	$\overline{g}(\lambda)$	$\overline{b}(\lambda)$	R(λ)	$\overline{r}(\lambda)$R(λ)	$\overline{g}(\lambda)$ R(λ)	$\overline{b}(\lambda)$R(λ)
400	0.00030	−0.00014	0.01214	56	0.01680	−0.00784	0.6798
420	0.00211	−0.00110	0.11541	46	0.09706	−0.05060	5.3089
440	−0.00261	+0.00149	0.31228	36	−0.09396	+0.05364	11.242
460	−0.02608	0.01485	0.29821	32	−0.83456	0.47520	9.5427
480	−0.04939	0.03914	0.14494	34	−1.67926	1.33076	4.928
500	−0.07173	0.08536	0.04776	35	−2.51055	2.98760	1.6716
520	−0.09264	0.17468	0.01221	49	−4.53936	8.55932	0.5983
540	−0.03152	0.21466	0.00146	63	−1.98576	13.52358	0.092
560	+0.09060	0.19702	−0.00130	72	+6.52320	14.18544	−0.094
580	0.24526	0.13610	−0.00108	75	18.39450	10.20750	−0.081
600	0.34429	0.06246	−0.00049	78	26.85462	4.87188	−0.038
620	0.29708	0.01828	−0.00015	78	23.17224	1.42584	−0.012
640	0.15968	0.00334	−0.00003	78	12.45504	0.26052	−0.002
660	0.05932	0.00037	0	78	4.62696	0.02886	0
680	0.01687	0.00003	0	77	1.29899	0.00231	0
700	0.00410	0	0	77	0.31570	0	0
Total	0.94564	0.94654	0.94136		R=82.112	G=57.852	B=33.836

to as the **horseshoe perimeter** of the chromaticity diagram.

– **The Boundary of the Horseshoe—Spectral Colors and the Line of Purples**

How is the boundary of the horseshoe of the chromaticity diagram determined? Note that the color matching functions themselves have a special significance: Consider a **monochromatic** source of light, with wavelength λ. Then the corresponding **color matching functions**, $\overline{r}(\lambda)$, $\overline{g}(\lambda)$, and $\overline{b}(\lambda)$ determine the **color coördinates** $\{r(\lambda), g(\lambda)\}$ of the monochromatic source of wavelength λ, as follows. Since I(λ) is non-zero only for the particular wavelength λ, **the three color coördinates for a given wavelength are nothing but the tristimulus values for a monochromatic spectral intensity with that wavelength!**

Thus, the color coördinates for the wavelength λ are

$$r(\lambda) = \frac{\overline{r}(\lambda)}{\overline{r}(\lambda) + \overline{g}(\lambda) + \overline{b}(\lambda)}$$
$$g(\lambda) = \frac{\overline{g}(\lambda)}{\overline{r}(\lambda) + \overline{g}(\lambda) + \overline{b}(\lambda)} \tag{15.11}$$

As we move from one wavelength to another, **the set of color coördinates $\{r(\lambda), g(\lambda)\}$ produce the curve that forms the boundary of the horseshoe of the chromaticity diagram.**

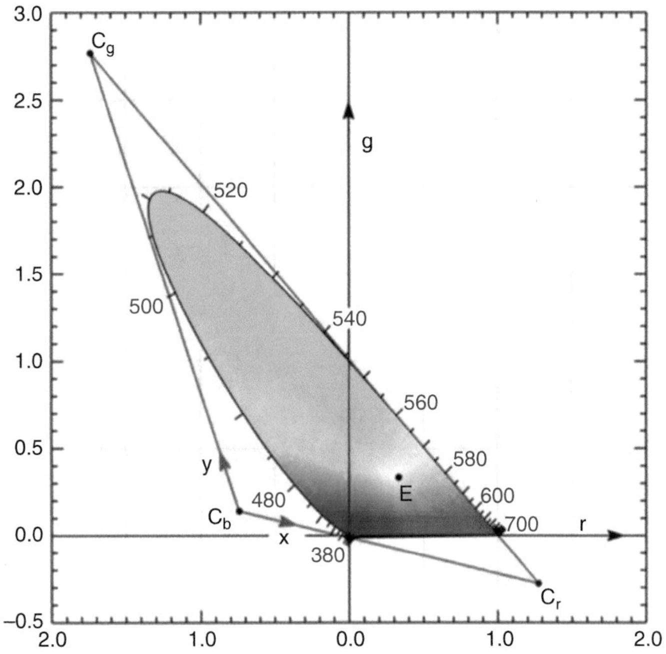

Fig. 15.5 Wright-Guild chromaticity diagram along with the CIE 1931 chromaticity space (source: http://upload.wikimedia.org/wikipedia/commons/1/16/CIE1931_rgxy.png)

Now let us recognize that a single source of light, with a given spectral intensity, is equivalent to adding an infinite number of independent monochromatic sources involving generally all wavelengths and with variable intensity, according to the given spectral intensity. Then, the calculation of the tristimulus values amounts to adding the corresponding tristimulus values for all wavelengths, weighted by the spectral intensity.

- **Gamut of Colors**

  The **gamut** of colors corresponding to the W-G primaries consists of all colors that lie within the small **triangle**, whose corners lie at {0,0}, {1,0}, and {0,1}. These points correspond to the wavelengths of the primaries: λ_B = B=435.8-nm, λ_G = 546.1-nm, and λ_R = 700-nm. All points outside this triangle have a negative color coördinate and cannot be matched by a mixture of primaries. For example, to the left of the vertical axis, $r < 0$. We will discuss the significance of such a color coördinate.

  Based upon the occupied area, the vast majority of colors is not matchable by the WG primaries! A different choice of primaries can be considerably better. See the chromaticity diagram and gamut for the set of primaries {436, 515, 700} in Fig. 15.11. It was produced using the calculation found in Appendix M.

- Equal Energy White **[E]** is at the **center** of the chromaticity diagram, not in the corner close to R, as is **W** in Fig. 15.3.
- **Line of Purples**

  There is a line connecting the blue primary with the red primary. None of the colors along this line are spectral colors, having no corresponding wavelength. This line is referred to as the **line of purples**, with the colors being various purple colors, among them **magenta**.

- **The Red Triangle**

  Later in this chapter we will discuss the **CIE Chromaticity Space**. It has a chromaticity diagram that is a standard for representing colors without any reference to a particular set of primaries. This space is defined by the triangle with red legs, having corners labeled C_r, C_g, C_b. Its gamut includes all colors.

15.6.4 Primary Units

Suppose we were to add equal intensities for each of the three primaries. For example,

$$I(\lambda_R) = I(\lambda_G) = I(\lambda_B) = 1 \qquad (15.12)$$

The resulting tristimulus values would be

$$R = \bar{r}_{\lambda_R} = 0.00410 \quad G = \bar{g}_{\lambda_G} = 0.215$$
$$B = \bar{b}_{\lambda_B} = 0.296 \qquad (15.13)$$

The color ccoördinates would be {0.0080, 0.4174, 0.5746}, far from being the Equal Energy White that we expect or hope for.

Note Before we continue, I would like to forewarn the reader that articles and textbooks are often imprecise and even inconsistent regarding important parameters, as demonstrated by inconsistent units. For example, in the wonderful book by S.J. Williamson and H.Z. Cummins,[14] both of whom were outstanding experimental physicists, the parameter "**spectral power distribution**," with the acronym **SPD**, is introduced. The SPD is our spectral intensity per unit wavelength. In the book the units given this parameter are given to be Watts, Watts/m^2, and Watts/m^2 per nm, corresponding, respectively, to Power, Intensity, and spectral intensity. A major reason for this casual treatment of physical parameters is that people working in color vision and color reproduction focus first and foremost on the faithful reproduction of color across devices. The need to modify and control the brightness takes last place. We will also omit the issue of reproducing the spectral intensity by mixing primaries.

So let us suppose that we want to mix the primaries so as to produce Equal Energy White. Let the spectral intensities needed to produce Equal Energy White be $I(\lambda_R)$, $I(\lambda_G)$, and, $I(\lambda_B)$ without regards to brightness of total intensity. The tristimulus values would then be

$$\boxed{\begin{aligned} R &= \bar{r}_{\lambda_R} I(\lambda_R) \\ G &= \bar{g}_{\lambda_G} I(\lambda_G) \\ B &= \bar{b}_{\lambda_B} I(\lambda_B) \end{aligned}} \qquad (15.14)$$

If we choose the spectral intensities to be the **inverses** of the corresponding matching functions, the tristimulus values will all equal unity, corresponding to [E].

Let us define the values of these spectral intensities as the **primary units** of the three primaries —the red unit intensity u_R, the green unit intensity u_G, and the blue unit intensity u_B. If we wanted to be precise we would know what the dimensionality of these parameters are. We won't bother to do so since we are interested only in reproducing the color produced by the spectral intensity by mixing the primaries. We have

$$\boxed{u_R = \frac{1}{\bar{r}_{\lambda_R}} \qquad u_G = \frac{1}{\bar{g}_{\lambda_G}} \qquad u_B = \frac{1}{\bar{b}_{\lambda_B}}} \qquad (15.15)$$

Let us isolate the color matching functions for the primaries themselves. See Table 15.4. The color matching functions for the primaries vanish except for their corresponding wavelengths. Note that the array—usually referred to as a **matrix**—of numbers form an **anti-diagonal matrix**. That is to say, the only non-zero numbers lie along a line running from the lower left to the upper right. The matrix is anti-diagonal as opposed to **diagonal** because we have generally followed the custom to list the primaries in the order of Red to Green to Blue, while wavelengths **increase** from Blue to Green to Red. Note that the non-zero values of the color matching functions for a given primary correspond to the wavelengths of the corresponding primaries.

From this table we obtain the **primary units**:

$$\boxed{\begin{aligned} &u_R = \frac{1}{0.00410} = 244 \quad u_G = \frac{1}{0.215} = 4.64 \\ &u_B = \frac{1}{0.293} = 3.42 \end{aligned}} \qquad (15.16)$$

[14] S. J. Williamson and H. Z. Cummins, *Light and Color*, (John Wiley and Sons, New York, 1983)

Table 15.4 Values of the color matching functions for the spectral primaries themselves [435.8 nm, 546.1 nm, and 700.0 nm]

λ(nm)	$\bar{r}(\lambda)$	$\bar{g}(\lambda)$	$\bar{b}(\lambda)$
435.8	0	0	0.293
546.1	0	0.215	0
700	0.00410	0	0

15.6.5 Mixing Primaries to Produce a Color

Suppose that we want to match the color of the spectral intensity I(λ). We use the TCMF to calculate the tristimulus values {R, G, B}. Then we mix the primaries as follows:

$$\mathrm{R} \cdot u_R \oplus \mathrm{G} \cdot u_G \oplus \mathrm{B} \cdot u_B \equiv \mathrm{I}(\lambda) \qquad (15.17)$$

In this equation:

- $\mathrm{R} \cdot u_R = 244\,\mathrm{R}\ldots.$
- The symbol ⊕ might not be an ordinary addition; one might have three separate sources of primaries that are added—for example, beams that are reflected off a wall with uniform reflectance over wavelengths.
- The symbol ≡ means that the addition produces a **match of colors**.
- While the mixture matches the color of I(λ) there is no reason to expect that it will be close to matching its brightness.

It is important to realize that only the ratios of the chosen unit intensities matters. This is because we are matching only the color and not the intensity. We have chosen to use unit intensities that are greater than those of Williamson and Cummins by a factor of 244 in order to simplify their relationship to the so-called color matching functions to be introduced later.

15.6.6 Properties of a Chromaticity Diagram

In Fig. 15.6 we exhibit a **schematic** representation of the chromaticity diagram based upon the primaries. All chromaticities are represented by color coördinates (r, g) within the **horseshoe shaped perimeter of the chromaticity diagram**. The monochromatic primaries have color coördinates at the respective corners of the triangle: Thus the red primary is at (1,0), the green primary is at (0,1), and the blue primary is at the origin (0,0).

The reader should note that the actual diagram (Fig. 15.5) that corresponds to the Wright-Guild primaries differs considerably from this schematic diagram.

We summarize here the important characteristics of the chromaticity diagram:

1. All monochromatic (spectral) sources are represented by color coördinates which lie along the curved, upper part of the horseshoe perimeter.
2. We see the straight line at the lower boundary, called the **line of purples**, representing a mixture of the two monochromatic sources 400-nm and 700-nm.
3. Any point that lies *outside* the triangle with corners at the points (0, 0), (1, 0), and (0, 1), will involve a negative color coördinate. Examples of such chromaticities are:

$$r = 0.60,\ g = 0.42,\ b = 1 - r - g = -0.02$$

$$r = 0.60,\ g = -0.02,\ b = 1 - r - g = +0.42$$

$$r = -0.13,\ g = 0.46,\ b = 1 - r - g = +0.67$$

4. **Color with a Negative Color Coördinate** What is the meaning of a negative color coördinate, given that there is no meaning to a negative intensity? First of all, it implies that any chromaticity with a negative color coördinate cannot be produced by a mixture of the three primaries.

   For a full answer as to how one interprets a negative color coördinate, let us consider the particular example above, with $r = -0.13$, $g = 0.46$, and $b = +0.67$. This point

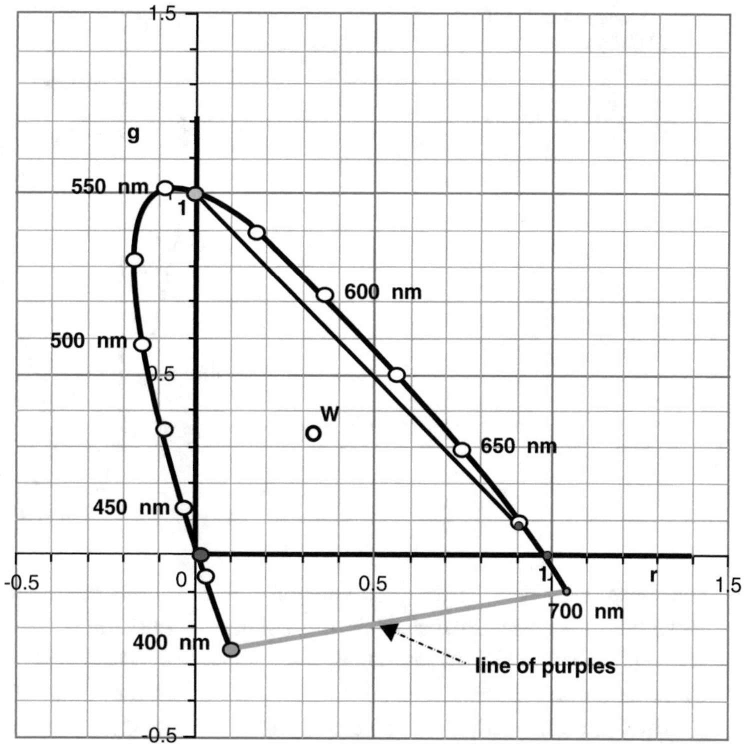

Fig. 15.6 Schematic chromaticity diagram

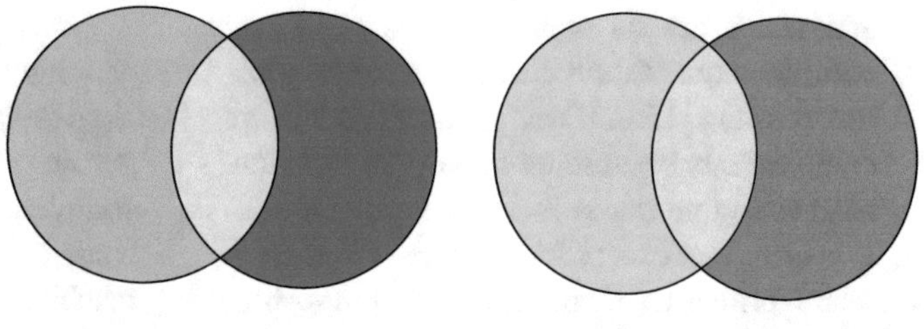

Fig. 15.7 Matching a color having a negative color coördinate

lies on the boundary of spectral colors and corresponds to monochromatic $\lambda = 480$-nm, which is a monochromatic cyan.

Note Note that all monochromatic spectra have one negative color coördinate, **except** for a **primary** that happens to be monochromatic.]

The above color coördinates mean that there is a mixture of the green and blue primaries that will match a mixture of the red primary and the spectral cyan color. The monochromatic cyan is **desaturated** by the addition of some red primary. In symbolic form:

$$\begin{gathered}0.46 \text{ units of green primary} \\ \oplus\, 0.67 \text{ units of blue primary} \equiv \\ 1 \text{ unit of monochromatic cyan}(\lambda = 480\text{-nm}) \\ \oplus\, 0.13 \text{units of red primary}\end{gathered}$$

The matching of colors is indicated in Fig. 15.7.

5. Pure, **equal energy white** is at the center of the horseshoe, with the color coördinates (1/3, 1/3). Recall that this is so because of the particular choice of the unit intensities of the primaries. [**White** is usually labelled with the letter **E**, for **equal energy**. We have labelled it with the letter "W."]

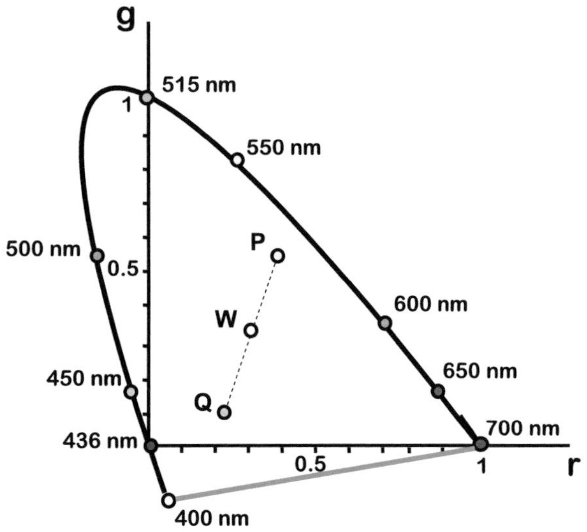

Fig. 15.8 Complements exhibited in a chromaticity diagram

6. The closer a point is to the perimeter of the horseshoe, the more **saturated** is the color. White is totally lacking in saturation.
7. **Complementary Colors**

   Any two points within the horseshoe which are on a line segment passing through W and which are on opposite sides of W are **complements** of each other. This idea is illustrated in Fig. 15.8. Here, points P and Q are complements because an appropriate mixture of P and Q produces W.[15]
8. It is clear that except for the three monochromatic primaries themselves, no monochromatic color coördinates lie along a line that joins two primaries.

   As we previously stated:

   The only fully saturated (i.e., monochromatic) chromaticities that can be produced by the full addition (without negative color coördinates) of the three primaries are the primaries themselves!
9. **A Numerical Characterization of Color—HUE and PURITY**

   The **hue** and degree of **saturation** of a color C are given *numerical values* as follows, using Fig. 15.9. Draw a line from W, through C—at the point (r, g)—to the perimeter of the horseshoe curve.

   Then the **hue** is defined as the wavelength at the point H on the perimeter. This wavelength is referred to as the **dominant wavelength** of the color C.

   Note that the point C' does not have a dominant wavelength since there are no wavelengths associated with points on the line of purples. Its hue is defined in terms of its complement, as follows. We extend the line from C' to the perimeter of spectral colors, reaching the point $\overline{H'}$. The point $\overline{H'}$ is the complement of H' since

$$\overline{H'} + H' \equiv W \tag{15.18}$$

   $\overline{H'}$ is referred to as the **complementary hue** of C'. The hue of C' is defined as the complement of the hue whose wavelength lies at $\overline{H'}$.

   The **degree of saturation** is expressed as the **purity** of the color. It is defined as follows:

$$\boxed{p \equiv \%\ \text{purity} = \frac{\overline{CW}}{\overline{HW}} \times 100\%} \tag{15.19}$$

   Note, in particular, that:

   (a) $p = 0$ at W.

   (b) $p = 100\%$ for a monochromatic and hence a fully saturated color.

[15] This situation is comparable to a see-saw, wherein one can balance the see-saw with unequal lengths ℓ_1 and ℓ_2 opposite the pivot point, as long as the weights satisfy the condition $\ell_1 w_1 = \ell_2 w_2$.

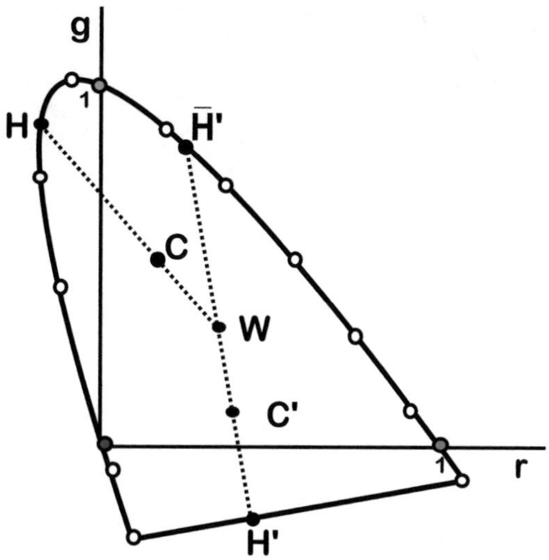

Fig. 15.9 Determination of the hue and saturation of a color

(c) Special case: Point C' is associated with the point H' on the line of purples. Then the purity is calculated using the point H' in Eq. (15.19). The hue is defined as discussed above.

10. What happens to the hue and purity of a filter if white light passes through two identical filters in sequence—as discussed in Sect. 14.5 of Chap. 14? First we expect the hue to be close in value to the hue of a single filter. Second, since the saturation increases, the purity should increase. Suppose that we express the purity as a fraction less than unity instead of as a percentage. Then, very crudely, the new purity will be approximately the square root of the purity of the single filter. Thus, a purity of 64% will lead to a purity of approximately $\sqrt{0.64} = 0.80$, or 80%.

15.6.7 Mixing Two Incoherent Sources of Light

Suppose that we have two **incoherent sources**, with spectral intensities $I_1(\lambda)$ and $I_2(\lambda)$. **Incoherence** means that the two sources consist of **wave packets** that have a random distribution of phase relations. Therefore, the total intensity of the light is a sum of the intensities of the individual sources.

They have sets of tristimulus values (R_1, G_1, B_1) and (R_2, G_2, B_2) and **color coördinates** (r_1, g_1) and (r_2, g_2), respectively. Now suppose that the eye receives these two sources. What will be the color coördinates (r, g) of the resulting color? The result is quite simple. The coördinates (r, g) lie along the line joining the two sets of coördinates (r_1, g_1) and (r_2, g_2), with the position dependent upon the relative strengths of the two sources. See Fig. 15.10 below.

Since the sources are incoherent the net **spectral intensity** received by the eye is the sum of the two spectral intensities.

$$I(\lambda) = I_1(\lambda) + I_2(\lambda) \qquad (15.20)$$

The relative *strengths* of the two sources expressed by the sums S_1 and S_2:

$$S_1 = R_1 + G_1 + B_1 \quad \text{and} \quad S_2 = R_2 + G_2 + B_2$$

In the next section, we will see how the tristimulus values and color coördinates are obtained from a given spectral intensity using a specific relation. Because the relation is linear, we add the tristimulus values. Thus, we have the simple result for the tristimulus values and strength of the mixture:

$$R = R_1 + R_2, \quad G = G_1 + G_2,$$
$$B = B_1 + B_2, \quad \text{and} \quad S = S_1 + S_2 \qquad (15.21)$$

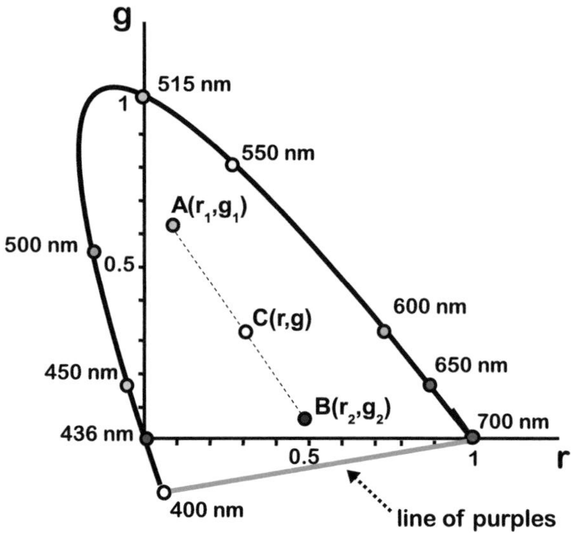

Fig. 15.10 Color coördinates for a mixture of two sources

The **color coördinates** will then be

$$r = \frac{\mathrm{R}}{\mathrm{S}}, \quad g = \frac{\mathrm{G}}{\mathrm{S}}, \quad \text{and} \quad b = \frac{\mathrm{B}}{\mathrm{S}} \tag{15.22}$$

Given that $r_1 = R_1/S_1$ and $g_1 = G_1/S_1$, we obtain

$$\begin{aligned} r &= \frac{\mathrm{R}_1 + \mathrm{R}_2}{\mathrm{S}} = \frac{\mathrm{S}_1}{\mathrm{S}} r_1 + \frac{\mathrm{S}_2}{S} r_2 \\ g &= \frac{\mathrm{G}_1 + \mathrm{G}_2}{\mathrm{S}} = \frac{\mathrm{S}_1}{\mathrm{S}} g_1 + \frac{\mathrm{S}_2}{S} g_2 \end{aligned} \tag{15.23}$$

The color coördinates (r, g) can be shown to lie along the line joining the two sets of color coördinates, (r_1, g_1) and (r_2, g_2). The position along the line segment depends upon the weights S_1/S and S_2/S.

It can also be shown that

$$\frac{\overline{AC}}{\overline{AB}} = \frac{\mathrm{S}_2}{\mathrm{S}} \quad \text{and} \quad \frac{\overline{BC}}{\overline{AB}} = \frac{\mathrm{S}_1}{\mathrm{S}} \tag{15.24}$$

Thus, if the strengths are equal, with $S_1 = S_2$, then $\overline{AC} = \overline{AB}/2$ and (r, g) is at the midpoint between (r_1, g_1) and (r_2, g_2). Generally, the greater the admixture of, say, $\mathrm{I}(\lambda)_2$, the closer will the point C be to the point B.

The result is similar to the see-saw we mentioned earlier[16]: In order to balance two unequal weights, one places the fulcrum at a position such that the ratio of the distances from the fulcrum to the two weights is **inversely proportional** to the ratio of the two weights. The formulas apply, with the strengths replaced by the weights.

[16] See also Sect. 10.4.

15.7 Using a Different Set of Primaries

In Color Vision experiments, one often uses three primaries that are not monochromatic. Examples would be primaries from color filters or from the pixels used in a computer or a color TV monitor.

How can we determine the color coördinates corresponding to this new set of primaries? The new set of color coördinates will certainly be different from the first set. Must we repeat the same tests on individuals that were used to create the above table? Or, can we use the above table to create a new table that can be used to determine the color coördinates for the new set of primaries just as we did for the original primaries?

It turns out that the original table of color matching functions contains all the information we need to know in order to deal with the new primaries: The spectral intensities of the new primaries determine a **transformation between two sets of primaries** that allows us to calculate

Table 15.5 Values of the color matching functions for the wavelengths of the primaries of Stiles and Burch based upon the primaries of Wright-Guild

λ(nm)	$\bar{r}(\lambda)$	$\bar{g}(\lambda)$	$\bar{b}(\lambda)$
444.44	0.01832	0.01036	0.30849
526.32	−0.07897	0.19269	0.00796
645.16	0.12975	0.00228	−0.00002

each member of the new table of color matching functions. This transformation consists of **nine numbers**. If the new primaries are also spectral, these nine numbers can be calculated from **nine numbers** taken from the color matching table: the three color matching functions of each of the three new primaries taken from the table of the old primaries. See Table 15.5.[17] We have chosen the set of primaries that were used by Stiles and Burch (SB) to produce a table of color matching functions for these primaries by studying the vision of a group of people with normal vision, as did Wright and Guild.[18] All of the SB primaries are monochromatic. $\lambda_{B'} = 444.44$ nm, $\lambda_{G'} = 526.32$ nm, and $\lambda_{R'} = 645.16$ nm.

The details for determining a TCMF of a set of monochromatic primaries from a TCMF of another set primaries are presented in Appendix M. I found that both the Wright-Guild primaries and Stiles-Burch primaries lead to a chromaticity diagram with a large region requiring negative red coördinates. By playing around with other choices of the negative monochromatic primary, I arrived at a set that does extremely well in producing small regions of negative color coördinates. The wavelengths are:

$$\boxed{\lambda_{B''}=436\text{-nm}, \quad \lambda_{G''}=515\text{-nm}, \quad \lambda_{R''}=700\text{-nm}} \tag{15.25}$$

The horseshoe perimeter of the chromaticity diagram for these primaries is shown in Fig. 15.11. [It is not a schematic diagram.] We see that by merely reducing the wavelength of the green primary to 515-nm we obtain a gamut that includes most of the green region of colors that are missing in both the WG and SB gamuts.

15.7.1 General Features of a Different Set of Primaries

In Fig. 15.12, we exhibit the position of the color coördinates of the new set of primaries—**here not spectral primaries**—based upon a schematic chromaticity diagram of the first set of primaries. Note that, in order to be general, we have chosen primaries whose coördinates are not on the perimeter of the horseshoe; therefore; the primaries are not monochromatic. Because the primaries are not spectral, the gamut does not extend to the horseshoe perimeter of the original set of primaries.

Note that the **gamut** of colors for the new primaries is defined by triangle $\overline{r'g'b'}$. It is clear that by *not* using spectral primaries, we reduce the number of colors in the gamut of colors. Furthermore, if the new primaries are spectral, the gamut would change: some colors matchable by the old primaries would not be matchable by the new primaries, and vice versa.

Next, we lay out the new color coördinates for all colors of interest in the r'-g' plane. Some thought will make it clear that the positions of the color coördinates of the new primaries with respect to the new primaries themselves will lie at the vertices of the triangle shown in Fig. 15.13 below.

There is a nice way to understand and appreciate the significance of the above transformation. Suppose that the original color coördinates were laid out with respect to a set of r-g axes drawn on a piece of elastic material. The above

[17] The values in this table were obtained by interpolation, using Table 15.2.

[18] See Stiles, Walter Stanley & Birch, Jennifer M. (1958), *N.P.L. colour matching investigation: final report*. Optica Acta 6: 1–26. See also the website: http://cvrl.ioo.ucl.ac.uk/database/text/cmfs/sbrgb2.htm.

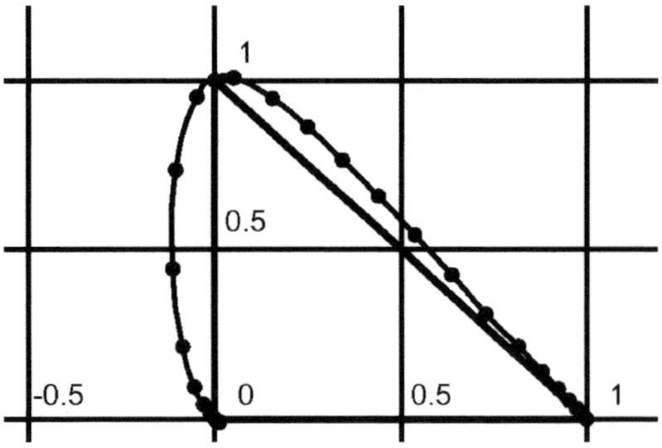

Fig. 15.11 Horseshoe perimeter for the primaries 436nm, 515nm, and 700nm

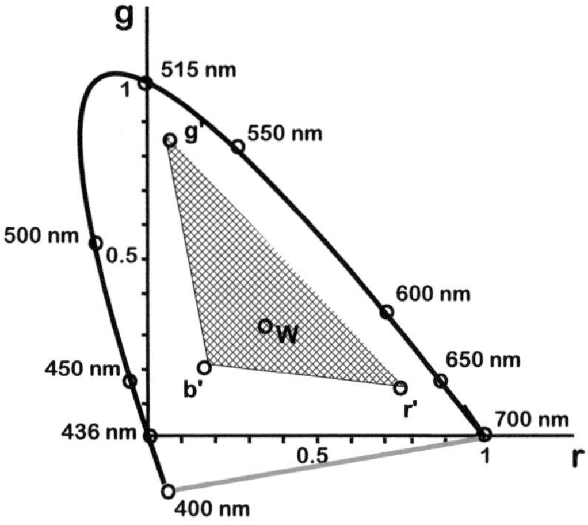

Fig. 15.12 Gamut of colors for the new primaries in the chromaticity diagram of the original spectral primaries

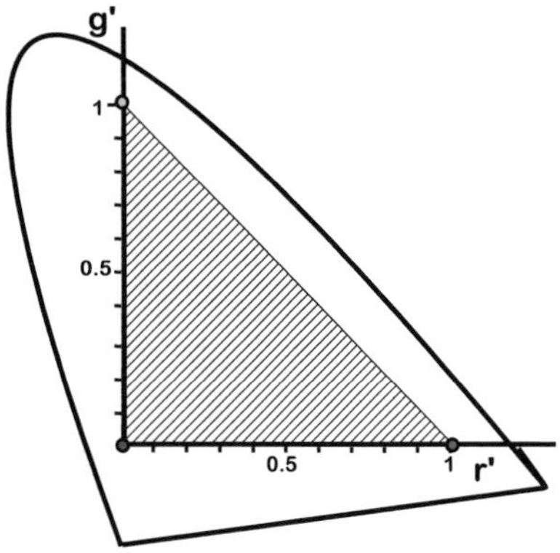

Fig. 15.13 Gamut of colors for the new primaries—the shaded triangle—within the full gamut of colors bounded by the horseshoe perimeter

transformation amounts to distorting the elastic material by a combination of a rotation, and a stretching and/or a compressing. In particular, the color coördinates of the new primaries would move to their respective corners of the triangle.

The triangle with corners at the points r', g', and b' contains the gamut of the primaries. One strives to choose primaries that at least have W included, as shown in the figure and as saturated as possible. With all other factors being equal, it is best to choose primaries that are monochromatic.

15.8 Comments About Tables of Color Matching Functions

1. Color matching functions were determined by testing many individuals with normal vision. As in the case of equal loudness curves, the functions reflect averages of their responses. The width of the distribution of responses is not readily available to the author.
2. Responses might exhibit variations from day to day and depend upon the ages of individuals.
3. Recall that spectral intensities are metamers with respect to each other if they generate the same color coördinates (hence color) from the given TCMF for a set of primaries.

15.9 Brightness and the Luminous Efficiency

In Sect. 11.1 we discussed the subject of assigning a number to the **loudness** of a sound of given frequency. We learned that loudness was far from being proportional to the intensity. Nor was the loudness proportional to the logarithm of the intensity. We introduced the **equal loudness curves** that were labeled by the loudness parameter called the **phon**—ϕ. However, the phon still maintains the logarithm of the sound intensity. Finally we learned in Sect. 11.2 that the **sone**—**s**—was a more accurate measure of loudness. It is a complex parameter that is essential a fractional power of intensity for $\phi \geq 40$-phons.

$$s \propto 2^{\phi} \approx 10^{0.3\times I'} \qquad (15.26)$$

where I' is the intensity of a 1000-Hz sound that matches the loudness of a sound with its frequency.

In the case of light our goal is to obtain a measure of the brightness of a spectral intensity. We begin by carrying out a study that is analogous to determining the equal loudness curves.

We choose a fixed source with wavelength λ_1, and measure its spectral intensity $I(\lambda_1)$. We then use a source whose wavelength λ we can vary, measuring the corresponding spectral intensity $I(\lambda)$. We find the wavelength that requires the **least spectral intensity** to match the given fixed spectral intensity $I(\lambda_1)$. Experiments lead to a wavelength λ_0= 555-nm, independent of the choice of the fixed wavelength λ_1. Clearly, the eye is most sensitive to the wavelength 555-nm.

Finally we define the **luminous efficiency** $V(\lambda)$ to be

$$\boxed{V(\lambda) \equiv \frac{I(\lambda_0)}{I(\lambda)}} \qquad (15.27)$$

Since for all wavelengths λ, $I(\lambda) \geq I(\lambda_0)$, the luminous efficiency varies from zero to unity. We see a plot of the luminous efficiency in Fig. 15.14.[19]

The plot indicates that in daylight, in order to match the brightness of a red light or a blue light with the brightness of a green light, the intensities of the red or blue lights must be greater than the intensity of the green light.

Suppose that we have a quasi-monochromatic source of light with wavelength λ and a spectral intensity $I(\lambda)$. According to the above, a source of light having a wavelength 555-nm and a spectral intensity $V(\lambda)I(\lambda)$ will be equally bright. Similarly, with sound: If we have a sound of frequency f and sound level SL, we can use the equal loudness curves to find the sound level of a sound source with frequency 1000-Hz that will be equally loud. The two sources have the same phon level.[20]

Note The product $V(\lambda)I(\lambda)$ for a given wavelength is the analog of the phon ϕ for a given frequency.

[19] Source: https://en.wikipedia.org/wiki/Luminous_efficacy#/media/File:CIE_1931_Luminosity.png. The reader should note that the luminous efficiency is symbolized by either $V(\lambda)$ or $\overline{y}_{\lambda}$. In more extensive studies, the two parameters are different but barely distinguishable.

[20] In the case of light, if we double $I(\lambda)$ we can maintain equal brightness with a source of wavelength λ_0=555-nm merely by doubling $I\lambda_0)$. Let us consider sound: Suppose that we have a sound of frequency f and sound level SL. If we double the sound intensity, the SL will increase by 3-dB. To match the change in loudness, we cannot generally double the intensity of the 1000-Hz sound, which would correspond to an increase of the phon level by 3-phons. Loudness appears to be more complex than brightness.

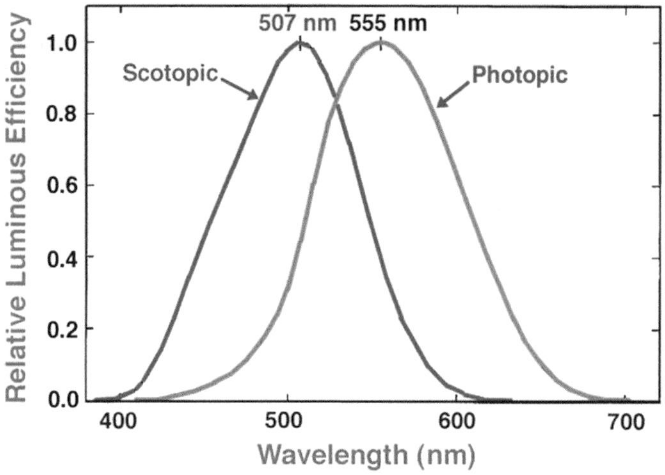

Fig. 15.14 Relative luminous efficiency for photopic (day vision) and scotopic (night vision)

Note For the luminous efficiency, as defined, to be a fixed function, the ratio of intensities that produces a match of brightnesses must be independent of the intensity $I(\lambda_0)$. For example, if two intensities produce equal brightnesses, doubling each will produce brighter light yet with equal brightness. While such is not exactly the case, the function $V(\lambda)$ is still used as a standard.

Later in this chapter, we will learn that the perception of brightness is not proportional to the intensity nor is it proportional to the logarithm of the intensity—as was the case for sound. I will not reveal the actual relation between intensity and brightness now.

15.9.1 Luminous Efficacy and Brightness

Having dealt with comparing light sources that are monochromatic, we can move on to examine the efficiency of a general spectral intensity to excite the eye. The parameter that is used is called the **luminous efficacy**. Let us see how this is determined.

The total intensity I is given by

$$\boxed{I = \text{area under the curve } \; I(\lambda)} \tag{15.28}$$

and has units of W/m^2

We define the **relative illuminance** Y.

$$\boxed{Y = \text{area under the curve } \; V(\lambda)I(\lambda)} \tag{15.29}$$

This sum takes into account the variation in brightness over the visible wavelengths. Brightness is the qualitative **photopic** (day vision) perception of light.

The **luminous efficacy K** is the ratio of the relative illuminance Y and the total intensity I. It is a measure of the overall efficiency of the intensity of the light to produce brightness. Thus,

$$\boxed{K = \frac{Y}{I}} \tag{15.30}$$

We will next describe the meaning of relative illuminance.

The **luminous flux** is the photopic parameter corresponding to the unit of **power**—we might refer to it as the "light-power." Its common unit is the **lumen**, which is well known to the layperson. A light bulb wrapping indicates the total electric power the light bulb consumes, e.g. 100-W, as well as the lumens of light emitted. Typically the efficiency of a light bulb is about 3%. Thus the light-power of a 100-W light bulb might be about 3-W.

At maximum sensitivity, when the light is monochromatic, with a wavelength of 555-nm, the relation between the lumen and the watt is

$$\boxed{\text{one-Watt} = 683\text{-lumens}} \qquad (15.31)$$

Thus, if the 100-W bulb emitted were monochromatic 555-nm light, we would expect the illuminance to be about 3-W×683-(lumens/W)=2,049-lumens. I have 100-W light bulb that is rated at 1280-lumens. Therefore, I would expect the luminous efficacy to be about 1280/2049=0.62, or 62%.

The **illuminance** is analogous to the intensity. It can be expressed in units of lumens/m^2, which is also referred to as the **lux**(plural: lux or luxes).[21]

We obtain the illuminance from the relative illuminance Y through the equation

$$\boxed{\text{Iluminance} = 683\text{-(lumens/W)} \times Y} \qquad (15.32)$$

Lightness

Recall from Chap. 11 that the loudness of a sound in phons is not directly related to the perceived loudness; the latter is measured by the **sone** level **s**; we learned that $s \propto I'^{0.3}$, where I' is the intensity of a sound of frequency 1000-Hz that has the same loudness in phons as the given sound. In the case of light, the perceived brightness for a given wavelength is not proportional to the intensity. The actual perceived brightness with respect to intensity is expressed by what is referred to as the **lightness L***, wherein approximately, $L^* \propto Y^\beta$, and where β is an exponent. Some sources claim that $\beta = 0.3$. [See http://en.wikipedia.org/wiki/Lightness_(color).] However, others point out that the value of β varies, depending upon whether or not the eye has adapted to the level of light intensity; specifically, the claim is made that β can vary from about 0.4 for the dark adapted eye to about 0.5 for the light adapted eye.

[21]We have introduced a number of photopic parameters that characterize a light beam: luminance, relative illuminance, luminous flux, and luminous efficacy. There are many more important parameters that are used by experts who need to characterize lighting very precisely. The reader is forewarned that there are often differences of the terminology for the parameters.

15.10 The Standard Chromaticity Diagram of the Commission Internationale de l'Éclairage (C. I. E.)[22]

In principle, any set of primaries can be used to produce a table of color matching functions that maps any color onto a set of color coördinates. To the extent that the experiments used in testing the color vision of a group of people have been carried out with care, all reflect color vision reliably and contain the same information! any set can be used to establish a standard table for labeling a color. All tables of color matching functions map onto one another with a one-to-one correspondence. With the advent of the results of Wright and Guild, a number of people studying color vision decided to establish a **universal standard table of color matching functions**, the **C.I.E. table**, that is independent of any set of primaries that one might choose to use. The table maps any spectral intensity onto a set of **CIE tristimulus values**—labeled **X**, **Y**, and **Z**.

In place of the color matching functions $\overline{r}(\lambda)$, $\overline{g}(\lambda)$, and $\overline{b}(\lambda)$ corresponding a particular set of primaries, we have the color matching functions $\overline{x}_\lambda$, $\overline{y}_\lambda$, and $\overline{z}_\lambda$.

Note

- The mapping was chosen so that **all CIE tristimulus values are positive**. This can be so because the CIE system does not correspond to a set of primaries that have specific intensities.
- The color matching function $\overline{y}_\lambda$ is equal to the luminous efficiency function V(λ).
- The tristimulus value Y is the **relative illuminance** that corresponds to the spectral intensity of a light source.
- The tristimulus values do not represent any linear measure of the intensity of any primary. They cannot since they are not specific to any set of primaries.

[22]See Wikipedia (1-6-2011): http://en.wikipedia.org/wiki/CIE_1931_color_space.

- The color coördinates that replace the set r=R/(R+G+B), g=G/(R+G+B), and b=G/(R+G+B) are

$$\boxed{\begin{aligned} x &= \frac{X}{X+Y+Z} \\ y &= \frac{Y}{X+Y+Z} \\ z &= \frac{Z}{X+Y+Z} \end{aligned}} \tag{15.33}$$

  Obviously, only two color coördinates specify a color since z=1-x-y. And none (x, y, or z) are negative.
- The coördinates $x = y = z = 1/3$ represents equal energy white.
- Since both the RGB values and the XYZ values are determined by a given spectral intensity, it should be no surprise that there is a straightforward transformation to obtain one set of values from the other without having to specify the spectral intensity from which they were calculated. Remember too, that we have **metamerization**, so that neither set of tristimulus values determines a specific spectral intensity!

The result is the standard C.I.E. chromaticity diagram shown in Fig. 15.15. The C.I.E. Color Matching Functions are found in Table 15.6.[23]

The first such set of C.I.E. color matching functions and chromaticity diagram were obtained from the Wright-Guild data. However, the results of later improved testing have led to modified C.I.E. standards.

Note that the horseshoe region is shifted and distorted, but lies entirely within the first quadrant, with coördinates ($x = 1, y = 0$), ($x = 0, y = 1$), and ($x = 0, y = 0$).

We will now see how Fig. 15.15 is mapped onto Fig. 15.5. The corners of the right triangle in Fig. 15.15 determined by the coördinates, ($x = 1, y = 0$), ($x = 0, y = 1$), and ($x = 0, y = 0$), correspond to the points labeled C_r, C_g, and C_b, respectively, in Fig. 15.5. [This triangle is cut off towards the right in the figure.] The line from C_b to C_r corresponds to the line segment along the x-axis, from $x = 0$ to $x = 1$, along which $y = 0$ and therefore Y=0. The line Y=0 corresponds to **zero relative illuminance** and is referred to as the **alychne**.[24]

Application to Butter

Suppose that equal energy white with a spectral intensity of light of 1-mW/m^2-nm is reflected off the surface of a block of butter. In Sect. 15.6.2 we introduced the reflectance of butter. By multiplying the reflectance by the incident spectral intensity we obtain the spectral intensity of the reflected light. In Table 15.7 we see a CIE table of color coördinates with this spectral intensity inserted into the table as I(λ) in units of mW/(m^2-nm).

We can obtain the relative illuminance from the sum 4.00-mW over the column y(λ)I(λ) by multiplying this number by the spacing 20-nm between the listed wavelengths:

$$\text{Y} = 4.00\text{-mW}/(\text{m}^2\text{-nm}) \times 20\text{-nm} = 80\text{-mW}/\text{m}^2 \tag{15.34}$$

The illuminance is given by

$$\begin{aligned} \text{Illuminance} = 683 \times \text{Y} &= 683\text{-(lumens}/W) \\ &\times 80\text{-mW}/\text{m}^2 = 55\text{-lumens}/\text{m}^2 \end{aligned} \tag{15.35}$$

We can also obtain the luminous efficacy:

$$K = \frac{\text{Y}}{\text{I}} = \frac{80}{193} = 0.41 \tag{15.36}$$

[23]For a source of many tables of color matching functions as well as comments about their reliability, see the website of the Color Vision Research Laboratory with the link http://www.cvrl.org.

[24]The word is pronounced **á-lichnee**, with an accent on the "a" and with "ch" pronounced as in the Scottish word for lake, "loch." The Greek word $\lambda\upsilon\chi\nu o$, pronounced "lichno," means "light." The term is based upon an ancient Greek word meaning "no" light, since the prefix "a" means "without." It is understood to have been coined by the great twentieth century theoretical physicist, Erwin Schrödinger.

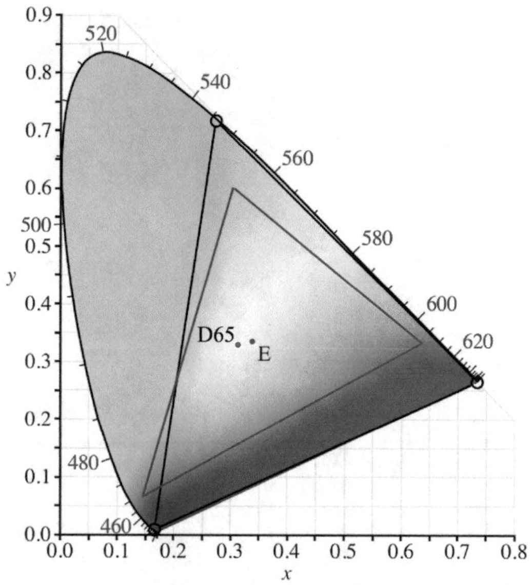

Fig. 15.15 Standard CIE chromaticity diagram with the gamuts of the Wright-Guild primaries (black triangle) and the sRGB primaries commonly used in color monitors (red triangle). Their respective white points are equal energy E (the red dot at x=1/3 and y=1/3) and D65 (the green dot at x=0.3127 and y=0.3290). The number "65" is associated with the chosen white having a spectral intensity of black body radiation at a temperature of 6500K. See Sect. 6.8. Source: https://upload.wikimedia.org/wikipedia/commons/6/60/CIE1931xy_CIERGB.svg with the sRGB gamut added by the author

15.10.1 sRGB Primaries

In the chromaticity diagram we see a highlighted triangle. At the corners are the color coördinates of the **primary illuminants** corresponding to **sRGB primaries**, which are the standard used in many color monitors. Note that the corners of the triangle are not on the perimeter of the limits of the horseshoe because they are not monochromatic. As a consequence, the gamut of colors (the triangle) produceable with the sRGB primaries is limited. The region that is most omitted is towards the "green corner." It happens that in this region, the eye is poor at seeing color differences with respect to changes in color coördinates. The **"white"** for sRGB primaries has coördinates labeled as "D65" and is close to a black body radiation of temperature 6500K. The coördinates of D65 are $x = 0.3127$ and $y = 0.3291$, which are supposed to correspond to the color of the sky in Europe at midday. Since x and y are not equal to 1/3, **D65 is not equal energy white.**

15.11 From Computer RGB Values to Color

We have noted that a computer stores colors identified by the {**RGB**} coördinates, each of which having values from 0 to 255. The 256 possible values store 8-bits of computer space: $2^8 = 256$. With three primaries, we need $(2^8)^3 = 2^{24} = 16{,}777{,}216$—corresponding to **24-bits** of computer space for each sub-pixel.

I am using boldface letters for the computer coördinates. I will use unbolded coördinates for the tristimulus values {RGB} that are associated with primaries. A common set of the latter are the sRGB primaries. In this section we will show how we can obtain the values {RGB} from the set of values {**RGB**} and conversely the values {**RGB**} from the set of values {RGB}. Our focus will be on **color monitors**. However, the essential issues raised apply to color printers as well. We will assume the use of 24-bit color.

Table 15.6 Table of color matching functions—C.I.E. 1964 (source: http://www-cvrl.ucsd.edu/cmfs.htm)

λ-nm	$\bar{x}_\lambda$	$\bar{y}_\lambda$	$\bar{z}_\lambda$	I(λ)	$\bar{x}_\lambda$I(λ)	$\bar{y}_\lambda$ I(λ)	$\bar{z}_\lambda$I(λ)
390	0.0023616	0.0002534	0.0104822				
400	0.0191097	0.0020044	0.0860109				
410	0.084736	0.008756	0.389366				
420	0.204492	0.021391	0.972542				
430	0.314679	0.038676	1.55348				
440	0.383734	0.062077	1.96728				
450	0.370702	0.089456	1.9948				
460	0.302273	0.128201	1.74537				
470	0.195618	0.18519	1.31756				
480	0.080507	0.253589	0.772125				
490	0.016172	0.339133	0.415254				
500	0.003816	0.460777	0.218502				
510	0.037465	0.606741	0.112044				
520	0.117749	0.761757	0.060709				
530	0.236491	0.875211	0.030451				
540	0.376772	0.96199	0.013676				
550	0.529826	0.99176	0.003988				
560	0.705224	0.99734	0.0				
570	0.878655	0.955552	0.0				
580	1.01416	0.868934	0.0				
590	1.11852	0.777405	0.0				
600	1.12399	0.658341	0.0				
610	1.03048	0.527963	0.0				
620	0.856297	0.398057	0.0				
630	0.647467	0.283493	0.0				
640	0.431567	0.179828	0.0				
650	0.268329	0.107633	0.0				
660	0.152568	0.060281	0.0				
670	0.0812606	0.0318004	0.0				
680	0.0408508	0.0159051	0.0				
690	0.0199413	0.0077488	0.0				
700	0.00957688	0.00371774	0.0				
710	0.00455263	0.00176847	0.0				
720	0.00217496	0.00084619	0.0				
730	0.00104476	0.00040741	0.0				
	SUM=23.3	SUM=23.3	SUM=23.3		X	Y=SUM ×10-nm	Z

- the chromaticity of each of the three primaries—this is expressed in terms of their CIE {xy} coördinates, resulting in six numbers that are used to carry out a transformation between the two sets of coördinates. For the sRGB primaries the xy color coördinates are: {x_r=0.64, y_r=0.33}, {x_g=0.30, y_g=0.60}, and {x_b=0.15, y_b=0.06}.
- the chromaticity—hence the {xy} coördinates—of the color corresponding to **R = G =B**, which provide us with two more numbers for the transformation. The intensities of light for each primary are related to the tristimulus values so as to produce the chosen white. Typically, the chosen white is not **equal energy white**. For the so-called

Table 15.7 Spectral intensity of **Butter** in a CIE table of color coördinates

λ-nm	x (λ)	y(λ)	z(λ)	I(λ)	x(λ)I(λ)	y(λI(λ)	z(λ)I(λ)
400	0.0023616	0.0002534	0.0104822	0.56	0.001322	0.000141904	1.39E-05
420	0.0191097	0.0020044	0.0860109	0.46	0.00879	0.000922024	0.000756
440	0.204492	0.021391	0.972542	0.36	0.073617	0.00770076	0.071596
460	0.383734	0.062077	1.96728	0.32	0.122795	0.01986464	0.241572
480	0.302273	0.128201	1.74537	0.34	0.102773	0.04358834	0.179377
500	0.080507	0.253589	0.772125	0.35	0.028177	0.08875615	0.021757
520	0.003816	0.460777	0.218502	0.49	0.00187	0.22578073	0.000409
540	0.17749	0.761757	0.060709	0.63	0.074182	0.47990691	0.004504
560	0.376772	0.96199	0.013676	0.72	0.271276	0.69263136	0.00371
580	0.705224	0.99734	0	0.75	0.528918	0.748005	0
600	1.01416	0.868934	0	0.78	0.791045	0.67776852	0
620	1.12399	0.658341	0	0.78	0.876712	0.51350598	0
640	0.856297	0.398057	0	0.78	0.667912	0.31048446	0
660	0.431567	0.179828	0	0.77	0.332307	0.13846756	0
680	0.152568	0.060281	0	0.77	0.117477	0.04641637	0
700	0.0408508	0.0159051	0	0.77	0.031455	0.012246927	0
				9.63		**4.006187635**	
			I=	**193-mW/m^2**	**Y=**	**80-mW/m^2**	

sRGB primaries we have $\{x_w=0.3127, y_w=0.3291\}$; this white, with the name **D65**, corresponds to the color of blackbody radiation at a temperature of 6500K and is reported to be close to the color of a clear midday sky in Western Europe.It is **NOT** an "equal energy white," that is, equal intensity white.

– the last parameter, referred to as the **"gamma value"**—symbolized by γ—reflects the fact that the tristimulus values {RGB} are not necessarily proportional to the computer's respective {**RGB**} values. The history of this situation is complex. A good approximation to the relationship is what is referred to as a simple power law:

$$R=\left(\frac{\mathbf{R}}{255}\right)^{\gamma},\quad G=\left(\frac{\mathbf{G}}{255}\right)^{\gamma},\quad B=\left(\frac{\mathbf{B}}{255}\right)^{\gamma} \tag{15.37}$$

Note that this scaling of the tristimulus values results in a range from zero to unity, so that maximum luminance or brightness is achieved with a value of unity. Note too that if γ were unity, we would have proportionality; however, typically its value lies between one and three. A reason for the non-linearity is that the cathode ray tubes (CRTs) that produce color have a light intensity I that is far from linear with respect to the strength of the electric signal—the so-called **voltage** V—that is responsible for the light. Approximately, $I \propto V^{\gamma}$. The {RGB} values are approximately proportional to the voltage, thus resulting in Eq. (15.37).[25]

Common values of gamma are 1.8 or 2.2 . For PC monitors the value is usually $\gamma = 2.2$. MAC monitors have fluctuated between both values.

Note There are important consequences of non-linear mapping from {**RGB**} to RGB.

[25]The subject of gamma and its related **gamma correction** is very complex. As a result it is extremely difficult to find resources that are reliable. Articles abound with contradictory information. For what I consider a very reliable reference I highly recommend Charles Poynton, *Video and HDTV* [Morgan Kaufmann Publishers and Elsevier Science, San Francisco, 2003].

- Suppose that all **RGB** values are multiplied by a constant **c**. Then we can see from Eq. (15.37) that all values of RGB are multiplied by the constant $\mathbf{c}^{\gamma}$. For example, let **R** change to 2**R**. Then

$$R=\left(\frac{\mathbf{R}}{255}\right)^{\gamma} \rightarrow \left(\frac{\mathbf{2R}}{255}\right)^{\gamma}=2^{\gamma}\left(\frac{\mathbf{R}}{255}\right)^{\gamma}=2^{\gamma}R \qquad (15.38)$$

Similarly, we will obtain $G \rightarrow 2^{\gamma}G$ and $B \rightarrow 2^{\gamma}B$. We then have no change in the *chromaticity* because all three RGB values change by the same factor.[26]
- The effect of non-linearity is dramatic when we don't add {**RGB**} values in the same proportion. For example, suppose that we add $\mathbf{R_2} = 150$ to $\mathbf{R_1} = 100$ and double the **G** and **B** values from 100 to 200. We will have $\mathbf{R} = 100 + 150 = 250$, so that the resulting value of R will be

$$R' = \left(\frac{\mathbf{250}}{255}\right)^{\gamma} = 2.5^{\gamma}R \qquad (15.39)$$

The result is that R will change by a factor **different** from that of G and B, so that the chromaticity will change: R changes by a factor 2.5^{γ} vs. a factor of 2^{γ} for G and B. This result is not surprising; we expect the chromaticity to change. On the other hand, more importantly, while $\mathbf{R_1} + \mathbf{R_2} = \mathbf{R}$, $R_1 + R_2 \neq R$:

$$R_1 + R_2=\left(\frac{\mathbf{100}}{255}\right)^{\gamma}+\left(\frac{\mathbf{150}}{255}\right)^{\gamma}\neq\left(\frac{\mathbf{250}}{255}\right)^{\gamma}=R \qquad (15.40)$$

See the last problem at the end of the chapter for an interesting effect of gamma non-linearity.

Suppose that you want to see the color associated with a given spectral intensity on a color monitor. A straightforward way to do so is to find the tristimulus values for the associated primaries and chosen white point and then obtain the computer **RGB** values.

In Table 15.8 we present the color matching functions for sRGB with a D65 white point.[27] Once you obtain the sRGB tristimulus values, you can proceed to calculate the color coördinates {r,g,b}, as we've seen earlier in the chapter. Finally, we can use the equations inverse to those of Eq. (15.37) to obtain the **RGB** values. They are given by

$$\boxed{\mathbf{R} = 255\ r^{1/\gamma} \quad \mathbf{G} = 255\ g^{1/\gamma} \quad \mathbf{B} = 255\ b^{1/\gamma}} \qquad (15.41)$$

15.12 How Many Colors Are There?

The answer to this question depends upon how we define colors. Whatever the choice, we have to be able to assign a numerical value. We can't be vague here. In the final analysis we need to make use of our knowledge of the chromaticity diagram as well as color vision studies. Thus, it is very important to avoid making rapid conclusions. Recently (April, 2010) the Sharp Electronics Corporation announced to the public that it is offering a **color monitor** that has a fourth primary. This monitor could increase the range of colors that the monitor will be able to display over that of a color monitor with three primaries. There are various reports as to how many more colors will be displayable. Most of the blogs are full of nonsense because the authors don't know enough about color vision.[28]

So let us start to examine the question very slowly. Here are some possible choices of what we can mean by countable colors.

---

[26] Tests of some monitors have revealed that their three **RGB** values do not have the same value of gamma. In this case, the chromaticity will change.

[27] The table was produced by using transformation matrices between the CIE table of color matching functions and the RGB coördinates for the sRGB primaries.

[28] For example, in the website (1-12-2011): http://www.gizmag.com/sharp-4-primary-color-tvs-enables-trillion-colors/13823/ we read: "By adding yellow to the colors red, green and blue, the televisions are capable of rendering nearly all the colors a human eye can discern."

Table 15.8 Table of color matching functions for sRGB primaries with D65 white point

λ-nm	$\bar{r}(\lambda)$	$\bar{g}(\lambda)$	$\bar{b}(\lambda)$	λ-nm	$\bar{r}(\lambda)$	$\bar{g}(\lambda)$	$\bar{b}(\lambda)$
390	0.0020379	−0.0013774	0.0111593	565	1.0624985	1.0735629	−0.1562685
395	0.0061633	−0.0042311	0.0344334	570	1.3786552	0.9410231	−0.1460794
400	0.0159679	−0.0111828	0.0915671	575	1.6757260	0.7950021	−0.1338111
405	0.0354432	−0.0254042	0.2098490	580	1.9509934	0.6471963	−0.1208752
410	0.0670300	−0.0495022	0.4144850	585	2.2124935	0.5076572	−0.1086960
415	0.1061226	−0.0818657	0.6990658	590	2.4299409	0.3743422	−0.0964009
420	0.1449626	−0.1176064	1.0349829	595	2.5687956	0.2520187	−0.0838849
425	0.1732094	−0.1478947	1.3643045	600	2.6307181	0.1456766	−0.0718077
430	0.1858490	−0.1678059	1.6516346	605	2.6169719	0.0584916	−0.0605933
435	0.1863771	−0.1788303	1.9107849	610	2.5280954	−0.0082826	−0.0504098
440	0.1673589	−0.1736197	2.0880869	615	2.3713247	−0.0550566	−0.0413530
445	0.1277173	−0.1503345	2.1491184	620	2.1632857	−0.0831681	−0.0335935
450	0.0693082	−0.1084812	2.1108656	625	1.9246978	−0.0946749	−0.0272949
455	0.0004766	−0.0539885	2.0064321	630	1.6625984	−0.0956921	−0.0218334
460	−0.0876709	0.0201495	1.8355095	635	1.3833738	−0.0904241	−0.0168117
465	−0.1866384	0.1050043	1.6264932	640	1.1222411	−0.0809174	−0.0126898
470	−0.3076486	0.2126340	1.3657585	645	0.8983389	−0.0700685	−0.0094939
475	−0.4228504	0.3271911	1.0514122	650	0.7041793	−0.0581450	−0.0070380
480	−0.5139261	0.4298260	0.7688802	655	0.5373194	−0.0457007	−0.0052031
485	−0.6087477	0.5423271	0.5441134	660	0.4017969	−0.0347817	−0.0038145
490	−0.6760153	0.6378142	0.3706395	665	0.2958794	−0.0260298	−0.0027567
495	−0.7419776	0.7493351	0.2392183	670	0.2144757	−0.0191002	−0.0019692
500	−0.8049760	0.8698089	0.1371703	675	0.1530033	−0.0137450	−0.0013898
505	−0.8462604	0.9884878	0.0607874	680	0.1079449	−0.0097546	−0.0009733
510	−0.8672447	1.1065961	−0.0032616	685	0.0756554	−0.0068610	−0.0006791
515	−0.8638713	1.2205595	−0.0489710	690	0.0527167	−0.0047904	−0.0004720
520	−0.8197702	1.3174593	−0.0846822	695	0.0365982	−0.0033320	−0.0003269
525	−0.7267116	1.3787319	−0.1128393	700	0.0253230	−0.0023074	−0.0002259
530	−0.5942649	1.4139555	−0.1332074	705	0.0174647	−0.0015906	−0.0001559
535	−0.4445743	1.4390806	−0.1497857	710	0.0120362	−0.0010948	−0.0001076
540	−0.2646612	1.4400910	−0.1608415	715	0.0083127	−0.0007546	$-7.452E{-}05$
545	−0.0503985	1.4052614	−0.1668914	720	0.0057481	−0.0005205	$-5.169E{-}05$
550	0.1904443	1.3472022	−0.1686456	725	0.0039784	−0.0003592	$-3.592E{-}05$
555	0.4600521	1.2772972	−0.1684127	730	0.0027597	−0.0002483	$-2.502E{-}05$
560	0.7523205	1.1875067	−0.1642469	735	0.0019210	−0.0001722	$-1.75E{-}05$

- the number of spectral intensities. We have pointed out that this number is a high level of infinity—labeled by mathematicians as $\aleph_0^C$.
- number of chromaticities (sets of color coördinates, each ranging continuously from 0 to 1). This number is also infinite but at a lower level of infinity. The mathematical infinity for the number of points in a finite area (the horseshoe of the chromaticity diagram) is $\aleph_0$.
- the number of colors that are visually distinguishable—estimated at about 26,000.
- the number of sets of **RGB** of color coördinates stored in a computer and associated with colors on a monitor. The settings are variable on a given monitor. For a three-color, **24-bit** monitor setting, we have $2^8 = 256$ (8-bits) different values for each subpixel, **R**, **G**, and **B** , ranging from 0, 1, 2, …, 255. The total

number of combinations is then $256 \times 256 \times 256 = 2^{24} = 16,777,216$.[29,30]

Another option is 36-bit color, which translates to a total of $2^{12} = 4096$ settings for each of **R**, **G**, and **B**, giving a total of 2^{36} ~70-billion total number of settings.

In all cases, the number of different possible tristimulus values produced by a monitor—but not the number of chromaticities.

– the number of different chromaticities produced by a given color monitor. This number can be calculated too. See the sample problem below.

Sample Problem 15.1 Why is the number of colors as defined in the context of color monitors not the same as the number of colors we would deduce using our definition of color?

Solution We defined **color** in terms of hue and saturation but did not include brightness or intensity. The 24-bit=16,777,216 different values of **RGB** of a monitor determine the hue, saturation, and illuminance (which is an objective measure of brightness).

For our future discussion, in order to avoid any mistake as to usage, we will refer to **hs-colors** as referring to the hue and saturation; on the other hand, we will refer to **hsb-colors** as referring to the hue, saturation, and brightness.

A crude estimate of the number of color coördinates we obtain from the **RGB** values of a monitor is to treat them as tristimulus values. [For technical reasons, the tristimulus values are not proportional to the monitor's **RGB** values.]

The number of different hs-colors according to our definition of color would be the number of distinct fractions r = **R/(R+G+B)** and **g=G/(R+G+B)**. We expect this number to be certainly less than 16,777,216, but still on the order of a few million. Bruce Boghosian estimated this number at about 14-million and then computed it to be exactly 13,936,094.

In Fig. 15.16 we see the color coördinates produced by all possible combinations of **R** $= 0$ to 31, **G=0** to 31, and **B**=0 to 31 in a chromaticity diagram. The monitor then has 15-bit color. The triangle represents the gamut of colors by the primaries of the monitor. There are $32 \times 32 \times 32 = 2^{15} = 32,768$ colored dots placed within the black triangular background of the figure. Each colored dot represents a possible producible color. Note that there are "avenues" of regions where colors cannot be produced!

Finally, we have arrived at the ultimate way to count colors. However, we might also want to know how well a person can distinguish among all possible colors as defined by their color coördinates: What matters as far as color vision is concerned—is what we see!

While the number of hsb-colors and hs-colors produced by a 24-bit monitor is in the millions, the question remains as to whether we can **distinguish** among all these colors. In concrete terms, we can have the two sets of color coördinates (r,g) in the chromaticity diagram, one for each of two colors, so close together that we cannot tell the colors apart.

Consider two points in the CIE Chromaticity Diagram that correspond to the spectral intensity of blackbody radiation at two different temperatures, 2000K and 3000K. [See Sect. 6.8.] For this author, the two colors are barely distinguishable in the figure. I have concluded that the two colors in the figure do not represent the actual colors represented by the color coördinates. You can check whether or not this is so for you by varying the color coördinates **(R,G,B)**, each ranging from

[29]Each element, e.g. a microscopic particle with two possible orientations, can store one bit of information. Two particles can store 2^2=4 possible bits of information. Eight particles can store $2^8 = 256$ bits of information; and so on.

[30]Adding a fourth color with 8-bits will increase this number by a factor of 256, so that we would have over 4-billion different combinations! In a recent (May, 2010) website of the SHARP Corporation, it was claimed that their monitor would produce trillions of colors. It is incomprehensible to understand how they can arrive at such a number. See SHARP website: http://www.sharpusa.com/AboutSharp/NewsAndEvents/PressReleases/2010/January/2010_01_06_Booth_Overview.aspx.

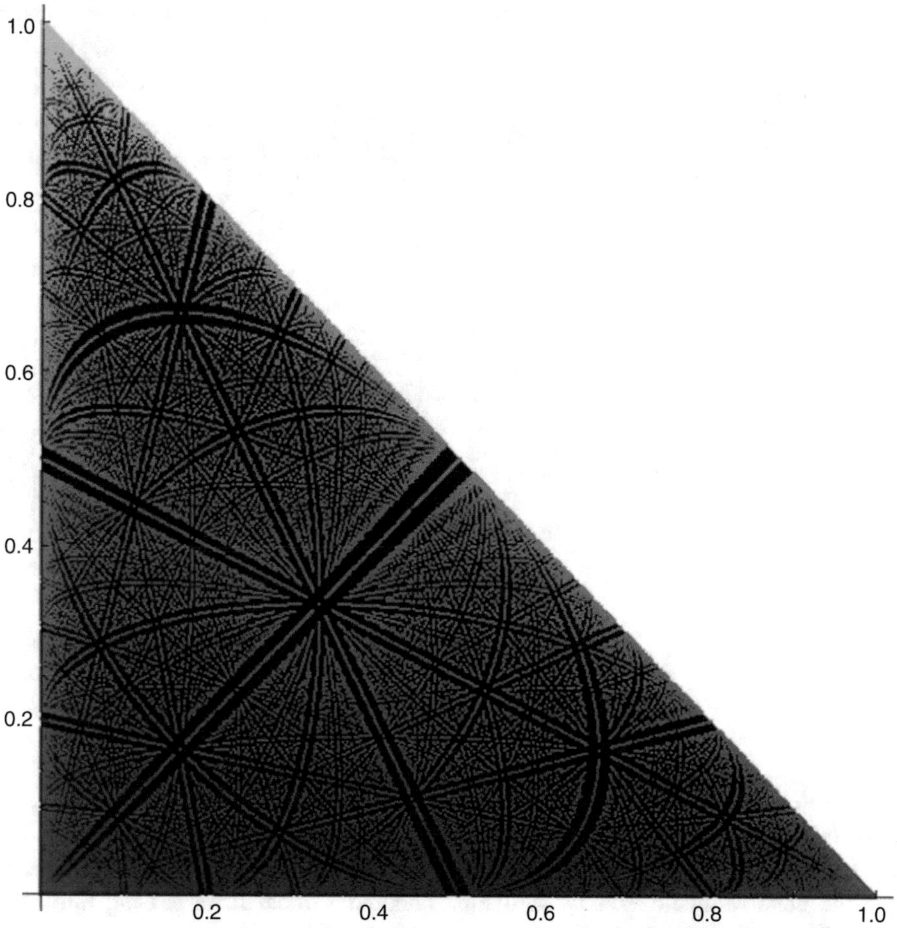

Fig. 15.16 Colors produced by three primaries on a 15-bit color monitor with **RGB** parameters having values from zero to 31 each—instead of the common zero to 255 each

0 to 255, on your color monitor and determining how well you can discriminate colors.

For any point in the diagram we can draw a small ellipse such that all points within the ellipse produce colors that are not distinguishable. We see a number of such ellipses in Fig. 15.17 below drawn on the standard CIE Chromaticity Diagram. The ellipses are drawn ten times their actual size.[31]

It is possible to draw a set of ellipses that fills the horseshoe with minimal overlap. The total number of these ellipses will be a good measure of the **number of distinguishable colors**. It is often stated that the number of distinguishable colors is on the order of 10-million.[32] However, this number refers to the number of hsb-colors, which takes into account varying brightness, and not hs-colors. According to recent studies, the

[31]Note that the ellipses are largest in the green region, indicating that the eye does not discriminate changes in chromaticity well in the green. On the other hand, the ellipses are much smaller towards the blue region. We can see this variation in discrimination in the CIE chromaticity diagram of Fig. 15.15.

[32]See D. B. Judd and G. Wyszecki (1975), *Color in Business, Science and Industry, Wiley Series in Pure and Applied Optics (3rd ed.). New York: Wiley-Interscience. p. 388.*

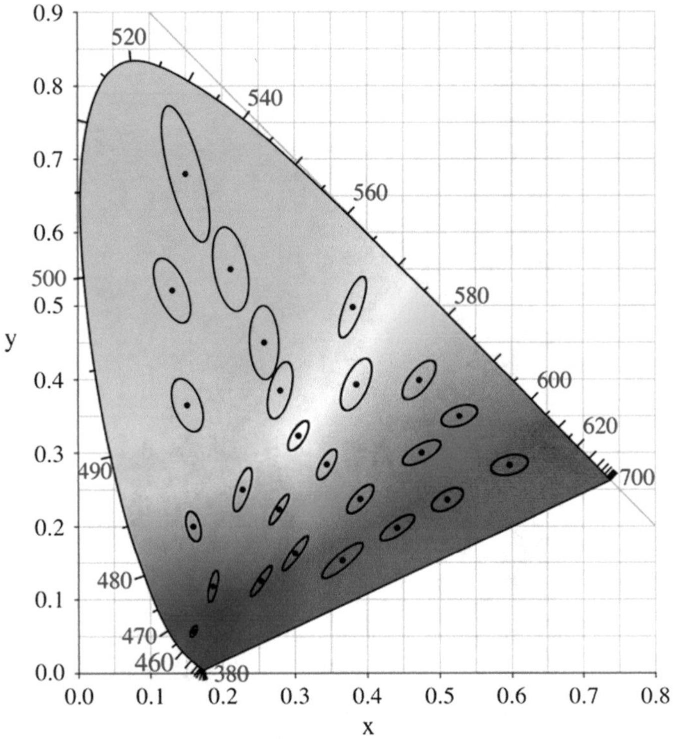

Fig. 15.17 Ellipses representing the area of colors that are indistinguishable from the color at the center of each respective ellipse—drawn ten times to actual scale (source: http://en.wikipedia.org/wiki/File:CIExy1931_MacAdam.png, attributed to David MacAdam)

number of distinguishable hs-colors is on the order of 26,000.[33,34]

In Fig. 15.18 we see a chromaticity diagram based upon the monochromatic primaries 436nm, 515nm, and 700nm.[35] I will refer to this diagram as the **lRGB gamut**.

The red triangle has corners at the points (0,0),(1,0), and (0,1). Thus all the hs-colors that can be matched with these monochromatic primaries have coördinates that lie within this triangle. Only a small fraction of all colors lies outside this triangular gamut. The monochromatic colors lie on the blue perimeter. Also displayed are the color coördinates for a standard set of primaries known as **sRGB**. They lie at the corners of the black triangle. **The sRGB primaries are not monochromatic.** If you look closely you will note that the gamuts are displayed on a grid. There are one-hundred horizontal lines and one-hundred vertical lines passing through the red triangle. The number of squares in the triangle is therefore $100^2/2$, or 5000. Since there are about 26,000 distinguishable hs-colors, a single square is associated with about five distinguishable hs-colors.

Current advertisements of color monitors brag about their gamut of colors—all points lying within the black triangle—in terms of the sRGB standard. A company might claim that their monitor has a gamut of 117% of the sRGB standard. From the fact that the human eye can distinguish about 10-million hsb-colors it is reasonable to assume that the 16.8 million monitor values of RGB in the sRGB gamut can produce essentially all the **distinguishable colors** that lie within the sRGB gamut.

We note that the sRGB standard is quite poor in the green area. If only the green primary at the top were moved along the line from R to G all the way up towards the perimeter, the gamut would be a good fraction of the optimum lRGB gamut!

[33] J. M. Linhares, et. al, J Optical Society of America, volume **25**, p. 2918 (2008).

[34] This number is just under twenty times the number of distinguishable pitches of pure tones, which has been found to be about 1400. See Wikipedia (1-7-2011): http://en.wikipedia.org/wiki/Pitch_(music).

[35] In Appendix M I show how this set of primaries is close to producing the largest possible gamut of colors.

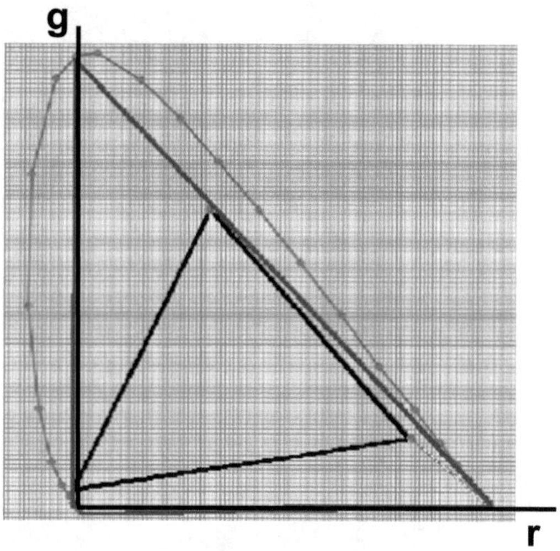

Fig. 15.18 sRGB coördinates (at the corners of the black triangle on a Chromaticity Diagram with monochromatic primaries 436nm, 515nm, and 700nm). The blue dots along the horseshoe perimeter are monochromatic colors

So how would the addition of a fourth primary in a color monitor increase the set of distinguishable colors? The answer is dealt with in problem 14.24. It should be clear that the number of distinguishable hsb-colors is not increased by even a factor of 256.

15.12.1 Limitations of a Broadened Gamut of a Monitor

We recognize that Sharp's four-color monitor will broaden the gamut of colors that can be produced. However, can it **faithfully reproduce** a broader gamut of colors? Suppose that an image or video shown by the monitor was produced by a camera with a smaller gamut? How would the monitor handle the input? What might be the gain?[36] To answer these questions we need to know how the camera handles a color that lies outside its gamut. The typical response of the camera is to **replace** a color outside its gamut by another that has the same hue but lies at the boundary of the gamut. As a result, the output will be a color that has no relation to how saturated the original color was. Sharp's monitor cannot reproduce a replication of the original; all it can do is to increase the saturation of the color by an amount not determined by the original color. Our conclusion is that the Sharp monitor can enrich the color of the image but **not faithfully**.

15.13 A Simple Physiological Basis for Color Vision

We mentioned that there are three different types of cones, or receptors of light.[37] The genes for the three cones have been identified in the laboratory.[38] They were originally hypothesized to exist as a means of understanding the results of color matching that are summarized in the chromaticity diagram.

We suppose that the three types of cones are distinguishable by their different spectral response curves, that is, the curves which describe how the rate at which each the individual type of cone produces nerve impulses depends upon the wavelength of a monochromatic source. These response curves are proportional to the respective **absorption spectra** of the respective **pigments** in the cones: Thus, R-cones have R-pigment, G-cones have G-pigment, and B-cones have B-pigment.

[36] I am grateful to Raymond Soneira for communication on this subject. You are invited to see his extremely informative website (1-27-2011): http://www.displaymate.com/eval.html.

[37] For more details, see http://en.wikipedia.org/wiki/Retina and http://webvision.med.utah.edu/sretina.html.

[38] See the article by Jeremy Nathans, who first identified the genes: Scientific American, volume 260, pp. 42–49 (1989).

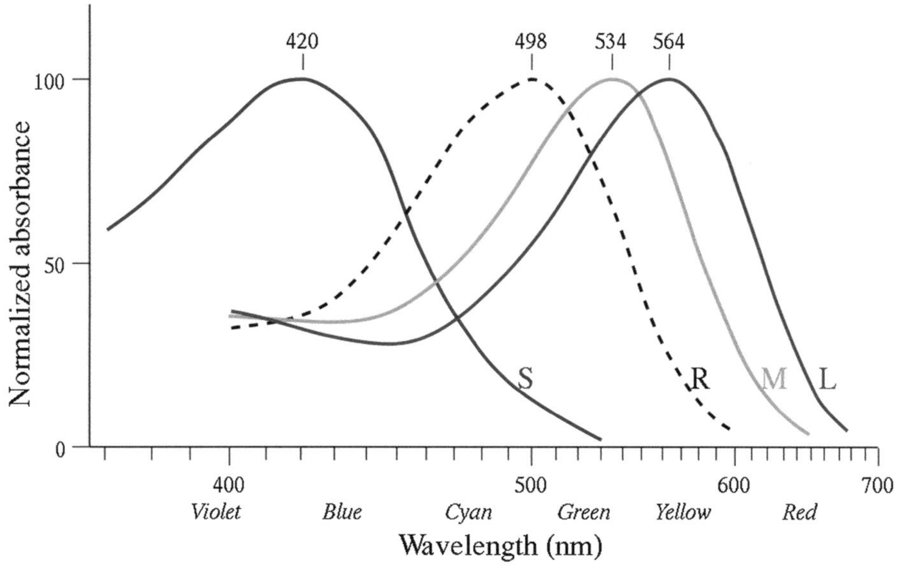

Fig. 15.19 Absorption by the rods (R) and by the three color pigments: long wavelength red cones (L), midrange wavelength green cones (M), and short wavelength blue cones (S). The curves are "normalized" so that all peak values are set to 100% (source: http://en.wikipedia.org/wiki/photoreceptor_cell)

The absorption spectrum tells us how the fraction of monochromatic incident light intensity that is absorbed depends upon the wavelength λ. The presumed spectra are shown in the Fig. 15.19.[39] The dashed curve is the absorption of the rods. As for the cones, "S" stands for "short wavelength," "M" stands for "medium wavelength," and "L" stands for "long wavelength."

Note that all the curves have a maximum value of 100. This is because the curves represent the variation of each cone with respect to wavelength; the actual *relative* absorption of one cone to another is quite different. Absorption by the blue cone is far weaker than the other two.

Consider, for example, the absorption spectrum of the L-cones. The curve has a peak at $\lambda =$ 564 nm. Notice how broad the peak is; the width is on the order of 150 nm, corresponding to about $150/440 = 0.34$, or 34% of the wavelength at the peak. Recall, however, that the width of a spectral line produced by a single atomic or molecular transition is very narrow. For atomic transitions, the width is on the order of **one part in ten million** of the frequency of the photon emitted.[40]

The reason that the absorption peak is so broad in the case of pigment is that the energy level diagram consists of a band of excited states. See Fig. 15.20 below. There are many excited states that correspond to a large number of transitions. Each transition has its own narrow absorption peak. Since these peaks overlap, the sum total of the peaks produces a single broad peak in the absorption spectrum.

When photons of light impinge upon a pigment molecule that is in its ground state (see Chap. 6, for a review of this subject), it has the possibility of exciting the molecule into one of the many quantum states that lies in the energy band. Now any system that is excited from its ground state by absorption of photons has the process of photon emission available to it as a means of returning to its ground state. However,

[39]The figure is based upon Bowmaker J.K. and Dartnall H.J.A., "Visual pigments of rods and cones in a human retina." J. Physiol. 298: pp501–511 (1980).

[40]If we plot the absorption spectrum as a function of the frequency, we would obtain a peak for the L-cone that has a width in frequency that is about 34% of the frequency at the peak.

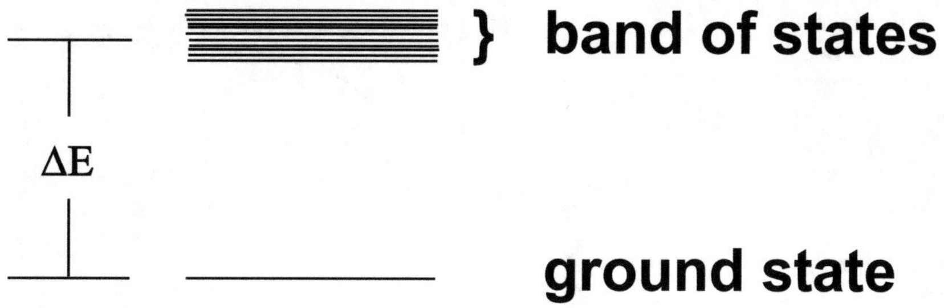

Fig. 15.20 Absorption of a photon by a molecule in a cone

additional processes are available to it for a return to the ground state.[41]

For example, a molecule in a gas can return to its ground state from an excited state by transferring its excitation energy to another molecule during a collision with that molecule. However, in the case of a molecule of cone pigment, in the process of returning to its ground state, excitation energy is used to produce a nerve impulse.

In the figure, ΔE is the difference between the energy of the ground state and the middle of the band of energy levels. It is equal to hf, where h is Planck's constant and f is the frequency ($= c/\lambda$). This frequency corresponds to the wavelength $\lambda \sim 430$ nm, where $\lambda = c/f$.

Sample Problem 15.2 Find the photon frequency and ΔE corresponding to $\lambda = 430$ nm.

Solution We have

$$f = \frac{c}{\lambda} = \frac{3 \times 10^8}{430 \times 10^{-9}} = 7.0 \times 10^{14}\ \text{Hz}$$

$$\Delta E = hf = (6.6 \times 10^{-34}\ \text{Joule-s})(7.0 \times 10^{14}\ \text{Hz})$$
$$= 4.6 \times 10^{-19}\ \text{J} \tag{15.42}$$

The **response of the cone**, in emitting nerve impulses, is proportional to its absorption spectrum. In practical terms:

nerve impulse rate of a cone at λ
∝ spectral absorption at λ ×
intensity I(λ) incident upon the cone

[41] See a full discussion of this subject in the section "Complex Scenarios of Absorption and Emission" in Chap. 6.

We are now in a position to present a simplified theory of a physiological basis for accounting for the chromaticity diagram. We introduce the following symbols:

Let

$$\mathrm{N}_R(\lambda) \equiv \text{nerve impulse rate of R-cone}$$

and

$$S_R(\lambda) \equiv \text{response function of R-cone for monochromatic } \lambda$$

The response functions $S_R(\lambda)$, $S_G(\lambda)$, and $S_B(\lambda)$ are proportional to the absorption spectra shown in Fig. 15.19.

Then

$$\mathrm{N}_R(\lambda) = S_R(\lambda) \times \mathrm{I}(\lambda) \tag{15.43}$$

Similarly for G and B we write:

$$\begin{aligned} \mathrm{N}_G(\lambda) &= S_G(\lambda) \times \mathrm{I}(\lambda) \\ \mathrm{N}_B(\lambda) &= S_B(\lambda) \times \mathrm{I}(\lambda) \end{aligned} \tag{15.44}$$

Now suppose that we have a general non-monochromatic source with spectral intensity I(λ). The spectral intensity tells us what the intensity is for each of the component wavelengths. For each component wavelength, we can compute the numbers $N_R(\lambda)$, $N_G(\lambda)$, and $N_B(\lambda)$. We next add up the set of these numbers for each primary. Thus obtaining the total nerve impulse rate emitted by each of the three cones:

$$\boxed{\begin{aligned} N_R &= \text{Sum of } \mathrm{N}_R(\lambda)'s \\ N_G &= \text{Sum of } \mathrm{N}_G(\lambda)'s \\ N_B &= \text{Sum of } \mathrm{N}_B(\lambda)'s \end{aligned}} \tag{15.45}$$

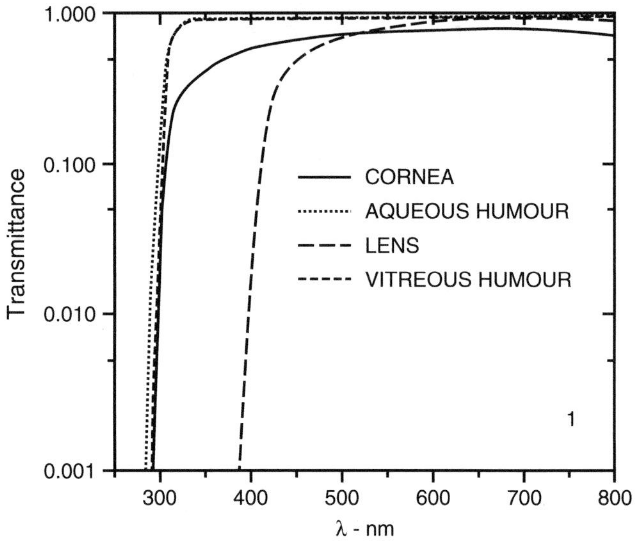

Fig. 15.21 Transmittance of the individual parts of the eye. The transmittances of the aqueous humour and vitreous humour are at the far left and overlapping; the humours have negligible absorption from 300nm to 700nm. Note the log scale of the transmittance, which indicates an even sharper drop at about 300nm than is shown. (source: Based upon a figure in the article by W. Ambach, et al., *Documenta Ophthalmologica*, Volume 88, pp. 165-173, 1994)

This is analogous to the mathematical process we carry out to obtain the tristimulus values R, G, and B. The response functions replace the color matching functions.

According to the simple theory of color vision, the signals N_R, N_G, and N_B are analyzable as distinct signals, so that they can be processed and interpreted by the brain to produce the following perceptions:

$$\boxed{\text{brightness} \propto N_R + N_G + N_B} \tag{15.46}$$

Hue and saturation are determined by the fractions

$$\boxed{\begin{aligned} n_R &= \frac{N_R}{N_R + N_G + N_B} \\ n_G &= \frac{N_G}{N_R + N_G + N_B} \\ n_B &= \frac{N_B}{N_R + N_G + N_B} \end{aligned}} \tag{15.47}$$

These three fractions reflect the color coördinates of the chromaticity diagrams. We do not expect the color coördinates, r, g, and b to be proportional to the respective fractions. However, we do expect that they will increase together: if one increases, so should the other.[42]

Comment It is important to keep in mind that testing of color vision involves light that must pass through various components of the eye on its way to the retina. All these components—the cornea, aqueous humour, lens, and vitreous humour—have a transmittance that falls off towards the ultraviolet, low wavelength range. Thus, we have

$$\boxed{\mathrm{I}_{\text{retina}}(\lambda) = \mathrm{T}_{\text{eye}}(\lambda) \cdot \mathrm{I}_{\text{incident}}(\lambda)} \tag{15.48}$$

where $\mathrm{T}_{\text{eye}}(\lambda)$ is the total transmittance of all these parts. The responses of the cones is to the ultimate light incident upon the cones, $\mathrm{I}_{\text{retina}}(\lambda)$.

In Fig. 15.21 we see the transmittance of the individual parts of the eye.[43]

Note The natural filtering out of UV radiation—especially by the lens—is beneficial in protecting the retina. With age, filtering increases but unfortunately moves into the visible region. Finally, cataract surgery, which involves removal of the

[42] In mathematics, we say that r is a **monotonically increasing** function of n_R, and so on.

[43] Omitted is the absorption of the macula, which contains the fovea. See the Wikipedia site (1-26-2011): http://en.wikipedia.org/wiki/Macular_degeneration, wherein it is pointed out that while "the macula comprises only 2.1% of the area of the retina ... almost half of the visual cortex [in the brain] is devoted to processing macular information."

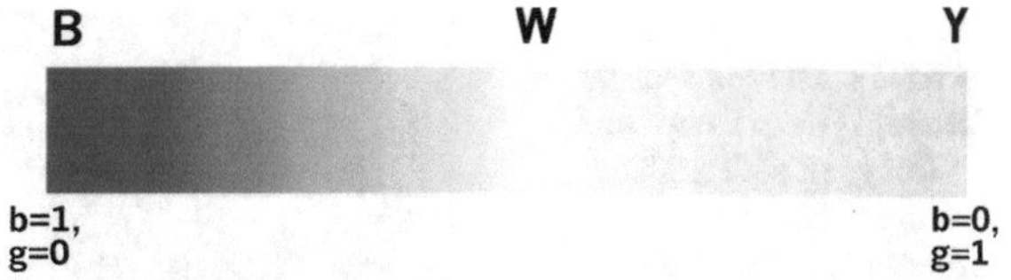

Fig. 15.22 Chromaticity diagram for Deuteranopes—the Blue-White-Yellow line

lens, strongly reduces UV filtering and increases potential damage to the retina.

15.14 Color-Blindness

We will briefly discuss the simplest kind of color-blindness, that of the **dichromats**.[44] Such color-blind individuals are missing one cone. If the green cone is missing, the condition is called **deuteranopia** and the individual is referred to as a **deuteranope**. If the red cone is missing, the condition is called **protanopia** and the individual is referred to as a **protanope**. And finally, for the rarest dichromacy, **tritanopia**, the blue cone is missing. Having only two cones, the dichromat cannot perceive the full range of colors associated with normal vision. They can still perceive the three characteristics—hue, saturation, and brightness; however, they can perceive only two hues.

It is reasonable to wonder how anyone could tell what colors they perceive. How can we know if a person is color blind? How can we compare the mappings of their sensations with the mappings under normal vision? The answer is that there are individuals who have color blindness in one eye and normal vision in the other, referred to as **unilateral dichromats**.[45] These individuals can make a mapping of the vision of their color blind eye onto the vision of their normal eye. Recently, eyeglasses have been developed to greatly improve the ability of deuteranopes distinguish to red from green.[46]

Below are some of its interesting features that are revealed by testing unilateral dichromats:

1. Their chromaticity diagram is reduced to a **line**.
2. There being but two pigments, a point on the line represents the fractional response **f** to light of a single pigment, say the B-pigment. Of course the fractional response to the second pigment would be **(1-f)**.
3. Equal energy white is towards the middle of the line segment. This white sensation can be produced by a single monochromatic source that has a wavelength of about 495 nm!
4. **Protanopes** and **deuteranopes** perceive a line of colors ranging from blue to yellow. See Fig. 15.22 above. They cannot distinguish among various shades of red, yellow, and green.[47] For further details, the reader is referred to the references listed in footnote number one at the beginning of this chapter.

In Fig. 15.23 we see at the top the spectral sensitivities of the blue and green cone of a protanope. At bottom is shown how a protanope would see a rainbow of colors. Note the gray band close to 500 nm; at the center of the band is a low intensity white.

[44]See the following website for a wonderful resource on color-blindness https://en.wikipedia.org/wiki/Color_blindness. It includes a fascinating set of figures that displays how the rainbow of colors appears for various types of colorblindness. It also discusses anomalous dichromacy, wherein there are three cones, but one of them is defective. Most interesting is the article's claim that a remote ancestor of the primates was a **tetranope**. In addition, the article points out that mothers of male dichromats tend to have a fourth color receptor in the green region.

[45]For studies of unilateral dichromats, see Martin Bodian's article "What do the Color Blind See?" in the **American Journal of Ophthalmology**, volume 35, p. 1471 (1952) and the article by Kurt Feig and Hans-Hilger Ropers in the journal **Human Genetics**, volume 41, p. 313 (1978).

[46]One approach, used by **EnChroma**, is to enhance the saturation of both red and green hues. See https://www.technologyreview.com/s/601782/how-enchromas-glasses-correct-color-blindness/. Another approach is the **ColorCorrection** *System*™ of Dr. Thomas Azman that uses a system of color filters.

[47]See the websites http://www.neitzvision.com/content/home.html and http://www.handprint.com/HP/WCL/color1.html#dichromat for details.

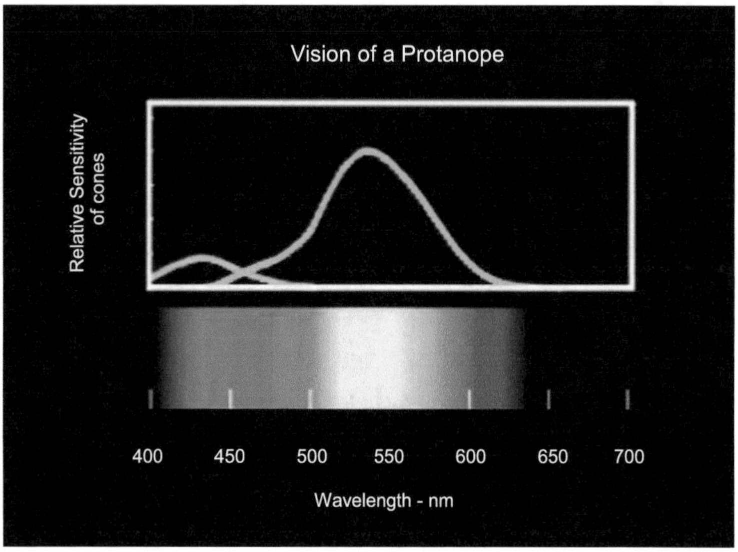

Fig. 15.23 Protanope: spectral sensitivities of the Blue and Green Cones and the Image of the Rainbow of Colors. (Courtesy Neitz Laboratory; http://www.neitzvision.com/content/excuseme.html)

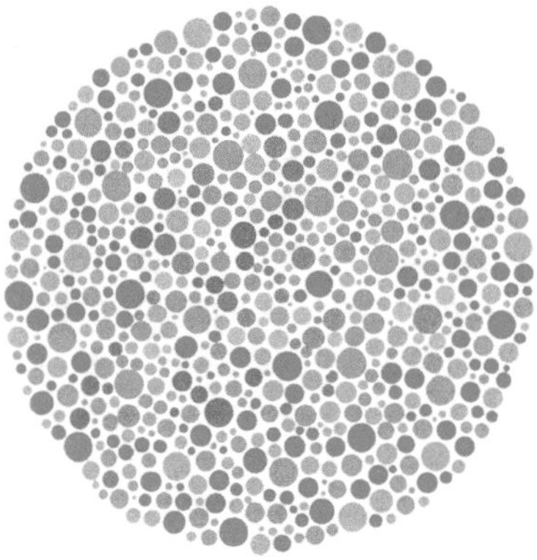

Fig. 15.24 Ishihara test for color-blindness (source: http://en.wikipedia.org/wiki/File:Ishihara_9.png)

You can test your own vision using one of the Ishihara tests for color blindness, shown in Fig. 15.24. The dots in the pattern and in the background are metamers for some observers but not for others. Individuals with normal vision see the number **74**. According to the Wikipedia Commons source of the image, deuteranopes might see the number **21**.

15.15 After-Images

Suppose that we produce a color patch of blue light over a white background in Fig. 15.25. Thus, when the blue light is removed, we will have white light everywhere.

Fig. 15.25 Blue light on a white background

If you stare at the patch, without moving our eyes much, for about thirty seconds and then stare at a blank area, there will appear a patch of yellow light in place of the blue patch. The *yellow* patch is referred to as the **after-image** of the blue patch. Our model of color vision can account for the phenomenon as follows.

When a cone pigment absorbs light and emits nerve signals, the pigment becomes "fatigued"—that is, it has a reduced ability to respond further to light by emitting nerve impulses. Such a process of fatigue also occurs in rod pigment, whose ground state color of purple is bleached (i.e., turned to white). The **recovery time**—that is, the time needed to regain full sensitivity—is only about 1-1/2 min for cones, in contrast to about 25-30 min for rods.

What happens in the above experiment is that the blue patch fatigues the blue receptor cones that lie on the image area of the retina. Upon removing the blue light, that retinal area suddenly *receives* white light. Since the rate of emission of nerve impulses from the blue cones is less than normal, the bluish component of incident white light is reduced, leaving its **complement**, yellow.

Interestingly, if you look carefully at a blue patch, you may notice a yellow halo around its boundary. A detailed explanation appears to be lacking. Nevertheless, this phenomenon is one of many that indicate that there is interference between the nerve impulses emitted by neighboring cones. A related phenomenon is the appearance of Mach bands, which are described and discussed in Chap. 11 in the context of pitch perception of sound. A second source of the halo may be the after-images produced by the erratic, so-called **saccadic** movements of the eye. (See **R. L. Gregory**, op. cit.)

15.15.1 Questions for Consideration

1. What after-images would result from removing the following patches from a white background?
   (a) a yellow patch?
   (b) a green patch?
   (c) a black patch?
2. Stare at the complementary image of the US flag. Identify the pairs complementary colors. For this exercise, I suggest that you choose a specific star and focus your attention on it. Doing so will help you keep your eyes fixed and improve on the effect of fatigue.

Fig. 15.26 Complement of the US flag

15.16 Terms

- additive primary
- after-image
- alychne
- blue cone
- brightness
- cathode ray tube (CRT)
- chromaticity
- chromaticity diagram
- coherent light
- color coördinate
- color matching functions
- color-blindness
- color monitor
- complementary colors
- deuteranope
- deuteranopia
- dominant wavelength
- D65 white
- gamut of colors
- green cone
- horse shoe boundary of chromaticity diagram
- hue
- incoherent light
- non-linearity
- line of purples
- luminance
- metamer
- metamerization
- nerve impulse from retina
- objective property
- primary colors
- primary illuminants
- protanope
- protanopia
- purity
- recovery time of a cone cell
- red cone
- response function of cones
- **RGB** coördinates
- saccadic movement
- saturation
- spectral color
- spectral intensity with respect to frequency
- spectral intensity with respect to wavelength
- spectral primary
- sRGB primaries
- subjective property
- subtractive primary
- table of color matching functions
- tristimulus values
- unilateral dichromat
- voltage fed to a color pixel in a color monitor

15.17 Important Equations

Color coördinates in terms of the intensities of the primaries used to produce the color:

$$\begin{aligned} r' &= \frac{I_R}{I_R + I_G + I_B} \\ g' &= \frac{I_G}{I_R + I_G + I_B} \\ b' &= \frac{I_B}{I_R + I_G + I_B} \end{aligned} \tag{15.49}$$

Color coördinates based upon the tristimulus values R, G, and B that are calculated from the spectral intensity:

$$\begin{aligned} r &= \frac{R}{S} \\ g &= \frac{G}{S} \\ b &= \frac{B}{S} \end{aligned} \tag{15.50}$$

where

$$S = R + G + B \tag{15.51}$$

Color coördinates for a light source that is a mixture of two incoherent light sources:

$$\begin{aligned} r &= \frac{R_1 + R_2}{S} = \frac{S_1}{S} r_1 + \frac{S_2}{S} r_2 \\ g &= \frac{G_1 + G_2}{S} = \frac{S_1}{S} g_1 + \frac{S_2}{S} g_2 \end{aligned} \tag{15.52}$$

where for the first source

$$r_1 = \frac{R_1}{S_1}$$
$$g_1 = \frac{G_1}{S_1} \tag{15.53}$$
$$b_1 = \frac{B_1}{S_1}$$
$$S_1 = R_1 + G_1 + B_1$$

and similarly for the second source.

Color coördinates for monochromatic light of wavelength λ:

$$r(\lambda) = \frac{\overline{r}(\lambda)}{\overline{r}(\lambda) + \overline{g}(\lambda) + \overline{b}(\lambda)}$$
$$g(\lambda) = \frac{\overline{g}(\lambda)}{\overline{r}(\lambda) + \overline{g}(\lambda) + \overline{b}(\lambda)} \tag{15.54}$$

15.18 Problems on Chap. 15

1. Discuss briefly how an infant might be taught to distinguish among the colors of objects. How does the infant get to appreciate what the significance of color is as distinct from among the other various attributes that an object can have?
2. (a) The complement of magenta is _______;
   (b) The complement of cyan is _______.
3. The chromaticity diagram shows (choose one)
   (a) the relative response of each type of cone to various wavelengths of light.
   (b) a comparison of how rods and cones react to light.
   (c) a way of plotting all colors in terms of two variables.
   (d) all the colors that are complementary pairs.
   (e) which three colors are the additive primaries.
4. (a) What color is complementary to blue?
   (b) What is **subtractive mixing** of colors? What are the primaries of subtractive mixing?
   (c) What are **metamers**?
   (d) What kind of color mixing do you suppose color television uses?
   (e) What are the two attributes of color?
   (f) What colors have zero saturation?
5. When we view a magenta light (choose one)
   (a) blue and red cones in the retina are stimulated.
   (b) blue and green cones in the retina are stimulated.
   (c) red and green cones in the retina are stimulated.
   (d) magenta cones in the retina are stimulated.
   (e) since green is the complementary color to magenta, only green cones in the retina are stimulated.
6. A **deuteranope** (choose one)
   (a) sees all objects as shades of red and green.
   (b) cannot distinguish between red and green.
   (c) has lost all red and green vision and sees the world in shades of blue.
   (d) is nearsighted for red light and far-sighted for green light, or vice versa.
   (e) reverses reds and greens.
7. There is evidence that some women are **tetrachromats**, meaning that they have four different color cones instead of three. How could you test for **tetrachromacy**?
8. (a) On a schematic chromaticity diagram such as Fig. 15.6, draw the triangle within which lie all colors that have a purple hue. We can refer to these colors as the set of **purple colors**.
   HINT: The point W is at one of the corners of the triangle.
   (b) Does this definition of purple colors depend upon the choice of primaries?
9. There are animals that have more than three color receptors; that is, they have more than three different cones. Animals with four different cones, such as doves, are called **tetrachromats**. Interestingly, there is evidence that some *women* have four receptors—those, in particular, who have the recessive gene for a dichromat.

See the following website for more information: http://www.freerepublic.com/forum/a3a24199b1ef8.htm.

Given what we have discussed about color perception for both **dichromats** and **trichromats**, discuss what changes you might expect for **tetrachromats**. Consider the range of perception of colors, in particular. How many primaries might one need to match any perceived color? How about the ability to discriminate between two spectral intensities? Might one expect improvement?

10. Might two people be found to need two different sets of color coördinates to match a given spectral intensity?
11. In practice, the primaries of various devices, such as a computer's color monitor or primaries of color printers that print color images, or primaries of digital cameras, are not spectral colors; that is, the primaries are not monochromatic. Furthermore, the actual sets of primaries vary from monitor to monitor and from color printer to color printer.
    (a) Comment on the consequences for the reliable reproduction of color images.
    (b) How might we deal with this situation? What we would we have to know about each of the sets of primaries used in the various devices in relation to a standard such as the C.I.E. table of color matching functions?
    (c) Name some other devices involving color reproduction that would require analysis in order to determine the relation between the image produced and the input of light.
12. Two monochromatic sources, D and E, are projected onto a screen, so as to appear *equally bright*. A photocell indicates that their intensities are **100 units and 10 units, respectively**. Qualitatively compare the eye *sensitivities* to the *two sources*.
13. Suppose that for a certain set of primaries,

$$0.1\ \mathrm{W/m^2}\ \text{of B} \oplus 0.1\ \mathrm{W/m^2}\ \text{of G} \oplus 1.0\ \mathrm{W/m^2}\ \text{of R}$$

produces a match with W (white). Suppose also that we choose 1-W/m^2 to be the *unit intensity* for all *three primaries*.
    (a) What would be the color coördinates of W?
    (b) Why would such a choice *not* be practical?
14. Use the schematic chromaticity diagram in Fig. 15.6 for the following problem:
    Four units of R, two units of G, and two units of B are mixed together.
    (a) Find the color coördinates of the mixture.
    (b) Find the dominant wavelength and purity of the mixture.
    (c) Describe the color.
    (d) Find the dominant wavelength of the complementary hue of this chromaticity.
15. Consider a table of color matching functions such as the one we are using that is based upon the spectral primaries $\lambda_B = 435.8$ nm, $\lambda_G = 546.1$ nm, and $\lambda_R = 700.0$ nm.
    (a) What would be the effect on the tristimulus values of a given spectral intensity if the values of all of the color matching functions were doubled?
    (b) What would be the effect on the color coördinates of a given spectral intensity if the values of all of the color matching functions were doubled?
    (c) What would be the effect on the color coördinates of a given spectral intensity if the values of all of the color matching functions were multiplied by any given number beside two?
16. In this problem you will produce two metamers. Open the following website http://www.cs.brown.edu/exploratories/freeSoftware/repository/edu/brown/cs/exploratories/applets/spectrum/metamers_java_browser.html. See Fig. 15.1 for an example of what you should see. Note that the graph is a plot of the spectral intensity vs. frequency and not wavelength.

(a) By sweeping your mouse across each of the frequency axes of each window, produce two **different** spectral intensities that are **orange** in hue and try to produce metamers as best you can. For one of the two metamers, arrange for the spectrum to have a gap in the orange region. Note that the two boxes at the bottom give you the tristimulus values of each of the colors. If you are using this book for a course, print out the result of your applet and hand it in as part of the homework set.

(b) Recall that in Chap. 9 we learned that the origin of the blueness of the sky is that scattering of sunlight is inversely proportional to the fourth power of the wavelength. As a consequence, the spectral intensity at 400-nm is $(7/4)^4 \sim 9$ times the spectral intensity at 700-nm.

Explain why the spectral intensity increases as the fourth power of the frequency.

(c) In the metamer applet, produce with your mouse a spectral intensity that increases from the red end to the blue end by about nine-fold and thus produce the resulting color patch. You need not be fussy about the precise shape of the spectral intensity.

17. Suppose that the two spectral intensities, $\mathrm{I}_1(\lambda)$ and $\mathrm{I}_2(\lambda)$ are metamers and we add the spectral intensity $\mathrm{I}_3(\lambda)$ to each, resulting in two new spectral intensities

$$\mathrm{I}_1'(\lambda) = I_1(\lambda) + \mathrm{I}_3(\lambda) \quad \textit{and}$$
$$\mathrm{I}_2'(\lambda) = \mathrm{I}_2(\lambda) + \mathrm{I}_3(\lambda) \qquad (15.55)$$

Are these two new spectral intensities metamers? To answer this question, consider how the resulting color coördinates would compare. Remember how these numbers are obtained by adding the color matching functions together.

18. The goal of this problem is to produce the chromaticity diagram for the Wright-Guild primaries corresponding to the color matching functions in Table 15.3. The table is also accessible as an EXCEL file from the course website: https://sites.tufts.edu/pmclg/links/. CLICK on "LINKS" and you will see the link entitled "Table of Color Matching Functions for Wright-Guild primaries." You can download the EXCEL file by clicking the filename "TCMF Wright Guild" shown in blue letters.

(a) Determine the color coördinates $\{r(\lambda), g(\lambda)\}$ for monochromatic sources of wavelength λ. See Eq. (15.23).

HINT: Note that for monochromatic a 400-nm source, $\mathrm{I}(\lambda)$ vanishes for all but 400-nm. Setting $\mathrm{I}(400\text{-nm})=I$, we obtain $\mathrm{R}=0.00030I$, $\mathrm{G}=-0.00014I$, and $\mathrm{B}=0.01214I$, from which one can obtain r_{400}, g_{400}, and b_{400}.

(b) Plot the coördinates $(r(\lambda), g(\lambda))$ on graph paper and connect the points, thereby obtaining the horseshoe shaped perimeter of the chromaticity diagram corresponding to Table 15.3. Label the points on the perimeter with the corresponding wavelengths.

19. (a) Use Fig. 15.4 to fill in the reflectance of a **tomato** in the Wright-Guild TCMF, as shown in the EXCEL in this chapter or in the EXCEL file on the course website: https://sites.tufts.edu/pmclg/links/

(b) Determine the tristimulus values and color coördinates of a tomato.

(c) Place a dot in the chromaticity diagram in problem 18.

20. (a) Determine the tristimulus values and the color coördinates of the spectral intensity depicted in Fig. 15.27.

(b) Determine the dominant wavelength and the purity using Fig. 15.6 and Eq. (15.19), respectively.

21. White light is passed through a filter that has a transmittance with a spectral intensity having the tristimulus values

$$\mathbf{R} = \mathbf{30}, \ \mathbf{G} = \mathbf{60}, \ \mathbf{B} = \mathbf{250}$$

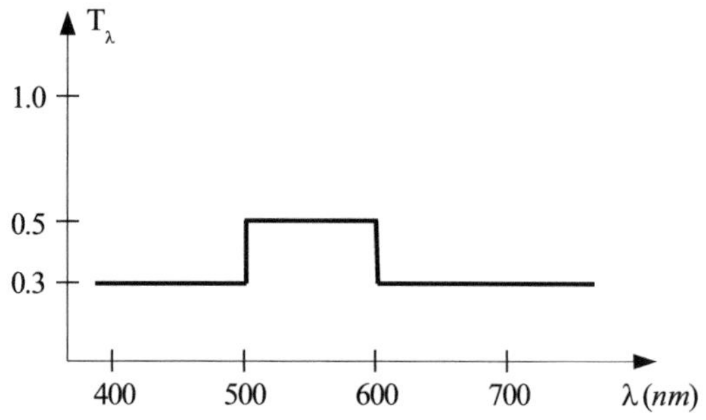

Fig. 15.27 Spectral intensity for Problem 14.21

(a) Calculate the color coördinates for this transmittance.
(b) Display the color coördinates in the chromaticity diagram shown in Fig. 15.28.

22. On a Chromaticity diagram, draw the region within the horseshoe that has hues that must be expressed in terms of their complement.
23. **Unmatchable Color**

Suppose that a digital camera were to store data according to the chromaticity diagram shown below. In Fig. 15.29 we see a color (identified by a red arrow pointing to a black dot) that is unmatchable by mixing the primaries.

(a) Determine the hue of the color.
(b) Why is the color not matchable?
(c) Determine from the diagram what color coördinates might actually be stored. EXPLAIN your procedure in detail.

Note: A common replacement is a color on the green axis that has the same hue as the given color.

24. (a) Consider Fig. 15.18, which displays the primaries of sRGB in an lRGB chromaticity diagram. Suppose you wanted to add a fourth primary to these three primaries. We would have a 4-primary color set for producing colors. Note that it is perfectly fine to have four primaries as sources of light even though a normal eye has only three different cones. Approximately where would you put the color coördinates so that the gamut of colors that lies within the resulting **quadrangle** encompasses the largest gamut of colors? It should be clear that the color of this fourth primary would not be yellow! The SHARP Corporation claims that in its 4-color monitor, it has added a Yellow primary. Perhaps the other three primaries are far from sRGB.
(b) The added fourth point will define a second triangle. Explain why the additional primary adds only colors that lie within this triangle.
(c) Explain why for any point within the quadrangular gamut, there is more than one way to produce a color by mixing the four primaries. The factor of 256 to the monitor-count of colors will thus be redundant.
(d) An additional redundancy is produced by the fact that the density of points in the gamut will be much greater than the density in the 3-primary gamut. Give an argument as to why the number of distinguishable colors should be increased by only about 50% at best.

25. Suppose that the spectral intensity of problem 14.20 is produced by white light passing through a filter. That is, the above spectral intensity is the *transmittance* of the filter. (See Sect. 14.5.) In order to increase the saturation of the light beam, two identical such filters are *stacked* (i.e. are placed back to back).

(a) What is the resulting transmittance of the stacked filters? Exhibit your answer as a graph of transmittance vs. wavelength.

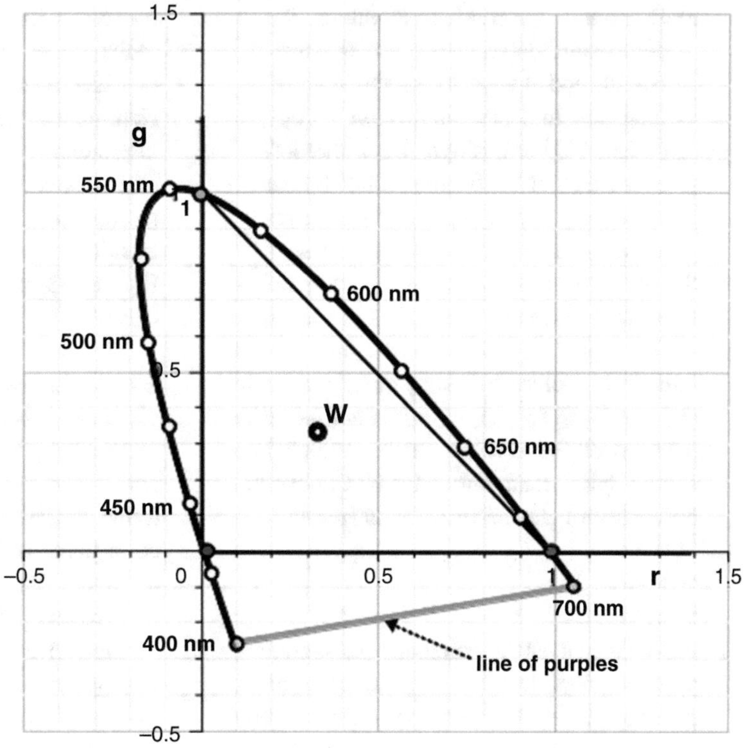

Fig. 15.28 Color coördinates of a filter

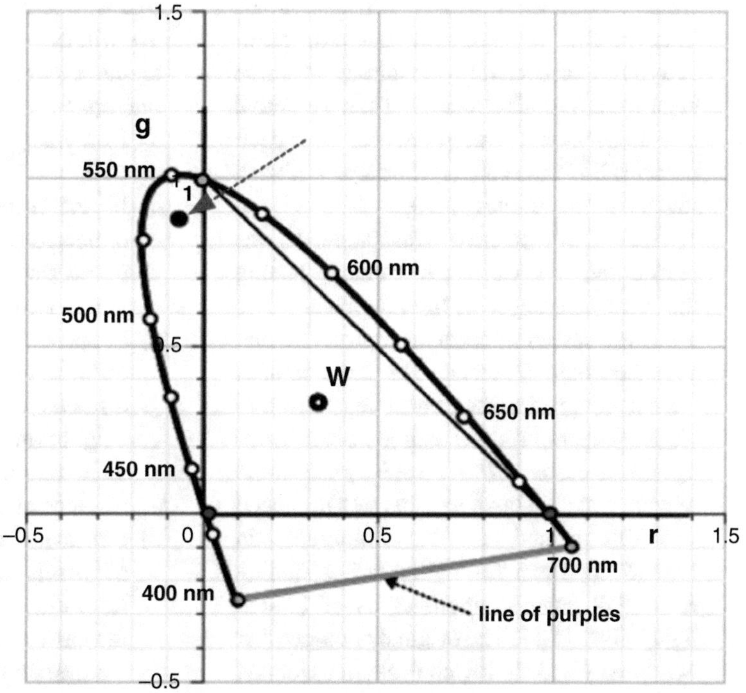

Fig. 15.29 For the problem of replacing the color coördinates of an unmatchable color

(b) Determine the tristimulus values, the color coördinates, the dominant wavelength, and the purity of the spectral intensity. Indicate roughly on a copy of Fig. 15.6 or a rough *sketch* thereof, how the color has been modified by the stacking process. Has the stacking resulted in an increase in the saturation?

26. Suppose you are adding two color patches so as to produce various shades of red, from white to pink. You have one source of red light that is close to being monochromatic and you want to use the *minimum intensity* possible of the second source. Which hue should that second source have? Choose the best answer below. *Explain your answer.*
   (a) white
   (b) blue
   (c) cyan
27. Before the advent of color film photography, photographers learned to produce a color photograph by the following trick: A scene was photographed three times in black and white (BW), each with a different color filter in front of the camera—red, green and blue, respectively—resulting in three negative BW transparencies. Then a positive BW transparency was produced from each negative BW transparency. See a pair of such transparencies in Fig. 15.30.

    Each BW positive transparency was placed into its dedicated camera, having the corresponding color filter in place. White light was passed through each projector and projected onto a screen. The three images were overlapped to the utmost degree of precision so as to produce a single distinct image on the screen. We see a color photograph of Mohammed Alim Khan that was produced with this method of color photography in Fig. 15.31. There follow the corresponding three BW positive photographs of filtered light in Fig. 15.32. The original photographer was Sergey Mikhaylovich Prokudin-Gorsky.

    Explain how this process works.
28. What color is the after-image of a bright green light? Why?
29. In Sect. 15.15.1 it was suggested that you focus on a specific star of the U.S. flag in order to enhance the effects of fatigue. Comment on why focusing might be of help in producing a clear after-image.
30. The impressionist painter Georges Seurat is famous for his paintings using **pointilism** mentioned earlier in this chapter. In Fig. 15.33 is a print of his painting "Le Chahut."[48]

    The painting is produced by laying down a huge number of dots of paint. From the distance, the dots are not visible and the viewer perceives a smooth variation of colors and textures. Since any cone on the retina receives light from many dots, the color perceived is a mixture of the colors of the dots. The purpose of this problem is to explore this phenomenon.
31. In Fig. 15.34 is a magnified image of the display on an iPhone. For simplicity, we will treat the rectangles as squares that touch each other. Each pixel consists of three **sub-pixels**, one for each of the three primaries. We will assume that there are 200 sub-pixels along a horizontal (or vertical) length of one-cm.[49]
   (a) How many pixels are there per square cm?
   (b) If the screen of the iPhone has dimensions 10-cm×6-cm, how many **pixels** does the screen have? Express your answer in **megapixels**.
   (c) Suppose that the iPhone is held at a distance of 25-cm. **CALCULATE** the distance between the centers of the neighboring images of the sub-pixels on a retina with an effective image distance of d_{ie}=24-mm.
   (d) Comment on whether the density of pixels has reached the limit of resolution by the eye. That is, would increasing the density lead to an increase in the resolution of the image the eye sees?

[48] https://en.wikipedia.org/wiki/Georges_Seurat.

[49] See http://prometheus.med.utah.edu/~bwjones/wp-content/uploads/2010/06/iPhone-4-Display_.jpg].

Fig. 15.30 Left: Negative Black and White Transparency of Richard Feynman; Right: Positive Transparency Thereof

Fig. 15.31 Color photograph of Alim Khan—using three black and white filtered photographs. (source: http://en.wikipedia.org/wiki/File:Mohammed_Alim_Khan_cropped.png)

32. In Fig. 15.35 we see two colored circles. The upper circle consists of a checkerboard of two colors. The lower circle has a uniform color. If one were to stand far enough away from the figure, the upper circle will appear uniformly colored and indistinguishable from the lower one. The basis for this phenomenon is that the distance between the centers of neighboring squares in the image of the checkerboard on the retina is comparable to or less than the limit of resolution of the retina. If the cones behaved independently of one another, the limit of resolution should be about equal to the distance (between the centers) of neighboring cones. In fact, as we have noted,

Fig. 15.32 Three black and white color filtered photographs that produce the color photograph of Alim Khan (courtesy: Anandajoti Bhikkhu)

cones communicate with each other, so that the distance between cones can be expected to be a bit smaller than the limit of resolution. We expect that the tristimulus values of the apparent color of the checkerboard should be at least close to the average of the tristimulus values of the red and green squares.

Exercise

(a) Measure the distance between neighboring squares in the upper disc.

(b) Look at the figure from a distance and determine the **minimum distance** at which the upper disc has a uniform color.

(c) Assuming that the distance from your lens to your retina is 2.5 cm, determine the distance between neighboring squares in the image on your retina. Is the value reasonable?

(d) At a close distance the squares are observed to be distinct, with different colors. A single particular cone will receive light from only one square on the disc. Another cone will receive light from a different square. At a great distance, each cone will receive light from many squares. Thus, all cones receive the same spectral intensity. As a consequence, the upper disc in the figure appears uniform in color.

Tom Cornsweet, who produced the image in Fig. 15.35, chose the spectral intensities of the individual squares and of the uniform disc so as to produce a match of the color of the uniform disc and the perceived color of the checkerboard when the check boarded disc is viewed from afar.

Fig. 15.33 *Le Chahut* by Georges Seurat

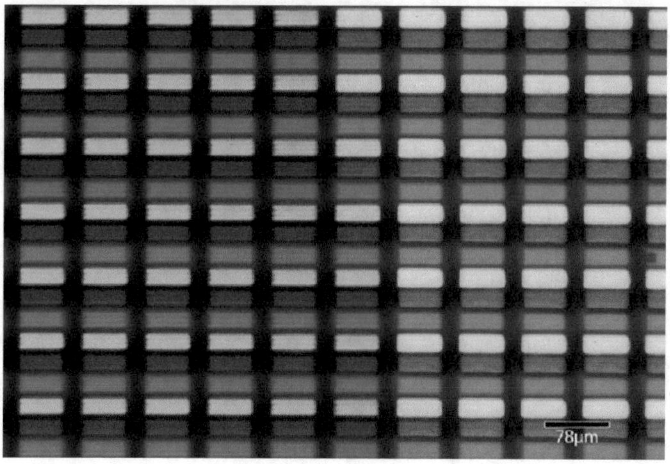

Fig. 15.34 iPhone4 display

Let us label the squares with the numbers 1 and 2. At **closeup**, a single cone might receive light from a square #1 with tristimulus values (R_1, G_1, B_1). Another cone will be in a position on the retina that receives light from a square #2 with tristimulus values (R_2, G_2, B_2). The corresponding color coördinates from square #1 are $\{r_1, g_1\}$ while the color coördinates from square #2 are $\{r_2, g_2\}$.

Fig. 15.35 Two discs that appear alike at a great distance—a checkerboard and a uniform patch (source: Courtesy, Tom Cornsweet, , *Visual Perception*, (Academic Press, 1970))

At a great enough distance from the checkerboard disc, each cone will receive light from many squares. The light falling on a cone is the sum of the intensities received from many squares.

i. Suppose that the intensity light on a cone from a square #1 is reduced by a factor of 1,000 in viewing the disc from afar. By what factor will the intensity of light on a cone from a square #2 be reduced in viewing the disc from the same greater distance?
ii. Each type of square produces light with three tristimulus values. By what factor will the six tristimulus values of the light on a cone, three from each type of square, be reduced?
iii. How will the color coördinates of the light on a cone contributed from each of the two types of squares change? Explain your answer.
ddsdsdccc
iv. Show that the resulting color coördinates of the disc viewed at a great distance, when the disc appears uniform in color, are given by

$$\boxed{\begin{aligned} r &= \frac{S_1 r_1 + S_2 r_2}{S_1 + S_2} \quad , \\ g &= \frac{S_1 g_1 + S_2 g_2}{S_1 + S_2} \end{aligned}} \tag{15.56}$$

where

$$\boxed{\begin{aligned} S_1 &= R_1 + G_1 + B_1 \quad , \\ S_2 &= R_2 + G_2 + B_2 \end{aligned}} \tag{15.57}$$

See Sect. 15.6.7 and Eqs. (15.21), (15.22), and (15.23) therein.

The resulting color coördinates (r,g) are said to be a **weighted average** of the two sets of color coördinates (r_1, g_1) and (r_2, g_2).

Note that when $S_1 = S_2$, the color coördinates are a simple average of the original two.

$$r = \frac{r_1 + r_2}{2} \quad , \quad \frac{g_1 + g_2}{2} \qquad (15.58)$$

(e) Do the color coördinates (r,g) depend upon the distance from the disc?

(f) Use the results of Eq. (15.40) in Sect. 15.11 to show that if the color patches match at a distance when viewed with one monitor, they will not necessarily match when viewed with another monitor. The same holds true for color printing.

(g) Finally, consider the question of viewing the figures from a printed copy. If the patches match when viewed with one light source reflected off the page, will there necessarily be a match if you use a different light source having a different spectral intensity?

(h) In Fig. 15.36 we see a checkerboard made up of red and green squares. What hue would you expect the full checkerboard to have when viewed at a great distance? Explain. Check out your prediction.

Note that on my monitor the squares have color monitor coördinates close to pure red and pure green, respectively. However, at a great distance, you might not see a uniform patch that is as bright as you might hope for.

33. **From spectral intensity to monitor color—the Blue Sky**

The purpose of this problem is to give you experience in producing the color on your monitor that corresponds to a given spectral intensity. You will focus on the spectral intensity of sunlight that is scattered by a clear sky. We learned in Chap. 9 that theory predicts that the spectral intensity is inversely proportional to the fourth power of the wavelength. Thus, we write $I(\lambda) \propto 1/\lambda^4$. The goal is to produce the corresponding color on your monitor, as discussed in Sect. 15.11.

First, determine the color coördinates, R, G, and B for **sRBG primaries**. To do so, you can use the EXCEL file for the Table of Color Matching Functions (TCMF) on the website for the Tufts course on the Physics of Music and Color.[50] In the intensity column, you can enter the function $(1/\lambda^4)$, where the wavelength is listed in the first column.

To determine the color coördinates, **R**, **G**, and **B** of a monitor, you will need to choose a value for gamma. If you cannot determine gamma for your monitor, use the value of 2.2 to complete the problem.

Note that the color on your monitor is likely to have a low brightness compared to what we are used to seeing in the sky. You can resolve this problem by dividing each of the color coördinates {r, g, b} by **b**. You will obtain a new set of numbers—{r/b, g/b, 1}—which you will substitute into Eq. (15.41). As a result, you will have **B** = 255.

34. In Chap. 15 we learned that on account of the Doppler effect, the observed color of an object depends upon the state of motion of the object with respect to the observer. In the acclaimed book, "Art and Science," by Leonard Shlain, [Harper Perennial, NY, 2007] the author claims that the Doppler Effect indicates that the color of an object is subjective—not objective.

Is it correct to say the color of an object changes when the object moves?

35. **Successive Contrast of Pairs of Colors**

The yellow color patches at the bottom of Fig. 15.37[51] have identical colors. Stare

---

[50] http://sites.tufts.edu/pmclg/. The table is to be found as a Link. You may need the password: pmc2012. The labels of the color matching functions are primed r, g, and b, as opposed to overbarred letters.

[51] Source: http://upload.wikimedia.org/wikipedia/commons/thumb/e/e2/Successive_contrast.svg/2000px-Successive_contrast.svg.png.

Fig. 15.36 A red and green checkerboard

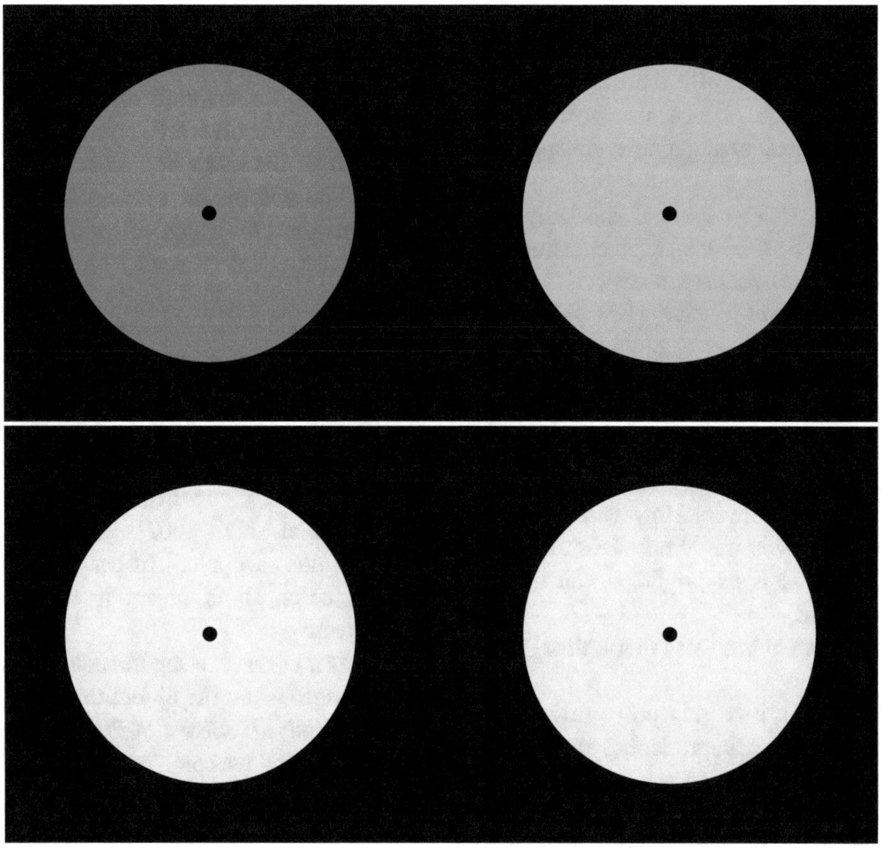

Fig. 15.37 Successive color contrast

at the top half of the image for about 20 s. Then quickly look at the bottom half. For people with normal vision, the two lower patches will appear to have two different colors.

Explain the reason for this phenomenon.

36. **Copyrights of Colors**

The Coca Cola Corporation has a patent on its color red, as do a number of other companies—for example the green and yellow color combination of the John Deere Corporation, a manufacturer of farm and construction equipment. [See https://www.businessinsider.com/colors-that-are-trademarked-2012-9 and https://www.deere.com/en/our-company/news-and-announcements/news-releases/2017/corporate/2017oct17-deere-wins-trademark-lawsuit/.]

Discuss how color coordinates can be used to define the patent and the issue of precision in characterizing the patented color.

Correction to: The Physics of Music and Color: Sound and Light

Leon Gunther

Correction to:
L. Gunther, *The Physics of Music and Color*,
https://doi.org/10.1007/978-3-030-19219-8

The original versions of chapters 8 and 11 were revised and the corrections for these chapters have been updated in the book.

The updated original version for these chapters can be found at
https://doi.org/10.1007/978-3-030-19219-8_8
https://doi.org/10.1007/978-3-030-19219-8_11

L. Gunther, *The Physics of Music and Color*,
https://doi.org/10.1007/978-3-030-19219-8_16

Manipulating Numbers

A

A.1 Units

Physics introduces many concepts that help us understand and analyze physical processes—sound and light phenomena, in particular. However, ultimately we come to trust the claims of Physics by comparing observed **numerical values** of important parameters—such as sound or light intensity and sound or light frequency—to how they are related to each other according to the laws of Physics. We must note that a numerical value depends upon the choice of **units**. For example, my height is about 5-ft 10-in, or 70-in. In the metric system, my height is 1.78-m. Another unit of length is the Ångstrom, abbreviated with the letter Å. Given that 1-Å=10^{-10}-m, my height is 1.78×10^{10}-Ångstroms. The relationship between various units for a given physical parameter is listed in Appendix D on the conversion of units.

A.2 Order of Magnitude

We begin by noting that while Physicists would like to know a value as precisely as possible, they are also extremely concerned with what is referred to as the **order of magnitude** of a parameter. Typically it is expressed by giving the value to the nearest power of ten. Thus, for example, any value ranging from 5 to any value less than 50 is rounded off to ten. We say that a value of 21 has an order of magnitude of ten. An adult has a height on the order of magnitude of one meter, if we choose to use the meter as our unit of length. However, if we choose to use the foot as our unit of length, most adults have a height that is on the order of magnitude of ten feet. Yet 10-ft is approximately 3 meters.

An atom has a diameter on the order of magnitude of 1-Ångstrom $\equiv$ 1-Å$\equiv$ 10^{-8}-cm. The size of the observable universe is 10^{27}-cm. The weight of most adults is an order of magnitude of 100-lbs. The number of hairs on an adult with a full head of hair can have an order of magnitude as high as 100-thousand.[1]

A.3 Significant Figures

The order of magnitude is a very crude estimate of the value. We often have to express *how precisely we know the value of a parameter.* For this purpose, let us now introduce the term **significant figures**. A value of my height as 70-in provides me with two significant figures. A height expressed as 184-cm has three significant figures. Over the course of one day, my height certainly wouldn't vary by as much as a centimeter. Therefore, I might wake up with a height of 181-cm and go to bed with a height of 180-

[1]We can arrive at this number as follows: The dimensions of a scalp is on the order of 30-cm by 25-cm, corresponding to an area of 750-cm^2. For a thick head of hair, the distance between hairs might be about 1-mm. Each hair takes up an area of about 1-mm^2, equivalent to $(10^{-1}\text{cm})^2 = 10^{-2}$-cm^2. Therefore the number of hairs is $750/10^{-2} = 75,000$. On the other hand, if the distance between neighboring hairs in 3-mm, the area per hair would be 9-mm^2; in this case, the number of hairs would be $75,000/9 \sim 8,300$ and the order of magnitude would be 10,000.

L. Gunther, *The Physics of Music and Color*,
https://doi.org/10.1007/978-3-030-19219-8

cm due to the compression of the disks in my spine. If I were to express my height to only two significant figures, I wouldn't be taking into account this change.

Here is an ambiguity that can arise. Consider the following: If I were to tell you that my height is 180-cm, you would not know whether or not the "0" digit is significant. The use of **scientific notation** takes care of this problem as follows: If the zero is significant, I would express my height as 1.80×10^2-cm; this expression provides my height to three significant figures. If the zero is **not** significant, I would express my height as 1.8×10^2-cm and we see my height expressed to two significant figures. One foot is about one-third meter, which is, to one significant figure, 0.1-m.

Rounding off to a given number of significant figures can lead to some curious numbers. Consider the following: Normal body temperature is said to be 98.6^0 F. However, a body temperature that is anywhere between about 97^0 F and 99^0 F is regarded as "normal" or acceptable. Given this fact, how could such a strange number arise? Why should the temperature be specified to three significant figures!?

To answer this question note that ^{0}C=(5/9)(^{0}F-32). Then we find that 98.6^{0}F = 37^{0}C to three significant figures. In fact, doctors realize that people's body temperatures are not expected to be precisely 37^{0}C. This value was chosen as a rounded off number.

Here is another example of how a choice of units introduces uncanny constants. The so-called **body mass index** (BMI) is defined in British units (feet for length, seconds for time, and pounds for weight) as

$$703 \times \text{weight in lbs/(height in inches)}^2.$$

Here again, we have a strange number "703" occurring with three significant figures. It wouldn't seem to matter had the number been chosen to be 700 precisely. Using metric system, with kilograms for weight and meters for length, changes the coefficient to unity (one).

Problem Confirm the above result. NOTE: One kilogram is equivalent to about 2.20-lbs and one meter equals about 39.37 inches.

The lesson we learn from these examples is that we must be careful not to take seriously the number of significant figures listed for an important parameter.

A.4 Relative Changes

The relative change of a parameter gives us a sense of the magnitude of the change that is independent of the units. Specifically

$$\boxed{\text{relative change} \equiv \frac{|\text{change in value}|}{|\text{original value}|}} \tag{A.1}$$

Here the vertical bars surrounding a number represent the **absolute value** of the number.[2]

Sample Problem A.1 Above we discussed the change of my height over the course of a day. My height changed from 181-cm in the morning to a height of 180-cm at the end for the day. What is the relative change in my height?

Solution Letting the heights be h_1 and h_2, and Δh be the absolute value of the change in height, we have to one significant figure

$$\text{relative change} = \frac{\Delta h}{h_1} = \frac{181 - 180}{180} = 0.0.006 \tag{A.2}$$

or about 0.6%.

Suppose that you know that the frequency of a sound has changed by 4-Hz. Is this change significant? The relative change will be extremely informative.

Suppose that the original frequency is f_1 and the new frequency is f_2. Then

$$\text{relative change} \equiv \frac{|f_2 - f_1|}{|f_1|} \tag{A.3}$$

[2]Thus, $|2| = |-2| = 2$.

If we set $\Delta f = f_2 - f_1$, we can express the relative change of the frequency as

$$\boxed{\text{relative change} \equiv \frac{|\Delta f|}{|f_1|}} \tag{A.4}$$

Thus if the original frequency is 400Hz, the **relative change** is 4/400 or one part in a hundred (0.01). [Note that the relative change as we have defined it would be 0.01 whether the change is ±2 Hz, that is, from 400-Hz to 396-Hz or 400-Hz to 404-Hz, since our definition of the relative change includes the absolute value.]

In the context of music, if the relative change is much less than unity, the relative change in sound frequency is closely related to the change in **pitch**: As we will learn in Chap. 12, a change in frequency by about 6% or one part in 16 corresponds to a change by what is referred to as a **half-step**.[3]

One might ask whether a change of one part in a hundred significant. The answer depends upon the ear of the listener. For a person with a "good ear," the change would certainly be significant. For many people the change would not be noticed.

Problem Suppose that the change in frequency is 4-Hz and the original frequency is that of the lowest note on a double bass, about 41-Hz. Determine the relative change in frequency.

A.4.1 Proportionalities

Suppose that a variable y is proportional to a variable x. We have

$$y \propto x. \tag{A.5}$$

Alternatively we can write

$$y = bx, \tag{A.6}$$

where b is a constant.

[3]For example, going from note B to note C or from A to A#.

Here is an example wherein we can use algebra to simplify a calculation. Suppose that we know x_1 and its corresponding y_1, so that $y_1 = bx_1$. Next, we know another value of x, say x_2, and want to know its corresponding value y_2. We could determine the constant b and then calculate y_2. Alternatively, we can use the following trick:

We have

$$\frac{y_2}{y_1} = \frac{bx_2}{bx_1} = \frac{x_2}{x_1} \tag{A.7}$$

or,

$$\frac{y_2}{y_1} = \left(\frac{x_2}{x_1}\right) \tag{A.8}$$

and finally

$$y_2 = x_1\left(\frac{x_2}{x_1}\right). \tag{A.9}$$

For example, if $x_1 = 4$, $x_2 = 12$, and $y_1 = 2$, $y_2 = (12/4) \times 2 = 6$.

We might want to know the relative change in y in terms of the relative change in x. We will now prove that **if the two are proportional to each other, the relative changes are equal.**

Then

$$\frac{\Delta y}{y} = \frac{\Delta(bx)}{bx} = b\frac{\Delta x}{bx}. \tag{A.10}$$

Thus

$$\boxed{\frac{\Delta y}{y} = \frac{\Delta x}{x}.} \tag{A.11}$$

Sample Problem A.2 Here is a simple common example. Suppose that two people run the same distance d at different velocities. Given that the relative change in the velocities is 2%, what is the relative change in the times of travel?

Solution The time of travel is proportional to the velocity: Since distance = velocity × time, or $d = vt$, we have

$$t = \frac{\text{v}}{d}. \tag{A.12}$$

Distance is then proportional to velocity (at fixed distance). Therefore, the relative change in the time is also 2%.

Now consider a string on a stringed instrument whose tension $\mathcal{T}$ is changed. In Chap. 2, we will learn that the frequency of a wave is proportional to the wave velocity v.

$$f \propto v. \tag{A.13}$$

In addition, for a given string under variable tension, as with a string on a violin, the wave velocity is proportional to the square root of the tension T. Therefore, the frequency is proportional to the square root of the tension:

$$f \propto \sqrt{\mathrm{T}}. \tag{A.14}$$

In terms of exponents, we have

$$f \propto \mathrm{T}^{1/2}. \tag{A.15}$$

Sample Problem A.3 Suppose that we have two strings identical except that they have different tensions, T_1=100-N and T_2=110-N. The frequency of string #1 is 300-Hz. Find the frequency of the second string.

Solution We can use the relation $f = bT^{1/2}$.

Then

$$\frac{f_2}{f_1} = \frac{bT_2^{1/2}}{bT_1^{1/2}} = \frac{T_2^{1/2}}{T_1^{1/2}}, \tag{A.16}$$

or,

$$\frac{f_2}{f_1} = \left(\frac{T_2}{T_1}\right)^{1/2}. \tag{A.17}$$

Finally, we have

$$f_2 = f_1\left(\frac{T_2}{T_1}\right)^{1/2} = 300 \times \left(\frac{110}{100}\right)^{1/2}$$

$$= 300 \times 1.05 = 315\text{-Hz}. \tag{A.18}$$

Suppose that the tension is changed by a relatively small degree. That is, the relative change is much less than unity. What is the relative change in the frequency?

It can be shown that the result is extremely simple.[4]

$$\frac{\Delta y}{y} = \frac{1}{2}\frac{\Delta \mathrm{T}}{\mathrm{T}}. \tag{A.24}$$

For example, if the tension is reduced by 2%, the frequency will be reduced by 1%.

Generally, for small changes, if $y \propto x^n$,

$$\boxed{\frac{\Delta y}{y} \approx |n|\frac{\Delta x}{x}}.$$

Sample Problem A.4 In Chap. 4, we discuss the **intensity** I of a source of sound or of light at various distances r. We pointed out that the intensity is inversely proportional to the square of the distance:

$$I \propto \frac{1}{r^2} \tag{A.25}$$

What is the relation between the corresponding relative changes?

Solution In this case, $n = -2$. Therefore,

$$\frac{\Delta I}{I} \approx \frac{1}{2}\frac{\Delta r}{r} \tag{A.26}$$

---

[4] We have

$$y = bx^{1/2}. \tag{A.19}$$

where b is a constant.

Then

$$\Delta y = bx_2^{1/2} - bx_1^{1/2} = b(x_2^{1/2} - x_1^{1/2}). \tag{A.20}$$

Now $x_2 = x_1 + (x_2 - x_1) \equiv x_1 + \Delta x$. And, if $\Delta x << x_1$, we can use the binomial theorem to show that

$$x_2^{1/2} \sim x_1^{1/2} + \frac{1}{2}\frac{\Delta x}{x_1^{1/2}}. \tag{A.21}$$

Then

$$\frac{\Delta y}{y} = \frac{b(x_2^{1/2} - x_1^{1/2})}{bx_1^{1/2}} = \frac{(x_2^{1/2} - x_1^{1/2})}{x_1^{1/2}}. \tag{A.22}$$

$$\frac{\Delta y}{y} \sim \frac{1}{2}\frac{\Delta x}{x^{1/2}} \times \frac{1}{x^{1/2}} = \frac{1}{2}\frac{\Delta x}{x}. \tag{A.23}$$

Note that in this case the relation holds only for relatively small changes in distances.

Kepler's Third Law

In the late 1500s the astronomer, **Tycho Brahe,**[5] compiled extremely precise data for the orbits of the then known six planets about the Sun—Mercury, Venus, Earth, Mars, Saturn, and Jupiter. After Brahe's death, his data became available to his assistant, **Johannes Kepler**, so that he was able to analyze the data with extreme diligence, obviously without the help of a computer or a pocket calculator. He discovered that the data can be summarized in terms of three mathematical statements. These statements are referred to as **Kepler's Three Laws**. However, it is important to keep in mind that these statements were not derived on the basis of any then-known theory. It was Isaac Newton's Theory of Gravity that allowed Newton to give Kepler's Laws a theoretical basis.

We will consider only **Kepler's Third Law**. It deals with the relationship of the **planetary period** P of a planet's orbit around the Sun and its distance R from the Sun.[6] As an example, the earth's period is one-year, while its distance from the Sun is 150-million=1.5×10^8-km.

Kepler's Third Law can be expressed as a proportionality:

$$P \propto R^{3/2} \tag{A.27}$$

Alternatively, we can write $P = bR^{3/2}$, where b is a constant.

Sample Problem A.5 Consider the orbit of the planet Mars. Its distance from the Sun is 2.28×10^8-km. Find the period of its orbit.

[5]The Wikipedia article on Brahe life and work is incredibly fascinating. See https://en.wikipedia.org/wiki/Tycho_Brahe.

[6]The orbit is an ellipse, though typically very close to being circular. Here, R refers to what is called the "semi-major axis" of the ellipse. We need not be exact here.

Solution We use the same approach as in Eq. A.18. We have (with four place accuracy)

$$P_2 = P_1\left(\frac{R_2}{R_1}\right)^{3/2} = 1\text{-yr} \times \left(\frac{227.9}{149.6}\right)^{3/2} = 1.88\text{-yr} \tag{A.28}$$

A.5 Problems for Appendix on Numbers

1. Normal body temperature is said to be 98.6^0F. It is well known that in fact there is a range of "normal" body temperatures −a range that is considered quite acceptable. Given this fact, how is it that the temperature is given to three significant figures?

   To answer this question express this temperature in degrees Celsius, given that ^{0}C=(5/9)(^{0}F-32). Note how the number "98.6" gives us the incredibly incorrect sense of what is a normal range of healthy body temperature.
2. Here is another example of how a choice of units introduces uncanny numbers. The so-called **body mass index—BMI** is defined in British units (feet for length, seconds for time, and pounds for weight) as 703 × weight in lbs/(height in inches)^2.

   Here again, we have a strange number "703" occurring with three significant figures. It wouldn't seem to matter had the number been chosen to be 700 precisely. However, using kilograms for "weight" and meters for length changes the coefficient to unity (one). **Confirm** the above result.

   NOTE: One kilogram is equivalent to about 2.20-lbs and one meter equals about 39.37 inches.
3. Consider a pen that has a reservoir of ink containing 1-cm^3 of ink. Estimate **how many words you can write with the pen**, given the following information:
   - The width of a line of writing is 0.7-mm.
   - The thickness of a line of writing is 2.5-microns.
   - A letter has an average length of 1-cm.

- The average number of letters in a word is five.

  How long a line of writing, including all punctuation, can the pen produce?

4. The radius of the planet Jupiter is 7.786×10^8-km. Find the period of its orbit.

Symbols

B

- Å Ångstrom
- α attenuation per distance in dB per km
- γ gamma (used to characterize ultra high frequency electromagnetic radiation)
- λ wavelength
- μ linear mass density
- ϕ loudness in phons
- ρ mass density
- σ optical activity (angle of rotation of axis of polarization per distance through medium)
- τ time of measurement of measurement
- θ angle
- θ_c critical angle for the absence of refraction
- θ_{inc} angle of incidence of a ray of light
- θ_{rfl} angle of reflection of a ray of light
- θ_{rfr} angle of refraction of a ray of light
- a acceleration
- a_0 Bohr radius (~ 0.53Å)
- A area *or* amplitude of a wave or of oscillation
- b blue color coordinate
- $b(\lambda)$ blue color coordinate for monochromatic light of wavelength λ
- B bulk modulus *or* magnetic field *or* blue tristimulus value of a spectral intensity
- B_{ind} induced magnetic field
- c speed of light in vacuum
- C musical interval in cents
- d distance between two sources of a wave
- dB decibel
- d_{ie} image distance for an eye (distance from the center of the effective lens of the eye and the retina)
- d_{im} diameter of an image as a result of diffraction
- d_i image distance for a lens
- d_{min} minimum diameter of an image as a result of diffraction
- d_{np} near point of vision
- d_o object distance for a lens
- E energy *or* electric field
- EM electromagnetic
- E_f final energy of a quantum system
- E_{ind} induced electric field
- E_i initial energy of a quantum system
- E_{ph} energy of a photon
- f frequency
- f focal length of a lens
- f_B beat frequency
- f_n frequency of the nth mode
- F force
- g green color coordinate
- $g(\lambda)$ green color coordinate for monochromatic light of wavelength λ
- G green tristimulus value of a spectral intensity
- h Planck's constant ($\sim 4.15 \times 10^{-15}$eV/Hz)
- h_i height of image
- h_o height of object
- electric current
- I intensity
- I_0 lowest audible sound intensity
- $\mathbf{I}(f)$ spectral intensity with respect to the frequency
- I(λ) spectral intensity with respect to the wavelength
- I_B intensity of blue primary
- I_G intensity of green primary
- I_R intensity of red primary
- k spring constant

L. Gunther, *The Physics of Music and Color*,
https://doi.org/10.1007/978-3-030-19219-8

- KE kinetic energy
- L horizontal distance from source(s) of a wave
- ℓ a length variable, not specific
- m mass
- M magnification of an object by a lens
- $\mathcal{M}$ magnifying power of a lens
- max d_o maximum object distance for the eye
- max f maximum focal length of the eye
- min f minimum focal length of the eye
- n index of refraction
- N_B signal to the brain from the blue cones
- N_G signal to the brain from the green cones
- N_R signal to the brain from the red cones
- p pressure
- p_s sound pressure
- p_0 lowest audible sound pressure
- P power
- PE potential energy
- q electric charge
- r radius *or* red color coordinate corresponding to a spectral intensity *or* distance from a point source
- $r(\lambda)$ red color coordinate for monochromatic light of wavelength λ
- R reflectance **or** red tristimulus value of a spectral intensity
- RT reverberation time
- s loudness in sones
- S sum of tristimulus values ($S = R + G + B$)
- SL sound level in dB's
- t time *or* time interval
- T period of a sine wave or sinusoidally oscillating object
- T for tension in a string
- T(λ) transmittance of a color filter as a function of wavelength
- v speed or velocity
- v wave velocity
- V volume
- x displacement **or** CIE color coödinate
- X CIE tristimulus value
- y displacement of an SHO **or** CIE color coödinate
- Y CIE tristimulus value
- z CIE color color coödinate
- Z impedance **or** CIE tristimulus value

Powers of Ten—Prefixes

C

- 10^3—one-thousand **kilo** as in **kilogram** (kg) or kilometer (km) or kilohertz kHz)
- 10^6—one-million **mega** as in **megahertz** (MHz) [frequency for WGBH FM radio waves is 89.7 MHz]
- 10^9—one billion **giga** as in **gigahertz** (GHz) [the frequency of microwaves in microwave ovens is 2.5 GHz]
- 10^{-2}—one-hundredth **centi** as in **centimeter** (cm)
- 10^{-3}—one-thousandth **milli** as in **millimeter** (mm)
- 10^{-6}—one-millionth **micro** as in **micrometer** (μm); 1 μm $\equiv$ 1**micron** [monochromatic red light has a wavelength of $\sim$ 0.7 μm]
- 10^{-9}—one-billionth **nano** as in **nanometer** (nm) = 10 **Ångstroms** = 10 $\AA$ (the size of an atom is typically a few tenths of a nanometer)
- 10^{-12}—one-trillionth **pico** as in **picogram** (pgm) or **picosecond** (psec) [a bacterium has a mass of about 250 pgm; there exist chemicals such that merely 1 pgm can be fatal(!); fiber optics signals can be made as short as a psec in duration]

Note that a cube, one-cm on a side, has a volume of 1 cm^3=(10 mm)^3=1000 mm^3.

Problem Suppose that an elemental device storing each BIT of a computer hard drive has a volume of 1 nm^3, with the bits stored in a compact way. Suppose, too, that the total volume occupied by the bits is 1 cm^3. How many bits are stored by this hard drive?

L. Gunther, *The Physics of Music and Color*,
https://doi.org/10.1007/978-3-030-19219-8

D Conversion of Units and Special Constants

Constants

$\pi = 3.14159\ldots$
$e = 2.7183\ldots$
$c = 2.998 \times 10^8$ m/s ... speed of light in vacuum
$h = 4.14 \times 10^{-15}$ eV per Hz = 6.63×10^{-34} J per Hz **Planck's constant**

Length

1 Å ngstrom (Å) = 10^{-8} centimeter (cm)
1 micron (μ) = 10^{-6} meter (m)=10^{-4} cm
1 cm = 0.39370 inch (in)
1 in = 2.540 cm
1 foot (ft) =30.480 cm
1 mile (mi)=5280 ft=1.61 kilometer (km)

Time

1 day (d)=86,400 s
1 year (yr)=3.15×10^7 s

Speed

1 mph=0.448 m/s
1 m/s=2.23 mph

Area

1 sq-in = 6.4516 sq-cm

1 sq-ft = 929.03 sq-cm

Volume

1 liter (lit)= 1000 cu-cm
1 gallon = 3.785 lit

Angle

1 radian (rad) = 57.3 degrees (deg)
1 deg=60 minutes
1 minute (min)=60 seconds (sec)

Force

1 Newton (N) = 0.224 pound (lb)

Weight equivalents (*symbol* $\doteq$) on the Earth's surface

1 lb$\doteq$454 gram (gm)
1 kilogram (kg)$\doteq$2.2 lb
1 ounce (oz)$\doteq$28.350 gram (gm)

Pressure

1 atmosphere (atm)= 1.0×10^5 Pascals (Pa) = 14.7 lb/sq-in

Energy

1 joule (J) = 10,000,000 ergs
1 electron-volt (eV)=1.6×10^{-19} J
1 calorie (cal)= 4.19 joule
1 Calorie=1 kilocalorie (kcal)=1000 cal=1 food calorie
1 foot-pound ft-lb)= 1.3549 J
1 British Thermal Unit (Btu) = 252.00 cal = 778 ft-lb

Power

1 horsepower (hp) = 746 watts (W)

L. Gunther, *The Physics of Music and Color*,
https://doi.org/10.1007/978-3-030-19219-8

References for The Physics of Music and Color

- Backus, J., *The Acoustical Foundations of Music*, [W.W. Norton, Inc., N.Y., 1977]
- Benade, A. H., *Foundations of Musical Acoustics*, [Oxford University Press, New York, 1976]
- Berg, R. and Stork, D., *The Physics of Sound*, [Prentiss-Hall, Englewood Cliffs, N.J., 1982]
- Center for Computer Research in Music and Acoustics (CCMRA), [https://en.wikipedia.org/wiki/String_(music)]
- Committee on Colorimetry, *The Science of Color* [Optical Society of America]
- Cornsweet, Tom N., *Visual Perception* , [Academic Press, N.Y., 1970]
- Deutsch, Diana, *Phantom Worlds and other Curiosities*, CD with audio illusions, [Philomel Records, Inc, La Jolla, CA, 1995]
- Falk, D., Brill, D. and Stork, D., *Seeing the Light*, [Harper & Row, Publishers, N.Y. 1986]
- Feynman, Richard, *Surely You're Joking Mr. Feynman*, [W.W. Norton, Inc., N.Y., 1985]
- French, A. P., *Vibrations and Waves*, [M.I.T. Press, Cambridge, MA, 1971]
- Gilbert, P. U.P. A. and Haeberli, W., *Physics and the Arts*, [Academic Press, N.Y., 2008]
- Gregory, R. L., *Eye and Brain* , 5th ed. [Princeton University Press, 1997]
- Hall, D., *Musical Acoustics*, [Brooks/Cole Publishing, Wadsworth, Inc. Belmont, CA, 1991]
- Hecht, E., Optics, 5th ed. [Pearson Education Ltd, Harlow, England, 2017]
- Heller, Eric, *Why You Hear What You Hear*, [Princeton University Press, 2013]
- Helmholtz, H., *On the Sensations of Tone*, [Dover Publications, Mineola, NY, 1954]
- Houtsma, AJ; Rossing, TD and Wagenaars, WM: *Auditory Demonstrations*, (available on Amazon, August, 2018), [Inst. for Perception Research, Eindhoven, The Netherlands, 1989]
- Isacoff, Stuart, *Temperament*, [Random House, Inc., N.Y., 2003]
- Judd, D. B. and Wyszecki, G., *Color in Business, Science and Industry*, [Wiley Series in Pure and Applied Optics (3rd ed.). New York , 1975]
- Hsien-Che Lee, *Introduction to Color Imaging Science*, [Cambridge University Press, 2011]
- Le Grand, Y., *Light, Color, and Vision*, [Dover, N.Y., 1957]
- Levitin, Daniel, *This Is Your Brain On Music,* [Penguin Group, N.Y., 2006]
- Loy, Gareth, *Musimathics*, Volumes 1 and 2, [M.I.T. Press, Cambridge, MA, 2006]
- Moles, A., *Information Theory and Esthetic Perception*, [University of Illinois Press, Urbana, IL, 1969]
- Moravcsik, M., *Musical Sound of Tones and Tunes* [Paragon House Publishers, N.Y., 1987]
- Overheim, R. and Wagner, D., *Light and Color*, [John Wiley & Sons, N.Y., 1982]
- Pierce, John R., *The Science of Musical Sound*, [Scientific American Library, N.Y., 1983]
- Pirenne, M. H., *Vision and the Eye*, [Associated Book Publishers, London, 1987]
- Poynton, Charles, *Video and HDTV*, [Morgan Kaufmann Publishers and Elsevier Science, San Francisco, 2003].

L. Gunther, *The Physics of Music and Color*,
https://doi.org/10.1007/978-3-030-19219-8

- Raichel, Daniel, T*he Science and Applications of Acoustics*, [Springer, N.Y., May 11, 2000]
- Rigden, J., *Physics and The Sound of Music*, 2nd ed., [John Wiley & Sons, New York, 1985]
- Roederer, J., *Int. to Physics and Psychophysics of Music 4th ed.*, [Springer-Verlag, New York, 2008].
- Rossing, T.D., *The Science of Sound* [Addison-Wesley, Reading, MA, 1982]
- Rossing, T.D., *Resource Letter:MA-2, Musical Acoustics*, American Journal of Physics, Volume 55, p. 589 (1987)
- Rossing, T.D, and Chiaverina, C. J, *Light Science*, [Springer, N.Y., 1999]
- Rossing, T.D. et al, editors, *Springer Handbook of Acoustics*, [Springer, N.Y., June 21, 2007]
- Rossing, T.D., Editor, *The Science of Stringed Instruments*, [Springer, N.Y., 2010]
- Schwartz, Steven, *Visual Perception: A Clinical Orientation*, [McGraw-Hill Medical; 4 ed., November 20, 2009]
- Scott, D., *The Physics of Vibrations and Waves*,[Merrill Publishing Company, Columbus, OH, 1986]
- Seashore, C.E., *Psychology of Music*, [Dover Publications; 1st Edition, 1967]
- Sethares, William, *Tuning, Timbre, Spectrum, Scale, 2nd ed.*, [Springer-Verlag, London, 2005]
- Shlain, Leonard, *Art and Science*, [Harper Perennial, NY, 2007]
- Stokes, M, et al, *A Standard Color Space for the Internet—sRGB* [https://www.w3.org/Graphics/Color/sRGB]
- Strutt, J.W., Lord Rayleigh, *Theory of Sound*, 2nd edition, MacMillan and Co. London, 1894]
- Taylor, C.A., *Exploring Music: the Science and Technology*, [Institute of Physics Publishing, Bristol, England, 1994]
- Tobias, J.V., *Foundations of Modern Auditory Theory*, [Academic Press, N.Y., 1970]
- Tongren, Mark C., *Overtone Singing*, Revised Second Edition, [Fusica, Amsterdam, 2004]
- von Bekesy, G., *Experiments in Hearing*, [McGraw-Hill, N.Y., 1960] QP461.V64
- Walker, Jearl, *The Flying Circus of Physics*, [John Wiley & Sons, N.Y., 2006]
- Waldman, G., *Introduction to Light*, [Dover Publications, Revised edition, 2002]
- White, H. and White, D., *Physics and Music*, [Saunders College/Holt, Rhinehart and Winston]
- Williamson, S. and Cummins, H., *Light and Color*, [John Wiley & Sons, N.Y., 1983]
- Wright, W.D., *The Perception of Light and Colour. Chapter 7 in: Dawson W.W., Enoch J.M. (eds) Foundations of Sensory Science.*, [Springer, Berlin, Heidelberg, 1984]
- Wyszecki, G.W. and Stiles, W.S., *Color Science: Concepts and Methods, Quantitative Data and Formulas* [John Wiley & Sons, N.Y., 1967]
- Yost, William, *Fundamentals of Hearing*, 5th ed. [Emerald Group Publishing, Ltd., 2006]

A Crude Derivation of the Frequency of a Simple Harmonic Oscillator

F

We start with the three Laws of Dynamics that Isaac Newton (1642–1727) proposed in order to account for the observed motion of the planets about the sun, the motion of the moon about earth, and the motion of projectiles (like bullets or baseballs) just above the earth's surface. To these three laws he had to add his law for the gravitational force.

Newton's First Law If an object experiences no net force, its velocity will remain constant, be it zero or otherwise.

Newton's Second Law If an object does experience a net force, its velocity will change, being reflected by a rate of change of velocity with respect to time—the **acceleration**—that is given by

$$a = \frac{F}{m} \quad \text{or} \quad F = ma \tag{F.1}$$

Note that acceleration is to velocity as velocity is to position:

$$\begin{aligned} \text{velocity} &= \frac{\text{change in position}}{\text{time interval}} \\ \text{acceleration} &= \frac{\text{change in velocity}}{\text{time interval}} \end{aligned} \tag{F.2}$$

Newton's Third Law When an object exerts a force on a second object, the second object automatically must be exerting a force on the first object with a force of equal magnitude but opposite in direction.

Thus, if I am pushing on a wall with a force of 450 N ($\sim$ 100lbs), the wall is pushing back on me with a force of 450 N. Likewise, if the mass of an SHO is pulling on the spring with a force F downwards, the spring is exerting a force F on the object upwards.

We can now combine Hooke's Law with Newton's 2nd and 3rd Laws as follows: Because the force on the massive object of an SHO is opposite to the direction of the displacement, we insert a minus sign and write Hooke's Law as

$$F = -ky \qquad \text{Hooke's Law} \tag{F.3}$$

In this equation, F is the force on the object. It is called the **restoring force** of the SHO because it tends to bring the object back towards the equilibrium position. Since $F = ma$,

$$ma = -ky \quad \text{or} \quad a = -\frac{k}{m}y \tag{F.4}$$

This equation can be analyzed mathematically. The analysis reveals that once the position and velocity is given at some time, the initial time, the motion is determined forever after. This characteristic of Newtonian dynamics is referred to as **determinism**. In particular, the equation can

L. Gunther, *The Physics of Music and Color*,
https://doi.org/10.1007/978-3-030-19219-8

be shown to lead to the sinusoidal behavior of the SHO.[1]

Now we return to our expression for the period T. During the first quarter cycle, the total displacement is A, while the speed changes from zero to a maximum value v_m. The average speed over $1/4$ of a cycle is given by

$$\langle v\rangle = \frac{\text{displacement}}{\text{time interval}} = \frac{A}{(T/4)} = \frac{4A}{T} \tag{F.5}$$

Since the maximum speed is smaller that the average speed, we will use the estimate:

$$v_m \sim 2\langle v\rangle \tag{F.6}$$

Thus,

$$v_m = \frac{8A}{T} \tag{F.7}$$

Next, the average acceleration over $1/4$ of a cycle is given by

$$\langle a\rangle = \frac{\text{change of velocity}}{\text{time interval}} = \frac{v_m}{(T/4)} = \frac{4v_m}{T} \tag{F.8}$$

The maximum acceleration can be obtained from Eq. (F.4) by setting $y = A$.

We ignore the minus sign since we are interested only in magnitudes. We then obtain

$$a_m = \frac{k}{m}A \tag{F.9}$$

We estimate that $a_m \sim 2\langle a\rangle$. Thus we obtain

$$\frac{k}{m}A = a_m \sim 2\langle a\rangle = 8\frac{v_m}{T} = \frac{64A}{T^2} \tag{F.10}$$

We finally arrive at the approximate relation

$$T^2 \sim 64\frac{m}{k} \tag{F.11}$$

so that

$$T^2 \sim 64\frac{m}{k} T \sim 8\sqrt{\frac{m}{k}} \tag{F.12}$$

This result compares very favorably with the exact relation $T = 2\pi\sqrt{m/k}$.

The two expressions differ only by the numerical prefactor, 2π vs. 8, respectively. Their ratio is $2\pi/8 \sim 0.8$. Most important is the agreement between the two expressions with respect to the mass m and the spring constant k.

[1]The reader might be interested in carrying out the exercise in the appendix to this chapter, entitled "Numerical Integration of the Equation of Motion of an SHO." In simple terms, the initial displacement y determines the initial acceleration a through Eq. (F.4). The initial acceleration determines the change in the velocity v from its initial value and hence its value soon after. The initial velocity determines the change in the displacement and hence the displacement soon after. This cycle is repeated on and on to yield the displacement, velocity, and acceleration for all future times.

Numerical Integration of Newton's Equation for a SHO

G

This mathematical procedure for analyzing Newton's equation of motion (the Second Law) for an SHO shows us dramatically one of the most important characteristics of classical physics, namely, that nature is such that if we are given all the information about a system at some instant of time, the behavior of the system in the future is fully determined. In the case of the SHO, the initial position x and velocity v of the mass of the SHO determine the future behavior of x and of v. We call this property **determinism**.

We seek to show how the equation

$$a = -\frac{k}{m}y \tag{G.1}$$

generates a sinusoidal function of time. For simplicity, we'll set k/m = 0.1/s^2, so that

$$a = -0.1y \tag{G.2}$$

where y is in meters and a is in m/s per second. This value of k/m can be shown to correspond to a period of oscillation equal to $2\pi\sqrt{10} \cong 19.9$ s.

We will drop the units in what follows. We will assume that initially, when the time $t = 0$,

$$y_0 = 0 \text{ and } v_0 = 1.00 \qquad \text{e.g., } v = 1 \text{ m/s} \tag{G.3}$$

We have put a subscript 0 next to the letters, so as to refer to $t = 0$.

Now suppose we want to know y and v after one second. Recall that

$$\text{velocity} = \frac{\text{change of displacement}}{\text{time interval}} \tag{G.4}$$

This will be strictly so only if the velocity is constant. Otherwise, the expression gives us the average velocity over the one second time interval.

Because one second is small compared to the period (though not very much smaller) the velocity doesn't change very much in one second. Then the above equation becomes a reasonable first approximation, albeit a crude one.

We then have (with a subscript 1 referring to a time $t = l$s):

$$v_0 \cong \frac{y(\text{at } 1\ s) - y_0}{1\ s} = y_1 - y_0 \tag{G.5}$$

or

$$y_1 \cong y_0 + v_0 = 0 + 1 = 1 \tag{G.6}$$

Next, we seek the velocity v_1 after one second. Recall that

$$\text{acceleration} = \frac{\text{change in velocity}}{\text{change in time}} \tag{G.7}$$

This will be strictly so if the acceleration is a constant. Otherwise, the expression gives us the *average* acceleration over the one second interval. However, if the acceleration doesn't change much during that interval, we can use Eq. (G.7) as an approximation.

L. Gunther, *The Physics of Music and Color*,
https://doi.org/10.1007/978-3-030-19219-8

The initial acceleration can be expressed approximately as

$$a_0 \cong \frac{v_1 - v_0}{1\text{ s}} = v_1 - v_0 \tag{G.8}$$

so that

$$v_1 \cong v_0 + a_0 \tag{G.9}$$

But

$$a_0 = -\frac{y_0}{10} = 0 \tag{G.10}$$

so that

$$y_2 \cong y_1 + v_1 \tag{G.11}$$

Similarly, after 2-s

$$v_1 = v_0 - \frac{y_0}{10} = 1 - 0 = 1 \tag{G.12}$$

and

$$v_2 = v_1 + a_1 = v_1 - \frac{y_1}{10} \tag{G.13}$$

After 3-s:

$$y_3 \cong y_2 + v_2 \tag{G.14}$$

and

$$v_3 \cong v_2 - \frac{y_2}{10}, \tag{G.15}$$

etc.

You see how knowledge of y and v at any time allows you to find y and v one second later. This process is known as **numerical integration**.

For homework:

a. Make a table, listing y and v after 1 s, 2 s, etc., at least up to the time when you obtain $1 - 1/2$ oscillations. Round off all numbers to the nearest hundredth. Prepare your table as follows:

Time (s)	y (m/s)	v (m/s^2)
0	0	1.00
1	1.00	1.00
2	2.00	0.90
3	2.90	?
4	?	?

WARNING: Any error you make is *propagated* on to the numbers which follow. So, be careful.

b. Make a graph of your results for both y vs. t and v vs. t. Compare your results with the plots below (Figs. G.1 and G.2).

Below is a plot of the displacement vs. the time n.

Next we plot the velocity vs. n.

c. Note that because of the approximations made, the numerical integration is unstable. That is, the displacement y oscillates with

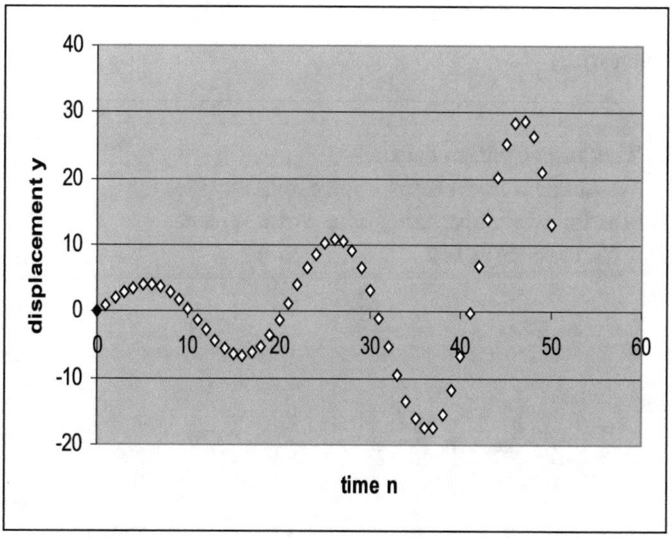

Fig. G.1 Resulting plot of the displacement: a crude numerical integration

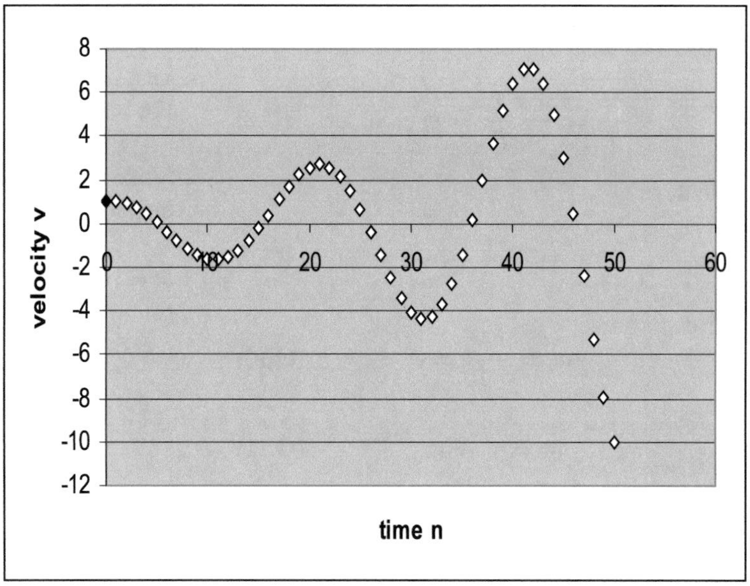

Fig. G.2 Resulting plot of the velocity: a crude numerical integration

ever increasing amplitude. The displacement ultimately diverges to infinity.

An Improved Approximation

Let n be the time in seconds. That is, $n = 1$ refers to one second, $n = 2$ refers to two seconds, etc. Our previous approximation can be expressed as

$$y_n \cong y_{n-1} + v_{n-1} \tag{G.16}$$

and

$$v_n \cong v_{n-1} + a_{n-1} \tag{G.17}$$

along with

$$a \cong -\frac{y_n}{10} \tag{G.18}$$

This last equation is exact.

We obtain a *much* better approximation if we use the following approximation, which replaces Eqs. (G.16) and (G.17):

$$y_n \cong y_{n-1} + \frac{v_{n-1} + v_n}{2} \tag{G.19}$$

and

$$v_n \cong v_{n-1} + \frac{a_{n-1} + a_n}{2} \tag{G.20}$$

Question: What do you suppose is the basis for these equations?

These equations, along with $a_n = -y_n/10$, may be solved for y_n and v_n. They lead to:

$$y_n \cong 0.95y_{n-1} + 0.98v_{n-1} \tag{G.21}$$

and

$$v_n \cong 0.95v_{n-1} - 0.096y_{n-1} \tag{G.22}$$

Tabulate and graph these equations. Use the initial conditions

$$y_0 = 0 \qquad \text{and} \qquad v_0 = 1.00.$$

Just to show you how the above equations work. We have:

$$\begin{aligned} y_1 &= 0.95y_0 + 0.98v_0 \\ &= 0 + 0.98 = 0.98 \end{aligned} \tag{G.23}$$

$$\begin{aligned} v_1 &= 0.95v_0 - 0.096y_0 \\ &= 0.95 - 0 = 0.95 \end{aligned} \tag{G.24}$$

$$\begin{aligned} y_2 &= 0.95y_1 + 0.98v_1 = 0.95(0.98) \\ &\quad +0.98(0.95) = \ldots \end{aligned} \tag{G.25}$$

$$v_2 = 0.95v_1 - 0.096y_1 = \ldots \tag{G.26}$$

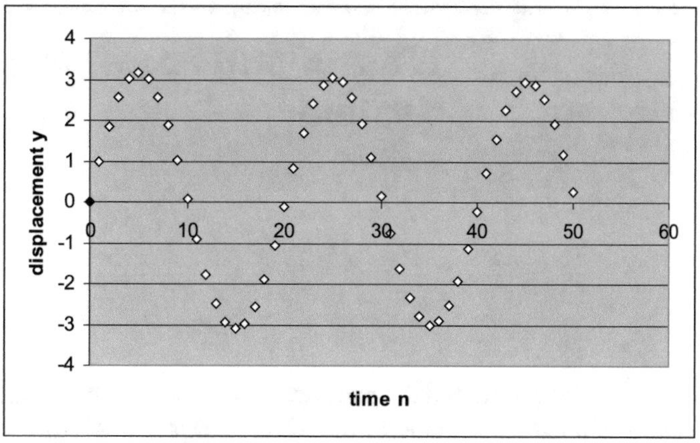

Fig. G.3 Resulting plot of the displacement: an improved numerical integration

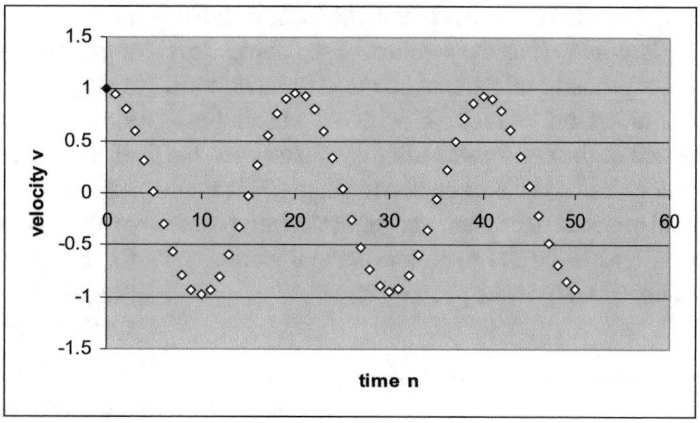

Fig. G.4 Resulting plot of the velocity: an improved numerical integration

Compare your results with the plots shown in (Figs. G.3 and G.4), which were obtained using the above equations.

Then **repeat** the above with the initial values: $y_0 = 1.00$ and $v_0 = 0.0$ (Figs. G.3 and G.4).

Problems with Numerical Integration and Oscillators of the velocity vs. n.

1. In this problem you will check the independence of the frequency of an SHO with amplitude. To do so, use the initial conditions $y_0 = 0$ and $v_0 = 4.00$. Use the refined set of equations to produce a plot of the displacement vs. time and compare.
2. In this problem you will study an oscillator whose restoring force is not Hooke's Law. Instead, let the force be proportional to the cube of the displacement. Thus, the acceleration is now given by

$$a = -0.1y^3 \tag{G.27}$$

Find the displacement vs time for the two initial conditions used for the SHO.

For both Hooke's Law (linear restoring force) and the cubic restoring force, the force increases with increasing displacement. However, the cubic restoring force increases faster with increasing displacement. This leads to a suggestion: Before obtaining your graphs, try to predict whether the frequency for a "cubic oscillator" should increase or decrease with increasing amplitude.

Magnifying Power of an Optical System

H

In Chap. 9 we introduced the parameter **magnification** of a lens as the ratio of the optical image size of a lens to the object size. However, the magnification depends on both the focal length of the lens as well as the position of the object. It is not a property of the lens alone. The maximum magnification is realized with an object positioned at the **near point** of an eye. However, the near point varies from person to person. It is standard practice to define the **magnifying power** of a lens to be the maximum magnification, as achieved with a near point of 25-cm.

If we have a compound lens system, such as a telescope or a microscope, the magnification would be the ratio of the final optical image size to the input object size. On the other hand, the ultimate purpose of an optical instrument such as a magnifying glass (a single converging lens), a telescope, or a microscope is to increase the size of the image of the object on the retina over the size of the image on the retina in the absence of the instrument. The maximum possible ratio for an optical instrument is referred to as the **magnifying power**, to which we will assign the symbol $\mathcal{M}$:

$$\mathcal{M} = \frac{\text{image on the retina with the instrument}}{\text{image on the retina in the absence of the instrument}} \tag{H.1}$$

By increasing the image size on the retina, we have two effects:

(1) The object appears to be larger.
(2) The details of the object are clearer: there is increased **resolution**.

An instrument that provides magnification produces an "effective object" for the eye to view. This lens object is called the **ultimate eye object**. The greatest magnification is achieved

L. Gunther, *The Physics of Music and Color*,
https://doi.org/10.1007/978-3-030-19219-8

when the ultimate eye object is at the **near point of the eye**, henceforth given the label **np**.[1] This is the closest that an object can be to the eye and still be in focus. For the so-called normal eye, this distance is about 25-cm, which is the value that is usually used as a standard for evaluating and characterizing optical instruments.

H.1 Image with the Naked Eye and with a Magnifying Glass

In Fig. H.1 we see the image produced on the retina R by a specific object at O of height h_o when the object is at the near point d_{np}. As the object is brought closer to the eye, the height h_{ie} of the image on the retina increases. The angle θ is referred to as the **angle subtended by the center of the lens of the object**. Clearly, this angle is a direct measure of the height of the image on the retina, since the distance d_{ie} is fixed. We also exhibit the central ray (dashed) from the object to the retina when the object is further from the eye than the near point; we see clearly that the image height h_{ie} is reduced.

The maximum height of the image produced on the **naked eye**, or **"unaided eye,"** is obtained when the object is located at the **near point**. The expression for the maximum height is simple to obtain. We have

$$\frac{h_i}{h_o} = \frac{d_{ie}}{d_o} \tag{H.2}$$

We set $d_o = d_{np}$ for the naked eye, leading to

$$\boxed{\max h_i = h_o \frac{d_{ie}}{d_{np}}} \tag{H.3}$$

If we try to increase the image on the retina by moving the object closer to the eye than the near point, we have the problem of not being able to bring the image into focus. However, by inserting a converging lens between the object and the eye, the lens can create an image at the near point with an increased height h'_o. This image serves as the object for the eye to view—that is the ultimate object for the eye-lens system. The lens is referred to as a **magnifying glass**. See Fig. H.2.

How much magnification we obtain depends upon the distance between the lens and the eye. One specific case is shown in Fig. H.3, wherein we see the magnifying glass held up against the eyeball. We also see the image of the magnifying glass, which becomes the **ultimate eye object** (in black), placed at the position of the near point. For comparison, we see the actual object repositioned in blue at the near point, where it would have to be placed in the absence of the magnifying glass in order to be seen clearly. Thus, we can appreciate the ability of the magnifying glass to "magnify" in the applied sense. Essentially, the magnifying has enlarged the ultimate eye object located at the near point from h_o to the image height of the magnifying lens h_i. This height becomes the new effective object height h'_o (shown in the figure) for the eye.

The magnifying power of the magnifying glass is given by

$$\mathcal{M} = \frac{h_o/d_o}{h_o/d_{np}} = \frac{d_{np}}{d_o} \tag{H.4}$$

Sample Problem H.1 Show that the magnifying power of a magnifying glass that is held up against the eye is given by

$$\mathcal{M} = \frac{d_{np}}{\mathrm{f}} + 1 \tag{H.5}$$

where f is the focal length of the magnifying glass.

Solution We use the thin lens equation to obtain an expression for the object distance d_o in terms of f and the near point distance d_{np}. Noting that

$$d_i = -d_{np} \tag{H.6}$$

we obtain

$$\frac{1}{d_o} = \frac{1}{\mathrm{f}} - \frac{1}{d_i} = \frac{1}{\mathrm{f}} + \frac{1}{d_{np}} \tag{H.7}$$

[1] See Chap. 13 for details.

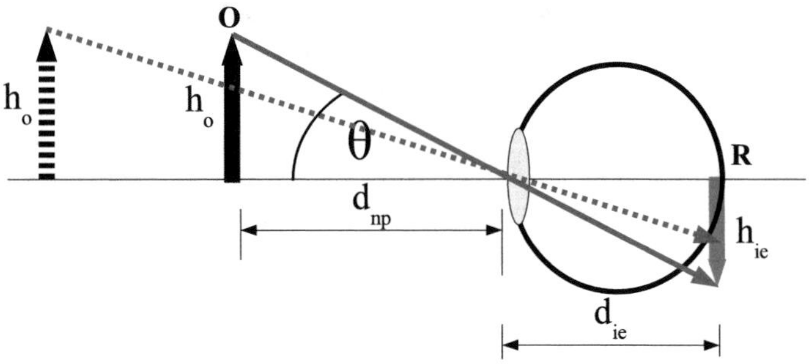

Fig. H.1 The image on the retina for various object distances with the naked eye

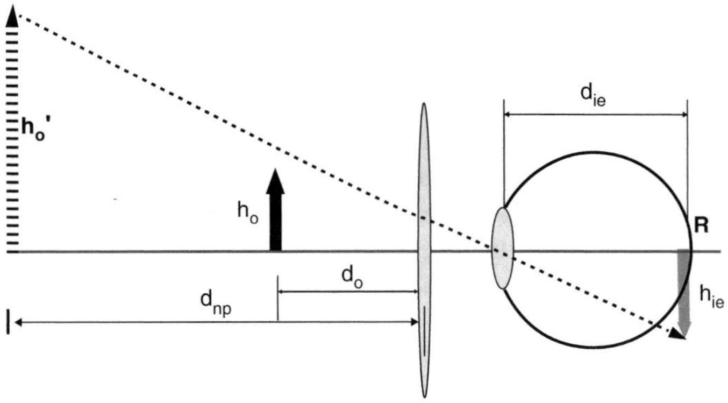

Fig. H.2 Magnification of the potential image on the retina by a lens

Substituting into Eq. (H.4) we obtain

$$\mathcal{M} = \frac{d_{np}}{d_o} = d_{np}\left(\frac{1}{\mathrm{f}} + \frac{1}{d_{np}}\right) = \frac{d_{np}}{\mathrm{f}} + 1 \quad \text{(H.8)}$$

Note The symbol × is used to identify magnifying power. Thus, a magnifying lens that has a magnifying power of 40 is characterized by the symbol 40×. It is understood that the near point is 25-cm and that the lens is held up to the eye.

Problem Use Fig. H.3 to show that the magnifying power is given by

$$\mathcal{M} = \frac{h_i}{h_o} \quad \text{(H.9)}$$

Here h_i is the image produced by the lens.

Sample Problem H.2 Assuming that d_{np}=25-cm, calculate the magnifying power of a magnifying glass that has a focal length of 25-cm.

Solution From Eq. (H.8) we have

$$\mathcal{M} = \frac{d_{np}}{\mathrm{f}} + 1 = \frac{25}{25} + 1 = 2 \quad \text{(H.10)}$$

Sample Problem H.3 Determine the focal length necessary to produce a magnifying power of 40×.

Solution We have

$$\mathcal{M} = 40 = \frac{d_{np}}{\mathrm{f}} + 1 = \frac{25}{\mathrm{f}} + 1 \quad \text{(H.11)}$$

Then

$$\frac{25}{\mathrm{f}} = 40 - 1 = 39 \quad \text{(H.12)}$$

from which we obtain

$$\mathrm{f} = \frac{25}{39} = 0.64\text{-cm} \quad \text{(H.13)}$$

H.2 The Microscope

Whereas the purpose of a telescope is to produce magnification of a typically huge object that is extremely far away, the purpose of a microscope

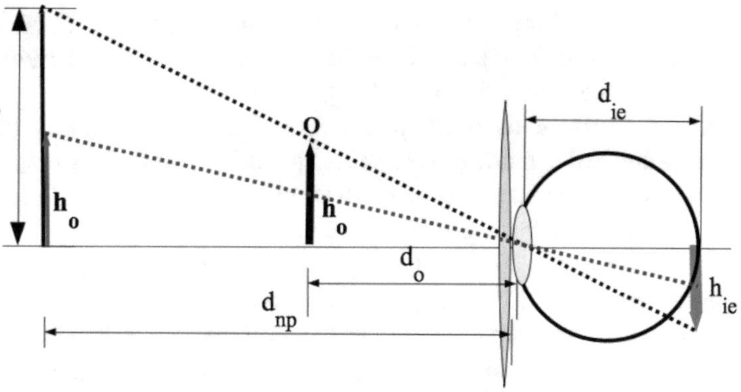

Fig. H.3 Magnifying glass up against the eyeball

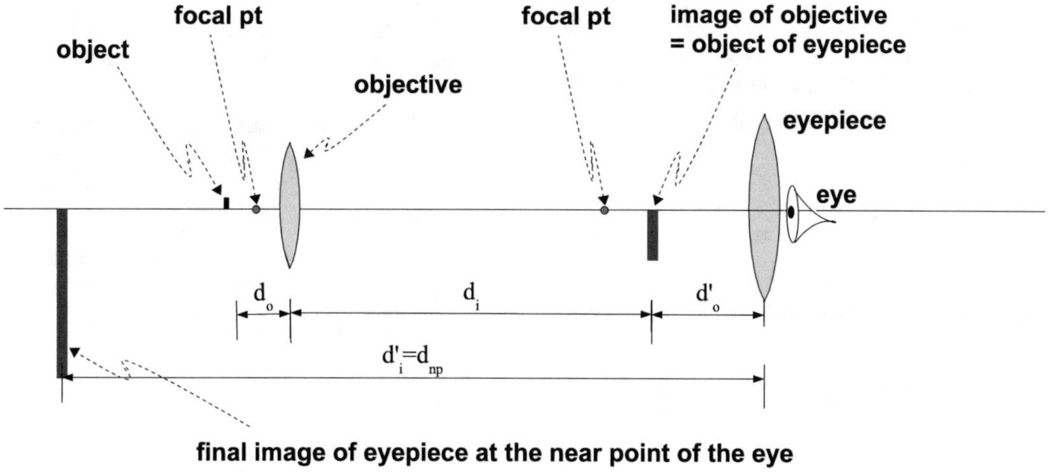

Fig. H.4 Schematic of a microscope

is to produce magnification of an extremely small object that can be brought extremely close to our eyes. For both devices there are two lens systems, with the final lens serving as a magnifying glass.

Figure H.4 is a schematic of a microscope. The first lens of a microscope is referred to as the **objective**. It has an extremely small focal length f_{ob} that allows us to bring the lens very close to the **microscopic** object. The second lens, with a focal length f_e, is the **eyepiece** or **ocular**, situated close to our eye.

We see that the objective of the microscope produces a real image that is situated just within the focal length of the **eyepiece**. We have magnification by the objective: This first image is much larger than the object. Next, the eyepiece serves as a magnifying glass of the first image, producing a virtual image that is the ultimate "object" of the eye itself. One can obtain maximum magnification by having the ultimate object located at the near point of the eye, as shown in the figure.

The overall magnification of the microscope is the **product** of the magnification of the objective and the magnifying power of the eyepiece:

$$\mathcal{M}_{\text{microscope}} = \frac{d_i}{d_0}\left(\frac{d_{np}}{\mathrm{f}_e}+1\right) \qquad \text{(H.14)}$$

Since $d_i \gg \mathrm{f}_{ob}$, according to the thin lens formula,

$$\frac{1}{d_o} = \frac{1}{\mathrm{f}_{ob}} - \frac{1}{d_i} \approx \frac{1}{\mathrm{f}_{ob}} \qquad \text{(H.15)}$$

Thus, $d_0 \approx \mathrm{f}_{ob}$ and we can rewrite the above magnification as

$$\mathcal{M}_{\text{microscope}} \approx \left(\frac{d_{np}}{\mathrm{f}_e}+1\right)\frac{d_i}{\mathrm{f}_{ob}} \qquad \text{(H.16)}$$

Unfortunately, there is a problem that we have to confront: People don't all have the same near point! Therefore, with the above design, the location of the image of the object would have to vary from individual to individual. How can we design a microscope that takes this fact into account?

The answer is that we can change the design of the microscope so that the objective of the microscope produces an image [which serves as the object for the eyepiece] that lies between the viewer and the eyepiece, but very close to the focal point of the eyepiece. In this case the final image (produced by the eyepiece) moves very far away and is therefore easily visible to any "reasonable eye". The object height is also huge. It can be shown that the magnifying power of the eyepiece is then:

$$\mathcal{M} = \left(\frac{d_{np}}{f_e} + 1 \right) \Rightarrow \frac{d_{np}}{f_e} \tag{H.17}$$

The overall magnification of the microscope will then be

$$\mathcal{M}_{\text{microscope}} \approx \frac{d_{np}}{f_e} \frac{d_i}{f_{ob}} \tag{H.18}$$

H.3 Problems on Magnifying Power

1. Assuming that d_{np}=25-cm and that the focal length of a magnifying lens is equal to 2.5-cm, calculate the corresponding magnifying power.
2. The magnifying power of a magnifying glass is typically defined in terms of a standard near point of 25-cm. However, in actuality, the near point depends upon the individual.

   Suppose that a magnifying glass is labeled as having a magnifying power of 10×. My near point has settled down to a stable value of 60-cm.

   Calculate the effective magnifying power of the magnifying glass corresponding to my eyes by first solving Eq. (H.5) for the focal length of the magnifying glass.
3. Consider Eq. (H.17). Is the change from one expression to the other significant if $\mathcal{M} = 10\times$? What about $\mathcal{M} = 40\times$?

Threshold of Hearing, Threshold of Aural Pain, General Threshold of Physical Pain

If we look carefully at the set of equal loudness curves in Chap. 11, we find two curves that bound the curves from above and below. The curve at the top is referred to as the **threshold of hearing**; the curve at the bottom is referred to as the **threshold of pain**. The former represents the minimum intensity of a pure tone that can be heard as a function of frequency. The latter corresponds to intensities that produce a sense of pain as opposed to sound. We will refer to this latter curve as the **threshold of *aural* pain** so as to distinguish it from a threshold of a more general sort of physical pain that one might experience in a body structure such as one's leg or back and normally find annoying. This curve is therefore a boundary between two types of sensation.[1] See Fig. I.1 below.

We will begin our discussion by characterizing a bit the threshold of hearing in physical terms. We have pointed out in Chap. 3 that sound corresponds to a variation of pressure in a medium. In the case of air, this pressure is produced by a huge rate of collisions of molecules of air against a surface. At a pressure of one atmosphere, there are about one trillion-trillion $(1,000,000,000,000,000,000,000,000)$ collisions each second on an eardrum, which has a surface area of about one square centimeter. The sound that we hear reflects a **difference** in the forces on the two sides of an eardrum.

In Fig. I.2 we see the results of a computer simulation of the pressure fluctuations. To the left we see the positions of an ensemble of dots representing molecules within a box. To the right is plot versus time of the pressure produced by the collisions of the molecules with the surface of the walls of the box.

In Fig. I.3 we depict the force on a sub-microscopic area of an ear drum due to these collisions over a short interval of time. Each spike represents a collision of a single molecule in the air, lasting about one-trillionth of a second.[2] In Fig. I.3a we see the force on the outside of the eardrum, while in Fig. I.3b we see the force on the inside of the eardrum. The latter force is shown downwards in a negative direction to represent the fact that the above two forces have opposite directions.

In Fig. I.3c we see the two forces together on one graph. Because each collision is so minuscule in its effect and because this rate over the entire eardrum is so huge, we experience a force that is uniform over the surface of the eardrum and is extremely steady over time. To appreciate this fact, consider a short time interval of one-hundred trillionths of a second. Imagine that there were one hundred collisions during this time interval and that they were spread uniformly over this time interval. The spikes would then touch each other, as shown in Fig. I.4. [Since collisions are actually random, many would actually overlap each other.]

[1] The closest analogous threshold for a large body part that I can think of is the boundary between a tickle and an ache.

[2] In order to get a feeling as to what the number one-trillionth is, suppose we were to cover an area the size of a football field with the dots over the letter "i" in this print. There would be about one-trillion such dots. Therefore, each dot takes up one-trillionth of the area of a football field.

L. Gunther, *The Physics of Music and Color*,
https://doi.org/10.1007/978-3-030-19219-8

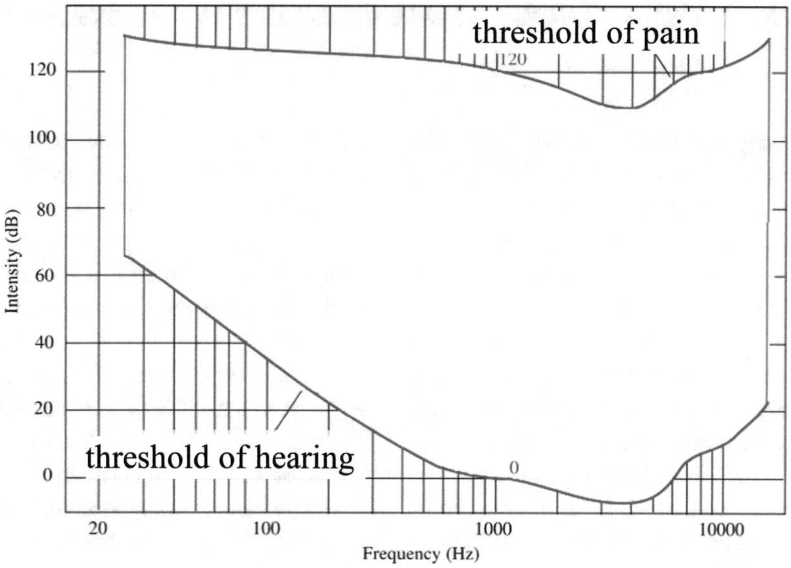

Fig. I.1 Hearing thresholds enclosing the set of equal loudness curves

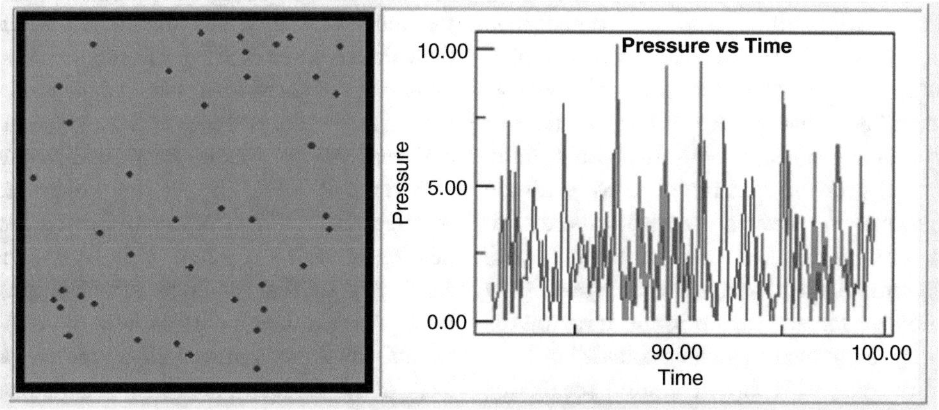

Fig. I.2 Computer simulation of the fluctuations of the pressure on a surface
(source: Wolfgang Christian and Gregor Novak, http://webphysics.davidson.edu/applets/Molecular/Pressure.html)

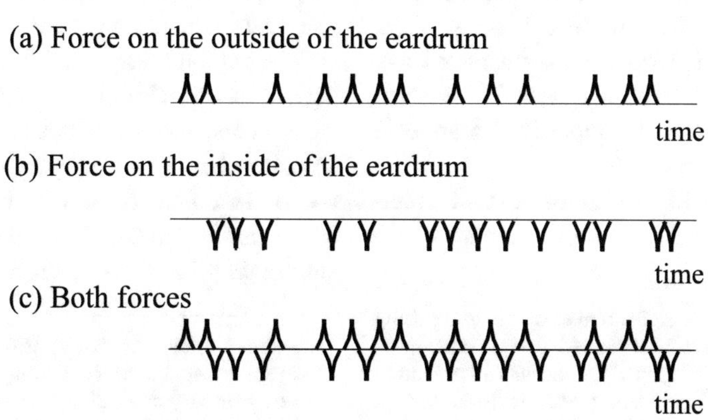

Fig. I.3 Force due to collisions of individual molecules on an eardrum

Fig. I.4 A segment of adjacent collisions

In fact, during an interval of one-hundred trillionths of a second there are one hundred trillion collisions, not merely one hundred. Imagine how dense the spikes actually are! It is then easy to appreciate why the force is extremely close to being constant in time.

Because of the discreteness of the collisions, the force produced is not exactly zero or constant in time. In Fig. I.3, representing the collisions over an extremely minute area of an eardrum, we see that in that short time interval there are thirteen collisions on the outside along with fourteen collisions on the inside, making a difference of one collision. The situation for the force on an entire eardrum is different. Over a period of about one second, we might have a total of about one trillion–trillion collisions on each side of the eardrum. The **difference** in the number of collisions will be on the order of one part in one trillion—that is the still huge number of one trillion collisions! The resultant variation in the overall pressure is referred to as **pressure fluctuations**.[3] As a result, even when there is no sound wave present, the forces on the two sides of an eardrum do not cancel each other. How does this net force compare to the force that is necessary to produce an audible sound?

In principle, with sufficient aural sensitivity, we should well wonder whether we can hear the individual collisions of molecules! In fact, the sensing apparatus for hearing is designed so that we cannot hear these collisions. Their presence is in fact mirrored in the fluctuations of nerve impulses mentioned in chapter 10. We hear sounds that produce nerve impulses that are over and above these fluctuations.

Still, we would like to get back to the question as to how the fluctuations compare to the sounds that are at the threshold of hearing. It is not too difficult to calculate the sound that is produced as a result of the variation of the force on the eardrums as a result of the discreteness of the collisions. There is a frequency spectrum to this force that ranges from zero frequency to a frequency corresponding to the duration of a single collision ($\sim$ one-trillionth of a second)—therefore, to a frequency of about one-trillion Hz. The sound is a uniform mixture of frequencies (as in the case of white light) and is often referred to as **white noise**. It is not straightforward, if at all possible to compare the resulting intensity with the intensities corresponding to the threshold of hearing among the equal loudness curves since these curves correspond to sounds having single frequencies. Ideally, one should test people for their threshold of hearing white noise. The best I can think of doing is to calculate the total intensity of the white noise over a range of audible frequencies—say in the most audible range of 1000Hz to 3000Hz. The result is a total intensity of about 10^{-12}W/m^2, equal to the threshold of hearing at 2000Hz. Thus, it is likely that the sensitivity of the ear is as small as possible, being close to being able to hear the collisions of individual molecules of air on the ear drums! The *inaudible* sound due to the randomness of the collisions is referred to as **background noise**. Our entire nervous system is wired up so that normally we don't sense the background noise that tends to excite our nerves.[4]

What is the pressure at the threshold of hearing? It is a minimum at a frequency to which we are most sensitive: about one-ten-billionth of an atmosphere at a frequency of about 3000 Hz. **Georg von Békésy**, in his study of cats' ears, found that at the threshold, the eardrum has an amplitude of about one-tenth of an Ångstrom,

[3]For comparison sake, imagine rain drops colliding with a window pane and pitter-patter sound they produce. Now imagine what would happen if the density of raindrops were to increase greatly and their rate of collision increase greatly. Ultimately, we would describe the sound produced as a steady continuous sound.

[4]The same situation holds for vision. The eye is capable of detecting the incidence of only about one-hundred photons over a period of a few seconds. Any greater sensitivity could lead to problems due to noise within the nervous system for vision.

a size that corresponds to about one tenth the diameter of a single molecule!

What about pain in general? Recently, I visited a physiatrist to deal with back pain that had developed over the past few months—another one of my many episodes. The main issues were whether the source of the pain was due to a relatively simple problem of sprained ligaments or strained muscles or a more serious problem of a herniated disc of the spine, or an even more serious disease such as cancer. The fact was that I had no recollection of having had any incident that might have done damage to my back.

My visit to his office informed me of a fascinating phenomenon of the human body. The brain can reset the level of background noise that can produce a sense of pain. My physiatrist's diagnosis to my backache was simple: I am getting old! My discs are wearing down. Old MRIs of my back were evidence that my level of deterioration might well be belated in life since they revealed at my then current age of fifty-five the back of a forty year old. What should be my response? I will paraphrase his response. "Simple," he said. "Stop worrying about your back. People are fortunate to have a unique organ in the brain called the **amygdala**. One of its functions is to control the response in your brain to nerve impulses that can generate sensations of pain. When we have a steady or continual input of such impulses, the amygdala can change the brain's response to these sources of pain by causing the nervous system to treat them as our new level of background noise! And, if we are fully fortunate, we will eventually not sense any pain. If you dwell upon the pain, you will weaken the amygdala's ability to perform this function and therefore prolong your pain. So ignore the pain and move on with your life."

MAPPINGS as a Basis for Arriving at a Mutually Agreed-Upon Description of Our Observations of the World—Establishing "Truths" and "Facts"

This book addresses the subject of physics, sound and light along with their relationship to our own experience of sound and light. We have introduced many concepts and equations that provide relationships among various physical quantities. Physics is all about relationships. And so is a piece of music or work of fine art: These are the relationships we perceive about the components of a given piece of music or of a given work of art. In addition, there are relationships of these components and their sum total that produce the full composition with our personal emotional responses to these compositions. All of these relationships are examples of what are more generally called **mappings**.

All forms of communication involve mappings. Moreover, physical laws are mappings of observations—that we share—onto mathematical equations. Satisfactory communication, as well as satisfactory laws of physics, requires agreement among those who share them. In this appendix, we will explore this subject a bit and relate mappings to the complex philosophical questions of truth and fact. According to my colleague George Smith of the Department of Philosophy at Tufts University, who is an expert on Isaac Newton, my orientation towards the nature of scientific investigation is within the framework of Newton's proposed system thereof.

I was led to consider mappings seriously because of my study of **color vision**, as this subject compels us to think with great clarity about the nature of mappings. One of the first sets of words parents teach their babies is colors. This process provides us with a wonderful example of how people learn how to share a common mapping of human experience. The parent shows the baby an object with a uniform surface of color, points, and says the word for the color of the surface—for example, *red*. The term color is technically better referred to as the **hue**, the term that we will henceforth use in this appendix. The parent then points to another object and says *green*. The baby must learn that it is the hue that the word is distinguishing and not another aspect of the object such as its shape or size. How is this aspect provided?—by using a number of objects that hopefully differ essentially in all ways except for the hue of their surfaces.

Note that, as we pointed out earlier in Chap. 15, there is no way to tell how the actual sensations compare among people. The same would be true for the baby and the parent.

There are cases when a baby will be confused: The parent shows the baby two surfaces, e.g. one red, the other green, and assigns these two different hues to the surfaces. The baby, on the other hand, seems to jump around, randomly assigning one or the other hues to both surfaces. The baby doesn't seem to differentiate. As you might guess, the baby is color blind. How does the baby handle this confusion?[1] I bring up color

[1]In color blindness, two words, red and green, are perceived to represent the same experience—perceived color. Ultimately, the child will be told that the two words represent different colors that he/she is incapable of distinguishing. I had problems of confusion of two words of a different sort in hearing Yiddish as a child. In a number of cases, two different pronunciations or even words were randomly assigned to what appeared to me

L. Gunther, *The Physics of Music and Color*,
https://doi.org/10.1007/978-3-030-19219-8

blindness here just to point out that there are situations wherein people are not always able to establish mappings that they can agree on. Imagine what the situation would be like if the prevalence of various types of colorblindness were close to 100%!

Note Suppose that an infant is fitted with a device placed over its eyes that inverts all images throughout infancy. Consider how the infant would map observations onto language:

1. Can you think of situations where there might be confusion in communication having to do with up and down?
2. How would the child draw itself as it sees itself in a mirror. Would the child draw an image that is upside down to us?

The important message that we learn from our study of color vision is that a chromaticity diagram tells us only how various color sensations are distinguishable. The diagram is a **mapping** of color sensations onto a set of pairs of numbers, the color coordinates. In technical terms it is more specifically referred to as a **one-to-one mapping** of color sensations onto the diagram; "one-to-one" means that for each color sensation there is a specific point in the diagram and for each point in the diagram there is a specific color sensation.

to represent the same idea. I was confused and blamed my difficulty on my own inability to remember or learn the correct pronunciation or to distinguish between two "different" words. For example, the word for the number "two" was pronounced as either "tsvay" (as in the English word "say") or "tsvy" (as in the English word "my"). I heard them as two different words. Ultimately, as an adult, I was told that the reason that my relatives were jumping back and forth between two pronunciations was that they naturally spoke with a "Galitsianer" accent (close to a German accent). However, Litvaks (from Lithuania), with their Litvak accent, were regarded as being more cultured. As a result, my relatives were sometimes embarrassed about their natural Galitsianer accent.

Of course there is a difference between the case of color blindness and a confusion between two dialects: While both involve a mapping of two different words onto what is conceived as representing the same experience, contrary to the latter situation, the former involves an intrinsic deficiency in perception that cannot be cured by explanation.

J.1 MAPPINGS being Central to Organizing Human Experience

Essentially all human experience is dominated by **mappings** of one kind or another. In the context of these notes, a mapping involves an association between two aspects of human experience.

Examples are:

1. written letters that spell words and their verbal counterpart as expressed words;
2. words that refer to classes of objects referred to as nouns;
3. images that we perceive in our consciousness and the scenes; that produce physical responses on the retina of an eye;
4. printed musical notes and the tones produced by a musical instrument or the human voice;
5. frequency and intensity of a pure tone and a sense of pitch;
6. spectral intensities and the corresponding sensations associated with color (hue and saturation) and brightness;
7. a sequence of positions of an object and the perception by the eye and brain as "motion" of the object;
8. the memory a person has of various perceptions of past inputs that correspond to actual physical inputs to a person's senses;
9. words that classify many objects that produce an experience that is common in some respect or respects—such as the appearance of tigers, lions, humans, or apples or love, or anger. Sometimes there is disagreement as to how objects are related to the words we ascribe to them. Severe arguments can arise, often merely as a result of people having different mappings. In these cases, ultimately what is important is how such classifications affect the way we use them—that is, how they are mapped onto other actions or attitudes. The important thing is for people to clarify as best they can the mappings they are using.[2]

[2]Recently (2008) the International Astronomical Union decided to demote Pluto to the status of being a "dwarf planet". See the article in the National Geographic

J.2 NUMBERS as a Mapping

A number of years ago, my wife, my then nine-year-old son Avi, and I were in Grenoble, France for one of my sabbaticals. Avi went to l'École Houille-Blanche, a public school whose student body was 50% French and 50% foreigners from all over the world. Avi was placed in a class with foreign children ranging in age from about 6 to 10, none of whom knew French. Few shared any particular language. How are such children to be taught and be prepared to join the rest of the student body in classrooms that used essentially only French? All I will mention here is the following: The very first subject that students were taught was mathematics—numbers being the first of this subject. Why was this so? Because it is relatively easy to teach and discuss the concept of numbers without using a particular verbal language. All one has to do is to present a number of objects such as one's fingers and assign a word to each finger: "One," "two," "three," "four," .... Or: "un," "deux," "trois," "quatre," .... We are observing the establishment of a one-to-one correspondence between an ordered set of objects (such as our fingers) and words expressed verbally or in written script. This numerical one-to-one correspondence is perhaps the simplest example of a "mapping."

Most of us would appreciate the probability that the first elements of communication between an earthling and an extraterrestrial would be the sharing of our "words" for numbers. The reason is the simplicity of this mapping and the small chance that the mapping will not be correctly communicated.

News, July, 2008. http://news.nationalgeographic.com/news/2006/08/060824-pluto-planet.html. It seems to me too ludicrous to regard astronomers of the past as having been mistaken in labeling Pluto as a planet. All we can say is that this new label allows astronomers to make statements about the now regarded "true" planets that will not be applied to Pluto.

J.3 The Concept of TIME as a Mapping

What is **time**? The first level of consideration and observation regarding **time** is the existence of an ordered sequence of observations. We refer to this observation as **time order**. This ordering is preserved in the patterns that our minds provide in what we call **memory**. Imagine what would happen if our brains destroyed the order or direction of this sequence!! The next level in establishing or characterizing our sense of the so-called "time" requires that the physicist observe a system behaving in a cyclic way: A pattern is observed to repeat itself again and again, with negligible observable change in the pattern. A sense of equality in the evolving pattern leads one to associate a time interval to a single occurrence of the pattern and to then assign a numerical value to an evolution of patterns—we number and order the patterns. The patterns are observed to be occurring simultaneously with other physical observations so that we can assign a value to the time interval of a sequence of physical observations. This special cyclic system becomes our "clock." Any time an event takes place, such as hearing a pulse of a sound or noting the position of a car on the highway, we can **correlate**, that is map, that event onto the numerical value of the number of cycles of a clock has made since we assigned an initial time. We can express the time interval between events by noting how many cycles took place between the two events. Here we have a mapping between two events and number of cycles of a clock.

Now consider that astronomers used the rotation of the earth and its revolution about the sun in order to measure time. These processes were believed to be periodic. People could thus count the days or years by making reference to the position of the sun or moon or stars in relation to the earth. Galileo is understood to have studied the motion of objects with respect to time at one point by relying on his trust of good time keeping by a musician acquaintance.[3] Later, Galileo used

[3]See Drake, S., *The Role of Music in Galileo's Experiments*, Scientific American, p. 98, June 1975. Also see

clocks that were still quite crude compared to what we would demand today. On the basis of the measurements of astronomers and scientists like Galileo, Newton was led to his three laws of mechanics and his Universal Theory of Gravitation. **Christiaan Huygens** was able to improve on the limitations of the **pendulum clock** by extending the observations of Galileo on an inclined plan through his contributions to **mathematics.**[4] Pure mathematics, along with Galileo's experiments, justified the trust he had in his clock.

Later developments in the improvement of clocks with respect to precision and accuracy depended upon the Laws of Mechanics. Their quality was based on **theory**. We now have the **Cesium clock**, which is understood to have an accuracy of one nanosecond (10^{-9} s) per day, or about one part in 100-trillion![5] The quoted accuracy is based upon an application of quantum theory.

The logic behind clocks is a bit confusing: Experiment based upon crude clocks led to theory; theory then led to more accurate clocks. Where lies the ultimate basis of evidence? Experiment or theory? Is the logic circular and therefore flawed?

It might seem as if we use theory as our ultimate judge, so that **circular reasoning** is not present. But that is not exactly so. The situation is more complex.

While Newton proposed his laws on the basis of a restricted domain of observations, his laws have been ultimately applied to a vast set of interconnected physical phenomena—for example, all the developments in engineering and medicine and in sending rockets to the moon. The Laws of Physics weave a network, an edifice, such that if any component were to fail to fit the theory, the structure would lose its reliability. It is because of the solidity of this edifice that physicists have such a high degree of faith in the laws of Physics—yes **faith** in the laws.[6] How did we end up with quantum theory? The answer is that new experiments destroyed our total trust by revealing that the edifice was flawed in a domain that takes into account the behavior of systems the size of atoms or smaller. Classical physics misses certain fine details and therefore had to be refined. Ultimately, Quantum Theory wove an intricate edifice that became the basis for a new level of trust as did the classical laws—hence the trust in the accepted accuracy of the Cesium clock. Nevertheless, physicists still use Classical Laws to account for or describe most behavior in the large, recognizing that corrections sometimes have to be made to take into account Quantum Theory.[7]

Note: I have raised the issue of time here because it is an example wherein the nature of a mapping can be complex and obscure.

J.4 Mappings as the Essential Goal of Physics

We observe the world about us. These observations are summarized by mappings within our brains. We communicate with others in words that represent these mappings, hoping to summarize these mappings in such a way that we can establish a one-to-one correspondence between our words and our observations that are shared among our fellow human beings. I will repeat a statement to be found in Chap. 5:

The essential goal of physics is to establish a theoretical framework for describing in a quantitative way what we decide to and are

the website (2-4-2011): http://www.joakimlinde.se/java/galileo/, which contains a beautiful applet that enables us to appreciate Drake's conjecture as to how Galileo might have used a musician to arrive at his law that when a ball rolls down an inclined plane, its speed increases linearly with time.

[4]A pendulum bob moves along a curved path that can be analyzed in terms of an infinite sequence of infinitesimal inclined planes having different angles of inclination.

[5](2-5-2001): http://en.wikipedia.org/wiki/Atomic_clock.

[6]Reader beware: the faith to which I am referring is not the same as the faith in religion, which has no such edifice and yet has its great benefits in helping some people handle the complexity of life's experience.

[7]I must warn the reader that the above view of Physics as being ultimately dependent upon faith is not shared by many physicists. Interestingly, this issue does not seem to arise among mathematicians because they recognize that a mathematical theory is dependent upon a set of axioms that is not provable.

able to measure. That framework makes use of models, concepts, and images. However, its ultimate content is a set of mathematical equations, which we call laws. The laws are as simple and all-encompassing as possible, and provide relationships among measurable quantities.

One of the most remarkable examples of such a mapping is the following: We observe an enormous variety of materials made out of a relatively small number of different kind of atoms (fewer than 100) arranged in a multitude of ways. We have millions of different organic compounds, metals, and alloys, complex materials like wood and so on. They have a variety of physical properties with respect to pliability, density, color, texture, and so on. And yet, it is understood by physicists that this entire variety of properties is describable, that is, can be *mapped* onto a set of a small number of mathematical equations.

For comparison sake: Note that a finite set of coupled algebraic equations are incredibly simple in content. For example, suppose that we have to solve the two equations, $x + y = 1$ and $x - y = 3$ for x and y. The solution is $x = 2$ and $y = -1$. Such equations cannot provide us with the richness of content that is associated with the behavior of materials.

Having discussed mappings, we need to answer the question of the relationship of mappings to questions of **truth** or **fact**. In my opinion, these issues are not subject to being defined by science. They are purely philosophical. In practice, I would say that we tend to use these terms in science to describe mappings for which there is essentially universal agreement under the rules that are used by scientists to test for acceptability. Any person has a *right* not to accept these rules, often to their own detriment. To add to this list of what I consider a **non-scientific issue** is the question of **reality**. "Do photons really exist?" I heard recently this question argued at a colloquium at Harvard University, during which Nobel prize winners couldn't agree![8] At best, we can say that the photon is a conceptual tool that is represented mathematically in physical laws that are mapped onto observations. Interestingly: While physicists might debate and disagree about the issues of truth, fact, and reality, these disagreements don't seem to affect their ability to conduct the discipline of physics.

A Final Remark

Let us recall the opening chapter of this book, wherein we exhibited a graph of a wave for a piece of music. It can be a joy to contemplate that this curve has all the content that maps onto our incredibly rich, sensual, and emotional experience when listening to the sound associated with the wave pattern. The graph maps onto a sound wave that ultimately produces nerve impulses that are processed in our programmed brains so as to produce our musical experience. The processing in the brain is quite complex. If **content** were to be measured by, say the number of megabytes stored on a computer, we would find that the ultimate content of our wonderful musical experience is vastly greater than the content of the input sound wave.[9]

[8] I ought to be specific: While Max Planck based his theory of Black Body radiation on the assumption that electromagnetic radiation is absorbed and/or emitted by atoms in discrete units, he didn't believe that the radiation itself was quantized. In 1905, Einstein produced a theory of the so-called **photoelectric effect**, which involves electromagnetic radiation knocking electrons out of a metal. Einstein's theory assumes that discrete units of radiation collide with the electrons. As a result it has been commonly understood that this experiment along with Einstein's theory provides proof as to the photon's existence. Someone in the Harvard audience asked whether the photoelectric effect does indeed prove the photon's existence. The vote was overwhelmingly in the negative, but not unanimously. Interestingly, no one who voted in the negative proposed an experiment that does prove its existence. Another example of existence questions has to do with atoms. Planck rejected their existence until as late as the early 1900s. Brownian motion (the motion of micron sized particles in a water suspension, due to collisions of water molecules with the particle) is given credit for **proving** their existence.

[9] The reader is invited to see the book by Abraham Moles entitled *Information Theory and Esthetic Perception*, [University of Illinois Press, Urbana, IL, 1969]. The author discusses the how one might assign a megabyte value to a piece of music. You can read a review of the book on this website: https://www.academia.edu/31546315/Information_Theory_and_Esthetic_Perception_Abraham_Moles.

The Mysterious Behavior of the Photon

K

Abstract
The goal of this Appendix is to demonstrate that the behavior of photons and all matter at the quantum level led to a decisive change in the discipline of Physics. Whereas prior to the quantum era, physicists were able to account for observations in terms of mathematical models as well as images, quantum systems revealed that we have to give up the use of images as part of a theory. The photon exhibits both wave and particle like behavior, each according to the experiment performed. We discuss in detail an experiment with photons that cannot be explained in terms of either a particle or a wave like nature. Our conclusion is that it is impossible to ascribe a nature to the photon that mirrors anything we can observe in our macroscopic world.

The goal of this Appendix is to demonstrate that the behavior of photons and all matter at the quantum level has led to a decisive change in the discipline of Physics. To understand what this change is about we need to review a bit of the history. Great physicists like Galileo and Newton and numerous others introduced a methodology that involved the following components:

1. **Measurements in the Laboratory**
   Choosing parameters that are measured by laboratory devices:
   e.g., the *position* of planets seen in a telescope; and the *position* of objects in motion; *clocks* that measure time by counting the number of repetitions of a measured parameter that are *assumed* to behave in a periodic way[1];
2. **Mathematical Relationships**
   Representing the measured parameters by mathematical variables such as x for position and t for time; then using the laboratory measurements to summarize relationships among the physical parameters;
   The combination of the above two components leads to **Laws of Physics**.
3. **Representation of Objects by Images**
   Note that Kepler did not see a planet move in an elliptical orbit about the Sun. He produced this image by examining the data he had for the position of a planet over time. Astronomers and astrophysicists accumulate data that enable astrophysicists to use theory to deduce the physical attributes of objects in outer space. For example, due to the detailed data accumulated about the Sun, we can deduce the physical state of the Sun extending from its surface down to its center. We can deduce the nature of a neutron star.

In the realm of quantum phenomena, which includes the behavior of photons, the third component, the ability to represent objects by images is denied to us. In order to demonstrate how this is so I will summarize the laboratory observations for a number of experiments involving photons.

[1] See Sect. J.3 for a discussion about the complexity of an essential device as a clock.

L. Gunther, *The Physics of Music and Color*,
https://doi.org/10.1007/978-3-030-19219-8

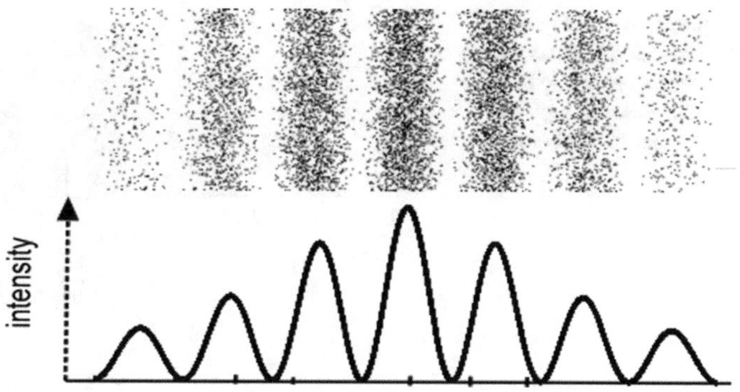

Fig. K.1 Interference with a low intensity beam of photons

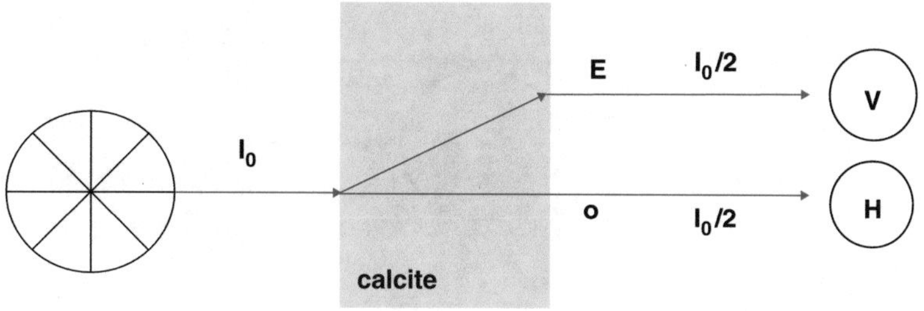

Fig. K.2 Schematic drawing of the action of a calcite crystal

The results are not deniable. The issue is what are we to conclude about the nature of the behavior of photons.

K.1 Experiments with Calcite Crystals

In Sect. 7.3.3 we described two-slit interference with such a low light intensity that one can detect individual photons. With a high intensity we see a continuous interference pattern on a screen. At very low intensity, photons strike the screen individually but in time, produce a speckled replication of the interference pattern (Fig. K.1). The behavior of photons is impossible to imagine in terms of what we observe at our macroscopic level.[2]

For this Appendix we will examine what experiments with calcite crystals reveal to us. In Sects. 8.6.2 and 8.6.3 we described the action of calcite crystals on a beam of light. In particular, we described the behavior of a beam of light that passes through a single calcite crystal as shown in Fig. K.2.

In this figure, we see the exit of the ordinary ray labeled with the letter **O** as well as the extraordinary ray, labeled with the letter **E**. I have modified the original image in Sect. 8.6.2 by replacing the angle 0^0 by a letter **H** indicating that the axis is **HORIZONTAL**. I have also replaced the angle 90^0 by the letter **V** for **VERTICAL**. Finally I have indicated the action of the calcite crystal on an incoming unpolarized beam. The incoming intensity I_0 is divided equally between the two outgoing polarized beams.

Suppose that we place two **photon detectors**, one in each of the paths of the two outgoing beams. We can then reduce the intensity to such a low level that only one detector registers a photon at a time. The result will be that the two detectors will detect photons in a random way—as in the tossing of a coin. On the average, the detectors will register an equal number of photons. In Fig. K.3 we see a beam in a so-called **DIAGONAL polarization state**, with the label **D**. The angle of polarization is 45^0. We see two detectors indicating that one or the other registers the detection of a photon.

[2] I strongly urge you to view the lecture by the great twentieth century physicist, Richard Feynman, on this website: http://www.cornell.edu/video/richard-feynman-messenger-lecture-6-probability-uncertainty-quantum-mechanical-view-nature. I cannot improve upon this presentation, unless I were to try to shorten it, albeit with a great loss.

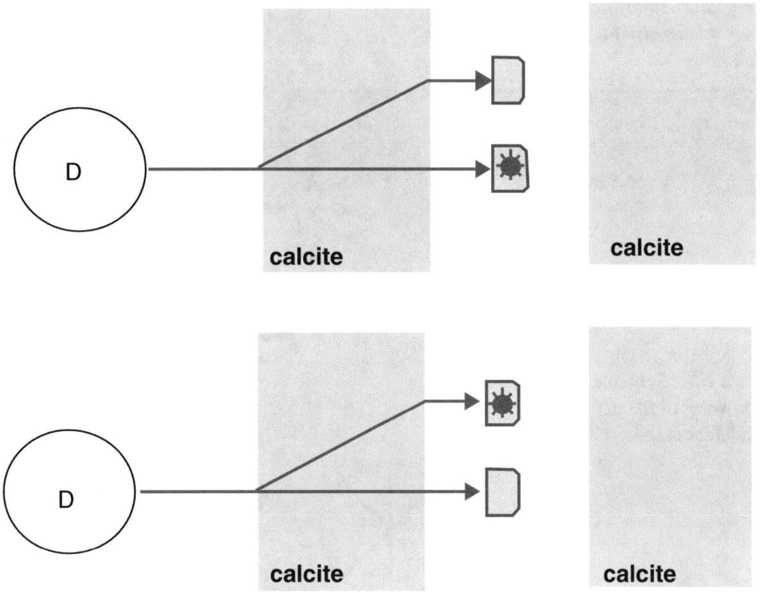

Fig. K.3 A beam in the diagonal polarization incident upon a calcite crystal detectors register the incidence of photons

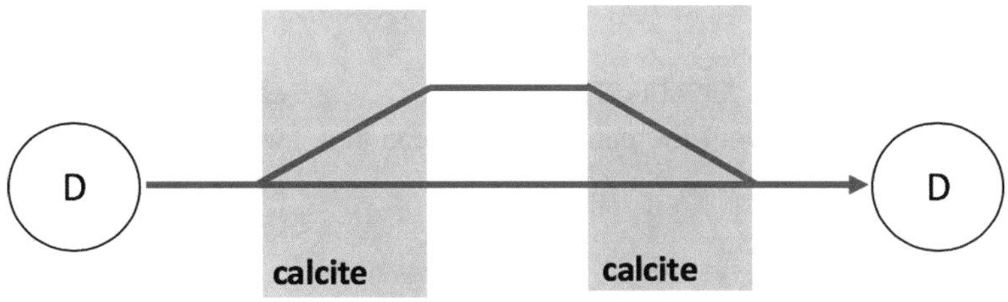

Fig. K.4 Calcite loop with a diagonally polarized incoming beam

The Calcite Loop

In Sect. 8.6.3 we described the behavior of a beam that passes through a **calcite loop**, as shown in Fig. K.4. We noted that the outgoing beam always has the same polarization state as the incoming beam. In the figure we show an incoming beam in the **DIAGONAL polarized state**. A fourth polarization state that we will refer to is the **SKEW polarized state**, designated by the letter **S**. Its angle of polarization is 135^0. The four polarization states that we will make use of in our study are shown in Fig. K.5.

Suppose that we have a diagonal polarized beam pass through a calcite, so that there is an outgoing diagonal polarized beam, as shown in Fig. K.6. We then place a SKEW polarizer in front of the outgoing beam. Since the DIAGONAL and SKEW polarizations are perpendicular to each other, there is no ultimate intensity.

At low enough intensity, the detectors will enable us to be certain that only one photon is in the loop at a time.

Let us now assume that a photon is incident upon the first calcite crystal and **behaves as a particle**, and therefore moves along a **well-defined path**, we would conclude that there are two possible scenarios, each with equal probability: First, the photon follows the path of the ordinary ray, having its polarization state changed from DIAGONAL to HORIZONTAL and emerging in the horizontal state. Or second, the photon follows the path of the extraordinary ray, having its polarization state changed from DIAGONAL to VERTICAL and emerging in the vertical state. There is no way to understand how the outgoing ensemble of photons is in the diagonal state. It is the wave theory of light that explains the state

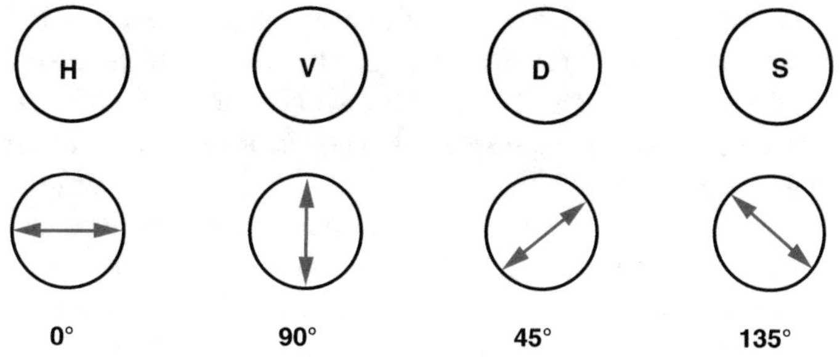

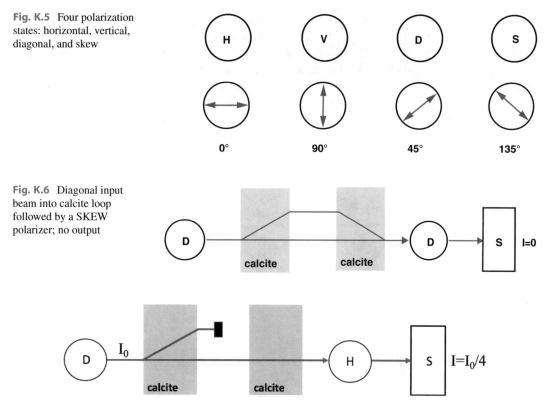

Fig. K.5 Four polarization states: horizontal, vertical, diagonal, and skew

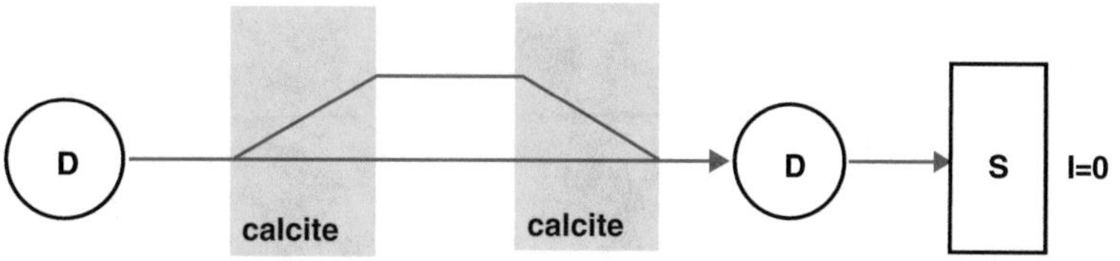

Fig. K.6 Diagonal input beam into calcite loop followed by a SKEW polarizer; no output

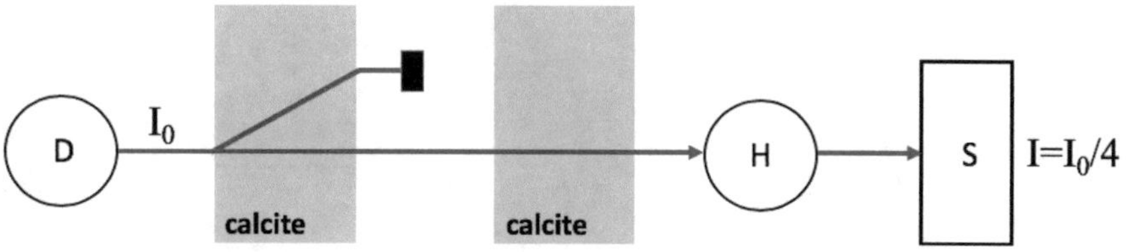

Fig. K.7 Destroying the calcite loop of Fig. K.6 by blocking the output from the vertical output of the first calcite crystal leads to an output from the SKEW polarizer

of the outgoing state. The ensemble of photons appears to reflect a mysterious incomprehensible combination of wave-like properties and particle-like properties.

A Paradox

We will assume that there is only one photon in a calcite loop at a time. Next, suppose that we block the path of any photon that exits the first calcite crystal in the extraordinary ray, as shown in Fig. K.7. In this case, all photons that reach the second calcite crystal will be in the horizontal state and pass out of the second calcite crystal in that state. The number of these photons will be half the number of photons incident upon the first calcite crystal. They will all pass through the horizontal polarizer. In turn, half of these photons, which strike the SKEW polarizer, will pass through this polarizer, since the angle between the horizontal and skew directions is 135^0.[3] The final result will be that **one-quarter** of the photons incident upon the first calcite crystal will exit the skew polarizer. We can refer to an intensity of the beam of photons and note that the ultimate output of photons has an intensity that is one-quarter of the input intensity. That is, $I = I_0/4$. The complete scenario is shown in Fig. K.7. Thus, by **reducing** the number of photons that reach the SKEW polarizer we will **increase** the outgoing intensity!

We might ask how a photon incident upon the skew polarizer can "know" that the path of the extraordinary ray from the first calcite crystal was blocked between the two calcite crystals, so that it ends up horizontally polarized and has a chance to pass through the skew polarizer? How can a photon be in two places at once? Our conclusion is that the photon is unknowable, or alternatively, indescribable, in terms of our own experience at the macroscopic level. This is also the case for the electron. In contrast to

[3] According to Malus' Law, the outgoing fractional intensity is $\cos^2(135^0) = \cos^2(45^0) = 1/2$.

the summary at the beginning of this Appendix about the nature of the discipline of physics as developed by Newton and Galileo, we have to give up our ability to represent objects by images in the quantum world.

There is an interesting comparison in the Torah, wherein it is written[4]: *Moses said to God, "Suppose that I go to the Israelites and say to them, "The God of your fathershas sent me to you" and they ask me, "What is his name?" What shall I tell them?" God said to Moses, "I AM WHO I AM. This is what you are to say to the Israelites: "I AM" has sent me to you."*

Similarly, suppose that we could speak to an electron and ask the electron to help us understand its seemingly incomprehensible behavior. The electron might reply by saying, "I am what I am."

[4]https://injil.org/TWOR/30.html.

Fusion of Harmonics: A Marvel of Auditory Processing

L

We take for granted that when a musical instrument plays a note we will hear the sound of but one source of sound. We have already learned in Chap. 2 that the frequency spectrum of the periodic wave of a musical instrument is a harmonic series, with the frequency of the wave. We have also learned that the timbre of an instrument is partly determined by the relative amplitudes and phases of the Fourier components associated with the instrument. On the other hand, more than three decades ago, I began to use a device made by the PASCO corporation for producing an electronic signal consisting of a periodic wave having up to nine harmonics, with a fundamental of 440 Hz. Such a device is called a **synthesizer**. The electronic signal was fed into an amplifier, which was connected to a loudspeaker. All listeners reported that they could hear the individual harmonics in the sound. Each harmonic was audibly separated, as if the harmonic had come from a separate source of sound. What is the difference between the wave produced by the Pasco synthesizer and the sound wave produced by a musical instrument? There is no visible characteristic on an oscilloscope trace that indicates a difference. Perhaps there is a significant difference that is too small to see on the trace with one's eyes.

We will discuss in this section the fact that the brain processes the sound so that we do not hear the individual harmonics. This process is referred to as **fusion of harmonics**.[1]

Fusion in the Taste of Food

It might be unclear to some readers what I mean by fusion of a mixture of harmonics. We can get some idea of fusion by considering the taste of a homogeneous dish of food. I recall many years ago, finding Lobster Cantonese an extremely delicious dish. When I first attempted to prepare the dish myself, I was amazed to learn that the essential ingredients were lobster and garlic. How well I knew the taste of garlic and yet how surprised I was that garlic was an essential ingredient in the recipe. Somehow, the blend of garlic and lobster produced a taste all its own—that of Lobster Cantonese—with the flavor of neither ingredient standing out. And so it seems to be with most superb dishes—as long as they are prepared properly. As another example, we can consider curries. Most are such that the ingredients are individually recognizable; there are fortunately some that have a wonderful homogenized flavor all their own. And so it seems to be with the fusion of a mixture of harmonics from a musical instrument!

It is interesting to consider and to try to perceive what the world of music would be like in the absence of fusion: The sound of a musi-

[1] I have been greatly helped in my attempt to weed out the known understandings of fusion by two audio-psychologists: Alan Bregman of McGill University and Brian Roberts of Aston University (Birmingham, England). Professor Bregman is the author of a book entitled *Auditory Scene Analysis* [MIT Press, Cambridge, 1994], in which he discusses how the brain processes an ensemble of sound inputs and organizes them according to sources. In particular, he explains how the brain is able to focus on one source of sound and ignore or become almost oblivious to other concurrent sources. As a result we are able to hear one person speak in the midst of a dense crowd at a party.

L. Gunther, *The Physics of Music and Color*,
https://doi.org/10.1007/978-3-030-19219-8

cal instrument would be heard as an ensemble of harmonics that would be superimposed with those of other musical instruments. We would lose our ability to separate out the sounds of the ensemble of instruments. Vibratos might lose their sweetness of tone. And so on.

Fusion of Harmonics is responsible for the rich beauty of musical instruments.

Here are some important interesting questions to be investigated.

1. What accounts for the ubiquitous fusion of the sound produced by musical instruments?
2. What are the conditions under which a sound wave with a superposition of harmonics will be fused, will be perceived as having a single source? Factors that seem to be important include:
   (a) the frequency of the fundamental—studies indicate that the lower the fundamental frequency, the greater is the degree of fusion;
   (b) the number of harmonics present and their relative amplitude;
   (c) the presence of "proportional modulation," sometimes referred to as "parallel modulation." Consider frequency modulation, which is characterized by two parameters. There is a rate at which the frequency is modulated—label it f_m. Next, there is an amplitude of variation of the frequency—label it Δf. Proportional frequency modulation would involve each harmonic being modulated with the same frequency of modulation f_m but with a variation f_a in proportion to the harmonic number n.

   Here are some possibly relevant sources of frequency modulation:

   Acoustic stringed instrument

   Normally we think of the vibrating string as having two fixed ends. However, the transmission of sound waves involves the string moving the bridge. Therefore, the string is not absolutely fixed at the bridge. Therefore, the string's length and its tension are modulated.

   Wind instrument such as a flute

   Here too—there is a modulation of the frequency due to the vibration of the mouth of the musician.
3. Are there strong variations in the auditory processing of people such that sounds that are fused for some people are not fused for others. It is reported that some people can sometimes distinctly hear the individual harmonics produced by a musical instrument.

L.1 Mathematica File

Below we provide a Mathematica file for producing a variety of waves that can be heard using the PLAY command of Mathematica. The reader can use this file to test his/her auditory processing a superposition of proportionally frequency modulated harmonics. When you run the commands a window will appear that allows you to listen to the wave form.

SYMBOLS:

Text: fn = central frequency for a given harmonic, n=1,2,3, . . . [f2 = 2 f1, . . . , f6 = 6 f1]

$\equiv \Delta f$ = maximum change in frequency

hn = $\delta fn/fn$ = **modulation index**

An = amplitudes

We begin with all modulation indices $hn \equiv \Delta f/fn$ the same:

$$h = fv/f1.$$

INPUT LINES

```
f1 = f
f = 440
f2 = 2*f
f3 = 3*f
f4 = 4*f
f5 = 5 f
f6 = 6 f
fv = 10
phi = f*t + (fv/(2 Pi f))*Cos[2 Pi f * t]
phi2 = f2*t + (fv/(2 Pi f2))*Cos[2 Pi* f2 * t]
```

```
phi3 = f3*t + (fv/(2 Pi f3))*Cos[2 Pi* f3 * t]
phi4 = f4*t + (fv/(2 Pi f4))*Cos[2 Pi* f4 * t]
phi5 = f5*t + (fv/(2 Pi f5))*Cos[2 Pi* f5* t]
phi6 = f6*t + (fv/(2 Pi f6))*Cos[2 Pi* f6* t]
A1 = 1
A2 = 1
A3 = 1
A4 = 0.01
A5 = 0.01
A6 = 0.001
```

INPUT COMMAND

```
Play[ A1*Sin[2 *Pi* phi] + A2*Sin[2* Pi * phi2]
+ A3*Sin[2* Pi * phi3] + A4* Sin[2* Pi * phi4]
+ A5* Sin[2* Pi * phi5] + A6*Sin[2* Pi * phi6],
t, 0, 5, SampleRate → 40000]
```

Next, the modulation indices are proportional to the harmonic:

$hn = n * fv/f1$

INPUT LINES

```
phi = f*t + (fv/(2 Pi f))*Cos[2 Pi f * t]
phi2a = f2*t + (fv/(2 Pi f))*Cos[2 Pi* f2 * t]
phi3a = f3*t + (fv/(2 Pi f))*Cos[2 Pi* f3 * t]
phi4a = f4*t + (fv/(2 Pi f))*Cos[2 Pi* f4 * t]
phi5a = f5*t + (fv/(2 Pi f))*Cos[2 Pi* f5* t]
phi6a = f6*t + (fv/(2 Pi f))*Cos[2 Pi* f6* t]
```

INPUT COMMAND

```
Play[ A1*Sin[2 *Pi* phi] + A2*Sin[2* Pi *
phi2a] + A3*Sin[2* Pi * phi3a] + A4* Sin[2* Pi
* phi4a] + A5* Sin[2* Pi * phi5a] + A6*Sin[2*
Pi * phi6a], t, 0, 5, SampleRate → 40000]
```

M

Transformation Between Tables of Color Matching Functions for Two Sets of Monochromatic Primaries *

In Chap. 15 we provided a Table 15.2 of color matching functions—henceforth referred to as **TCMF**—that was produced by Judd and Wyszecki by studying the color vision of a set of individuals with normal color vision for a specific set of primaries, $\lambda_B = 435.8$ nm, $\lambda_G = 546.1$ nm, and $\lambda_R = 700.0$ nm. The table consists of three columns of numbers, $\overline{r}(\lambda)$, $\overline{g}(\lambda)$, and $\overline{b}(\lambda)$—the **color matching functions**. What if we have a different set of monochromatic primaries? How should we mix these primaries so as to produce the same colors? In this appendix, we will derive a set of nine numbers that will allow us to determine a corresponding TCMF—starting with the original TCMF—that should allow the same set of individuals to match a spectral intensity using any other set of monochromatic primaries. We will label their wavelengths as $\{\lambda'_R, \lambda'_G, \text{and } \lambda'_B\}$.

In the TCMF, there are sixteen different wavelengths for each of the three primaries.[1] One might assume that each of the set of $3 \times 16 = 48$ numbers of the new TCMF depends upon the entire set of 48 numbers of the original TCMF. We would then need $48 \times 48 = 2304$ numbers to specify the relationship between the two TCMFs. We will see shortly that in fact, only $3 \times 3 = 9$ numbers are sufficient to determine the relationship.[2] The nine numbers are exhibited in the following three equations:

$$\begin{aligned}
\overline{r'}(\lambda) &= U_{RR}\overline{r}(\lambda) + U_{GR}\overline{g}(\lambda) + U_{BR}\overline{b}(\lambda) \\
\overline{g'}(\lambda) &= U_{RG}\overline{r}(\lambda) + U_{GG}\overline{g}(\lambda) + U_{BG}\overline{b}(\lambda) \\
\overline{b'}(\lambda) &= U_{RB}\overline{r}(\lambda) + U_{GB}\overline{g}(\lambda) + U_{BB}\overline{b}(\lambda)
\end{aligned} \qquad \text{(M.1)}$$

The nine numbers are represented by the symbols U_{RR}, U_{GR}, U_{BR}, U_{RG},U_{GG}, U_{BG}, U_{RB}, U_{GB}, and U_{BB}. They are commonly exhibited as an array of numbers, together forming what is referred to as a **matrix**. A common symbol for a matrix is $\mathbb{U}$, here shown for the letter U. The matrix $\mathbb{U}$ is exhibited below.[3]

$$\begin{pmatrix} U_{RR} & U_{GR} & U_{BR} \\ U_{RG} & U_{GG} & U_{BG} \\ U_{RB} & U_{GB} & U_{BB} \end{pmatrix}$$

[1]We realize that there are an infinite number of possible wavelengths spanning the range 400nm to 700nm. Our table essentially samples a given spectral intensity at a discrete value simply for convenience. In fact, our table was taken from another that had doubled the number of wavelengths, with values halfway between the values in our table. For ideal sampling, we would need color matching functions for the continuum of visible wavelengths. The tristimulus values would be integrals: $R = \int \overline{r}(\lambda)I(\lambda)d\lambda$, $G = \int \overline{g}(\lambda)I(\lambda)d\lambda$, and $B = \int \overline{b}(\lambda)I(\lambda)d\lambda$.

[2]In fact, if we don't care about maintaining a specific requirement on the unit intensities, we need only eight numbers.

[3]For those who are familiar with matrices, we will rewrite Eq. (M.1) in a simpler form. We introduce two vector functions, (λ) and $'(\lambda)$. They represent the color matching functions for the respective two sets of primaries which have the following components:

$$\overline{c}_R(\lambda) = \overline{r}(\lambda), \quad \overline{c}_G(\lambda) = \overline{g}(\lambda), \quad \overline{c}_B(\lambda) = \overline{b}(\lambda)$$

$$\overline{c'}_R(\lambda) = \overline{r}'(\lambda), \quad \overline{c'}_G(\lambda) = \overline{g}'(\lambda), \quad \overline{c'}_B(\lambda) = \overline{b}'(\lambda) \qquad \text{(M.2)}$$

We will use subscript notation, with subscripts $\alpha, \beta, ... = R, G, or B$. Then Eq. (M.1) can be written as

$$\overline{c'}_\alpha(\lambda) = \sum_{\beta=R,G,B} U_{\alpha\beta}\overline{c}_\beta(\lambda). \qquad \text{(M.3)}$$

L. Gunther, *The Physics of Music and Color*,
https://doi.org/10.1007/978-3-030-19219-8

The essential reason for this extraordinary simplification is *physiological*: color vision is based upon a *single* set of three receptors with three corresponding independent nerve impulse rates, as opposed to such a set for each possible set of primaries. This fact will be demonstrated in the last subsection of the appendix.

Explicit Expression for the Transformation Matrix $\mathbb{U}$

We will later show that the transformation matrix $\mathbb{U}$ can be expressed in terms of a **sub-matrix** obtained from the TCMF.
Suppose that the wavelengths of the second set of primaries are given by $\lambda_{R'}$, $\lambda_{G'}$, and $\lambda_{B'}$.

Then this sub-matrix, here symbolized by $\mathbb{V}$, is given by:

$$\begin{pmatrix} V_{RR} & V_{RG} & V_{RB} \\ V_{GR} & V_{GG} & V_{GB} \\ V_{BR} & V_{BG} & V_{BB} \end{pmatrix} \equiv \begin{pmatrix} \overline{r}(\lambda_{R'}) & \overline{g}(\lambda_{R'}) & \overline{b}(\lambda_{R'}) \\ \overline{r}(\lambda_{G'}) & \overline{g}(\lambda_{G'}) & \overline{b}(\lambda_{G'}) \\ \overline{r}(\lambda_{B'}) & \overline{g}(\lambda_{B'}) & \overline{b}(\lambda_{B'}) \end{pmatrix}$$

Note, for example, that $V_{RG} = \overline{g}(\lambda_{R'})$
We will show that

$$U_{\alpha\beta} = V_{\alpha\beta}^{-1} / u'_{\beta} \tag{M.4}$$

where the unit intensities are given by

$$u'_{\beta} = \sum_{\alpha=R,G,B} V_{\alpha\beta}^{-1} \tag{M.5}$$

so that

$$U_{\alpha\beta} = \frac{V_{\alpha\beta}^{-1}}{\sum_{\sigma=R,G,B} V_{\sigma\beta}^{-1}} \tag{M.6}$$

M.1 Application of the Transformation: Determining an Ideal Set of Primaries

We have tried to determine an ideal set of primaries starting with the TCMF produced by Judd and Wyszecki for the primaries 436nm, 546nm, and 700nm. Our goal is to minimize the total area that encompasses negative primaries. The process we will use involves a bit of trial and error. We first note that a change of primaries can be looked at as taking a piece of rubber on which we draw the TCMF and stretching it in various directions so that the points corresponding to the desired primaries lie at the corners of the triangle entirely encompassing positive color coordinates. The TCMF is best seen as a whole through the corresponding chromaticity diagram encompassed by the **horseshoe perimeter**. See Fig. M.1.

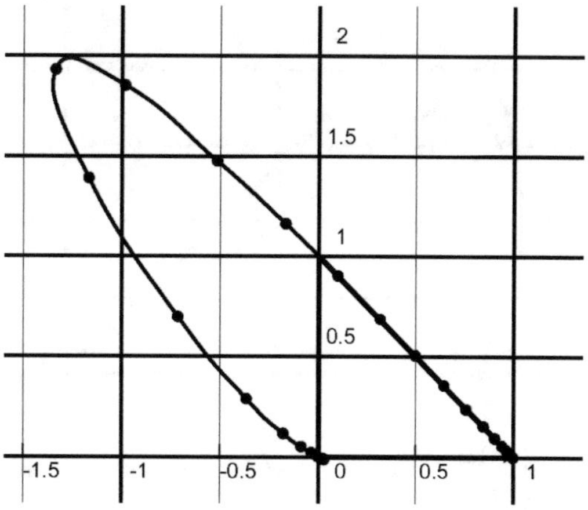

Fig. M.1 The horseshoe perimeter for the Judd and Wyszecki primaries 436nm, 546nm, and 700nm

We notice that the bulge to the upper left represents a large region having negative red coordinates. Consequently, this choice of monochromatic primaries is far from ideal in enabling one to match colors entirely with positive color coordinates. On the other hand, the perimeter from the green primary to the red primary is extremely close to being straight, so that there is an extremely small region having a negative blue primary coordinate. Finally, the red primary is at the end of the perimeter while the extreme blue end of the perimeter—at 400nm—is extremely close to the blue primary, so that we have an extremely small region having a negative green primary coordinate.

Since the extreme left end, with the greatest negative red coordinate, corresponds to a wavelength of about 510nm, it is reasonable to study a new set of primaries, with the green primary at a

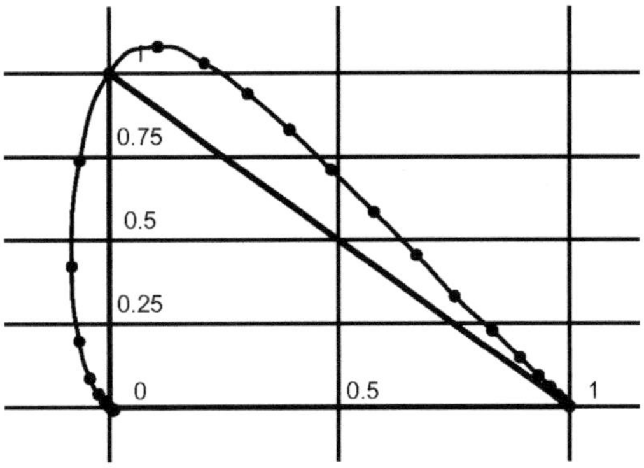

Fig. M.2 The horseshoe perimeter for the primaries 436nm, 510nm, and 700nm

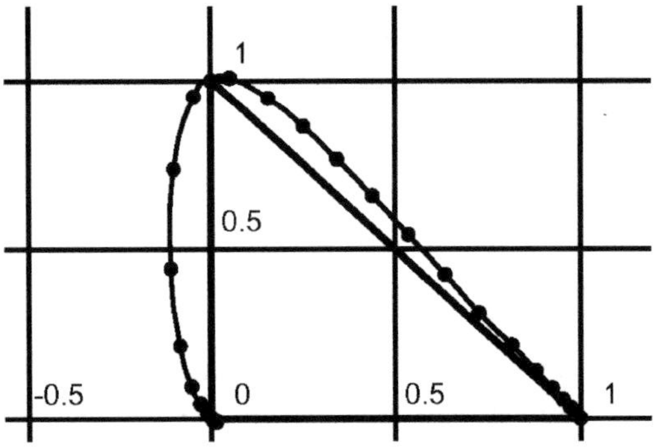

Fig. M.3 The horseshoe perimeter for the primaries 436nm, 515nm, and 700nm

wavelength of 510nm and with the same red and blue primaries. The resulting horseshoe is shown in Fig. M.2.

The improvement is dramatic. We have significantly reduced the area with a negative red coordinate. However, we now have a significant region with a negative blue coordinate. We therefore experiment with a green primary a bit closer to the original 546nm, hoping that the reduction in the region with a negative blue coordinate will not lead to a significant region with a negative green coordinate. We next switch to a green primary of 515nm. The resulting horseshoe is shown in Fig. M.3. The area of negative blue primary is now about the same as the area of negative red primary. With mathematical optimization techniques we could, perhaps, make further improvements; nevertheless, we will stop here and accept what we have now.

Procedure

We now summarize the procedure for obtaining the new TCMF, specifically for the set 436nm, 510nm, and 700nm. We first extract out of the original TCMF of Judd and Wyszecki those rows having to do with the new primaries.[4] We arrive at the 3×3 sub-matrix $\mathbb{V}$ that is shown in **light blue** in Fig. M.4. The wavelengths 700nm and 510nm are present in the TCMF. However, the wavelength 436nm of the blue primary, which is the same for both the original and the new primaries, is absent. That is not a problem for us since the red and the green color coordinates of the blue primary must be identically zero. The blue color coordinate can be obtained by interpolation between the wavelengths 420nm and 440nm.

Note that while the original TCMF was organized with the wavelengths running from 400nm at the top to the highest at 700nm, we have rearranged the three wavelengths in reverse order so that we have R,G, and B running from to bottom.

We next produce the **inverse of the matrix** $\mathbb{V}$, shown in **yellow** using the Excel.[5] The *unit intensities* of the new primaries are shown in magenta and are the respective sums of the columns above. Finally, we see the transform matrix, $\mathbb{U}$, in **light green**. It is the ratio, cell by cell, of the matrix $\mathbb{V}$ divided by the unit intensity corresponding to the column. Note that the sum of each of the three columns of $\mathbb{U}$ is unity.

M.2 Proof of Eqs. (M.1) and (M.6)

Let us begin by understanding better the content of the tables. The color of a spectral intensity is produced by mixing sources of the set of given primaries. Physiologically, the three tristimulus

[4] This text has only wavelengths that are multiples of twenty, while the original TCMF found in Williamson and Cummins has all multiples of ten.

[5] We highlight a 9×9 block of cells. We then type in the command line: **=MINVERSE(C7:E9)**, where (C7:E9) identifies the matrix $\mathbb{V}$ to be inverted - here C7 is the cell ID of V_{RR} and E9 is the cell ID of V_{BB}. Of course, your cell IDs might be different.

λ(nm)	R	G	B	R	G	B	R	G	B
700	0.0041	0	0	243.9	0	4E-17	0.591	0	0
510	-0.089	0.1286	0.027	168.82	7.776	-0.21	0.409	1	-0.2655
436	0	0	1	0	0	1	0	0	1.2655
			u'=	412.72	7.776	0.7902			

Fig. M.4 Matrices for transforming from the Judd-Wyszecki primaries 436nm, 546nm, and 700nm to the primaries 436nm, 510nm, and 700nm

calculated for that spectral intensity are NOT simply proportional to the corresponding rates at which nerve impulses are sent to the brain by the cones. In Chap. 15 we introduced the following functions:

The response functions $S_R(\lambda)$, $S_G(\lambda)$, and $S_B(\lambda)$ are the respective rates at which the cones, R, G, and B emit nerve impulses per *unit* intensity of wavelength λ.
The nerve impulse rate for a given spectral intensity $I(\lambda)$ are: $N_R(\lambda) = S_R(\lambda)I(\lambda)$, $N_G(\lambda) = S_G(\lambda)I(\lambda)$, and $N_B(\lambda) = S_B(\lambda)I(\lambda)$.

For a given spectral intensity, the total nerve impulse rates from the respective cones are given by[6]

$$N_R = \sum_\lambda S_R(\lambda)I(\lambda)$$
$$N_G = \sum_\lambda S_G(\lambda)I(\lambda) \qquad \text{(M.8)}$$
$$N_B = \sum_\lambda S_B(\lambda)I(\lambda)$$

where the right-hand sides are sums over all the wavelengths.

To produce a match by mixing the primary sources, the sources have to produce the same set of nerve impulse rates. Therefore we next need to obtain expressions for how these rates depend upon the primary sources. We note that generally, each of the three primaries produces nerve impulse rates from all three cones. We therefore introduce the following nine quantities:

$$\begin{aligned}
S_{RR} &= \text{Nerve impulse rate of the R-cones per unit intensity of R-primary}\\
S_{RG} &= \text{Nerve impulse rate of the R-cones per unit intensity of G-primary}\\
S_{RB} &= \text{Nerve impulse rate of the R-cones per unit intensity of B-primary}\\
S_{GR} &= \text{Nerve impulse rate of the G-cones per unit intensity of R-primary}\\
S_{GG} &= \text{Nerve impulse rate of the G-cones per unit intensity of G-primary}\\
S_{GB} &= \text{Nerve impulse rate of the G-cones per unit intensity of B-primary}\\
S_{BR} &= \text{Nerve impulse rate of the B-cones per unit intensity of R-primary}\\
S_{BG} &= \text{Nerve impulse rate of the B-cones per unit intensity of G-primary}\\
S_{BB} &= \text{Nerve impulse rate of the B-cones per unit intensity of B-primary}
\end{aligned} \qquad \text{(M.9)}$$

[6]See Chap. 15. With matrix and vector notation, we have

$$N_\alpha = \sum_\lambda S_\alpha(\lambda)I(\lambda). \qquad \text{(M.7)}$$

The nine quantities can be treated as a 9 × 9 matrix $\mathbb{S}$, exhibited below:

$$\begin{pmatrix} S_{RR} & S_{GR} & S_{BR} \\ S_{RG} & S_{GG} & S_{BG} \\ S_{RB} & S_{GB} & S_{BB} \end{pmatrix}$$

The nerve impulse rates depend upon the **tristimulus values**, **R**, **G**, and **B** and the matrix $\mathbb{S}$:

$$\begin{aligned} N_R &= \mathbf{R}\, S_{RR} + \mathbf{G}\, S_{RG} + \mathbf{B}\, S_{RB} \\ N_G &= \mathbf{R}\, S_{GR} + \mathbf{G}\, S_{GG} + \mathbf{B}\, S_{GB} \\ N_B &= \mathbf{R}\, S_{BR} + \mathbf{G}\, S_{BG} + \mathbf{B}\, S_{BB} \end{aligned} \tag{M.10}$$

We can now use the equations from Chap. 15 for the dependence of the tristimulus values on the color matching functions, namely,[7]

$$\begin{aligned} \mathbf{R} &= \sum_\lambda \overline{r}(\lambda) I(\lambda) \\ \mathbf{G} &= \sum_\lambda \overline{g}(\lambda) I(\lambda) \\ \mathbf{B} &= \sum_\lambda \overline{b}(\lambda) I(\lambda) \end{aligned} \tag{M.12}$$

These expressions, substituted into Eq. (M.10), provide us with an expression for the nerve impulse rates in terms of the matrix $\mathbb{S}$. Both this set of equations and the set of Eq. (M.9) in terms of the functions $S_R(\lambda)$, $S_G(\lambda)$, and $S_B(\lambda)$ must hold for any spectral intensity and therefore **must hold for any specific wavelength**. This fact will allow us to show how visual physiology reduces the number of independent variables necessary to relate matching with one set of primaries with matching by another set of primaries, as expressed by Eq. (M.1).

To clarify the above, let us join the first Eqs. (M.9) and (M.10) and all three equations of Eq. (M.12). We obtain

$$\begin{aligned} N_R &= \sum_\lambda S_R(\lambda) I(\lambda) = \mathbf{R}\, S_{RR} + \mathbf{G}\, S_{RG} + \mathbf{B}\, S_{RB} \\ &= \sum_\lambda \overline{r}(\lambda) I(\lambda)\; S_{RR} + \sum_\lambda \overline{g}(\lambda) I(\lambda)\; S_{RG} \\ &\quad + \sum_\lambda \overline{b}(\lambda) I(\lambda)\; S_{RB} \end{aligned} \tag{M.13}$$

We obtain a similar equation for N_G and N_B. Ultimately, we have for each wavelength[8]

$$\begin{aligned} S_R(\lambda) &= \overline{r}(\lambda) S_{RR} + \overline{g}(\lambda) S_{RG} + \overline{b}(\lambda) S_{RB} \\ S_G(\lambda) &= \overline{r}(\lambda) S_{GR} + \overline{g}(\lambda) S_{GG} + \overline{b}(\lambda) S_{BG} \\ S_B(\lambda) &= \overline{r}(\lambda) S_{BR} + \overline{g}(\lambda) S_{BG} + \overline{b}(\lambda) S_{BB} \end{aligned} \tag{M.15}$$

There are, correspondingly, nine quantities for the second set of primaries, the matrix $\mathbb{S}'$ as well as the second set of color matching functions, $\overline{r'}$, $\overline{g'}$, and $\overline{b'}$. We also have a set of equations parallel to Eq. (M.15):

$$\begin{aligned} S_R(\lambda) &= \overline{r'}(\lambda) S'_{RR} + \overline{g'}(\lambda) S'_{RG} + \overline{b'}(\lambda) S'_{RB} \\ S_G(\lambda) &= \overline{r'}(\lambda) S'_{GR} + \overline{g'}(\lambda) S'_{GG} + \overline{b'}(\lambda) S'_{GB} \\ S_B(\lambda) &= \overline{r'}(\lambda) S'_{BR} + \overline{g'}(\lambda) S'_{BG} + \overline{b'}(\lambda) S'_{BB} \end{aligned} \tag{M.16}$$

[7]In matrix notation, we can define the vector representing the three tristimulus values as $C_\alpha = (\mathbf{R,G,B})$. Then

$$C_\alpha = \sum_\lambda \overline{c}_\alpha(\lambda) I(\lambda) \tag{M.11}$$

[8]With matrix notation we have

$$S_\alpha(\lambda) = \sum_{\beta=R,G,B} S_{\alpha\beta} \overline{c}_\beta(\lambda). \tag{M.14}$$

Therefore,[9]

$$\begin{aligned}
\overline{r'}(\lambda)S'_{RR}+\overline{g'}(\lambda)S'_{GR}+\overline{b'}(\lambda)S'_{BR} &= \overline{r}(\lambda)S_{RR}+\overline{g}(\lambda)S_{GR}+\overline{b}(\lambda)S_{BR}\\
\overline{r'}(\lambda)S'_{RG}+\overline{g'}(\lambda)S'_{GG}+\overline{b'}(\lambda)S'_{BG} &= \overline{r}(\lambda)S_{RG}+\overline{g}(\lambda)S_{GG}+\overline{b}(\lambda)S_{BG}\\
\overline{r'}(\lambda)S'_{RB}+\overline{g'}(\lambda)S'_{GB}+\overline{b'}(\lambda)S'_{BB} &= \overline{r}(\lambda)S_{RB}+\overline{g}(\lambda)S_{GB}+\overline{b}(\lambda)S_{BB}
\end{aligned} \tag{M.18}$$

We see above that the two matrices, $\mathbb{S}$ and $\mathbb{S}'$, determine the relationship between the two sets of color matching functions and that the relationship is identical for each wavelength. The algebra of matrices leads to an expression for the transformation matrix $\mathbb{U}$ of Eq. (M.1) that involves the so-called **inverse matrix** of the matrix $\mathbb{S}'$:

$$\mathbb{U} = \mathbb{S}\mathbb{S}'^{-1} \tag{M.19}$$

For those who aren't familiar with these symbols, we will exhibit one of the matrix elements of $\mathbb{U}$:

$$U_{RG} = S_{RR}S'^{-1}_{RG} + S_{RG}S'^{-1}_{GG} + S_{RB}S'^{-1}_{BG} \tag{M.20}$$

Here, for example, S'^{-1}_{RG} is the RG element for the matrix $\mathbb{S}'^{-1}$.

Note When the two sets of primaries are identical, we expect the transformation matrix to yield the same TCMF as the original TCMF so that $\mathbb{U}$ should be the so-called **identity matrix**. Then all the diagonal elements (U_{RR}, U_{GG}, and U_{BB}) are unity while the remaining six elements vanish. Equation (M.19) confirms this result since in this case, $\mathbb{S} = \mathbb{S}'$

We will next prove Eq. (M.4)

$$\mathbb{U}_{\alpha\beta} = \mathbb{V}^{-1}_{\alpha\beta}/u'_{\beta} \tag{M.21}$$

where the parameters u'_{β} are the unit intensities for the second set of primaries, given by

$$u'_R = V^{-1}_{RR} + V^{-1}_{GR} + V^{-1}_{BR} \tag{M.22}$$

with corresponding expressions for the other two unit intensities.

Thus, the second TCMF is determined by the 9×9 sub-matrix $\mathbb{V}$ of the original TCMF.

Proof

First let us recall (M.1), which we rewrite here:

$$\begin{aligned}
\overline{r'}(\lambda) &= U_{RR}\overline{r}(\lambda)+U_{GR}\overline{g}(\lambda)+U_{BR}\overline{b}(\lambda)\\
\overline{g'}(\lambda) &= U_{RG}\overline{r}(\lambda)+U_{GG}\overline{g}(\lambda)+U_{BG}\overline{b}(\lambda)\\
\overline{b'}(\lambda) &= U_{RB}\overline{r}(\lambda)+U_{GB}\overline{g}(\lambda)+U_{BB}\overline{b}(\lambda)
\end{aligned} \tag{M.23}$$

In Chap. 15 we pointed out that if we sum any color matching function, $\overline{r}(\lambda)$, $\overline{g}(\lambda)$ or $\overline{b}(\lambda)$ over all of the wavelengths, we must obtain the same number so that a constant spectral intensity will produce equal energy white. This fact obviously holds true for the second set of primaries too, except that the constant common to the three sums can be different. We now recall that if we were to multiply every color matching function in a TCMF by the same number, all tristimulus values are multiplied by that number but the color coordinates are unchanged. We therefore are free to choose the sums to be equal for the two different sets of primaries.

If we carry out this sum in each of the above three equations, we will obtain the

---

9

$$\sum_{\beta=R,G,B} \overline{c}'_{\beta}(\lambda)S'_{\alpha\beta} = \sum_{\beta=R,G,B} \overline{c}_{\beta}(\lambda)S_{\alpha\beta}. \tag{M.17}$$

following three equations for the transformation matrix[10]:

$$U_{RR} + U_{GR} + U_{BR} = 1$$
$$U_{RG} + U_{GG} + U_{BG} = 1 \qquad \text{(M.25)}$$
$$U_{RB} + U_{GB} + U_{BB} = 1$$

Next, from Eq. (15.15) in Chap. 15 we obtain

$$\bar{r}(\lambda_R) = \frac{1}{u_R}, \quad \bar{g}(\lambda_G) = \frac{1}{u_G}, \quad \bar{b}(\lambda_B) = \frac{1}{u_B} \qquad \text{(M.26)}$$

with the remaining functions, e.g. $\bar{r}(\lambda_G)$, vanishing.

Let us introduce the matrix $\mathbb{C}$ defined by

$$\mathbb{C} = \begin{pmatrix} \bar{r}(\lambda_R) & \bar{r}(\lambda_G) = 0 & 0 \\ 0 & \bar{g}(\lambda_G) & 0 \\ 0 & 0 & \bar{b}(\lambda_B) \end{pmatrix}$$

Similarly, we have a corresponding matrix for the second set of primaries, λ'_R, λ'_G, and λ'_B , we have

$$\overline{r'}(\lambda'_R) = \frac{1}{u'_R}, \quad \overline{g'}(\lambda'_G) = \frac{1}{u'_G}, \quad \overline{b'}(\lambda'_B) = \frac{1}{u'_B} \qquad \text{(M.27)}$$

with the remaining functions, e.g. $\bar{r}(\lambda_G)$, vanishing. We also define the matrix

$$\mathbb{C}' = \begin{pmatrix} \overline{r'}(\lambda'_R) & \overline{r'}(\lambda'_G) = 0 & 0 \\ 0 & \overline{g'}(\lambda'_G) & 0 \\ 0 & 0 & \overline{b'}(\lambda'_B) \end{pmatrix} = \begin{pmatrix} 1/u'_R & 0 & 0 \\ 0 & 1/u'_G & 0 \\ 0 & 0 & 1/u'_B \end{pmatrix}$$

According to Eq. (M.1) we have

$$C'_{RR} = \overline{r'}(\lambda'_R) = \frac{1}{u'_R} = U_{RR}\bar{r}(\lambda'_R) + U_{GR}\bar{g}(\lambda'_R) + U_{BR}\bar{b}(\lambda'_R)$$
$$C'_{GG} = \overline{g'}(\lambda'_G) = \frac{1}{u'_G} = U_{RG}\bar{r}(\lambda'_G) + U_{GG}\bar{g}(\lambda'_G) + U_{BG}\bar{b}(\lambda'_G) \qquad \text{(M.28)}$$
$$C'_{BB} = \overline{b'}(\lambda'_B) = \frac{1}{u'_B} = U_{RB}\bar{r}(\lambda'_B) + U_{GB}\bar{g}(\lambda'_B) + U_{BB}\bar{b}(\lambda'_B)$$

These equations can be rewritten so as to exhibit the matrix $\mathbb{V}$:

$$C'_{RR} = V_{RR}U_{RR} + V_{RG}U_{GR} + V_{RB}U_{BR}$$
$$C'_{GG} = V_{GR}U_{RG} + V_{GG}U_{GG} + V_{GB}U_{BG}$$
$$C'_{BB} = V_{BR}U_{RB} + V_{BG}U_{GB} + V_{BB}U_{BB} \qquad \text{(M.29)}$$

These equations can be expressed as a multiplication of matrices:

$$\mathbb{C}' = \mathbb{V}\mathbb{U} \qquad \text{(M.30)}$$

or

$$C'_{\alpha\beta} = \sum_{\sigma=R,G,B} V_{\alpha\sigma}U_{\sigma\beta} \qquad \text{(M.31)}$$

We know so far only the matrix $\mathbb{V}$. We can obtain the unit intensities u'_α) as follows. First, we solve Eq. (M.31) for the matrix $\mathbb{U}$:

$$\mathbb{U} = \mathbb{V}^{-1}\mathbb{C}' \qquad \text{(M.32)}$$

In matrix element form we have

$$U_{\alpha\beta} = \sum_{\sigma=R,G,B} V^{-1}_{\alpha\sigma}C'_{\sigma\beta} = V^{-1}_{\alpha\beta}/u'\beta \qquad \text{(M.33)}$$

Next we recall Eq. (M.25). It can be written as

$$\sum_{\alpha=R,G,B} U_{\alpha\beta} = 1 \qquad \text{(M.34)}$$

[10]The three equations can be expressed in matrix notation as

$$\sum_\alpha U_{\alpha\beta} = 1 \qquad \text{(M.24)}$$

for all β.

for all β. Therefore,

$$u'_\beta = \sum_{\alpha=R,G,B} V^{-1}_{\alpha\beta} \qquad \text{(M.35)}$$

Finally we have our expression for the transformation matrix $\mathbb{U}$ in terms of the sub-matrix of the original TCMF, Eq. (M.6):

$$U_{\alpha\beta} = \frac{V^{-1}_{\alpha\beta}}{\sum_{\sigma=R,G,B} V^{-1}_{\sigma\beta}} \qquad \text{(M.36)}$$

M.3 Problems on the Transformation of TCMFs

1. Suppose that the matrix $\mathbb{V}$ has only diagonal matrix elements. We say that the matrix is a **diagonal matrix**.
   Explain why the matrix $\mathbb{U}$ is diagonal, with all its matrix elements equal to unity. What does this imply about how the new set of primaries is related to the old set?
2. Below is the TCMF produced by Stiles and Burch [See Chap. 15] for the primaries $\lambda_{B'} = 444.44$ nm, $\lambda_{G'} = 526.32$ nm, and $\lambda_{R'} = 645.16$ nm (Table M.1).
   a. According to the theory of color vision that we have presented, the data collected by the two groups, Stiles and Burch vs. Judd and Wyszecki, should be consistent. For example, if we were to apply our transformation matrices to the TCMF of Stiles and Burch, we should obtain the TCMF of Judd and Wyszecki. Carry out this process and compare the TCMFs. Beware that the two might differ by a constant factor that multiplies all the coordinates. Why is this so? On the other hand, the color coordinates do not depend upon such a constant factor, so compare the horseshoe perimeters.
   b. Derive the transformation matrix $\mathbb{U}$ for the set of primaries 436nm, 520nm, and 700nm that is based upon the Stiles and Burch TCMF.

Table M.1 Table of color matching functions based upon the spectral primaries 444.44 nm, 526.32 nm, and 645.16 nm

λ (nm)	$\bar{r}(\lambda)$	$\bar{g}(\lambda)$	$\bar{b}(\lambda)$
400	0.0089	−0.0025	0.04
410	0.035	−0.0119	0.1802
420	0.0702	−0.0289	0.467
430	0.0763	−0.0338	0.6152
440	0.0561	−0.0276	0.8778
450	−0.0044	0.0024	1.0019
460	−0.097	0.0636	0.9139
470	−0.2235	0.1617	0.7417
480	−0.3346	0.2796	0.472
485	−0.3776	0.3428	0.3495
490	−0.4136	0.4086	0.2564
500	−0.4452	0.5491	0.1307
510	−0.414	0.7097	0.058
520	−0.2845	0.8715	0.02
530	−0.0435	0.9945	0.0007
540	0.3129	1.0375	−0.0064
550	0.7722	1.039	−0.0094
560	1.271	0.9698	−0.0097
570	1.8465	0.8571	−0.0087
580	2.425	0.6953	−0.0073
590	2.9151	0.5063	−0.00537
600	3.1613	0.336	−0.00357
610	3.1048	0.1917	−0.00208
620	2.7194	0.0938	−0.00103
630	2.17	0.0371	−0.00044
640	1.5179	0.0112	−0.00014
650	1.007	0.000078	0
660	0.5934	−0.001988	0
670	0.3283	−0.002006	0
680	0.1722	−0.001272	0
690	0.0853	−0.000683	0
700	0.0408	−0.000337	0

   Find the TCMF for this second set of primaries and the corresponding horseshoe perimeter.
   Compare this perimeter with the one we derived by a transformation from the Judd and Wyszecki TCMF of the set of primaries 436nm, 546nm, and 700nm.
3. Prove that the tristimulus values C'_α for a second set of monochromatic primaries is related

to the tristimulus values C_α of the first set by the equation

$$C'_\alpha = \sum_\beta U_{\alpha\beta} C_\beta \qquad \text{(M.37)}$$

4. a. Prove that the sum $S = \sum_\alpha C_\alpha$ is independent of the set of monochromatic primaries.

   b. Prove that the color coordinates satisfy the equation

$$\frac{\overline{c'}_\alpha}{C'_\alpha} = \frac{\overline{c}_\alpha}{C_\alpha} \qquad \text{(M.38)}$$

Hommage to Pierre-Gilles de Gennes—Art and Science

N

In July of 2007 I learned that the physicist **Pierre-Gilles de Gennes** (Fig. N.1) had passed away on May 18, 2007. I received the news with deep sadness since de Gennes was my supervisor when I was a young Post-Doc at the Laboratory of the Physics of Solids in Orsay, France in the academic year 1966-67 and I had crossed paths with him many times since then. While only seven years older than me, he had already achieved fame as leader of a group of physicists studying superconductivity. De Gennes had a great influence on all those working in the field of Condensed Matter. Whatever subject he touched, he transformed. Frequently he attacked problem areas for which physicists and chemists felt that they had reached the limits thought possible conceptually. De Gennes discovered sets of concepts that produced a new exciting level of activity of research in the fields.

One meeting I had with him stands out in my memory: I was working on a problem and was stuck because of not knowing one of the essential parameters of the problem. **de Gennes** listened to me attentively with all due patience. He then proceeded to ask me one simple question, which I will recall with clarity: "What is the approximate wavelength of the electrons in bismuth in comparison to the roughness of the surface of the sample." The reader need not understand the meaning of the question. It was the utter simplicity and directness of his question that mattered. For that question was all I needed to hear in order for me to proceed to complete my study. Our meeting was very brief and yet extremely productive.

Of what relevance is this encounter to this book, in which our goal is to appreciate the connection between physics and the arts? It is that some people have an extraordinary gift to see order and simplicity in the complex. de Gennes was one of those who applied this ability to many fields, whose numerous familiar applications depend upon the results of his research. This ability to create order from a suspended blank state of our normal awareness is manifest in artists as well. De Gennes is quoted as describing the deep influence that the painter **Pablo Picasso** had on his own scientific studies. To appreciate this point, I am including an excerpt of one of a number of obituaries written after his death. For details, see: http://www.nature.com/nature/journal/v448/n7150/pdf/448149a.pdf.

The obituary makes reference to a film on Picasso that left a great mark on de Gennes' research as a physicist. You can view an excerpt of this film on the following YouTube site: http://www.youtube.com/watch?v=5tn5uTTCRCg.

Excerpt of an Obituary of de Gennes by Françoise Brochard-Wyart

"Chacun en nous a son trésor d'images entrevues dans un instant mais jamais oubliées. Un exemple pour moi: Picasso peignant grands traits blancs sur une vitre et filmé par Clouzot. Tout ce que j'ai essayé de dessiner laborieusement plus tard est né de ces moments."

"Every one of us has a treasure of images caught in glimpses but never forgotten. A personal example: Picasso painting white lines on glass using large strokes,

L. Gunther, *The Physics of Music and Color*,
https://doi.org/10.1007/978-3-030-19219-8

Fig. N.1 Pierre-Gilles de Gennes (source: http://authors.library.caltech.edu/5456/1/hrst.mit.edu/hrs/materials/public/DeGennes/DeGennesintro_fr1252.htm)

filmed behind the glass by Clouzot. Everything that I tried painting laboriously later was born from such glimpses."
(Pierre-Gilles de Gennes, from L'émerveillement by Thibaut de Wurstemberger, Saint-Augustin, 1998.)
With his strikingly simple yet pioneering ideas, Pierre-Gilles de Gennes drew "white lines in large strokes" that defined the physics of soft matter—liquid crystals, polymers, colloids and surfactants. He died on 18 May.
Educated at the École Normale Supérieure in Paris during 1951-55, de Gennes learned theoretical physics from the greatest masters of his time. He obtained his PhD in 1957 while at the French Atomic Energy Commission, specializing in magnetism and neutron scattering. During a stay at the University of California, Berkeley, in 1959 he studied with the solid-state physicist Charles Kittel, who taught him how to communicate ideas in physics using plain language, so avoiding the use of daunting equations. . . .
De Gennes fostered a collective research effort that is scarcely imaginable today. Papers were signed not with the names of individuals, but with the name of the group. Theoreticians would spend half their time contemplating liquid crystals under the microscope and discussing practical experiments. Researchers would often arrive in the morning to find a note from de Gennes that would launch them in yet another groundbreaking direction. Calling on his vast knowledge of physics, de Gennes drew analogies between different fields. For example, he realized that laws developed to describe superconductivity phenomena could be used to understand phase transitions in liquid crystals.
. . .
De Gennes pursued his research with extraordinary imagination, insatiable curiosity and an ability to grasp facts rapidly. But he also gave his time to others and helped them develop their ideas. A keen ambassador of science to the public, he generated passionate debates on subjects as diverse as "Physics and Medicine," "Inventors" and "Primo Levi." He inspired generations of students to pursue careers in physics and played an active role in establishing the L'Oréal-UNESCO Awards for Women in Science.
. . .

End of excerpt of the Obituary

Here is another expression of a likening of de Gennes revelations to the product of an artist in an excerpt of a description of de Gennes' book on Soft Interfaces: "[The book provides us with] an impressionistic tour of the physics of soft interfaces by Nobel Laureate Pierre-Gilles de Gennes. Full of insight and interesting asides, it not only provides an accessible introduction to this topic, but also lays down many markers and signposts for interesting new research possibilities."

I will end this essay with a personal remark:

Over the years, de Gennes and I corresponded and saw each other only once every few years. Yet, in spite of his fame and his multitude of activities and acquaintances, he remembered even the most trivial of our encounters. Most vivid in his mind when I last saw him at a talk he gave at Harvard was about a dinner that his wife and he had prepared for my family in 1967. They had forgotten to buy fresh cheese for our dessert; this mistake required deep apologies and remained in his mind for 40 years.

Index

L. Gunther, *The Physics of Music and Color*,
https://doi.org/10.1007/978-3-030-19219-8

Printed in the United States
by Baker & Taylor Publisher Services